1007493578

Weimar Culture
and Quantum Mechanics

Selected Papers by Paul Forman and
Contemporary Perspectives on the Forman Thesis

Weimar Culture and Quantum Mechanics

Selected Papers by Paul Forman and
Contemporary Perspectives on the Forman Thesis

Editors

Cathryn Carson
University of California, Berkeley, USA

Alexei Kojevnikov
University of British Columbia, Canada

Helmuth Trischler
Deutsches Museum, Germany

Published by

Imperial College Press
57 Shelton Street
Covent Garden
London WC2H 9HE

and

World Scientific Publishing Co. Pte. Ltd.
5 Toh Tuck Link, Singapore 596224
USA office: 27 Warren Street, Suite 401-402, Hackensack, NJ 07601
UK office: 57 Shelton Street, Covent Garden, London WC2H 9HE

British Library Cataloguing-in-Publication Data
A catalogue record for this book is available from the British Library.

Cover image: "Homage to the Uncertainty Principle: A Device to Aid in Locating Electrons in an Atom If There Were a Means to Look for Them," sculpture by Kenneth Snelson, 1964, height 22 inches (56 cm). Reproduced with the artist's permission. For elucidation of this device, see Kenneth Snelson, *Forces Made Visible* (Lenox, MA: Hard Press Editions, 2009), p. 145

WEIMAR CULTURE AND QUANTUM MECHANICS
Selected Papers by Paul Forman and Contemporary Perspectives on the Forman Thesis

Copyright © 2011 by Imperial College Press and World Scientific Publishing Co. Pte. Ltd.

All rights reserved. This book, or parts thereof, may not be reproduced in any form or by any means, electronic or mechanical, including photocopying, recording or any information storage and retrieval system now known or to be invented, without written permission from the Publisher.

For photocopying of material in this volume, please pay a copying fee through the Copyright Clearance Center, Inc., 222 Rosewood Drive, Danvers, MA 01923, USA. In this case permission to photocopy is not required from the publisher.

ISBN-13 978-981-4293-11-2
ISBN-10 981-4293-11-3

Typeset by Stallion Press
Email: enquiries@stallionpress.com

Printed in Singapore by B & Jo Enterprise Pte Ltd

Contents

Acknowledgments	ix
Contributors to this Volume	xi
The Forman Thesis: 40 Years After *Cathryn Carson, Alexei Kojevnikov and Helmuth Trischler*	1
Cold War Culture, History of Science and Postmodernity: Engagement of an Intellectual in a Hostile Academic Environment *J.L. Heilbron*	7
Selected Chronological Bibliography of Paul Forman's Scholarship	21

Part I. The Forman Thesis

Scientific Internationalism and the Weimar Physicists: The Ideology and its Manipulation in Germany after World War I *Paul Forman*	26
The Financial Support and Political Alignment of Physicists in Weimar Germany *Paul Forman*	57
Weimar Culture, Causality, and Quantum Theory, 1918–27: Adaptation by German Physicists and Mathematicians to a Hostile Intellectual Environment *Paul Forman*	85
Kausalität, Anschaulichkeit, and *Individualität*, or, How Cultural Values Prescribed the Character and the Lessons Ascribed to Quantum Mechanics *Paul Forman*	203

The Reception of an Acausal Quantum Mechanics in Germany and Britain   221
   Paul Forman

Part II. Quantum Physics in its Historical Contexts

Paul Forman and the Environment and Practice of Quantum History   263
   David C. Cassidy

Culture and Mechanics in Germany, 1869–1918: A Sketch   277
   Richard Staley

The Establishment of a Network of Reactionary Physicists in the Weimar Republic   293
   Stefan L. Wolff

Philosophical Rhetoric in Early Quantum Mechanics 1925–27: High Principles, Cultural Values and Professional Anxieties   319
   Alexei Kojevnikov

'The Shackles of Causality': Physics and Philosophy in the Netherlands in the Interwar Period   349
   Kai Eigner and Frans van Lunteren

Crisis, Measurement Problems and Controversy in Early Quantum Electrodynamics: The Failed Appropriation of Epistemology in the Second Quantum Generation   375
   Anja Skaar Jacobsen

Causality in Physics and in the History of Physics: A Comparison of Bohm's and Forman's Papers   397
   Olival Freire Jr.

Part III. Science and Culture: Cross-Disciplinary Debates

Forman Reformed, Again   415
   M. Norton Wise

From *Kosmos* to *Koralle*: On the Culture of Science Reading in Imperial and Weimar Germany   433
   Arne Schirrmacher

Living Ambiguity: Speculative Bodies of Science in Weimar Culture   453
   Cornelius Borck

Science and Politics: Pathology in Weimar Germany (1918–33)   475
Cay-Rüdiger Prüll

Jordan alias Domeier: Science and Cultural Politics in Late Weimar   487
Conservatism
Richard H. Beyler

The Causality Debates of the Interwar Years and Their Preconditions: 505
Revisiting the Forman Thesis from a Broader Perspective
Michael Stöltzner

Modern or Anti-modern Science? Weimar Culture, Natural Science   523
and the Heidegger-Heisenberg Exchange
Cathryn Carson

Acknowledgments

This book has its origin in the conference 'The Cultural Alchemy of the Exact Sciences: Revisiting the Forman Thesis', held at the University of British Columbia on 23 to 25 March 2007. We are grateful to all our colleagues who participated in that conference, delivered papers and offered commentaries.

For financial and logistical support of the conference we owe thanks to the University of British Columbia Department of History, Green College, St. John's College and the Peter Wall Institute for Advanced Studies. Further financial support for the conference and for the production of this volume came from the TransCoop Program of the Alexander von Humboldt-Stiftung and the U.S. National Science Foundation. This material is based upon work supported by the National Science Foundation under Grant No. SES-0621179. Any opinions, findings and conclusions or recommendations expressed in this material are those of the author(s) and do not necessarily reflect the views of the National Science Foundation.

For permission to reprint papers by Paul Forman we are grateful to the University of Pennsylvania Press, Transaction Publishers, Springer, the History of Science Society and the American Association for the Advancement of Science. A few papers in this volume appeared earlier in German in a special issue of the journal *Berichte zur Wissenschaftsgeschichte* 31, no. 4 (2008). We owe thanks to editor-in-chief Cornelius Borck for facilitating that publication. For assistance in assembling this volume we are indebted to Diana Wear and Lindsay Crawford of the Office for History of Science and Technology of the University of California, Berkeley. Translations from the German were undertaken by Gona Geyer and editing of the texts skilfully done by Paul Tyler. We also thank our editor at Imperial College Press, Romén Reyes-Peschl.

Many warm thanks go to J.L. Heilbron for much good advice in shaping this project. Our ultimate debt is to Paul Forman for intellectual inspiration and stringent criticism over the years.

Contributors to this Volume

Richard Beyler received his PhD in history of science at Harvard University in 1994. After receiving his PhD, he was a Walter Rathenau Postdoctoral Fellow in Berlin and later a Volkswagen Stiftung Fellow at the German Historical Institute in Washington, D.C. In 1996, he joined the faculty of Portland State University, first in PSU's University Studies program and then, starting in 1998, as a member of the History Department. In 2002, he returned to Berlin for a term as a guest researcher for the Presidential Commission of the Max-Planck-Gesellschaft on the History of the Kaiser-Wilhelm-Gesellschaft during National Socialism. He is currently Associate Professor of History at PSU, teaching in the areas of history of science, European intellectual history and German history; he also serves as chair of the university's Graduate Council. His research focuses on biophysics before the emergence of molecular biology, on the cultural and philosophical relations of physics in twentieth-century Germany and on the politics of German science across the thresholds of 1933 and 1945. Forthcoming publications include an essay on the 'ideology of non-ideology' in post-1945 German science; an essay in *Physics and Politics: Research and Research Support in Twentieth-Century Germany in International Perspective*, edited by Helmuth Trischler and Mark Walker; and a study of biophysical 'target theory' during the 1930s in *Creating a Physical Biology: The Three-Man Paper and the Origins of Molecular Biology*, edited by Brandon Fogel and Phillip Sloan.

Cornelius Borck is Professor for the History, Theory and Ethics of Science and Medicine and Director of the Institute for the History of Medicine and Science Studies of the University of Lübeck, Germany. A medical doctor by training and a philosopher with an habilitation in history of science, his appointments include the Karl Schädler Research Fellowship at the Max Planck Institute for the History of Science in Berlin; the directorship of the research group, Writing Life, Media Technologies and the History of the Life Sciences 1800–1900, in the Faculty of Media at the Bauhaus University in Weimar; and a Canada Research Chair in Philosophy and Language of Medicine at McGill University in Montreal. Among his research topics are the following: mind, brain, and self in the age of visualisation;

the epistemology of experimentation in art, science and media; and sensory prostheses and the human-machine relation between artistic avant-garde and technoscience. Recent publications include 'Recording the Brain at Work: The Visible, the Readable, and the Invisible in Electroencephalography', *Journal of the History of the Neurosciences* 17 (2008): 367–79; 'Blindness, Seeing, and Envisioning Prosthesis: The Optophone between Science, Technology, and Art', in *Artists as Inventors — Inventors as Artists*, ed. D. Daniels and B. Schmidt (2008); *Hirnströme: Eine Kulturgeschichte der Elektroenzephalographie* (2005); and *Psychographien*, ed. with Armin Schäfer (2006).

Cathryn Carson is Associate Professor of History at the University of California, Berkeley, where she also directs the Office for History of Science and Technology. Her training at the University of Chicago and Harvard University was in physics and the history and philosophy of science. She is the author of *Heisenberg in the Atomic Age: Science and the Public Sphere* (2010) and co-editor, with David A. Hollinger, of *Reappraising Oppenheimer: Centennial Studies and Reflections* (2005). Since 2008 she has chaired the editorial board of *Historical Studies in the Natural Sciences*, the current incarnation of *Historical Studies in the Physical Sciences*, in which Paul Forman's 'Weimar Culture' was published.

David C. Cassidy is Professor of Natural Sciences at Hofstra University, Hempstead, NY. He is the author of *Beyond Uncertainty: Heisenberg, Quantum Physics, and the Bomb* (2009), the successor to *Uncertainty*; as well as *J. Robert Oppenheimer and the American Century* (2005/2009), *Einstein and Our World* (2005) and papers on quantum history. He undertook postdoctoral work with J.L. Heilbron at the University of California, Berkeley and was a Humboldt Foundation fellow with Armin Hermann at the Universität Stuttgart, Germany. He is a former associate editor of the Einstein Papers Project.

Kai Eigner studied physics and philosophy of science at Utrecht University, specialising in the history of the natural sciences and in the philosophy of the cognitive sciences. In his master's thesis he explored the parallels between the Dutch intellectual climate of the interwar period and that of the Weimar Republic as described in Forman's thesis. He is presently affiliated with the faculty of philosophy of the VU University of Amsterdam, where he teaches philosophy of science and is finishing his doctoral dissertation on understanding in psychology. In this research project he elaborates and defends a new approach to scientific understanding.

Olival Freire's interests are concentrated on the history of the research into the foundations of quantum physics, the history of physics in Brazil and the use of history of physics in the teaching of science. His papers include 'Quantum

Dissidents: Research on the Foundations of Quantum Theory circa 1970' and 'The Origin of the Everettian Heresy' (co-authored by Stefano Osnaghi and Fabio Freitas), both in *Studies in History and Philosophy of Modern Physics* and 'Science and Exile: David Bohm, the Cold War, and a New Interpretation of Quantum Mechanics', in *HSPS*. He was a Senior Fellow at the Dibner Institute for History of Science and a Visiting Professor at the Université de Paris VII. He has a BSc in physics (Universidade Federal da Bahia – UFBa, Brazil) and a PhD in history (Universidade de São Paulo, Brazil). He is also an Associate Professor at UFBa and a Fellow [Bolsista de produtividade] for Brazilian CNPq. In 2000, he helped found the Graduate Studies Program in History, Philosophy and Science Teaching in Brazil.

John Heilbron, Professor Emeritus of History at the University of California Berkeley, is a long-time friend and one-time fellow student of Paul Forman. Like Forman, he used to work on the history of modern physics, but now prefers to write biography. His most recent attempt in this line is a study of Galileo (Oxford University Press, 2010). He is now working on parallels between the doing of physics and the writing of history in the early modern period.

Alexei Kojevnikov is Associate Professor in the Department of History at the University of British Columbia, Vancouver, Canada. He is the author of *Stalin's Great Science: The Time and Adventures of Soviet Physicists* (2004) and co-editor of *Intelligentsia Science: The Russian Century, 1860–1960*, vol. 23 of the *Osiris* series (2008).

Cay-Rüdiger Prüll studied medicine, history and philosophy at the University of Gießen in Germany. From 1992 to 2008 he was a lecturer at the Institute for the History of Medicine in Freiburg. He is currently a scientific assistant at the Institute for the History of Medicine at the University of Heidelberg, working on a project entitled, 'Patients, the Public and Medicine in Western Germany, 1945–1970', funded by the DFG. He also headed (together with Karl-Heinz Leven) a DFG-funded project entitled, 'War and Medical Culture: Patients and Physicians between 1914 and 1945'. His main research areas are the history of biomedicine in the nineteenth and twentieth centuries, especially the history of pathology, pharmacology and military medicine. He is particularly interested in the social and cultural history of medicine, as well as the relationship between medicine and politics.

Arne Schirrmacher worked as an historian of science at the Deutsches Museum in Munich for over a decade and is currently based at the Max Planck Institute for the History of Science in Berlin. He studied physics, biology and philosophy at Hamburg, Oxford and Munich and received a PhD in mathematical physics before turning fully to the history of modern mathematical sciences, which he pursued at Berkeley, Munich and Göttingen. His research interests cover the history of quantum theory and atomic representation; the relation of war and ideology to the development of modern science and the history of science communication in twentieth-century Europe. Recent publications include 'Looking into (the) Matter: Scientific Artifacts and Atomistic Iconography', in *Illuminating Instruments*, ed. Peter Morris and Klaus Staubermann (2009); a critical edition of the recollections of German physicist and Nobelist Philipp Lenard, *Erinnerungen eines Naturforschers* (2009); 'Nach der Popularisierung. Zur Relation von Wissenschaft und Öffentlichkeit im 20. Jahrhundert', *Geschichte und Gesellschaft* 34 (2008): 73–95 and *Wissenschaft und Öffentlichkeit als Ressourcen füreinander: Studien zur Wissenschaftsgeschichte im 20. Jahrhundert,* ed. with Sybilla Nikolow (2007).

Anja Skaar Jacobsen is postdoctoral scholar at the Niels Bohr Archive, Copenhagen, with a grant from the Carlsberg Foundation. She also teaches history of science at the Danish Institute for Study Abroad (DIS), Copenhagen. After having received her MA in physics and chemistry, she received her PhD in the history of science at the University of Aarhus in 2000 and has held postdoctoral positions at the University of Aarhus, Roskilde University, the Niels Bohr Archive and the Max Planck Institute for the History of Science in Berlin. Her main research interests include the history of early nineteenth-century science and Romanticism, and the history of twentieth-century theoretical physics, its epistemology and Marxism. She has published papers in *Ambix, Centaurus, Minerva, Historical Studies in the Physical and Biological Sciences* and *Science and Education*. She is currently the president of the Danish Society for the History of Science.

Richard Staley is Associate Professor in history of science at the University of Wisconsin–Madison and teaches the history of physics since Newton. His early interest in physics in Germany is evidenced by his 1992 dissertation on Max Born's early career, completed at the University of Cambridge, and by his participation in a collaborative project on innovation in Germany and Britain that culminated in two exhibitions at the Whipple Museum for the History of Science in Cambridge (1992–95). He is the author of several studies of physics circa 1900, including his recent *Einstein's Generation: The Origins of the Relativity Revolution* (2008).

Michael Stöltzner is Associate Professor of philosophy at the University of South Carolina at Columbia. He has studied mathematical physics, philosophy and history of science at Tübingen, Trieste, Vienna and Bielefeld, and held positions at the universities of Salzburg, Bielefeld and Wuppertal. He has been a visiting scholar and lecturer at the University of California at Irvine and the University of Notre Dame.

His main areas of research are the history and philosophy of physics and applied mathematics; the intersections between history of science and philosophical ideas, particularly those related to the development of twentieth-century physics; the history of logical empiricism; the core principles of mathematical physics, among them the principle of least action; the history and epistemology of universal encyclopedias and the philosophy of applied science, especially the role of models and ceteris paribus laws.

He is co-editor of *Wiener Kreis* (with Thomas Uebel) (2006); *Time and History* (with Friedrich Stadler), (2006); *Appraising Lakatos: Mathematics, Methodology, and the Man* (with G. Kampis and L. Kvasz) (2002); and *John von Neumann and the Foundations of Quantum Physics* (with M. Rédei), (2001). His papers include: 'Eine Enzyklopädie für das Kaiserreich: Die Kultur der Gegenwart im Kontext der Geschichte philosophischer Enzyklopädien', *Berichte zur Wissenschaftsgeschichte* 31 (2008): 11–28; 'On Optimism and Opportunism in Applied Mathematics', *Erkenntnis* 60 (2004): 121–45; 'The Least Action Principle as the Logical Empiricist's Shibboleth', *Studies in History and Philosophy of Modern Physics* 34 (2003): 285–318; 'Franz Serafin Exner's Indeterminist Theory of Culture', *Physics in Perspective* 4 (2002): 267–319; and 'Vienna Indeterminism: Mach, Boltzmann, Exner', *Synthese* 119 (1999): 85–111.

Frans van Lunteren studied physics at Utrecht University. In 1991 he received his PhD with a dissertation on theories of gravity from Newton to Einstein. He subsequently worked as a university lecturer at the Utrecht Institute for the History and Philosophy of Science. He is now Professor in the history of science, both at the VU University of Amsterdam (since 2002) and at Leiden University (since 2007). He has edited several volumes on the history of Dutch science in the nineteenth and twentieth centuries. His interests range from early nineteenth-century Humboldtian science to postwar radio-astronomy, from standardisation to discipline formation, and from natural theology to the place of science in modern society.

M. Norton Wise is Distinguished Professor of History and Co-director of the Center for Society and Genetics at UCLA. Methodologically, his work has focused on the cultural history of modern science, or the co-evolution of science and social life. It has ranged over the history of energy physics in Britain (with Crosbie Smith), quantum mechanics, the relation of political economy and natural philosophy in the nineteenth century, social resources for laboratory science in Berlin, the gender of time and of automata, English landscape gardens in Prussia (with Elaine Wise) and the changing character of explanation in recent science. He has PhD degrees in nuclear physics from Washington State University and in the history of science from Princeton University and has held positions at Auburn University, Oregon State University, UCLA, Princeton University and again, UCLA.

Stefan L. Wolff belongs to the Research Institute of Deutsches Museum, Munich and teaches history of physics at the University of Munich (LMU). He is co-editor of *Das Deutsche Museum im Nationalsozialismus: Eine Bestandsaufnahme* (2010). He does research in history of physics in the nineteenth and twentieth centuries, with special interests in kinetic theory and thermodynamics, the papers of Wilhelm Wien, physics during the era of National Socialism and the emigration of physicists at that time.

The Forman Thesis: 40 Years After

Cathryn Carson, Alexei Kojevnikov and Helmuth Trischler

Forty years ago, in 1971, Paul Forman published *Weimar Culture, Causality, and Quantum Theory, 1918–1927: Adaptation by German Physicists and Mathematicians to a Hostile Intellectual Environment*. His landmark study (too long, too thorough and too fundamental to be called simply an article) became immediately famous, and famously controversial. It has remained at the heart of debates about the historical relationship between science and culture ever since. The controversy surrounding the Forman Thesis was practically unavoidable, for Forman's work put forward and placed at the centre of a broader discussion the argument that the cultural values prevalent in a given place and time could influence the results of discipline-bound research, i.e. the very content of scientific knowledge. This idea, if still controversial, has since become commonly used in cultural studies of science, but at the time of its introduction it created uproar as it explicitly contradicted generally accepted and cherished beliefs about science. Yet tectonic shifts were already underway, if not always visible, that would eventually put those very beliefs into question. The Forman study both reflected and forwarded these shifts in our general perspectives on the nature and practice of science. Despite some heated objections to its findings, Forman's work has fundamentally changed directions of research in the history, sociology and philosophy of science and established itself as a classic in this group of fields, sometimes collectively called science studies. In subsequent decades it has been read and discussed in practically every graduate program that trains students in those fields, circulating in numerous copies and translated into many languages, while the original publication in the journal *Historical Studies in the Physical Sciences* has long become a bibliographic rarity.

Forman's seminal paper is reprinted in full in the first part of this volume, along with his related articles of the same period, so that the reader may gain a comprehensive introduction to the Forman Thesis. Here, a short summary of its main argument suffices to present the issues at stake. Forman described the cultural climate in economically depressed and socially volatile Germany after the nation's defeat in World War I and the collapse of the German Empire. A sense of spiralling social crisis affected all aspects of life, including science. It particularly inspired widespread discussion about the 'crisis in science', which encouraged some scientists to question the conceptual foundations of their respective disciplines. Fashionable philosophical and ideological treatises, including Oswald Spengler's *Decline of the West*, mobilised the educated public against ideas of rationality, progress, modernity and materialism popular prior to the war. In this new romantic intellectual atmosphere that elevated intuition and celebrated irrationalism, many academics wavered in their attachment to the rationalist values heretofore central to the practice of the exact sciences. Mechanical determinism, or the causality principle, came under special criticism as being too rationalistic and, indeed, mechanical. Responding to such hostile critiques, a number of prominent physicists and mathematicians expressed a readiness to accommodate their discipline to the *Zeitgeist* by abandoning or restricting the validity of causality in physics. Forman showed that proposals such as this predated 1925, the year of the invention of quantum mechanics. Once that revolutionary theory appeared, acausality was quickly ascribed to it and proclaimed the fundamental scientific principle of the new quantum mechanics of atoms and electrons.

As a methodological model, Forman's study has arguably been the most influential article ever published in the historical studies of science, with the possible exception of Boris Hessen's equally famous and controversial 1931 analysis of classical mechanics in *The Social and Economic Roots of Newton's Principia*. The two works have sometimes been confused as representing, in the eyes of some critics, the same 'externalist' approach to science. It is true that both upended, each in its own way, the essentially Platonic ideology of science as a pure intellectual activity, a noble search for abstract truth supposedly in control of its intrinsic scientific method and of the criteria of true knowledge. Instead, both approached science as an essentially human, and thus also earthly, social and cultural activity, and accepted the necessary epistemological consequences of such an assumption. Yet the differences between these two classic works, separated by another 40 years, are no less important than the similarities. Hessen developed a demonstrably Marxist argument that focused primarily on economic and technological influences on science. In Forman's analysis, culture plays a key role, mediating and channelling the impact of economic and social conditions. Hessen, writing in the midst of the revolutionary industrialisation of the Soviet Union, promoted an unabatedly optimistic view on science, counting it without reservation among the major forces of social and political progress. In Forman's day, in the era of DDT, napalm and Agent Orange, the question of science and scientists' political associations became less straightforward. His study found some leading proponents of the quantum revolution entering a pact with anti-rationalist conservative ideological currents whereas those physicists who upheld the values of causality and reason, often rhetorically dismissed at the time as scientifically 'conservative', were frequently aligned with the progressive forces supporting the Weimar Republic. Last, but not least, Hessen's essay was largely declarative and programmatic. It inspired and required further empirical justifications, including Robert K. Merton's *Science, Technology and Society in Seventeenth Century England* (1938). The Forman Thesis relied on a extensive foundation of primary sources, many heretofore unused, and came out of a broader empirical — archival and historical — study then underway.

The professional historiography of the quantum revolution, which was only emerging during the 1960s, stood at that time at the forefront of methodological innovation in the history of science. In preceding decades, with quantum mechanics still relatively recent science and its founders active and publishing, its history was dominated by participants' own accounts and physicists' popular writings. Entering the field as trained historians of science, Thomas Kuhn, John L. Heilbron, Paul Forman and Lini Allen embarked in 1961 upon the ambitious project of the Archive for the History of Quantum Physics (AHQP) — not an archive in the usual sense but a comprehensive effort to locate and catalogue an international body of manuscripts and correspondence of several hundred quantum scientists active between approximately 1900 and 1935. The resulting primary source database became the main foundation for practically every historian working in the field since, including AHQP team members' own research into the history of quantum ideas. Taking a proactive approach to sources, the AHQP project microfilmed many crucial collections, bringing them closer to researchers. It also pioneered the technique of oral history in the history of science by recording interviews with about 100 physicists, including Niels Bohr, Max Born and Werner Heisenberg.

As related in this volume by John L. Heilbron in his recollections of that *Sturm und Drang* era, contemporary events influenced historians' thinking about the past and encouraged them to ask questions beyond the traditional repertoire of the history of ideas. If the Vietnam War and the Cold War attracted their critical attention to scientists' social and political roles, the post-Sputnik transformation of science into a mass profession with an outsized infrastructure inspired inquiries about the scientific community as a whole, the institutions and funding that sustained it, and the social relations operating within it. A team effort by Forman, Heilbron and Spencer Weart produced by 1975 an international survey of physics as disciplinary practice circa 1900. Forman's own doctoral dissertation of 1967 analysed the German physics community from the angle of its placement within the economic, political and social situation immediately after World War I. The dissertation remained unpublished (unjustifiably so), but established the empirical foundation for the development of the Forman Thesis three years later and has since been used by many historians in the new area of research that it opened up. Two of Forman's articles closely linked to its content are reprinted below: on the international relations of German science during the postwar boycott and its struggles for funding and political alliances. Together, they describe the postwar academic community and the general social environment within which the ideas of quantum acausality would brew. The other two papers that complete Part I of this volume extend the original Forman Thesis on its own terms. One takes the argument beyond the issue of causality by analysing other culturally sensitive notions (individuality and *Anschaulichkeit*) that were important for Weimar physicists' thinking, while the other examines the reception of quantum acausality outside the German cultural sphere, particularly in Great Britain.

The history of quantum physics has since come into its own as a mature subject, with detailed studies of its technical formalism, philosophical interpretation and institutional settings as well as biographies and editions of the collected papers and letters of its major contributors. Forman's work served as a touchstone for new generations of researchers; the methodology it adopted and the questions posed continue to generate further inquiries and controversies. Historians developing the genre of the cultural history of science applied and extended Forman's argument further, adapting its conceptual language and questions to other cases and situations, and checking its applicability in different cultural milieus. In March 2007, a conference in Vancouver, British Columbia, brought together scholars whose research addresses historical and philosophical problems related to the Forman Thesis. Many working in the field today — whether they agree with Forman, disagree over details or even disagree profoundly — feel indebted to his inspiring ideas. The papers in Part II of this volume represent, first of all, the lines of argument developed in contemporary research on the history of quantum physics. They explore different cultural environments of physics from the *fin de siècle* to the mid-twentieth century, compare the situation in Germany with those in other countries, analyse the cases of some major contributors and their detailed interactions in the process of inventing quantum mechanics and look at those quantum physicists who, in different political and social circumstances, strove for a causal understanding of the theory.

The latter case deserves special comment, since attitudes toward the philosophical interpretation of quantum theory have changed dramatically in recent decades. When the Forman Thesis was published in the early 1970s, the acausality of the quantum laws was generally seen as part of the core scientific formalism of the theory in accordance with the dominant Copenhagen Interpretation. Due largely to the work of David Bohm and John Bell, physicists' views shifted, allowing more space for a philosophical pluralism within which different interpretations, including causal ones, are possible. Bell was aware of Forman's historical critique, which may have provided additional encouragement to his efforts in challenging the Copenhagen orthodoxy from the physics side. In recent years, more historians and philosophers of science have turned in their analyses to those physicists who disagreed with prevailing opinion and defended the causality principle in quantum mechanics, to whose previously neglected views Forman had called 'sympathetic attention'.

Historians have also extended Forman's argument to cases beyond physics and the exact sciences. The Weimar cultural atmosphere and the contemporary setting for the academic enterprise affected other fields of scholarship as well, if not in exactly the same way and without always framing causality as the main issue at stake. The papers in Part III of this volume explore similar ideological and political issues in cases involving the life sciences, the human sciences and philosophy, and in the ways that science was popularised and presented to the general public. Forman's ideas have also been tested for utility in a larger chronological frame of German history outside the Weimar period *per se*, in particular in inquiries into whether similar cultural influences continued during the Nazi period, or even to a certain degree into the post-World War II era. While many historians have built upon Forman's themes in the depiction of the cultural milieu, others have recorded variations and argued about non-uniformity of culture and its effects. An important and particularly contested problem for contemporary scholarship is the manner in which Weimar culture was heterogeneous, combining anti-modern with modernist trends.

Over time, perceptions of the Forman thesis have changed. Initially, most of the opposition came from technical challenges and, sometimes animating them, the modernist rejection of the possibility that scientific knowledge, supposedly universal, could be influenced by its local and idiosyncratic cultural setting. Yet as the paradigmatic example demonstrating such influence, the Forman Thesis was instrumental in the rise of new scholarly approaches to science during the 1970s and 1980s: namely, the sociology of scientific knowledge and cultural studies of science. With the growing number of examples involving various cultures and scientific disciplines, scholars who described them met with significantly less opposition than Forman initially did, until the understanding that science is produced locally in particular cultural and social settings has become widely accepted, almost to the point of hardly requiring a proof. Current assumptions about science, however, make it harder to explain how such locally produced knowledge manages to travel across cultures, establishing itself internationally; hence the importance of comparative studies related to the case described by Forman and similar ones. And while the central methodological lesson of the Forman Thesis has been broadly accepted, other aspects continue to cause disagreement as well as discomfort to contemporary postmodernist feelings. Forman's account of physicists succumbing to ideological currents of the day, in particular, flies in the face of contemporary currents that insist that individuals are free agents, even when they shop as prescribed by the latest TV commercial. Here, we are dealing with one of the basic contradictions of contemporary postmodernist culture, which may only be sorted out once this culture, too, becomes a thing of the past.

Paul Forman's subsequent work on topics beyond science in Weimar Germany cannot be adequately represented or discussed in this volume. However, some of his major publications are listed in the bibliography that follows this introduction. In a series of groundbreaking papers in the 1980s and 1990s, Forman subjected to similar critical analysis the Cold War era in American science, in particular the then dominant ideology of scientists who claimed to be able to pursue 'pure science' while receiving funds from the military establishment and its granting agencies. In case studies centred on the development of quantum electronics, the invention of the maser and atomic clocks, Forman displayed how military and Cold War agendas underwrote the development and directions of basic research in physics in the 1940s and 1950s. His analysis, sometimes called the 'second' Forman thesis, has been as influential in the field of history of American science as his works on causality and quantum physics were in the history of science in Germany.

Postmodernism, too, has already experienced its own history. With Forman's finely-tuned radar for pressing questions and problems, he has come full circle to identify postmodernity as a provocation for historians of science and technology. In a recent article, again of monograph-like character, Forman discussed the controversial question of the aims and direction of knowledge production in postmodern societies. He argues that the 1980s experienced a deep caesura in the relation between scientific and technical knowledge. During the period of modernity, science held priority relative to technology. Then, in a sudden and unexpected reversal in the 1980s, technology gained priority over science. For Forman, the subsumption of science under technology serves as a demarcator of postmodernity from modernity, and he makes no bones about the fact that from a normative stance he is highly critical of this change. To the contrary: in his article, which meanwhile has become another landmark in the history and philosophy of science, Forman takes to task his fellow historians of technology. They have ignored the epochal elevation of the cultural standing of technology — and he attributes their deliberate ignorance of that fact to the ideological character of their discipline. Without claiming to be visionary, it is easy to predict that Forman has yet again come up with a seminal publication that will keep cohorts of students in science studies busy in the years to come.

Cold War Culture, History of Science and Postmodernity: Engagement of an Intellectual in a Hostile Academic Environment

*J.L. Heilbron**

I thank the organisers of this welcome meeting for the opportunity to declare my debt to my old teacher Paul Forman. This is not an acknowledgment got up for this occasion — it occurs from time to time in my correspondence with him, from which, with his permission, I shall quote as I proceed.[1] The hostile intellectual environments to which my title refers are Berkeley, Weimar and Washington.

Berkeley

There was a professor in the picture when we were teaching one another in graduate school in Berkeley almost 50 years ago. He was T.S. Kuhn, then maturing his paradigms and, despite the opening to social history they were to give to others, still teaching and practising the history of science in the style of Alexandre Koyré. The three of us learned a lot about history making in the project we undertook between 1962 and 1964 to preserve the recollections and microfilm the papers of decaying quantum physicists. Although few of them could answer the ridiculously detailed questions about the intellectual history of quantum physics we put to them, many proved able to recollect much that was interesting and sometimes dramatic, and also plausible, about their early lives, training, teachers, colleagues and institutions.[2]

---

* April House, Shilton, Burford, Oxfordshire, OX18 4AB, England; john@heilbron.eclipse.co.uk.

 The following abbreviation is used: *HSPS, Historical Studies in the Physical Sciences* (later, *Historical Studies in the Physical and Biological Sciences*).

[1] J.L. Heilbron to Paul Forman, 23 July 1992: 'I always thought I learned more from you — your thesis, our collaboration, our correspondence, your papers — than from anyone else'. All correspondence cited in this essay is in the author's possession.

[2] Thomas S. Kuhn, J.L. Heilbron, Paul Forman and Lini Allen, *Sources for History of Quantum Physics: An Inventory and Report* (Philadelphia: American Philosophical Society, 1967), 4–5.

Of course, 40 or 50 years of intervening experience had distorted their recollections. We had to try to purge the alleged facts from the teller's fancy — as I recommend you do with what you hear from me today.

Paul was the only one of us who saw the point and possibility of testing our actors' testimony about the milieu against contemporary descriptions. He took Alfred Landé, one of the few respondents able to recollect the development of ideas as well as the character of the milieu, and who had preserved some important contemporary correspondence, as his primary probe and problem. In the original design of his dissertation, Paul intended to explore the routes to innovation in atomic physics during the 1920s through biographical studies of Landé and others who tried to puzzle out the Great Doublet Riddle, which Paul identified as the key crisis of the old quantum theory.[3] It happened that he wrote up his account of Landé's invention of the *g*-factor during the turbulent early years of Berkeley's Free Speech Movement.[4] The outcome was a change in plan. He directed his dissertation at environment rather than innovation. Perhaps he had convinced himself, as he later wrote, that a genesis account, in the sense of rational reconstructions of an individual's steps to a discovery, is 'an impossible and contradictory undertaking'.[5]

The question of the legitimacy of genesis accounts played a large part in the genesis of Forman's historiography, for its importance in its own right and for its relation to the key issue, for him, of the relations between historians of science and scientists. In 1969, Kuhn and I published a paper entitled 'The Genesis of the Bohr Atom', fighting words, almost, to Paul. He wrote me a nice note about the narrative but criticised our neglect of the prior problem, whether the goal ('a reconstruction of Bohr's *path*') was accessible in principle.[6] He had been trying to solve this problem in a new introduction to his then still unpublished paper on Landé's innovation. 'The reasoning now is no longer simple, but rather involved, and perhaps sophistical. I'm not sure I've convinced myself'. Perhaps a little clear thinking, as he put it, would 'return me to the repudiation of geneses'.[7] Perhaps the genesis of the Bohr atom had clouded his mind. As his more recent studies of Townes and Rabi evince, however, he never

---

[3] Paul Forman, 'The Doublet Riddle and Atomic Physics circa 1924', *Isis* 59 (1968): 156–74.

[4] Paul Forman, 'Alfred Landé and the Anomalous Zeeman Effect, 1919–1921', *HSPS* 2 (1970): 153–261.

[5] Paul Forman, *Fisici a Weimar: La cultura di Weimar, la causalità, la teoria quantistica*, trans. Tito M. Tonietti (Pistoia, Italy: Editrice CRT, 2002), 6–7.

[6] J.L. Heilbron and Thomas S. Kuhn, 'The Genesis of the Bohr Atom', *HSPS* 1 (1969): 211–90; Forman to Heilbron, 10 December 1969.

[7] Forman to Heilbron, 8 January 1970.

definitively rejected genesis; or rather, as will appear, he found out how to have genesis stories and independence too.[8]

The problem is not trivial. Can we just scrap the genesis stories told by scientists? If not, how can we use them? 'It's enough to make one tear one's hair, for how can we possibly avoid making use of such testimony, or know (without the documents that would make the testimony superfluous) which are tainted, and to what extent?'[9] Paul made an ingenious use of what he called the myths of origin of X-ray crystallography. P.P. Ewald, grand old man in the field and guardian of the myths, immediately dismissed Paul's paper as 'worthless except as an example to which conclusions the study of the literature under a biased point of view can lead'.[10] What better confirmation could one require that in this case genesis had become gospel? Or, as Clifford Truesdell, the maverick editor of the journal in which Paul's exposé appeared, put it, that 'physicists are as pig-headed and superstitious as any monk of the dark ages'?[11] At one stroke Paul had exploded a myth, confirmed his doubts about intercourse with scientists and showed that history can be an experimental field — at least in Bacon's sense of twisting the lion's tail.

Let us look for the forces in Forman's environment at Berkeley that might have prompted his turn from genesis accounts to a dissertation on the social, political and economic circumstances in which Weimar physicists worked. I've mentioned one — the Free Speech Movement (FSM) going on around him as he tried to place his actors in their environment. Another was Hunter Dupree, Forman's *Doktorvater*, who taught the histories of science and technology in America. Dupree began his classes by reading and ranting at a clipping he had taken from the day's newspaper. Paul's classnotes explain that discussing the clippings allowed Dupree 'to work off his righteous indignation [and] to convince himself that the lecture that follows is dealing with a subject worth our attention'.

---

[8] The papers in question are: '"Swords into Ploughshares": Breaking New Ground with Radar Hardware and Technique in Physical Research after World War II', *Reviews of Modern Physics* 67 (1995): 397–455; 'Into Quantum Electronics: The Maser as "Gadget" of Cold-War America', in *National Military Establishments and the Advancement of Science and Technology: Studies in Twentieth Century History*, ed. Paul Forman and José Sánchez-Ron (Dordrecht: Kluwer Academic Publishers, 1996), 261–326; 'Molecular Beam Measurements of Nuclear Moments before Magnetic Resonance: I. I. Rabi and Deflecting Magnets to 1938. Part I', *Annals of Science* 55 (1998): 111–60; 'Researching Rabi's Relics: Using the Electron to Determine Nuclear Moments before Magnetic Resonance, 1927–1937', in *Artefacts: Studies in the History of Science and Technology*, vol. 2, *Exposing Electronics*, ed. Bernard Finn (Amsterdam: Harwood Academic Publishers, 2000), 161–74; and 'Inventing the Maser in Postwar America', *Osiris* 7 (1992): 105–34, esp. 126–29.
[9] Forman to Heilbron, 2 November 1968.
[10] Paul Forman, 'The Discovery of the Diffraction of X-rays by Crystals: A Critique of the Myths', *Archive for History of Exact Sciences* 6 (1969): 38–71; P.P. Ewald to Stephen Brush, 1 March 1969.
[11] Clifford Truesdell to Forman, 10 March 1969.

Paul thought Dupree's combination of current events with historical analysis convincing, and his 'passionate engagement with the material' inspiring. It was not smooth sailing, however, with the passionate Dupree, a former Naval officer who would have liked to try the leaders of FSM for mutiny. Paul and Karl Hufbauer, then Dupree's teaching assistants, were sympathetic to FSM. Paul was 'caught between conflicting loyalties', between a professor he admired and a social movement he could not resist.[12]

In 1991, Dupree received the History of Science Society's Sarton Medal. Paul drew up the motivation. What he chose to praise can be read as autobiography. Dupree merited the medal because of his 'conviction ... that historical knowledge can be, must be, applied prescriptively to guide our future', his insistence that history of science is social history and his liberation of the 'historian of science to construe the world from his or her own point of view, to judge the better or the worse not by the authority of science, but by his or her own lights and insights'. By that time, 1991, Paul had made this liberation a theme and goal of his writing. He believed that he had come only recently to demand such freedom for himself; but in this he was mistaken.[13]

Forman had declared his freedom from science in his paper on Landé. He put it this way in the final, published version of the introduction: a genesis story should illuminate the practice of a community and not pretend to a reconstruction of a discovery. Thus the 'strength of the thought controls exercised by particular doctrines' would stand revealed and the internalist-externalist distinction be dissolved. And then came a flash of light. Forman observed that 'th[is] distinction [is] so foreign to contemporary historical scholarship that we must regard its persistence among historians of science as one of the more blatant ideological atavars testifying to our ... connection with the sciences'.[14] The unhappy distinction between the external and the internal is a consequence of our incomplete liberation from the scientists whose history we write: they made it, to preserve their claims to autonomy, and therefore, if for no other reason, we must discard it.

Paul's struggle for freedom became the more insistent as he began personally to feel the pressure of institutions. He writes from the University of Rochester that his colleagues there will not listen to his agitation on his and their behalf. '[They] simply do not want to pay the price of freedom and self-determination. Back to the old stance of ignoring and despising [those who will not think for themselves] ....

---

[12] Forman to Hunter Dupree, 22 December 1981.
[13] Paul Forman, 'Sarton Medal Citation' [for A.H. Dupree], *Isis* 82 (1991): 281–3; Paul Forman, 'Independence, Not Transcendence, for the Historian of Science', *Isis* 82 (1991): 71–86.
[14] Forman, 'Alfred Landé' (ref. 4), 156–57.

To Hell with them all!' He is Galileo against the Simplicios, the Royal Society against nincompoops who swear in the words of others. Shortly after taking up his curatorial position at the Smithsonian, he cries out: 'I want very much that my work feel like my own free activity, an expression of myself', and not the smothering vicarious limping life of a curator, 'for what else is it when one seeks to learn what other people — contemporary physicists — judge to be significant in their line of work, and then set about gaining possession of these objects and taking pleasure in having done so?'[15]

Forman's dissertation (not his 'thesis') was a revelation to me. Quite apart from its wealth of information on the great inflation, the consequences of the postwar isolation of Germany, intellectual life as a substitute for political and military power, anti-Semitism in the universities and so on, it demonstrated that, and also how, the powerful analytical tools and bibliographical resourcefulness of general historians, and also their results, could be brought to bear on the history of science. Like other historians, we have to read as widely as possible in the period of our interest, look at its art and architecture, listen to its music and try to understand its social and political life: not necessarily to practise the sociology of knowledge, but to bring ourselves closer to the experiences of the people we write about. Koyré had revealed to Kuhn and his generation that the history of science was not a chronicle of true belief, but of the battle of ideas required to arrive at useful concepts. Forman revealed to me that the history of science is history and that the writing of history was as demanding, creative and rewarding a discipline as the doing of science.

Weimar

You all know the story, the carefully constructed facts and their explanations. First we learn what no one knew before, that German-speaking physicists and mathematicians liked to express sympathy for acausal theories before general academic audiences and also before the invention of quantum mechanics. We learn further that this uncharacteristic discourse was a sort of *Gleichschaltung*, an alignment with a hostile intellectual regime. The speakers hoped thereby to raise their prestige and improve their resources.[16] There are of course other possible explanations,

---

[15] Forman to Heilbron, 'Saturday Afternoon', 1971 (?); and Forman to Heilbron, 23 December 1974. The latter complaint was motivated in part by Forman's concurrent effort to disentangle his life from his wife's and by his unsuccessful negotiation over editing the Einstein Papers.

[16] Paul Forman, 'Weimar Culture, Causality, and Quantum Theory, 1918–1927: Adaptation by German Physicists and Mathematicians to a Hostile Intellectual Environment', *HSPS* 3 (1971): 1–115, on 3, 7–8.

some more flattering to the historical actors than Forman's. For example, reference to matters of current interest is a standard way for a specialist to gain the attention of a general audience. Perhaps Forman's actors, or some of them, were not grantsmen or opportunists but perplexed specialists condemned to speak to a general university audience condemned to listen; and, as a *captatio benevolentiae*, they described recent developments in their field so as to awaken people presumed to have overdosed on Spengler and crisis talk.

Did Forman's actors sincerely advocate acausal physics or did they cynically pretend to place their field under the prevailing *Geist* so as to improve their social and financial situation? The question is perhaps secondary in comparison to the thesis that calls for acausality, and the crisis talk into which they fitted, contributed more than the troubles of the old quantum theory to the invention of quantum mechanics.[17] For students of Forman's historiography, however, the question, sincere or cynical, has priority. He accepts the question as legitimate — 'the historian ... must insist upon a causal analysis' — but, as he acknowledges, his choice of the sociological over the psychological method does not give him the tools to resolve it.[18]

Nonetheless, he lets us know indirectly where he stands. In the style of Dupree, he compares the situation in Weimar with that in America at the time of writing, 1970, when scientists were making 'a far reaching [and, in Forman's view, undesirable] accommodation of scientific ideology to a hostile intellectual milieu'.[19] (This reference to his immediate environment and hint about his attitude toward it is not whiggish insofar as 'all true history is contemporary history'.[20]) To continue: Forman criticises the rhetoric of *Gleichschaltung* as duplicitous, 'weak, vague, and disingenuous', and the talk of crises as 'an entrée, a ploy, to achieve instant "relevance"'.[21] Concessions of acausality are 'capitulations to Spenglerism' and 'repudiations [by physicists and mathematicians of] their own disciplines'.[22] Forman identifies with the opponents of the accommodationists, with the conservatives, with Planck and Einstein. Why? Because they were

---

[17] Ibid., 108, 110. Writing with Kuhn's *Structure* in mind, Forman makes even the '*possibility* of the crisis ... dependent upon the physicists' own craving for crises, arising from participation in, and adaptation to, the Weimar intellectual milieu' (quote in ibid., on 62).

[18] Ibid., 3, 114. Forman to Heilbron, 11 March 1971: '[S]ensitivity to the particular circumstances of the different individuals ... is so strikingly lacking in my "Weimar Culture ..." with its unsympathetic, and "mechanistic", treatment of the process of adaptation to intellectual environment'.

[19] Forman, 'Weimar Culture' (ref. 16), 5.

[20] Benedetto Croce, *Teoria e storia della storiografia* (Milan: Adelphi Edizioni, 1989), 14.

[21] Forman, 'Weimar Culture' (ref. 16), 50–51, 58, 91.

[22] Ibid., 86, 48, 55.

'uncompromising and courageous' defenders of reason, men faithful to their discipline, true *Wissenschaftler* who accepted causality as the precondition of all science.[23]

Paul did not regard his temperate defence of discipline, which became a passion, or his exposure of the moral faults of physicists, which would become a favourite pastime, as the primary programmatic product of his time in Weimar. Rather, it was the demonstration that the autonomy of modern science, conducting its own affairs according to the paradigm of the moment, was a fiction. The behaviour of the Weimar physicists, if they really did make, or accept, an acuasal physics in response to the wider cultural environment, knocked down our old professor's *Structure of Scientific Revolutions*. Paul found that gratifying.[24] He also found gratifying the corollary, which I put in this way in a letter in 1980:

> As to physics in the hostile milieu, I think it is the most important and original work on 20th century physics by an historian. It is distinguished by method, content, and "above all" (as you would say) by freedom, by its declaration and exemplification of independence from physicists' history …. Which is not to say that I accept its thesis or would allow it to fall into the hands of children.

To which he replied, 'I was struck by your criterion: distance from physicists' history. And I realized that, for psychological reasons, nearly all my work has aimed at just that. But you're absolutely right that it should not be put into the hands of children'.[25] Please, dear colleagues, take note of this warning!

I have a fatherly feeling toward the Weimar paper that may not jump out from the remarks just quoted. Paul decided to unveil it at the Christmas meeting of the History of Science Society in 1970. He also decided that I should unveil it. He had not been feeling very energetic and, as the time approached, 'felt less and less like seeing people, talking, throwing myself into the maelstrom'. Anticipating the maelstrom, I proposed some small changes, which Paul graciously accepted, and one big one that he did not. I urged that he state explicitly that he did not claim, or claim to have demonstrated, that quantum mechanics is acausal because Spengler was popular. To this reasonable request he returned a noteworthy reply:

> I am reluctant to retrench my claims explicitly; I would like the audience to *worry* rather than be reassured. If I knew *just* what I thought to be the

---

[23] Ibid., 91, 111.
[24] Forman, *Fisici* (ref. 5), 12.
[25] Heilbron to Forman, 23 May 1980, and reply, undated. There is also this in Heilbron to Forman, 2 October 1969: '["Weimar Culture"] can't help but be a *model* for studies in HS outside of the 20th century as well as in it, for (1) coming to grips with the abused int-ext distinction (2) new information (3) Urkundenmeisterei … (4) generally goosing up standards (5) etc'.

implications of my "discovery" I would perhaps be ready to detail my claims and what I do not claim. But in general I have the feeling that physicists *tend* to find the kind of thing they are looking for, and even more generally in the negative: physicists tend not to find the kind of thing they do not want to find.[26]

At least we can say that we would have a bigger burden of explanation if Victorian physics were acausal and quantum mechanics causal.

After I returned from the maelstrom, Paul said that he had a positive reason for throwing me into it. 'I thought it just as well for you to have the opportunity to see how you liked taking responsibility for my irresponsible assertions .... I mean in anticipation of our collaboration'.[27] This referred to an elaborate project conducted in Berkeley from 1971 to 1973.[28] In harness with Spencer Weart and myself, Paul produced the unreadable monograph 'Physics circa 1900', which contains some valuable tables in a sea of footnotes into which more and more information was thrown even as the manuscript made its way through the press.[29] The galleys groaned under so many new numbers, names and notes that they had to be reset entirely. Forman's habits of overkill footnotes and revision up to the instant of printing were socially constructed in our competition to outdo one another in the discovery of recondite references.

The higher purpose of this project, of which only the first of three planned parts appeared, was to provide a firm and detailed benchmark for ourselves (for we intended to carry the work forward) and other historians of modern physics. The published part dealt with funding, workplaces, manpower and outputs; the second part was to describe physicists' institutions from the parochial (the local physical society) through national and international associations; the third was to discuss the self-evaluation of the physics profession, its ideology and prospects; only a piece of the third part saw print — an exercise by me in the style of the Forman thesis.[30]

---

[26] Forman to Heilbron, 11 December 1970.

[27] Forman to Heilbron, 14 January 1971.

[28] In Forman to Heilbron, 4 May 1969, after declaring his dislike of teaching, Paul proposed, 'Why don't you give a little attention to the problem of funding a research institute in the history of science? .... Perhaps we could write a joint proposal'. This was the seed of the collaboration.

[29] Paul Forman, J.L. Heilbron and Spencer Weart, 'Physics circa 1900: Personnel, Funding, and Productivity of the Academic Establishments', *HSPS* 5 (1975): 1–185.

[30] J.L. Heilbron, 'Fin-de-Siècle Physics', in *Science, Technology and Society in the Time of Alfred Nobel*, ed. C.G. Bernhard *et al.* (Oxford: Pergamon, 1982), 51–73.

Washington

The Smithsonian Institution, where Paul began in earnest after leaving Berkeley in 1973, was not the place for a scholar of his habits and interests. It became for him a hostile intellectual milieu. In a letter of 1974, he describes it as 'utterly banal and soporific'. He was able to put together editions of his Weimar papers but neither that accomplishment nor his curatorial duties fulfilled him. In the introduction to the Italian translation of his work written around 1978, he allows that the old essays were very good, not to brag but to admire their author, a person 'able to work and think so hard, so "consequently,"' from whom he felt very distant.[31] He gave himself over to preparing exhibits on atom smashers, parity, atomic clocks and other matters.

They took a lot out of him and also out of the profession. 'In the throes of winding up an exhibit, I feel completely alien to such scholarly concerns [as attending professional meetings]'. He felt himself growing ever more unsociable and a 'more or less frankly angry middle-age man looking for an angle that will justify indulgence of it [the anger?] in my scholarship'.[32] In 1984, he wrote a mutual friend that he no longer derived much satisfaction from his scholarly work; and that what did interest him was 'directed *entirely* toward the contemporary social situation and function of science. I really have very little interest any more in the content of science, present or past .... Indeed, I've largely ceased to believe in the relevance of history, i.e., of *anything* that happened more than 50 years ago'.[33] In July 1974, he began psychoanalysis.[34] Eight years later, he was still on the couch and saw no end in sight.[35] The accelerating hostility of the Smithsonian toward his conception of scholarly work and its place in the institution no doubt exacerbated his condition. In time, he felt compelled to sue the institution to obtain the recognition due him.

By the 1990s, Forman had resumed scholarly work, however hostile the environment. Overcoming his distaste of the vicarious, he curated exhibits, searched for artefacts and broadened his scholarship. In joining the Smithsonian he moved

---

[31] Forman to Heilbron, March 1974; Forman, *Fisici* (ref. 5), 14.
[32] Forman to Heilbron, 12 October 1982 and 25 January 1985.
[33] Forman to Erik Rüdinger, 7 July 1984.
[34] Forman to Heilbron, 1 December 1974. Immediately after the analysis began, Paul wrote this self-evaluation: 'My opinion of my own intrinsic capacities is not high, and no "achievements" seem able to raise it. At the same time my expectations of myself know no bounds, and so I am in a bind which can only be tightened by others sustaining high opinions (= expectations) of me. In almost all cases I can "protect" myself by holding a low opinion of the opinion of others'.
[35] Forman to Heilbron, 22 February 1982: 'When I'll be free of psychoanalysis I am no longer predicting; I am acting as if it will go on forever, and prepared for it to end next week'.

from Weimar to Washington, from German physicists with philosophical complexes to American physicists soaked in pragmatism and entrepreneurship. Like the émigré physicists of the 1930s, he experienced culture shock. He came to grips with the technical details of postwar physics and with the culture of American enterprise. I will say of Forman's truly pioneering studies of postwar radar, quantum optics and electronics, and, more recently, Rabi and molecular beams, only that they appeared in such strongholds of externalist-constructivist-doublethink history as the *Reviews of Modern Physics*, the *Proceedings of the Institute of Electrical and Electronic Engineers* and a publication of the Army Signal Research and Development Command.[36]

Well, I should also say that this work forced Forman to a definitive resolution of the problem of genesis accounts. The resolution: we must entertain causes that overdetermine outcomes. Freed from the necessity of declaring for a single pathway, the historian can 'recognize and wrestle with the very large number of characteristically very different material, social, and personal factors involved'.[37] This is not like assigning ten causes to the Crusades, as in my high-school history book, but, as I understand it, like summing the possible routes of an electron between two states.

As he studied American physics after World War II, Forman's moral juices heated up and boiled over. He indicted physicists for betraying their discipline, for cosying up to the military, for bending and warping their research to please their new paymasters. His lengthy exposé, 'Behind Quantum Electronics', displays his mastery of occult and manifest sources, and the power of argument and luminous simplification that make his writing memorable. Paul has no patience with the physicists' version of their relations with the federal paymaster — that they were in the driver's seat, that they could get everything they wanted because the military felt it needed to keep them and their students on standby in case of a national emergency. On their telling, physicists used their leverage to advance their science as well as themselves, which, as Forman had observed in his old exercise on Landé, was exactly how they do and ('private vices, public virtues') should behave.[38]

Forman prefers a tale of sin and insincerity, of which, no doubt, there were enough. Once again he energised himself by outrage at the contemporary scene, in

---

[36] Paul Forman, '"Atomic Clocks": Preview of an Exhibit at the Smithsonian', in *Proceedings of the 36th Annual Frequency Control Symposium* (US Army Signal Research and Development Command, 1982), 220–22; Paul Forman 'Atomichron®: The Atomic Clock from Concept to Commerical Product', *Proceedings of the Institute of Electrical and Electronic Engineers* 73, no. 7 (1985): 1181–204; Forman, 'Swords into Ploughshares' (ref. 8), 129.

[37] Forman, 'Inventing the Maser' (ref. 8), 129.

[38] Forman, 'Alfred Landé' (ref. 4), 156–57: 'How indeed can we avoid regarding an innovation as the outcome of the motivated acts of an acute man, working in a particular social environment for his own advancement as well as for the progress of science?'

this case, Reagan's Star Wars, the Strategic Defense Initiative, which was buying up the brains of scientists on the make. Similarly, according to Forman, during the 1950s senior academic physicists eagerly accepted money from the military although the proceeds distorted their science. They could do so — and this was their sin — by cultivating and compartmentalising a false consciousness. They worried about their purity and autonomy and managed to obtain perhaps 5 per cent of the military R&D budget for what they considered basic research; but they neglected to take into account the other 95 per cent, which supported their colleagues and students in applied projects of direct military interest.[39] They ignored, 'as scientifically irrelevant, the very large technical component that brought them their quota of scientific freedom'. That was a tragedy. 'Though they have maintained an illusion of autonomy, the physicists had lost control of their discipline'.[40]

The minuet between the funders and the physicists of the 1950s was subtler than this updated Weimar thesis suggests. Forman pointed out that the military categorised most of the projects it supported as applied, whereas the universities categorised them as basic.[41] Who was right? Both sides had constituencies to satisfy: the military had to convince Congress that it spent its money on its missions, while professors had to assure their peers that they solicited money only for basic research. Duplicity and conflict of interest were inevitable. Who got the better of the deal? How do we know how far military money drove scientists from the true path of science? and what was meant by such a path? Forman's story is less about science than about morality, about double-think, the refusal to see things as they are. What irked him appears from his gentle rebuke of the Nobel laureates of Columbia: 'Committed to a snobbish idealism of fundamentality [Columbia's elite relied] on compartmentalization to milk its military sponsors while keeping its own research lily pure'.[42]

Since 1991, in parallel at first with his studies of American science, Forman has focused his intense light on his colleagues. Previously he had done so in blinks, in incisive and not always friendly book reviews, which offered such compliments as 'pretentious and fatuous', 'ludicrous', 'shoddy and muddled', 'turgid and jejune'.[43]

---

[39] Paul Forman, 'Behind Quantum Electronics: National Security as Basis for Physical Research in the United States, 1940–1960', *HSPS* 18, no. 1 (1987): 149–229, esp. 149–50, 152, 198, 227.
[40] Ibid., 228–29.
[41] Ibid., 179–80, 186; Forman, 'Inventing the Maser' (ref. 8), 130–34.
[42] Forman, 'Inventing the Maser' (ref. 8), 133.
[43] Paul Forman, review of *Atomic Order: An Introduction to the Philosophy of Microphysics*, by Enrico Cantore, *Isis* 61 (1970): 535–36. Forman has not lost his zest; see Paul Forman, review of *Uncertainty: Einstein, Heisenberg, Bohr, and the Struggle for the Soul of Science*, by David Lindley, *Isis* 100 (2009): 180–81: 'blinkers ... blindness', 'intellectual arrogance', 'studied opinionatedness', 'sloppiness in reasoning and fact-checking'.

He could also make very helpful comments on manuscripts that the brave and the risk-prone submitted to his judgment. Now, however, his light has become a beacon and in books and papers recalls us to our craft and purpose. A golden text, 'Independence, not transcendence, for the historian of science', rebottled the old wine of Dupree. Our prophet now thunders against a more pernicious fiction than the discredited internal-external distinction, the fiction of the transcendental. Natural science, Forman warns, has required its cultivators to believe in and aim for transcendence, for formulations of knowledge valid for all times and places. Historians of science must not construe historical work as the recovery of old paving stones of the road to Transcendence. Rather, they should choose a subject without regard to the views of scientists; they should investigate whatever their independent judgments think significant and they must historicise, not transcendentalise, whatever they treat. The choice of subject by an historian is a moral one, that is, it depends on taste, character and free will. The choice of a scientist is amoral or worse because it depends on surrendering to the fiction of the transcendental.[44] As he once wrote me, 'the process whereby individuals *willingly*, yea *urgently*, flee their individuality, incorporating themselves into a coherent, coordinated body', had been a mystery to him since grammar school.[45] Paul's inability to incorporate is the source of his strength as an historian and of his conviction (mistaken in my view) that he is a loner.

Many of those who claim the noble title of historian of science fumble when exercising the independence Forman claims for them, fall victim to fads and betray their discipline. Now, in his old age, he sometimes indulges these transgressions as a postmodern inevitability. In postmodernity, he tells us, emphasis is on the local, idiosyncratic and contingent; disciplines fade as fast as the boundaries between them; professional values have little purchase, devotion to long-term projects wanes, standards decline. '[Postmodernity is] vacant of any utopias, vacant of any absolutes, ideals, destinies, or even destinations'.[46] The consequences are not only feckless scholars but also degraded institutions. You will not be surprised to know that the Smithsonian Institution entered postmodernity early by scrapping its 'scholarly integrity' to align with 'the utilitarian relativism of the postmodern political milieu'.[47]

---

[44] Forman, 'Independence' (ref. 13).
[45] Forman to Heilbron, 16 February 1975.
[46] Paul Forman, 'What the Past Tells Us about the Future of Science', in *La ciencia y la tecnologia ante el tercer milenio*, ed. José M. Sánchez-Ron (Madrid: Sociedad Estatal España Nuevo Milenio, 2002), 27–37, on 29.
[47] Ibid., 37.

Forman has a long and spirited paper in press under the provocative title, 'Regarding the Preposterous Primacy of Science Relative to Technology Prior to Postmodernity'. Its disciplined heap of facts evidences the prevalence in modernity of the trickle-down theory that science precedes technology, and in postmodernity of a role reversal that gives technology the priority. The typical postmodern demand for quick payoff and the progressive mechanisation of science have imposed engineering goals and schedules on scientists and crushed whatever was left of pure, long-term, non-mission-oriented research except in fields with little prospect of useful application, such as astrophysics and paleontology. This is a moral outrage. The sinners this time are not scientists, however, but historians, modernist historians of technology, who have missed the revaluation of their subject of study in postmodernity. Their inability to see a development of such consequence to themselves reveals that, as historians, they really are (as, according to Forman, they had believed themselves to be) inferior to historians of science.[48]

Another paper of Forman's, which appeared in a recent special issue of *HSPS* edited by Lew Pyenson in honour of Russell McCormmach, takes much the same tack. Here the facts concern Lewis Mumford, whose writing Paul has put through the finest mesh of his analytical engine, and the sinners are every pre-postmodern who has written about Mumford. Their sins are primarily against discipline: they have not lived up to the standards of historical investigation and analysis.[49] And on those more recent scholars, for whom Paul occasionally invokes the protection of a postmodern decay beyond their control, he confers a terrible epithet. They are 'unlovely'. The persevering reader of 'Preposterous primacy of science', which treats the advent of postmodernity with world-weary wisdom, will find this horrible condemnation in an oasis of a footnote in a desert of disciplinary dissolution. It reads:

> While I believe that no one convinced of the fact of postmodernity can hold any but the slimmest hope for the survival of disciplinarity, the alternative to pursuing scholarship 'as if' we remain fully committed to disciplinary objectives and constraints is a chaos of purposes and practices in which only unlovely characters can thrive.[50]

Historia magistra vitae. We must remain true to our discipline, to our sources, to our colleagues past and present who have laboured over many sources and years

---

[48] Printed as Paul Forman, 'The Primacy of Science in Modernity, of Technology in Postmodernity, and of Ideology in the History of Technology', *History and Technology* 23 (2007): 1–152.
[49] Paul Forman, 'How Lewis Mumford Saw Science, Art, and Himself', *HSPS* 37, no. 2 (2007): 271–336.
[50] Forman, 'Primacy' (ref. 48), 128–42.

to create the instruments and standards of our science — that is, of historical research. Our guru teaches us to avoid cooptation not only by scientists but also by the groups and fads in our own discipline. Historians who exercise independence in their choice of subject matter, honesty in their thought and faithfulness to the highest standards of their discipline have no need of the fiction of transcendence, but every need for a cultivated sense of moral responsibility. For it is well and truly said that although God cannot alter the past, historians can.

Selected Chronological Bibliography of Paul Forman's Scholarship

Thomas S. Kuhn, John L. Heilbron, Paul Forman and Lini Allen, *Sources for History of Quantum Physics: An Inventory and Report* (Philadelphia: American Philosophical Society, 1967).

Paul Forman, *The Environment and Practice of Atomic Physics in Weimar Germany: A Study in the History of Science*. (PhD diss.: University of California, Berkeley, 1967).

Paul Forman, 'The Doublet Riddle and Atomic Physics circa 1924', *Isis* 59 (1968): 156–74.

Paul Forman, 'The Discovery of the Diffraction of X-Rays by Crystals: A Critique of the Myths', *Archive for the History of Exact Sciences* 6 (1969): 38–71.

V. V. Raman and Paul Forman, 'Why Was It Schrödinger Who Developed de Broglie's Ideas?', *Historical Studies in the Physical Sciences* 1 (1969): 291–314.

Paul Forman, 'Alfred Landé and the Anomalous Zeeman Effect, 1919–1921', *Historical Studies in the Physical Sciences* 2 (1970): 153–261.

Paul Forman, 'Weimar Culture, Causality, and Quantum Theory, 1918–1927: Adaptation by German Physicists and Mathematicians to a Hostile Intellectual Environment', *Historical Studies in the Physical Sciences* 3 (1971): 1–115 (translations into German, Italian, Portuguese, Spanish, and Swedish).

Paul Forman, 'Scientific Internationalism and the Weimar Physicists: The Ideology and its Manipulation in Germany after World War I', *Isis* 64 (1973): 150–80.

Paul Forman, 'The Financial Support and Political Alignment of Physicists in Weimar Germany', *Minerva* 12 (1974): 39–66.

Paul Forman, 'Industrial Support and Political Alignments of the German Physicists in the Weimar Republic', in *Industrielles System und politische Entwicklung in der Weimarer Republik*, Hans Mommsen, Dietmar Petzina and Bernd Weisbrod (eds.), (Düsseldorf: Droste, 1974), 716–31.

Paul Forman, J. L. Heilbron and Spencer Weart, 'Physics circa 1900: Personnel, Funding, and Productivity of the Academic Establishments', *Historical Studies in the Physical Sciences* 5 (1975): 1–185.

Paul Forman, 'The Reception of an Acausal Quantum Mechanics in Germany and Britain', in *The Reception of Unconventional Science*, Seymour H. Mauskopf (ed.), (Boulder, CO: Westview Press, 1979), 11–50.

Paul Forman, 'Kausalität, Anschaulichkeit, and Individualität, or, How Cultural Values Prescribed the Character and the Lessons Ascribed to Quantum Mechanics', in *Society and Knowledge: Contemporary Perspectives in the Sociology of Knowledge and Science*, Nico Stehr and Volker Meja (eds.), (New Brunswick, NJ: Transaction Books, 1984), 333–48.

Paul Forman, 'Il Naturforscherversammlung a Nauheim del settembre 1920: una introduzione alla vita scientifica nella Repubblica di Weimar', in *La ristrutturazione delle scienze tra le due guerre mondiali*, vol. 1, *L'Europa*, Giovanni Battimelli, Michelangelo De Maria and Arcangelo Rossi (eds.), (Roma: La Goliardica, 1984), 59–78. Translated into German, with added notes, as 'Die Naturforscherversammlung in Nauheim im September 1920: Eine Einführung in das Wissenschaftsleben der Weimarer Republik', in *Physiker zwischen Autonomie und Anpassung: Die Deutsche Physikalische Gesellschaft im Dritten Reich*, Dieter Hoffmann and Mark Walker (eds.), (Weinheim: Wiley-VCH, 2007), 29–58.

Paul Forman, 'Atomichron®: The Atomic Clock from Concept to Commerical Product', *Proceedings of the Institute of Electrical and Electronic Engineers* 73, no. 7 (1985): 1181–204.

Paul Forman, 'Los propósitos de la historia de la ciencia', *Revista de Occidente* no. 64 (1986): 51–62.

Paul Forman, 'Behind Quantum Electronics: National Security as Basis for Physical Research in the United States, 1940–1960', *Historical Studies in the Physical Sciences* 18 (1987): 149–229.

Paul Forman, 'Social Niche and Self-Image of the American Physicist', in *Proceedings of the International Conference on the Restructuring of the Physical Sciences in Europe and the United States, 1945–1960*, Michelangelo De Maria, Mario Grilli and Fabio Sebastiani (eds.), (Singapore: World Scientific, 1989), 96–104.

Пол Форман, 'К чему должна стремиться история науки', *Вопросы истории естествознания и техники* # 1 (1990): 3–9 (Translation into Russian of 'At What Should the Historian of Science Aim?').

Paul Forman, 'Independence, Not Transcendence, for the Historian of Science', *Isis* 82 (1991): 71–86.

Paul Forman, 'Inventing the Maser in Postwar America', in *Science after '40*, Arnold Thackray (ed.), *Osiris* 7 (1992): 105–34.

Paul Forman, 'Física, modernidad y nuestra evasión de la responsabilidad', *Arbor — Ciencia, Pensamiento y Cultura* 147, nos. 577–578 (1994): 51–74.

Paul Forman, '"Swords into Ploughshares": Breaking New Ground with Radar Hardware and Technique in Physical Research after World War II', *Reviews of Modern Physics* 67 (1995): 397–455.

Paul Forman and José M. Sánchez-Ron (eds.), *National Military Establishments and the Advancement of Science and Technology: Studies in Twentieth Century History* (Dordrecht: Kluwer Academic Publishers, 1996) (Boston Studies in the Philosophy of Science, vol. 180).

Paul Forman, 'Into Quantum Electronics: The Maser as 'Gadget' of Cold-War America', in *National Military Establishments and the Advancement of Science and Technology: Studies in Twentieth Century History*, Paul Forman and José M. Sánchez-Ron (eds.), (Dordrecht: Kluwer Academic Publishers, 1996), 261–326.

Paul Forman, 'Recent Science: Late-Modern and Post-Modern', in *The Historiography of Contemporary Science and Technology*, Thomas Söderqvist (ed.), (Amsterdam: Harwood Academic Publishers, 1997), 179–213. Reprinted with a few revisions in *Science Bought and Sold: Essays in the Economics of Science*, Philip Mirowski and Esther-Mirjam Sent (eds.), (Chicago: University of Chicago Press, 2002), 109–48.

Paul Forman, 'Molecular Beam Measurements of Nuclear Moments before Magnetic Resonance. Part I: I. I. Rabi and Deflecting Magnets to 1938', *Annals of Science* 55 (1998): 111–60.

Paul Forman, 'Ciencia y humanidades en la postmodernidad', *Revista de Occidente* no. 213 (1999): 110–21.

Paul Forman, 'What the Past Tells Us about the Future of Science', in *La ciencia y la tecnología ante el tercer milenio*, vol. 1, José M. Sánchez Ron (ed.), (Madrid: Sociedad Estatal España Nuevo Milenio, 2002), 27–37.

Paul Forman, 'From the Social to the Moral to the Spiritual: The Postmodern Exaltation of the History of Science', in *Positioning the History of Science*, Kostas Gavroglu and Jürgen Renn (eds.), (Berlin and New York: Springer, 2007), 49–55 (Boston Studies in the Philosophy of Science, vol. 248, Festschrift for S.S. Schweber).

Paul Forman, 'The Primacy of Science in Modernity, of Technology in Postmodernity, and of Ideology in the History of Technology', *History and Technology* 23 (2007): 1–152.

Paul Forman, 'How Lewis Mumford Saw Science, and Art, and Himself', *Historical Studies in the Physical and Biological Sciences* 37 (2007): 271–336.

Paul Forman, '(Re)cognizing Postmodernity: Helps for Historians — of Science Especially', *Berichte zur Wissenschaftsgeschichte* 33 (2010): 157–175.

Part I

THE FORMAN THESIS

"Demostrate at home! As proof of his unshakable protest, every morning in Berlin Professor Patrius Rumbler gives the globe a couple of resounding smacks on its face" (Cartoon by Olaf Gulbransson from Simplicissimus, 1919, 24 : 40.)

Scientific Internationalism and the Weimar Physicists: The Ideology and Its Manipulation in Germany after World War I

By Paul Forman*

I. INTRODUCTION

THIS ESSAY IS NOT LIMITED to the Weimar physicists nor does it extend to a consideration of the institutional forms for international scientific relations established in the years following World War I.[1] It aims to achieve some conceptual clarity about the ideology of scientific internationalism and to trace the fate of that ideology as it interacted in the Weimar period with political circumstances and with other elements of academic ideology. For my efforts in this field the work of Brigitte Schröder-Gudehus has been an essential starting point, guide, and support,[2] and the result of these efforts is a mode of treatment far closer to Dr. Schröder's than I had anticipated at the outset.[3] Further, I owe to A. Hunter Dupree[4] and Karl Hufbauer[5] much stimulus and insight in examining the interrelations of nationalism and internationalism in science.

Structurally, the argument developed in the following sections has three main

---

Received Jan. 1972: revised/accepted Aug. 1972.

* Department of History of Science and Technology, National Museum of History and Technology, Smithsonian Institution, Washington, D.C. This essay is a revision and amplification of a paper read at the conference on "Science, Government, and Internationalism, 1900–1939," organized by Roger Hahn under the auspices of the Institute of International Studies, University of California, Berkeley, April 3–4, 1970. I have profited from the criticisms of Gerald Feldman, Joseph Haberer, Roger Hahn, John L. Heilbron, Karl Hufbauer, and Daniel J. Kevles.

[1] For institutional forms see Daniel J. Kevles, "'Into Hostile Political Camps': The Reorganization of International Science in World War I," *Isis*, 1971; *62*:47–60, and "The International Research Council, 1914–1931," to be published in the proceedings of the above conference; Brigitte Schröder-Gudehus, *Deutsche Wissenschaft und internationale Zusammenarbeit, 1914–1928* (Geneva: Dumaret & Golay, 1966); Harold Spencer-Jones, "The Early History of ICSU, 1919–1946," *ICSU Review*, 1960, *2*:169–187.

[2] Schröder-Gudehus, *Deutsche Wissenschaft*; "Charactéristiques des relations scientifiques internationales, 1870–1914," *Journal of World History*, 1966, *10* (No. 1):161–177; "Les professeurs allemands et la politique du rapprochement," *Annales d'études internationales* (Genève), 1970 (No. 1):23–44.

[3] *Cf.* my review of Schröder-Gudehus, *Deutsche Wissenschaft* in *Isis*, 1969, *60*: 589–590.

[4] A. Hunter Dupree, "Nationalism and Science —Sir Joseph Banks and the Wars with France," in *A Festschrift for Frederick B. Artz*, ed. D. H. Pinkney (Durham, N.C.: Duke University Press, 1964), pp. 37–51.

[5] Karl Hufbauer, "The Formation of the German Chemical Community, 1700–1795" (Dissertation, University of California, Berkeley, 1970; Ann Arbor: University Microfilms, 1970).

components: the substance of the ideology of scientific internationalism (Sections II and III); the manipulations of that ideology (Sections IV, V, and VII); the rationale underlying those attitudes and actions (Sections VI and VIII).

In Section II, I consider "internationalism" as an element of the ideology of scientists, stressing the essentially nationalistic foundations and functions of scientific internationalism while emphasizing that the classical formula for the participation of the nation in the scientist's fame spares the scientist any conflict between advancing his science and advancing the interests of his nation.

A most essential tenet of this ideology is the universality of scientific knowledge. In Section III, I consider whether the cold war in international scientific relations which followed World War I might have had its root in a repudiation of this tenet. But I argue that however divided, ambivalent, and inconsistent the German physicists and mathematicians may have been on the issue of universality, not merely they, but even the most nationalistic *Geisteswissenschaftler*, whose adamant rejection of all relations with the Allies was probably founded upon a strong distaste for the very notion of supranational knowledge, represented themselves in this context as the champions of true internationalism. This posture was in fact mandatory if political capital were to be made of German scientific and scholarly achievements.

Section IV begins the consideration of the manipulations of this ideology. The Wilhelmian *Gelehrten* endeavored to broaden the classical formula providing national prestige from scientific achievement to an equation between political and scientific great-power status, contending that a decline in one of these sources of "power" must adversely affect the other. In the Weimar period, however, following Germany's military and economic prostration, the *Gelehrten* shifted this rhetoric subtly and significantly to the proposition that scientific great-power status could function as a *substitute* for political great-power status—a notion which found wide appeal across the German political spectrum.

I then turn in Section V to inquire how indeed German science was wielded as a political instrument, with the discussion based upon a special definition of *Kulturpolitik* as a monopolization of the international relations of other nations by artificially multiplying bilateral ties. This preemption of international scientific relations—while paying lip service to the ideology of scientific internationalism—was generally expected not merely to fabricate prestige but also to provide a basis for influencing the behavior of both the scientists and the government of the country in question.

In order to make intelligible a combination of formal allegiance to the ideology of scientific internationalism with its effective subversion I then make an excursus upon the anti-political and "mandarin" ideologies characteristic of scientists and professionals generally, but especially highly developed among the German academics (Section VI). The conviction that politics is utterly incompatible with the objectivity of the scientist served both as a basis for rejecting the policies of the democratic-parliamentary Weimar regime and as the foundation of a self-conception which did not permit the *Gelehrten* to see their own behavior as political. Moreover, the mandarin inclination to take science and scholarship as the *raison d'être* of the state encouraged the Weimar *Gelehrten* to see themselves, not the politicians, as the representatives of the true interests of the German nation.

With this preparation I then turn in Section VII to the characteristic feature of the international relations of the Weimar academics: a readiness to subordinate the interests of science, even of German science, to the interests of the nation. Admission of the primacy of foreign politics, particularly when it demanded behavior at variance with the ideology of scientific internationalism, generated pressures for the individual scientist to surrender all responsibility and initiative to the legitimate representatives of his nation. The great majority of the German *Gelehrten* turned, however, not to their national government, but to their academic corporations. The small minority who sought governmental support for an effort to bring Germany's international scientific relations in line with the foreign policy of the Weimar regime were in fact leaning on a very weak reed. Finally (Section VIII) I point out that the intransigence of the German natural scientists at the level of formal international relations was possible only because of the existence of extensive informal relations with Allied scientists and scientific enterprises.

II. THE IDEOLOGY OF SCIENTIFIC INTERNATIONALISM

The quantitative growth and qualitative transformation of scientific activity and literature in late-sixteenth- and seventeenth-century Europe was paralleled by a heightening self-, class-, and national consciousness among the scientists themselves. Thus by the middle of the seventeenth century we have in highly developed form that apparently contradictory union of the notion of a republic of science—of an activity and body of knowledge which transcends national boundaries and loyalties—with the most acute consciousness of the national origin or affiliation of individual scientists and scientific achievements.[6] The contradiction may be real enough, but it is essential, just as essential as the tension between cooperation and competition at all other levels of scientific activity. A level of merit presupposes an equally high level of competition, and if honors are to be distributed at an international level, then there must be competition between national scientific champions.[7]

---

[6] Hufbauer, *ibid.*, p. 112, and in personal communications, has pointed out that while phrases such as "the republic of learning" and "the commonwealth of learning" are exceedingly common in the last third of the seventeenth century (e.g., Pierre Bayle's journal, *Nouvelles de la republique des lettres*, begun in 1684), they appear to be rare in the first third. To illustrate the extreme national consciousness, which from this period onward subsisted alongside the notion of a republic of letters, examples may be drawn *ad libitum* from Newton's correspondence in Louis T. More, *Isaac Newton* (1934; reprinted New York:Dover, 1962), pp. 146, 316, 384, 398: John Wallis to John Collins, "I would very fain that Mr. Hooke and Mr. Newton would set themselves in earnest for promoting the designs about telescopes, that others may not steal from us what our nation invents"; Henry Oldenburg to Isaac Newton, "They think it necessary to use some means to secure this invention from the usurpation of foreigners"; Edmond Halley to Newton, "so laudable a piece, so much to your own and the nation's credit"; Wallis to Newton, "You are not so kind to your reputation (and that of the nation) as you might be, when you let things of worth lie by you so long."

[7] The analogy between international science and international sport has been employed by Erwin Schrödinger. "Ist die Naturwissenschaft milieubedingt?"—expansion of a lecture to the Preussische Akademie der Wissenschaften, Berlin, Feb. 18, 1932, in *Über Indeterminismus in der Physik; Ist die Naturwissenschaft milieubedingt? Zwei Vorträge zur Kritik der naturwissenschaftlichen Erkenntnis* (Leipzig: J. A. Barth, 1932), pp. 35–36. This second lecture is further expanded, and divided into two essays, by James Murphy in a far too free translation, "Is Science a Fashion of the Times?" and "Physical Science and the Temper of the Age" in Erwin Schrödinger, *Science, Theory and Man* (New York:Dover, 1957), pp. 81–105, 106–132; the passage in question is on p. 97. On the tension between cooperation and competition in contemporary science: see Warren O. Hagstrom, *The Scientific Community* (New York:Basic Books, 1965), p. 100, *et passim*; John Ziman, *Public Knowledge. An Essay Concerning the Social Dimension of Science* (Cambridge:Cambridge University Press, 1968), p. 100, *et passim*.

An agonistic simile such as the Olympic games can be of some help here, especially in making clear the necessity of agreement about the bases of competition, that is, about scientific doctrine, data, methods; these must be substantially supranational if competition is to be possible at all. Thus the very possibility of a deeper understanding of the nature of atoms, just as the possibility of breaking the world hop, skip, and jump record (and thus, *a fortiori*, the possibility of deriving individual or national prestige from the achievement) depends upon the existence of supranational agreement on the ground rules.[8] Moreover, the competitive element inherent in our simile is also an essential element in the pursuit of science. The question of the quality of scientific performance—the major preoccupation of the scientist—has no meaning apart from (competitive) comparisons.

Who, however, is to judge that competition? Here, as elsewhere, it is praise from parties with a negative bias, from competing nations, which is regarded as most genuine and cogent; thus the great prestige carried by *foreign* honors. When, then, it is a question of measuring a nation's scientific achievement, the only standard legitimized by science itself is the relative amount of attention which that nation's scientific work attracts among foreign scientists.[9]

Nonetheless, there is in modern science also a cooperative element to which our agonistic simile does not do justice. For the competition is not merely between the scientist and his colleagues, but between all and nature, in which each success, if it comes up to *accepted* standards of performance, is applauded, although perhaps not very loudly. Thus scientists or scientific organizations which have committed themselves to a large investment of time and money in a particular line of research need not have the courage to "go for broke." They will usually prefer to restrain direct competition by cartelizing the field, trading a slim chance of carrying off all the laurels for a guarantee of a share in the recognition. Moreover, in certain areas of applied or cosmical physics requiring large quantities of data from diverse geographical sites, division of labor and profit sharing become practical necessities. But again, any such cartelization requires communication, personal contact, and coordination at the international level.

It is then to be expected that the scientists of all nations—but especially the leading scientists of the leading scientific nations—in order to assure themselves of these advantages will affirm and "verify" internationalism in and through scientific ideology,

---

[8] F. S. L. Lyons, *Internationalism in Europe 1815–1914* (European Aspects, Series C, No. 14) (Leiden: Sythoff, 1963), pp. 381–386. Thus a thorough-going sociological view of science would see the scientists' insistence that the laws of nature are indifferent to national boundaries as an element of scientific ideology and would stress the social and personal benefits which the scientists derive from adherence to it: "til Fremme af vor Forstaaelse af Naturen og til Aere for dansk Videnskab," as Niels Bohr put it in concluding his address at the inauguration of his Institut for teoretisk Fysik, Mar. 3, 1921. Archive for History of Quantum Physics, Bohr MSS, Microfilm 9; for descriptions, locations, etc. of this archive, see T. S. Kuhn, *et al.*, *Sources for History of Quantum Physics. An Inventory and Report* (Memoirs of the American Philosophical Society, Vol. 68) (Philadelphia, 1967); referred to hereafter as AHQP and *SHQP*.

[9] "Dass die deutsche Forschung auf dem richtigen Wege fortschreitet zeigt das hohe Ansehen auf dem ganzen Erdball, das sie geniesst...." Wilhelm Wien, *Universalität und Einzelforschung, Rektorats-Antrittsrede, gehalten am 28. November 1925*, Münchener Universitätsreden, Heft 5 (Munich, 1926), p. 5. This measure was, e.g., adopted by Alphonse de Candolle, *Histoire des sciences et des savants depuis deux siècles* (2nd ed., Geneva/Basel, 1885), and is implicit in the procedures for selecting, as well as the popular pastime of counting, Nobel laureates.

that is, in and through the corpus of generally accepted conceptions *about* the nature and products of scientific activity, in contradistinction to the generally accepted conceptions which constitute the scientific products themselves. The propositions and rhetoric asserting the reality and necessity of supranational agreement on scientific doctrine, of transnational social intercourse among scientists, and of international collaboration in scientific work are thus regarded here as tenets of scientific ideology; we may call these tenets the ideology of scientific internationalism.[10] This ideology assumes a substantial measure of national sentiment and organization among scientists. Its function is to control and exploit this "scientific nationalism" in the interest of the advancement of science and of scientists.

Moreover, to function smoothly and effectively this ideology must leave room for the expression and fulfillment of the nationalistic sentiments it channels. This it has done since the seventeenth century through the eminently simple formula that the fame and honor which the scientist wins accrues also to his nation and patron. As Lavoisier's colleagues urged in his defense while he was awaiting trial by the Revolutionary Tribunal: "the opinion of most of the scientists of Europe assigns to Citizen Lavoisier a distinguished place among those who have brought honor to France...."[11] According to this classical conception—largely due to and propagandized by the scientists themselves—the contribution of science to national prestige is an automatic and inevitable byproduct of scientific achievement. It does not require a choice on the scientist's part between serving the interests of science and serving the interests of his nation, between behaving like a good scientist and behaving like a good patriot.

Of course the scientist may be strongly inclined, especially in wartime, to manifest his patriotism by *applying* his knowledge and skills toward the solution of urgent practical problems. But that is quite a different case: (1) because the scientist's service to his nation is then measured not by scientific prestige abroad but by practical

---

[10] In maintaining a distinction, at least in nuance and by implication, between supranational, transnational, and international, I am following Schröder-Gudehus, *Journal of World History*, 1966, *10*:162.

[11] Quoted by Douglas McKie, *Antoine Lavoisier, Scientist, Economist, Social Reformer* (1952; reprinted, New York:Collier, 1962), p. 281. The Greeks' obsession with "noble fame" (Jacob Burckhardt, *Griechische Kulturgeschichte*, ed. R. Marx, Leipzig: A. Kröner, 1929, Vol. II, p. 27, Vol. III, p. 421, *et passim*) encourages one to look for antecedents of this formula in classical antiquity. Although examining a somewhat different problem, Edgar Zilsel's discussions of "Ruhmverleiher" in *Die Entstehung des Geniebegriffes* (Tübingen:Mohr, 1926), pp. 52–65, 103–104, 111–130, suggest, however, that in classical antiquity there was no clear conception of the literary man *reflecting* his own glory upon his patron, but only of his conferring of glory upon his patron by writing about that patron. This was evidently not an option open to the natural philosopher or mathematician, and in fact after Socrates the philosophers generally repudiated fame as a motive for scientific research—and implicitly for support thereof. In the Renaissance, on the other hand, *gloria* appears as a motive for literary work of all sorts, along with a notion of the scholar's patron, but apparently not the scholar's nation, participating in the literary fame won by the protégé. A large element of this conception persisted in the seventeenth and eighteenth centuries, particularly in Germany, where national consciousness was weaker and dependence upon the prince stronger. Leibniz, for example, in pressing potentates to support his scientific academies, confidently held out to them "gloriam immortalem vermittelst des incrementi scientiarum." (Quoted by A. Harnack, *Geschichte der ... Akademie der Wissenschaften zu Berlin*, Berlin:Reichsdruckerei, 1900, Vol. I, p. 19.) At the same time one can trace at least from the latter sixteenth century a growing emphasis by the scientist upon the honor he is conferring upon his *country*, even where it is the prince to whom he must turn for patronage (e.g., J. L. E. Dreyer, *Tycho Brahe*, 1890; reprinted New York:Dover, 1963, pp. 84, 116).

achievements at home; (2) the scientist must indeed choose to subordinate the interests of his science to those of his nation. Logically this need not be accompanied by any repudiation of the ideology of scientific internationalism, although with this reorientation of the scientist's goals that ideology loses much of its utility. In fact, however, the scientist's *persona*, his image of himself, usually makes him most reluctant to admit that he has chosen between his science and his nation. He will often prefer to equate the interests of his nation with those of science (and, perhaps, humanity), most easily and most often when his nation is both a leading scientific power and a leading military power. At the same time, in order to maintain that *persona* intact, he must take care to avoid any direct repudiation of the ideology of scientific internationalism. And those who have the most to lose from a cancellation of international competition and recognition are likely to take the greatest care.

The peculiarly interesting feature of the period following World War I is that while in contrast to the war years the political contribution of science was once again measured primarily in terms of prestige, with formal allegiance to the ideology of scientific internationalism, the German scientists—and in some measure the Allied as well—no longer conceived of their political role in the classical passive terms. Rather, they regarded themselves as agents, or even as bearers, of the foreign policy interests of their nation and as such were often obliged to sacrifice the interests of German *science*, and their personal interests as scientists, for the sake of patriotic political posturing.

III. THE UNIVERSALITY OF SCIENTIFIC KNOWLEDGE

A discussion of scientific internationalism in the interwar period, and especially in relation to Germany, immediately calls to mind a host of historical-intellectual phenomena whose common feature is their implicit or explicit challenge to the leading proposition in the ideology of scientific internationalism: the assertion of the universality of science.[12] And when one has learned that international scientific relations in this period were characterized by boycotts and counter-boycotts, exclusionism and separatism,[13] one is naturally led to suppose that there is an intimate connection between these two historical phenomena. Is not the deplorable state of international scientific relations, the division of science into "hostile political camps," only possible because the ideology of scientific internationalism has been undermined, in particular because the assertion of the intellectual universality of science has lost its cogency? That, however, was not the case—at least not for the German scientists. Neither was the intellectual universality of science repudiated—at least not in this context—nor

---

[12] I know of no extended study of these phenomena. Indications of their breadth and pervasiveness are given by Georg G. Iggers, "The Dissolution of German Historism," *Ideas in History. Essays Presented to Louis Gottschalk*, R. Herr and H. T. Parker, eds. (Durham, N.C.: Duke University Press, 1965), pp. 288–329, esp. p. 306; Kurt Sontheimer, *Antidemokratisches Denken in der Weimarer Republik* (Munich: Nymphenburger, 1962), pp. 53–60. Their common tendency may be indicated by a quote from the preface to their most characteristic expression, Oswald Spengler's *Decline of the West*, Vol. I (New York: Knopf, 1926), p. xiii: "I can then call the essence of what I have discovered 'true'—that is, *true for me*, and, as I believe, true for the leading [German] minds of the coming time; not true in itself as dissociated from the conditions imposed by blood and history, for that is impossible."

[13] Schröder-Gudehus, *Deutsche Wissenschaft*; Kevles, "'Into Hostile Political Camps.'"

would the history of international scientific relations in the Weimar period be intelligible had it been.

To be sure, from the time of the Franco-Prussian War there had been a variety of challenges to the intellectual universality of science. In Germany the struggle against English influence in physics forms a continuous thread from Friedrich Zöllner's attacks on Helmholtz as a Trojan Horse in the 1870s and 1880s through the manifestos of Philipp Lenard, Johannes Stark, and Willy Wien during World War I.[14] Yet even the views of such radicals show considerable ambivalence. Thus in Lenard's wartime and postwar agitation we find alongside distinctions between German physics and English physics (or, subsequently, Jewish physics) bitter complaints that German contributions are not receiving due recognition abroad (or, subsequently, by a domestic establishment dominated by Jews or the Jewish spirit).[15] As Lenard wrote to Stark in July 1915 congratulating him on his recent researches in atomic physics: "Knowledge about atoms thus makes good progress. And inasmuch as that occurs in Germany just now [i.e., during the war], even the 'perfectly neutral' people will not be able to put it down to the credit of the English."[16] One notes here the implicit assumption of a unique and thus universal "knowledge," and indeed it is only with such an assumption that the complaints about lack of recognition abroad make any sense.[17]

If one wishes to find during World War I (as during the Franco-Prussian War) strong and widespread denials of the universality of natural science by respectable spokesmen, one must look to the French and their Allies, not to the Germans. These repudiations of the ideology of scientific internationalism by Allied scientists constituted an important precondition to their formation of a comprehensive and ex-

---

[14] J. C. F. Zöllner, *Über die Natur der Cometen. Beiträge zur Geschichte und Theorie der Erkenntniss* (1st ed., Leipzig, 1872; 3rd ed., Leipzig, 1883). P. Lenard, *England und Deutschland zur Zeit des grossen Krieges* (Heidelberg: Winter, 1914). Letters from W. Wien to J. Stark, Dec. 21, 1914, Jan. 26, 1915, Feb. 26, 1915, in the Nachlass Stark, Staatsbibliothek Preussischer Kulturbesitz, Berlin-Dahlem.

[15] Helmut Heiber, *Walter Frank und sein Reichsinstitut für Geschichte des neuen Deutschlands* (Stuttgart: Deutsche Verlags-Anstalt, 1966), p. 592–594, 839, 970–972, illustrates this latter theme in the Nazi period. Its earlier subterranean course may be traced in Stark's correspondence; *cf.* Armin Hermann, "Albert Einstein und Johannes Stark," *Sudhoffs Archiv*, 1966, *50*:267–285. A dissertation on this topic is being prepared by Alan Beyerchen, Department of History, University of California, Santa Barbara.

[16] "Die Kenntnisse über die Atome schreiten also gut vor. Dass das jetzt bei uns geschehen kann, werden wohl selbst die 'neutralsten' Leute nicht als engl. Verdienst hinstellen können!" Philipp Lenard to Johannes Stark, Heidelberg, July 14, 1915, postcard; published in part in facsimile in A. Hermann, ed., *German Nobel Prizewinners* (Munich: Heinz Moos, 1968), p. 77.

[17] The feeling which Lenard expressed publicly and blatantly—that the contributions of the German physicists were being plagiarized or suppressed by foreigners—was in fact widely shared by his colleagues. As Arnold Sommerfeld wrote Bohr, Feb. 5, 1919, thanking him for acknowledging the work of Sommerfeld's school so liberally in his publications, "Dadurch werden wohl auch die Fachgenossen in den feindlichen Ländern, die sonst gern alle deutschen Leistungen unterschlagen möchten, gezwungen sein, einzusehen dass sich die deutsche Wissenschaft selbst im Kriege nicht unterdrücken lässt." (AHQP, Bohr Scientific Correspondence, Microfilm 7.) The feeling was certainly mutual; so, P. G. Nutting, "National Prestige in Scientific Achievement," *Science*, 1918, *48*:605–608: "Plagiarism and piracy were common practices, and from personal knowledge I doubt whether a third of even the more eminent German scientists were free from this taint." Yet such statements, charging the other side with cheating and gaining prestige by dishonest means, far from indicating a repudiation of the ideology of scientific internationalism, are rather indicative of basic adherence to it. This general point was made by R. K. Merton in his analysis of disputes over priority: "Priorities in Scientific Discovery: A Chapter in the Sociology of Science," *American Sociological Review*, 1957, *22*:635–659.

clusionist international scientific organization at the end of the war—the International Research Council. Yet such deprecation of the enemy's scientific men and scientific contribution, never characteristic of the Germans, receded very quickly among the Allies after 1918. Their only essential role in international scientific relations in the Weimar period was as unrepeated but unretracted insults to the Germans, rankling in much the same way as the unretracted manifesto of the ninety-three German scholars and artists of October 1914 continued to rankle among the Allies.[18]

Nevertheless, it is undeniable that in Germany the postwar intellectual atmosphere was saturated with sociologies and biologies of mind and knowledge and that restrictions upon the universality of scientific knowledge were more or less explicit in the bulk of such doctrines—certainly in those with substantial appeal to a wholly non-Marxist and largely non-Catholic educated middle class. The German academics, even the physicists, did not remain immune to such influences.[19] Notable in this connection is Erwin Schrödinger's contention,

> In a word, we all are members of our cultural milieu. So soon as the orientation of our interest plays any role whatsoever in a matter, the milieu, the cultural complex, the Zeitgeist, or whatever one wishes to call it, must exert its influence. In all areas of a

---

[18] The "Aufruf an die Kulturwelt" of Oct. 4, 1914, signed by ninety-three German scientists, scholars, and artists (published and republished in numerous places, e.g., Hans Wehberg, *Wider den Aufruf der 93! Das Ergebnis einer Rundfrage an die 93 Intellektuellen über die Kriegsschuld*, Charlottenburg: Deutsche Verlagsgesellschaft für Politik und Geschichte, 1920, 40 pp.; there is an English translation of the manifesto in Ralph H. Lutz, *The Fall of the German Empire, 1914–1918*, Palo Alto: Stanford University Press, 1932, Vol. I, pp. 74–78). The manifesto was widely construed outside Germany as a repudiation of the universality of science, although in fact there is nothing in the document to support such a construction. Disturbed by such "unzutreffenden Vorstellungen von der Gesinnung seiner Unterzeichner," Planck wrote an open letter to H. A. Lorentz in Mar. 1916 (Wehberg, *Wider den Aufruf der 93!*, pp. 19–20) declaring that "Was ich aber Ihnen gegenüber mit besonderem Nachdruck zu betonen wünsche, ist die feste, auch durch die Ereignisse des gegenwärtigen Krieges nie zu erschütternde Ueberzeugung, dass es Gebiete der geistigen und sittlichen Welt gibt, welche jenseits der Völkerkämpfe liegen, und dass ehrliche Mitwirkung bei der Pflege dieser internationalen Kulturgüter, wie auch nicht minder persönliche Achtung von Angehörigen eines feindlichen Staates, wohl vereinbar ist mit glühender Liebe und tatkräftiger Arbeit für das eigene Vaterland." *Cf.*, for contrast, Michael Pupin to George E. Hale, Oct. 20, 1917, "I heartily endorse the sentiments expressed by Monsieur Picard and by Mr. W. W. Campbell. Science is the highest expression of a Civilization. Allied Science is, therefore, radically different from Teutonic Science. It is true that the highest aim in Science is to disassociate itself from all its anthropomorphic elements, but we are still very far from that ideal goal. Man is intensely anthropomorphic, of course, and we see today more clearly than we have ever seen before, that Science can-not be disassociated from the varying moods and sentiments of man. I thank God it is so, for this brings Science very much closer to my heart than it ever was before, because I feel that scientific men are men first and scientists after that." (California Institute of Technology, Archive, Hale Papers, Microfilm 41, frame 140.) "Indeed it is a common saying that science is international. But we are beginning to revise our verdict." William Ramsay in *Nature*, Oct. 1914, as quoted by Roy M. MacLeod, "Into the Twentieth Century," *Nature*, 1969, *224*:457–461. See, also, Paul Gary Wersky, "The Perennial Dilemma of Science Policy," *Nature*, 1971, *233*:529–532.

[19] This paragraph has been revised and the following paragraphs in this section have been added in response to Prof. Gerald Feldman's query whether "important German scientists were ambivalent about or even hostile to Western science in the same way that their counterparts in the social sciences and humanities often asserted the separation of German *Kultur* from the civilization of the West." This most interesting question is part of the general problem which Fritz Ringer, *Decline of the German Mandarins . . . 1890–1933* (Cambridge, Mass.: Harvard University Press, 1969), has placed before historians of science, *viz.* how far did German natural scientists share the academic ideology of their colleagues the *Geisteswissenschaftler*?

culture there will exist common features deriving from the world view and, much more numerous still, common stylistic features—in politics, in art, in science (*Wissenschaft*).[20]

Indeed, Schrödinger himself was prepared to go even further and insist that physical conceptions be judged on the basis of their conformity to such extrinsic preoccupations; as he wrote Willy Wien in August 1926: "Physics consists not merely of atomic research, science not merely of physics, and life not merely of science. The purpose of atomic research is to fit our experiences from this field into the rest of our thought."[21]

Schrödinger's declarations are unusual, possibly unique—not however in substance, but in clarity and boldness. As I have pointed out elsewhere, in their public addresses Weimar physicists and mathematicians were as often ready to concede as to combat the notion that the doctrines and methods of their disciplines were decisively influenced by the *Zeitgeist*.[22] More common still, and more to the point, was a readiness to represent the race or nation as determinative of the scientific style or as endued with a special talent for some particular science. Planning its strategy for a public protest against the curricular reforms of the Prussian Ministry of Education, the Mathematischer Reichsverband, the central organization of the various German mathematical societies, chose to base its case upon the proposition that "mathematics belongs alongside the German language among the specifically German cultural studies."[23]

No doubt the mathematicians who led the Mathematischer Reichsverband would have represented their discipline somewhat differently in other circumstances and to other constituencies. Addressing an audience of humanistically educated officials and eminent *Geisteswissenschaftler* (and their wives) at a public session of the Prussian Academy of Sciences, they might rather emphasize, as did Max Planck, that "*die Wissenschaft*, just like art and religion, can in the first instance grow properly only on national soil. Only when such a basis has been established is a fruitful union of the nations in high-minded competition possible"[24]—thus reminding us strongly of Ernst Troeltsch's contrast between the Western "ideal of a final union of fundamentally equal human beings in a rationally organised community of all mankind," and the German "ideal of a wealth of national minds, all struggling together and all developing

---

[20] Schrödinger, "Ist die Naturwissenschaft milieubedingt?" (1932), pp. 37–38. The main thrust of Schrödinger's argument is that while the choice of the subject of scientific investigation may be "subjective" the resulting scientific knowledge is "objective." He therefore tended to restrict the influence of his contemporary cultural milieu to the "style" of physical science and did not explicitly extend it to the content of the theories which he and his colleagues were forming and adopting.

[21] Schrödinger to W. Wien, Aug. 25, 1926 (AHQP). This passage is included in the extract published in [W. Wien], *Aus dem Leben und Wirken eines Physikers* . . . (Leipzig: Barth, 1930), p. 74.

[22] Paul Forman, "Weimar Culture, Causality, and Quantum Theory, 1918–1927: Adaptation by German Physicists and Mathematicians to a Hostile Intellectual Environment," *Historical Studies in the Physical Sciences*, 1971, *3*: 1–115.

[23] "Mathematik gehört nebst Deutsch zur spezifisch deutschen Kulturkunde und kann nicht in Parallele zu den [foreign] Sprachen gesetzt werden." *Unterrichtsblätter für Mathematik und Naturwissenschaften*, 1924, *30*: 72. For the Mathematischer Reichsverband and its struggle "gegen die Herabsetzung des Intellektualismus" see Forman, "Weimar Culture, Causality . . . ."

[24] Max Planck, "Ansprache des Sekretars . . . vom. 1. Juli 1926," *Max Planck in seinen Akademie-Ansprachen. Erinnerungsschrift der Deutschen Akademie der Wissenschaften zu Berlin* (Berlin: Akademie Verlag, 1948), p. 94. Planck's biographer Hans Hartmann waxes quite ecstatic over "seine Auffassung von dem Verhältnis nationaler und 'internationaler' Wissenschaft . . . in sich so rein, so geschlossen, so organisch ausgebaut, dass keine höhere Form denkbar ist." *Max Planck als Mensch und Denker* (Frankfurt: Ullstein Buch No. 490, 1964), p. 38.

thereby their highest spiritual powers."[25] Or, conversely, before still other audiences they might, like Planck, present a picture of the individual researcher contributing his mite "to the great treasurehouse of international science."[26]

It must therefore be recognized and admitted that in some considerable measure the German physicists and mathematicians participated in—and in certain situations appealed directly to—the anti-universalist sentiments in their intellectual milieu. But for the purpose of our present inquiry the striking and important circumstance is that in other situations—notably, in the Weimar discussions of international scientific relations—not merely the physical scientists but even the *Geisteswissenschaftler* refrained from introducing the notion of qualitatively distinct national sciences.[27] And

---

[25] Ernest Troeltsch, "The Idea of Natural Law and Humanity in World Politics" (1922), in Otto Gierke, *Natural Law and The Theory of Society, 1500–1800*, trans. E. Barker (Boston: Beacon Press, Paperback No. 50, 1957), pp. 201–222, on p. 211.

[26] Max Planck, "Physikalische Gesetzlichkeit im Lichte neuerer Forschung," Vortrag gehalten am 14. Februar 1926 in den Akademischen Kursen von Düsseldorf, *Physikalische Abhandlungen und Vorträge* (Braunschweig: Vieweg, 1958), Vol. III, pp. 159–171, on p. 159. See also Planck's letter to Lorentz quoted above, n. 18. Among the very few outspoken opponents of the prevalent anti-universalism was Gerhard Hessenberg. In his *Antrittsrede* as professor of mathematics at Tübingen (*Vom Sinn der Zahlen*, Leipzig: Neue Geist, 1922), Hessenberg lampooned Spengler's utterly superficial knowledge of mathematics and challenged his basic assumptions. "The thesis: 'There is no mathematic, there are only mathematics' is for him the test of strength of his more far-reaching contention of the independence of the cultures. This latter thesis cannot be more easily refuted (and in fact is refuted *eo ipso*) than by the unity [*Einheitlichkeit*] of the mathematical knowledge of all civilized peoples [*Kulturvölker*]" (pp. 48/49). But precisely because of Hessenberg's unusually outspoken and uncompromising stand on the universality of mathematical knowledge, it is interesting to note that he himself was prepared to attribute the geometrical garb of Hellenic mathematics at least in part to "die besondere anschauliche Veranlagung der Rasse" (p. 30).

[27] Even while urging a counter-boycott and the rejection of all overtures, the radically nationalist *Geisteswissenschaftler*—to be found above all in the leadership of the Verband der Deutschen Hochschulen, Weimar's AAUP—represented themselves as the defenders of true internationalism in science. Thus the classical archaeologist Heinrich Karo claimed to stand for "that real and genuine internationality of science (*Internationalität der Wissenschaft*), which the German *Gelehrten* . . . have always regarded as a matter of course," even while he was doing his utmost to maintain the split in German-Allied scientific relations. (Heinrich Karo, "Der geistige Krieg gegen Deutschland," *Mitteilungen des Verbandes der Deutschen Hochschulen*, June 1924, *4*, Beilage, 22 pp., on p. 13. See, also, Karo, "Krieg oder Friede," *Mitt. V. D. H.*, Oct. 1925, *5*:165–169; "Ergänzungsbericht," *Mitt. V. D. H.*, Feb. 1926, *6*:25–29.) At its fourth congress in Jan. 1925 the Verband itself passed a resolution which declared that "science is not a subject for political conflict but an affair of humanity, for which only the truth is valid as supreme law," and then continued with the assertion that any invitation to individual Germans to participate in international scientific meetings must be rejected as an "insulting presumption" (*Mitt. V. D. H.*, Feb. 1925, *5*: 50).

If, however, we look closely at the notion of "that real and genuine internationality of science" held by these radical nationalists, we find precisely the anticipated distinction between German and Western culture; and there, of course, lies one of the basic motives for their utter intransigence in international scientific relations. In an unguarded moment Karo might deplore the "representatives of 'objective supra-national science'" ("Versöhnungsfanatiker, schlecht Unterrichtete und Vertreter der 'objektiven, über den Nationen stehenden Wissenschaft,'" *Mitt. V. D. H.*, 1926, *6*:25). Far more often, however, he and his associates would argue that "one can scarcely conceive of a greater distortion and falsification of the axiom that science is international than a decisive influence upon the world by French *Kulturpolitik*." For this reason he rather welcomed the "world boycott, which here can even be a blessing, because it leaves us complete freedom," above all freedom from French influence. ("Der geistige Krieg . . . ," p. 12.) Such arguments became particularly common in 1925–1926 when Germany's entrance into the League of Nations seemed imminent. Thus the chairman of the Auslandsausschuss of the Verband der Deutschen Hochschulen, Otto Franke, warned of subordination and subservience of German science to French civilization in "Geistige Zusammenarbeit," *Mitt. V. D. H.*, Oct. 1925, *5*:170–174; Wilhelm Riedner, "Völkerbund und geistige Zusammenarbeit," *Mitt. V. D. H.*, 1926,

this is not altogether surprising: the ideology of scientific internationalism, and especially the universality of the results of scientific research, was the essential prerequisite for any attempt to make political capital out of German scientific and scholarly achievement; it was implicit in the very notion of science as a *Macht-Ersatz*.

IV. SCIENCE AS A *MACHT-ERSATZ*

To make intelligible the intensity of the Weimar scientist's preoccupation with his national political role—the depth of his conviction of his national political mission—it is necessary to emphasize once again that proposition, that pervasive rhetoric, which Brigitte Schröder-Gudehus has so clearly analyzed: the notion that quite apart from any economic, technical, or military advantages to be derived from leadership in science, the very fact of being a scientific great power is an attribute comparable to, and in some vague sense interconvertible with, political great-power status.[28]

In the fall of 1909 Wilhelm II was induced to request from his favorite *Gelehrter*, the Director of the Prussian State Library, Adolf Harnack, a report on the desirability of establishing in Germany scientific research institutes independent of the universities. With the aid of the organic chemist Emil Fischer and the medical scientist August Wassermann, Harnack prepared a memoir urging the necessity of, and sketching the organization of, what then became the Kaiser-Wilhelm-Gesellschaft, now the Max-Planck-Gesellschaft.[29] The motivation for this radical shift in the patterns of research organization and support in Germany was primarily the American precedent and challenge, especially in the biomedical sciences. With such slogans as "The military and science (*Die Wehrkraft und die Wissenschaft*) are the two strong pillars upon which Germany's greatness rests," Harnack *et al.* were not referring to the economic importance of applied science, say, but argued that the threatened loss of Germany's leadership in natural science was "national-politically ominous," "because today, in contrast with earlier periods [!], with the extraordinarily heightened national feeling, a national stamp is affixed to every product of scientific research."[30]

Harnack's memoir was passed for analysis and comment to the responsible bureaucrats in the Prussian Ministry of Education and the Reich Interior Ministry, Friedrich Schmidt and Theodor Lewald. In their joint memorandum they summarized the argument of the savants as follows:

> For Germany the maintenance of its scientific hegemony is just as much a necessity for the state as is the superiority of its army. A decline in Germany's scientific prestige

---

6:85–92, stressed that "the true internationality of science rests upon the nationally and culturally freely developing cooperative scientific work of the *Völker*," but in the organizations dominated by French cultural imperialism "kann eine kulturelle Eigentümlichkeit nicht gedeihen" (p. 92); and in an article on "Locarno und die deutsche Wissenschaft," *Mitt. V. D. H.*, Dec. 1925, 5:213–216, Ludwig Bernhard, professor of political economy at the University of Berlin, warned that France's zealous efforts "to tackle the problems of science in the French spirit" constitute "an enormous political power from which in the long run no individual and no nation can withdraw,"
and urged the severest measures against his colleagues who, knowingly or unknowingly, acted in the interests of French cultural hegemony.

[28] Schröder-Gudehus, *Deutsche Wissenschaft*, pp. 181–182, 199.

[29] Published in *50 Jahre Kaiser-Wilhelm-Gesellschaft und Max-Planck-Gessellschaft zur Förderung der Wissenschaften, 1911–1961. Beiträge und Dokumente*, ed. Generalverwaltung der Max-Planck-Gesellschaft (Göttingen, 1961), pp. 80–94.

[30] *Ibid.*, pp. 81, 82, 89.

reacts upon Germany's national repute and national influence in all other fields, leaving entirely out of account the eminent importance for our economy of superiority in particular fields of science, such as chemistry, especially.[31]

What we have in this imagery and rhetoric is an attempt at a substantial extension of the classical nationalistic exploitation of scientific internationalism. The spokesmen for German science here advance beyond the concept of national honor and prestige, giving an imperialistic and anti-utilitarian twist to the Baconian-Comtian slogan "science is power." Nor was this merely rhetoric, merely for Prussian governmental consumption. On the contrary, while His Majesty received Harnack's report with "unqualified approval,"[32] the Wilhelmian bureaucrats Schmidt and Lewald, as also the Chancellor Bethmann-Hollweg, were uncomfortable with and unready to adopt the savants' conception of their own national-political importance. "The decisive consideration," they said, "is not whether German science has been overtaken by other countries; it is quite sufficient to establish that German science, for its own sake, requires for its full development large research institutes of the type proposed."[33] But even if not yet ready to recognize German scientific hegemony as indispensable to German political hegemony, the leading government officials would have joined the spokesmen for German science in subscribing to the converse proposition—the indispensability of Germany's political great-power status to her position as a scientific great power. It was to this conviction that the manifesto of the ninety-three gave expression.[34]

Without trying to trace the social bases of this thirst for power by the German academics, or the forms it took after the outbreak of World War I, let us jump forward four years to November 1918. Germany is militarily and economically prostrated and

---

[31] Transcribed from the Nachlass Schmidt-Ott, B LXXVI, No. 3, Vol. IV, Deutsches Zentralarchiv, Merseburg, by Günter Wendel, "Zur gesellschaftlichen Stellung und Funktion der Kaiser-Wilhelm-Gesellschaft zur Förderung der Wissenschaften e.V., dargestellt anhand ihrer Gründungsgeschichte und Entwicklung bis zum I. Weltkrieg (1911–1914)" (Inaugural dissertation, Phil. Fakultät, Karl-Marx-Universität, Leipzig, 1964; microfilm copy at the Center for Research Libraries, Chicago), "Dokumentenanhang," document No. 1: "Aufzeichnung, betreffend die Harnacksche Denkschrift. . . ." Apart from a few identifiable interpolations the "Aufzeichnung" was written by Schmidt. With the exception of this first section summarizing Harnack's "Grundgedanken" and various other less interesting passages, it is published in *50 Jahre Kaiser-Wilhelm-Gesellschaft*, pp. 96–103.

[32] ". . . den lebhaftesten uneingeschränkten Beifall Seiner Majestät. . . ."; v. Valentini to Harnack, Dec. 10, 1909, *ibid.*, p. 94.

[33] *Ibid.*, p. 97. Bethmann-Hollweg's report (*qua* President of the Prussian Staatsministerium) to Wilhelm II, likewise drafted by Schmidt, dilutes the argument from national prestige with the argument from full development: "Fasst man die Notwendigkeit voller Entfaltung unserer Wissenschaft und der Erhaltung ihres geschichtlich erworbenen Ruhmes in Auge, so. . . ." This passage, too, is omitted from the partial copy published, *ibid.*, pp. 103–105; the complete text is given as document No. 2 by Wendel, *op. cit.* In 1913 Bethmann-Hollweg publicly rejected the academics' call for a governmentally sponsored *Kulturpolitik* in a letter to the historian Karl Lamprecht, quoted from the *Vossische Zeitung*, by Georg Schreiber, *Deutsche Wissenschaftspolitik von Bismarck bis zum Atomwissenschaftler Otto Hahn*, Arbeitsgemeinschaft für Forschung des Landes Nordrhein-Westfalen, Geisteswissenschaften, Heft 6 (Cologne: Westdeutscher Verlag, 1954), p. 61; Schröder-Gudehus, *Deutsche Wissenschaft*, pp. 40, 48–49, cites the originals in the Politisches Archiv, Auswärtiges Amt, Bonn.

[34] *Cf.* Friedrich Meinecke, "Politik und Kultur," *Süddeutsche Monatshefte*, Sept. 1914, *11*:796–801: *Kultur* is the sap of the tree which is the state, and if the tree is struck at the root, *Kultur* must dry up. "Alle diejenigen unter uns, die von einer Kultur ohne Staat träumten, werden jetzt erwachen im Angesicht der Gefahr, die ihr droht." For Max Weber's similar opinions: Wolfgang J. Mommsen, *Max Weber und die deutsche Politik, 1890–1920* (Tübingen: Mohr, 1959), pp. 74–76.

in the throes of a political revolution which the academics almost universally deplored and whose issue, the Weimar Republic, few ever accept as the legitimate bearer of their own or their nation's interests. Their immediate and consistent reaction was to reach for the prewar "science as power" rhetoric and images; but now, adapted to a state of national impotence, it became the doctrine of science as a substitute for military and economic power. Addressing a plenary session of the Prussian Academy of Sciences on November 14, 1918, Max Planck declared:

> If the enemy has taken from our fatherland all defense and power, if severe domestic crises have broken in upon us and perhaps still more severe crises stand before us, there is one thing which no foreign or domestic enemy has yet taken from us: that is the position which German science occupies in the world. Moreover, it is the mission of our academy above all, as the most distinguished scientific agency of the state, to maintain this position and, if the need should arise, to defend it with every available means.[35]

Planck has here struck the keynotes of the German academics' view of international relations in the Weimar period: scientific and scholarly prestige is the sole great-power attribute remaining to the German nation; it is to be valued and defended on that account rather than as an end in itself or even as the index of genuine scientific achievement; it will, by some unspecified mechanism, function as a surrogate for the other, lost attributes of a great power. These basic themes may be heard repeated endlessly with but small variations in the following years. So the spokesmen for the Union of German Universities in justifying their adamantine line toward overtures from the International Research Council: "After the complete disarmament and the severe breakdown of the economy, German science is just about the only asset which Germany has to throw onto the scales."[36] So the physical chemist Fritz Haber, who had played a central role in the mediation of such overtures:

> I would guess that for my compatriots the chief question will be, in what sense is the invitation issued. We know perfectly well that we lost the war and politically as well as economically no longer sit on the board of directors of the world. But scientifically we believe we can still be numbered with those peoples which have a claim to be reckoned among the leading nations. Whether this claim is recognized. . . .[37]

---

[35] Max Planck, *Preussische Akademie der Wissenschaften, Berlin, Sitzungsberichte* (1918), p. 993. This *Ansprache* is not included in the collection cited above, n. 24, nor in the bibliography contained therein, but is quoted at length in Hartmann, *Max Planck*, pp. 32–33.

[36] "Die Referenten [O. Franke, Berlin, and G. Karo, Halle] führten aus, dass nach der völligen Abrüstung und nach dem schweren Niederbruch der Wirtschaft die deutsche Wissenschaft so ziemlich das einzige Aktivum sei, das Deutschland in die Wagschale zu werfen hatte." *Münchener medizinische Wochenschrift*, 1926, *73*, No. 7, Akademie der Wissenschaften zu Berlin, Archiv, VIa 17, Bd. 8, fol. 23. *Cf.* W. Riedner, *Mitt. V. D. H.*, 1926, *6*:92: "Die deutsche Wissenschaft . . . ist ein Machtfaktor geblieben, dessen unsere auswärtige Politik . . . um so weniger entraten kann, als andere Machtfaktoren verschwunden sind."

[37] Fritz Haber to H. R. Kruyt, July 7, 1926 (AHQP, H. A. Lorentz Papers, Microfilm 9): "Aber ich vermute, dass die Hauptfrage bei meinen Landsleuten sein wird, in welchem Sinne die Einladung ergangen ist. Wir wissen sehr genau, dass wir den Krieg verloren haben und politisch ebenso wie wirtschaftlich nicht mehr im Vorstande der Welt sitzen. Aber wissenschaftlich glauben wir noch zu den Völkern zu zählen, die einen Anspruch haben, unter die führenden Nationen gerechnet zu werden. Ob dieser Anspruch anerkannt wird oder nicht, vermögen wir schwer aus einer Einladung zu ersehen, die in gleicher Form und mit gleicher Einstimmigkeit auch früher an Siam ergangen ist." Similarly, Haber, "Über Staat und Wissenschaft," *Aus Leben und Beruf* (Berlin: Springer, 1927), pp. 158–166, quoted, along with much other pertinent material, by Brigitte Schröder-Gudehus, "The Argument for Self-Government and Public Support of Science in Weimar Germany," *Minerva*, 1972, *10*:537–570.

164   PAUL FORMAN

It is interesting and important that the cogency of this rhetoric was not limited to the Weimar *Gelehrten*, but found a substantial resonance right across the social and political spectra. Harnack introduced it into the budget debate in the national constitutional convention in February 1920 in an appeal for financial support for science, arguing that we must of course limit our expenditures to the vital necessities of the state, but "among the vital necessities of the state belongs also the conservation of the few large assets which it still possesses. Among these entries in the assets column a prominent place belongs to German science."[38] And to see how widely and firmly this rhetoric took hold of Weimar parliamentarians, note Julius Moses (who, even more than most Social Democrats, had reason to bear a grudge against the German academic establishment) urging his fellow members of the Reichstag to support a substantial appropriation for scientific research:

> The former Minister of the Interior Koch once said in the Reichstag: "German science is the one thing for which the world still envies us." Well, it seems to me that if you don't want the one thing for which the world still envies us to go miserably to ruin, then strike billions from the military budget, and re-employ these unproductive billions for the purposes of culture and of German science.

At which, the transcript notes, there was lively approval among the United Social Democrats.[39]

To go further in this direction would lead us into such questions as the nature and

---

[38] Quoted by Kurt Zierold, *Forschungsförderung in drei Epochen: Deutsche Forschungsgemeinschaft*... (Wiesbaden: Steiner, 1968), p. 4. *Cf.* Planck, as *Vorsitzender Sekretar* of the Prussian Academy of Sciences in the public session of July 3, 1919: "Denn die Wissenschaft gehört mit zu dem letzten Rest von Aktivposten, die uns der Krieg gelassen hat, den einzigen, denen auch die Begehrlichkeit unserer Feinde bisher nichts Wesentliches anhaben konnte." *Sitzungsberichte*, 1919, p. 548. Although this address, unlike that cited in n. 35, is included in *Max Planck in seinen Akademie-Ansprachen* (1948), pp. 29–31, the passage quoted is omitted from it without any editorial indication that the text had been altered.

[39] Reichstag, *Verhandlungen*, 268. Sitzung, Nov. 16, 1922, p. 9005 B-C. Moses was here refering to Koch's address at the "Parliamentary Evening," Nov. 23, 1920, unveiling the Notgemeinschaft der Deutschen Wissenschaft: "Die deutsche Wissenschaft, um die uns die Welt beneidet—es ist vielleicht fast das einzige, um das uns die Welt noch beneidet—, die deutsche Wissenschaft zu fördern, ist unsere Aufgabe." *Internationale Monatsschrift für Wissenschaft, Kunst und Technik*, 1920, *15*:97–100. Koch himself was drawing upon—if his speech was not actually drafted by—Friedrich Schmidt-Ott's lieutenant, Eduard Wildhagen: "Die Not der deutschen Wissenschaft," *Internat. Monatsschr.*, Oct. 1920, *15*:1–32, which opens with the proposition that "Von dem allgemeinen Zusammenbruche der dem unglücklichen Kriege folgte, schien zunächst ein gewaltiger Faktor deutscher Weltgeltung unberührt geblieben zu sein: die deutsche Wissenschaft... und ist heute vielleicht das einzige, um das die Welt Deutschland noch beneidet."

Similarly, in a letter of Nov. 24, 1919, from the Prussian minister of education, Social-Democrat Konrad Haenisch, to the *Rektor* and *Senat* of the University of Berlin we read that "Under the present circumstances, since economically we will stand for a long time... upon a field of ruins, we are especially dependent upon the aid of our intellectual (*geistigen*) forces. Fate has here allotted a huge task to German science." Original quoted from the Deutsches Zentralarchiv, Merseburg, by Siegfried Grundmann, "Der deutsche Imperialismus, Einstein und die Relativitätstheorie (1914–1933)," *Relativitätstheorie und Weltanschauung* (Berlin: VEB Deutscher Verlag der Wissenschaften, 1967), pp. 155–285; on p. 167. It is likely that this letter was drafted by Haenisch's *Staatssekretär*, the Islamicist Carl Heinrich Becker, whose *Kulturpolitische Aufgaben des Reiches* (Leipzig: Quelle und Meyer, 1919), pp. 15 and 18, included such slogans as "Bei seiner politischen und wirtschaftlichen Ausschaltung hat das deutsche Volk im Ringen der Völker nur noch seinen Ideengehalt als Einsatz," and "das Reich braucht in Ermangelung einer militärischen eine ideele Hausmacht."

sources of financial support for science in the Weimar Republic and then to the reasons for the flourishing of science, especially physics, in this environment. These questions, seemingly of quite a different order from those with which we began, are in fact intimately related to them through the rhetoric of science as a *Macht-Ersatz*.[40]

V. SCIENCE AS A POLITICAL INSTRUMENT – *KULTURPOLITIK*

What does all this remarkably vague talk about science as a surrogate for political and economic power come to in practice? How indeed does an individual or a nation wield science as a political instrument bending other nations to its will? The first halting steps beyond the purely passive classical relationship, in which the nation merely basked in the light of its scientific luminaries, had already been taken in the years before World War I. These measures, of which the German exchange professorships with Harvard, Columbia, and Wisconsin are perhaps the clearest examples, were tacitly based on the concept of an active *Kulturpolitik*, of a policy carefully planned and orchestrated by the state in which scientific internationalism is killed with kindness by artificially fostering innumerable bilateral contacts and institutions, thus forestalling the multilateral competition.[41]

The number of American students at German universities had reached a peak in the early 1890s and then began to decline with the rise of research-oriented graduate training in the American universities.[42] The exchange-professor programs represented, then, an attempt to hold onto that special bilateral relationship between the German and American academic elites—from which the German academics anticipated a far-reaching influence upon American public opinion.[43] One might suppose that the most thorough disappointment of these expectations at the outbreak of World War I would also have thoroughly discredited *Kulturpolitik* as an effective political instrument. But in fact, while before the war the German governments had rather reluctantly accepted

---

[40] Schröder-Gudehus, "Public Support," *Minerva*, 1972, *10*:537–570. Paul Forman, "Financial Supports and Political Alignments of the German Physicists in the Weimar Republic," *Minerva*, 1973, *11* (in press).

[41] I am here giving the term *Kulturpolitik* a very narrow and technical meaning related to, but I think more precise than, Barghoorn's concept of "cultural diplomacy." (F. C. Barghoorn, *The Soviet Cultural Offensive*, Princeton: Princeton University Press, 1960.) A large amount of material on the historical development of the concept and manifestations of *Kulturpolitik* in the broadest sense is cited by Manfred Abelein, *Die Kulturpolitik des Deutschen Reiches und der Bundesrepublik Deutschland . . .* (Cologne/Opladen: Westdeutscher Verlag, 1968). Richard Martinus Emge, *Auswärtige Kulturpolitik. Eine soziologische Analyse einiger ihrer Funktionen, Bedingungen und Formen* (Berlin: Duncker & Humblot, 1965), proves, notwithstanding its promising title, quite useless.

[42] Statistics given by F. Charles Thwing, *The American and the German University* (New York: Macmillan, 1928), pp. 42–43. See Jurgen Herbst, *The German Historical School in American Scholarship; A Study in the Transfer of Culture* (Ithaca: Cornell University Press, 1965), Ch. 1; Thomas N. Bonner, *American Doctors and German Universities; A Chapter in International Intellectual Relations, 1870–1914* (Lincoln: University of Nebraska Press, 1963).

[43] Schröder-Gudehus, *Deutsche Wissenschaft*, p. 41. This expectation, which seems so unreasonable, arose largely from a tenet of German academic ideology which held that the university teacher simply and solely by virtue of his influence upon university students, and the subsequent trickling down of this influence to the lower intellectual orders of German society, was the maker of the German public mind. See, e.g., Friedrich Paulsen, *The German Universities and University Study*, trans. from the 1902 German ed. by F. Thilly and W. W. Elwang (London: Longmans Green, 1906), pp. 5, 119–120, and Ringer, *Decline of the German Mandarins*, pp. 5–11, *et passim*.

this instrument pressed into their hands by the academics, in the Weimar period *Kulturpolitik* was enthusiastically adopted right across the political spectrum as one of the few instruments of an active foreign policy remaining to Germany. In this enthusiasm one sees yet another example of the remarkable success of the Weimar *Gelehrten* in selling their own academic ideology to a regime which by that very ideology they could not respect.[44]

Siegfried Grundmann, using East German state archives, has traced in some detail the great interest of the Prussian Ministry of Education and the German Ministry of External Affairs in Albert Einstein's travels abroad, the satisfaction which they took in his enormous popularity, their anxiousness to have him appear as a German (not Swiss or Jewish) scientist, and their consternation when this man "with whom we can carry on real culture-propaganda"[45] threatened to leave Berlin and Germany because he and his theory of relativity had become the focus of anti-semitic political agitation. In this they were at one with the principal eminences of theoretical physics in Germany—Max Planck and Arnold Sommerfeld—whose reactions on each such occasion betray as much concern over the political implications as over the personal or scientific implications. "To forsake Germany now," Sommerfeld wrote Einstein in the fall of 1920, "when it is so unspeakably mistreated from all sides—I couldn't regard you as capable of it."[46] Einstein stayed. Following the murder of Walther Rathenau in the summer of 1922, many feared for the life of this other equally prominent Jew who had served as an instrument of Rathenau's foreign policy. Einstein laid low, pretended to be out of town, and then went for several months to Japan.[47] In November 1923, at the time of Hitler's Beerhall Putsch, Einstein was evidently directly threatened, and he fled to Holland. Planck's first reaction was to write Einstein "expressing only the one, but heartfelt and most urgent, request: undertake no steps which would make your return to Berlin finally and permanently impossible."[48] A few weeks later Planck reported to Lorentz: "About Einstein's affair I am now somewhat reassured, although for several days I could not free myself from the feeling of indignation and shame that this man,

---

[44] Schröder-Gudehus, "Les professeurs allemands," likewise concludes that the problems in re-establishing international scientific relations provided academic groups access routes to both government and public opinion, routes which they then used to propagate political convictions which had only an indirect connection with the problems of international scientific relations.

[45] R. Sthamer, German *chargé d'affaires* in London, to the Ministry of External Affairs, Sept. 2, 1920. Published by Grundmann, "Der deutsche Imperialismus," p. 265. Further interesting documents on the insistence by the German ambassador in Stockholm that he, and not the Swiss ambassador, was entitled to transmit Einstein's Nobel Prize have been published by F. Herneck, "Über die deutsche Reichsangehörigkeit Albert Einsteins," *Forschungen und Fortschritte*, 1963, 37:137-140.

[46] Albert Einstein and Arnold Sommerfeld, *Briefwechsel*, ed. Armin Hermann (Basel: Schwabe, 1968), p. 68. (English translation in preparation by R. and H. Stuewer.)

[47] Grundmann, "Der deutsche Imperialismus," pp. 223-230; Ronald W. Clark, *Einstein: The Life and Times* (New York: World, 1971), pp. 292-305.

[48] "Ich bin ganz ausser mir vor Zorn und Wut über diese infamen Dunkelmänner, welche es wagten und fertig gebracht haben, Sie von Ihrem Hause, von der Stätte Ihrer Wirksamkeit zu trennen... Aber tiefer als meine Empörung geht mir der Schmerz darüber, dass Sie... keine Lust mehr hätten, zurückzukommen... so möchte ich Ihnen hier nur die eine, aber herzliche und dringendste Bitte aussprechen, jetzt keinen Schritt zu unternehmen, der Ihre Rückkehr nach Berlin endgültig und für allezeit unmöglich machen würde." Planck to Einstein, Nov. 10, 1923 (Einstein Papers, Institute for Advanced Study, Princeton, N.J.).

for whom the entire world envies us, could be caused by intrigues of the lowest sort to abandon his working places."[49]

Yet in Einstein's case we do not have a clear example of *Kulturpolitik* in our narrow sense, but rather of *Kulturpropaganda*. For here the mechanism still remains, despite the active interest of the Ministry of External Affairs, basically the classical one of reflected glory. A far clearer example of the attempt to stack the deck of scientific internationalism by promoting bilateral ties may be found in Germany's relations with Soviet Russia during the Weimar period. Probably no one could have been more astonished than Rathenau himself at the reaction of the politically so conservative German academic world to the announcement that he had met with the commissars at Rapallo on Easter Sunday 1922 and signed a treaty renouncing reparations and establishing diplomatic relations. Certainly the physical scientists were jubilant. On May 15 the German Chemical Society held a special Rapallo celebration at its Berlin headquarters, and an overflowing auditorium applauded the Russian chemists who happened to be in town.[50] During the summer of 1922 the German Physical Society suddenly developed a most sympathetic concern for the difficulties of their Russian colleagues, who were "almost all as good as cut off from German science," and the society appealed to its members to contribute multiple copies of offprints of their publications for distribution to the Russian physicists.[51] In December 1922 the society elected Orest Chwolson, professor in Petrograd, as their one and only honorary member, stating that this was meant as a visible expression of their satisfaction at renewing the ties with their Russian colleagues.[52] And in order to keep these colleagues securely tied one had recourse to such practices as maintaining lower standards for acceptance of articles submitted by Russians to German journals.[53]

The Russians themselves were not altogether pleased by the German scientists' efforts to saturate and so monopolize their international relations—as the Germans knew perfectly well. Otto Hoetzsch, *qua* President of the Society for the Study of

---

[49] "Ueber Einsteins Angelegenheit habe ich mich jetzt wieder etwas beruhigt, nachdem ich mehrere Tage lang das Gefühl der Empörung und der Scham nicht los werden konnte, das dieser Mann, um den uns die ganze Welt beneidet, durch Umtriebe der niedrigsten Art veranlasst werden konnte, seine Arbeitsstätte zu verlassen." Planck to Lorentz, Dec. 5, 1923 (AHQP, Lorentz Papers, Microfilm 9). It is noteworthy that Planck never claims that Einstein is a German, but only that he works in—and thus sheds his prestige upon—Germany.

[50] *Berichte der Deutschen Chemischen Gesellschaft*, 1922, 55A:107–109; V. N. Ipatieff, *The Life of a Chemist* (Palo Alto: Stanford University Press, 1946), pp. 355, 505; Horst Schützler, "Wissenschaftliche Beziehungen der Berliner Universität zur Sowjetunion in der Zeit der Weimarer Republik 1918 bis 1933," *Forschen und Wirken. Festschrift zur 150 Jahr-Feier der Humboldt-Universität zu Berlin* (Berlin: Deutscher Verlag der Wissenshaften, 1960), Vol. I, pp. 529–546, on p. 532; Eduard Winter, "Die deutsche Wissenschaft und Rapallo," pp. 153–159 of *Rapallo und die friedliche Koexistenz*, ed. Alfred Anderle (Berlin: Akademie-Verlag, 1963). Schützler and Winter both err in dating the meeting June 15.

[51] *Verhandlungen der Deutschen Physikalischen Gesellschaft*, Aug. 31, 1922, *3*:66; *Zeitschrift für Physik*, Aug. 14, 1922, *10*:352.

[52] *Verhl. D. P. G.*, Aug. 31, 1922, *3*:93. In the previous two years there had been a steady stream of Russian scientists through Germany, so that we must suppose that Rapallo was responsible not for the renewed ties but the satisfaction therein. In this same pre-Rapallo period various individual physicists, notably Paul Ehrenfest, had been trying, largely unsuccessfully, to arouse interest among the Germans in the work and plight of their Russian colleagues.

[53] Otto Lubarsch, *Ein bewegtes Gelehrtenleben* (Berlin: Springer, 1931), p. 369: toward such articles "I have been considerably more generous than toward those by Germans or by other foreigners, because I wanted to prevent their straying into French and English journals.... And that is true not only of Virchows Archiv, but also of other strictly scientific German journals."

Eastern Europe, acknowledged this in 1927 in requesting support from the Ministry of External Affairs for a strictly bilateral scientific fair. But, Hoetzsch explained, the Society nonetheless "considers it to be our duty to exploit in the field of science as well the head start won by the early assumption of political relations with Soviet Russia."[54]

As the rhetoric of science as a *Macht-Ersatz* would lead us to expect, this deliberate fostering of bilateral ties to the exclusion of multilateral was seen by the German *Gelehrten* not merely as a means for artificially maintaining the international prestige of German science, but also as providing them with a mechanism for directly influencing the policies and behavior of foreign scientists and even of foreign governments. Thus in the summer of 1925, abetted by the Ministry of External Affairs, the German *Gelehrten* threatened the Soviet government with a rupture of the close relations with their Russian colleagues, and in particular a refusal of the invitations to the celebration of the two-hundredth anniversary of the Russian Academy of Sciences, in order to influence the trial then in progress in Moscow of three alleged German terrorists.[55] Essentially similar was Fritz Haber's effort in 1930 to exploit the close relations between German and Japanese chemists which he had carefully cultivated in the preceding decade. Having learned that the Japanese representative body in the International Chemical Union had failed to support the German proposals, Haber informed the German Ministry of External Affairs that "if Japan sticks to this position, ill feeling will arise between German and Japanese chemistry . . . which will have the distressing result that Japanese chemists will have difficulty gaining admission to German

---

[54] Original quoted from DZA, Potsdam, by Ursula Kretzschmar, "Die Russische Naturforscherwoche in Deutschland (19.-25. Juni 1927)," *Jahrbuch für Geschichte der U.d.S.S.R. und der volksdemokratischen Länder Europas*, 1966, 9:97–119, on p. 102. In reply to O. Vogt's welcoming address at the Naturforscherwoche the geochemist A. E. Fersman, Vice President of the Soviet Academy of Sciences, made very clear the Russians' dissatisfaction with the Germans' attempt to monopolize their international scientific relations: Irene Strube, "Zur Publizierung der Ergebnisse der sowjetischen Naturforschung, insbesondere der Chemie, in Deutschland (1918-1930)," *NTM. Zeitschrift für Geschichte der Naturwissenschaften, Technik und Medizin*, 1968, 5 (No. 11):55–63.

Rudolf Ludloff, "Der Aufenthalt deutscher Hochschullehrer in Moskau und Leningrad 1925 anlässlich des 200 jährigen Bestehens der Russischen Akademie der Wissenschaften," *Wissenschaftliche Zeitschrift der Friedrich-Schiller-Universität Jena, gesellschafts- und sprachwissenschaftliche Reihe*, 1956/1957, 6:709–721, shows the German *Gelehrten*, both at the grass-roots level and in the Verband der Deutschen Hochschulen, acting on the policy that "Germany ought to be represented as strongly as possible at the celebrations," while the Ministry of External Affairs tried to restrict participation in order that the German delegation not be "conspicuously strong" compared with the Latin and Anglo-Saxon. Some further interesting material on this episode is given, seemingly in ignorance of Ludloff's work, by Conrad Grau, "Die deutschen Universitäten und die 200-Jahr-Feier der Akademie der Wissenschaften der UdSSR 1925," *Deutschland, Sowjetunion: Aus fünf Jahrzehnten kultureller Zusammenarbeit* (East Berlin:Humboldt-Universität, 1966), pp. 172–178. The seventy contributions to this volume give numerous examples of cultural diplomacy and show most clearly the enormous number of strictly *bi*lateral relations.

[55] Ludloff, "Der Aufenthalt." On Aug. 1, 1925, the Minister of Foreign Affairs addressed all Reich and Prussian ministries, pointing out that recently many Russians with letters of recommendation from Soviet authorities had been asking to visit scientific institutes in Germany: "Im Hinblick auf die Erfahrung des Moskauer Studentenprozesses, bemerke ich *vertraulich*, dass gegenwärtig kein Anlass zu besonderem Entgegenkommen . . . besteht und dass beabsichtigt ist, solange der Prozess eine unseren Forderungen entsprechende Lösung nicht gefunden hat, den von Sowjetbehörden unterstützten Wünschen solcher Kreisen um Besichtigung deutscher Institute gegenüber, sofern nicht im Einzelfalle eine Ausnahme angezeigt erscheint, grösste Zurückhaltung zu üben." (Auswärtiges Amt, VIB. 9981/17098.)

university laboratories. I assume that on the Japanese side the situation has not been clearly comprehended...." Haber recommended that the Ministry telegraph the German embassy in Tokyo, directing them to call in the key Japanese representative and explain the situation to him.[56]

VI. THE ANTI-POLITICAL AND "MANDARIN" IDEOLOGIES

The German-Soviet scientific relations show that the foreign *Kulturpolitik* of the Weimar period contains no explicit repudiation of the ideology of scientific internationalism, however much the readiness to substitute political purposes and criteria for scientific may in practice pervert the reality. Such a situation is not in itself surprising: it is the most common characteristic of anti-social behavior. What is surprising is that the German *Gelehrten* would have rejected with indignation any suggestion that their international scientific relations, let alone their scientific judgments, were affected by political considerations. In February 1925 the classicist Eduard Meyer—one of the most influential and also one of the more irrational of the Weimar academics—discussed in the Ministry of External Affairs the question of the German response to the anticipated overtures from the International Research Council:

> Our reserve can bring us in only advantages.... If we bite too soon we expose ourselves to the danger of prematurely surrendering an advantageous position. Also in the case of congresses reserve is already demanded by economic considerations, which one must throw up to the opposite side over and over again as their fault.... Science must work scientifically and hold itself free of all politics. Therefore... even outstanding foreign scientists who have displayed antagonism to Germany are not to receive any academic honor from Germany.[57]

---

[56] Fritz Haber to Ministerialdirigent Terdenge, June 13, 1930 (Politisches Archiv, Auswärtiges Amt, Bonn: Kulturabteilung, VI W, Institute und Vereinigungen, No. 42, Bd. 3). "Das japanische Votum ist, wie Sie erkennen werden, mit unseren Auffassungen in vollem Widerspruch. Hält Japan an diesem Votum fest, so entsteht zwischen der deutschen chemischen Wissenschaft und der japanischen eine Verstimmung, die ich bedauern werde, aber nicht hindern kann. Sie wird sich in peinlicher Weise dahin auswirken, dass die Aufnahme von Japanern an deutschen Hochschulstellen Schwierigkeiten finden wird. Ich möchte annehmen, dass man von japanischer Seite die Situation nicht klar übersehen hat. Ich würde es für zweckmässig halten, wenn von Seiten des Auswärtigen Amtes der Herr Botschafter in Tokio telegraphisch benachrichtigt und gebeten würde, sich Herrn Professor Setsuro Tamaru kommen zu lassen und den Gegenstand mit ihm zu besprechen."

[57] "Unsere Zurückhaltung könne uns nur Vorteile einbringen.... Wenn wir zu früh zugriffen, setzten wir uns der Gefahr aus, eine günstige Position vorzeitig preiszugeben. Auch bei Kongressen sei Zurückhaltung schon aus wirtschaftlichen Rücksichten geboten, welche man der Gegenseite als deren Schuld immer wieder vorhalten müsse.... Die Wissenschaft müsse wissenschaftlich arbeiten und sich von jeder Politik freihalten. Deshalb... könnten auch hervorragende Wissenschaftler des Auslandes, welche sich deutschfeindlich betätigt hätten, deutscherseits keiner akademischen Ehrung teilhaftig werden." (Politisches Archiv, Auswärtiges Amt, Bonn: VI B, No. 583, Bd. 1, 1921–1925, "Protokoll der Sitzung im Auswärtigen Amt vom 6. Februar 1925 betreffend Verhalten der deutschen Gelehrtenwelt gegenüber dem Auslande," p. 4.) Erhard Moritz, "Zu den Auslandsbeziehungen der Berliner Universität in der Zeit der Weimarer Republik," *Forschen und Wirken*, Vol. I, pp. 471–497, points out that "Das 'Eintreten für Deutschland' wurde an der Universität zum stillschweigenden [!] Kriterium für die Einladungs-würdigkeit eines Gastes. Das betonte auch Professor Otto Franke, der spätere Vorsitzende der Kommission für Gastvorlesungen." See Otto Franke, *Erinnerungen aus zwei Welten* (Berlin: W. De Gruyter, 1954), p. 163.

I have edited Meyer's remarks to bring his contradictions into immediate juxtaposition. But the fact remains that he, and his colleagues, were oblivious to these contradictions, precisely because it was a basic axiom of their academic ideology, and thus of their own self-conception, that science (= objectivity) and politics were wholly antithetical.[58]

To some degree this anti-political ideology has been common to scientists and scholars of all nations.[59] (The metaphor of the republic of science stems from an age in which parties were regarded as subversive of republics.) And the function of this ideology is to safeguard the interests of the science by denying the legitimacy of any route to status and power in the scientific community other than the ascribed status resulting from distinguished scientific contributions.[60] Thus the only legitimized politics in scientific associations is "deference politics."[61] But in early-twentieth-

---

In this connection it is interesting to note the rather ostentatious celebration of the Dutch spectroscopist Pieter Zeeman, who was generally considered "deutschfreundlich," and the considerable reluctance to celebrate his countryman H. A. Lorentz, by far the more significant physicist, who was widely regarded as "deutschfeindlich." Sommerfeld to Friedrich Paschen, July 24, 1921 (*SHQP* Microfilm 33); Karl Kerkhof, *Internationale wissenschaftliche Kongresse 1922–1923* (Berlin:Privately printed, 1923), p. 13; Planck to Einstein, Dec. 4, 1925 (Einstein Collection, Institute for Advanced Study), Lise Meitner to Max von Laue, Nov. 26, 1923, published by S. Grundmann, "Zum Boykott der deutschen Wissenschaft nach dem ersten Weltkrieg," *Wissenschaftliche Zeitschrift der Technischen Universität Dresden*, 1965, *14*:799–806.

[58] Klaus Schwabe, *Wissenschaft und Kriegsmoral: Die deutschen Hochschullehrer und die politischen Grundfragen des Ersten Weltkrieges* (Göttingen:Musterschmidt, 1969), pp. 182–183, observes with respect to Meyer's colleague Dietrich Schäfer that "his publicistic activity is all the more astonishing in as much as from his own point of view (*Selbstverständnis*) in the last analysis his stance remained unpolitical." German academic ideology is treated as such by Ringer, *Decline of the German Mandarins*, and Helene Tompert, *Lebensformen und Denkweisen der akademischen Welt Heidelbergs im Wilhelminischen Zeitalter*, Historische Studien, Heft 411 (Lübeck/Hamburg:Matthiesen, 1969), esp. pp. 59–82. Other studies of political attitudes of the German professorate are: Mommsen, *Weber*; Hans Peter Bleuel, *Deutschlands Bekenner: Professoren zwischen Kaiserreich und Diktatur* (Bern:Scherz, 1968); Kurt Töpner, *Gelehrte Politiker und politisierende Gelehrte. Die Revolution von 1918 im Urteil deutscher Hochschullehrer* (Göttingen: Musterschmidt, 1970), who also notes (p. 14) as "ein Spezifikum des akademischen Berufsstandes, dass Parteipolitik als suspekt angesehen wurde."

[59] E.g., Paul Gary Werskey, "*Nature* and Politics between the Wars," *Nature*, Nov. 1, 1969, *224*:462–472; Joseph Haberer, *Politics and the Community of Science* (New York:Van Nostrand, 1969).

[60] "Although rank and authority in science are *acquired* through past performance, once acquired, they tend to be *ascribed* (for an indeterminate duration)." Harriet Zuckerman and Robert K. Merton, "Patterns of Evaluation in Science: Institutionalisation, Structure, and Functions of the Referee System," *Minerva*, 1971, *9*:66–100, on pp. 81–82.

[61] I have borrowed the concept from Ronald Formisano, "Political Character, Antipartyism, and the Second Party System," *American Quarterly*, 1970, *21*:683–709. The scientists' malaise in a political environment and its connection with the politics of professional associations is noted by Anselm L. Strauss and Lee Rainwater, *The Professional Scientist: A Study of American Chemists* (Chicago:Aldine, 1962). Today, however, we are witnessing a revolt against the anti-political ideology and with it an attack on "deference politics" in scientific associations. E.g., Robert Golub, letter to the editor, *Physics Today*, March 1970:11: "When one examines the ballot for American Institute of Physics officers that has just been sent out, the reason that policies so harmful to the profession can be instituted becomes clear. We are told of the fine academic work and impressive research accomplishments of the various candidates, but we are not given any further information as to their views on the function of AIP and the American Physical Society, on the relationship of the physics community to society .... In the absence of any debate between the candidates the elections are a farce, a mere popularity contest...." The logical alternative, already realized in the American Chemical Society, is the development of "machine politics" in scientific associations: *Science*, Apr. 21, 1972, *176*:260–263.

century Germany this quasi-universal professional ideology was integrated with and reinforced by the anti-democratic socio-political ideology of those classes in Germany from which the scholars and scientists were drawn. The bureaucratic authoritarianism of the old regime, basing policy not on "politics" but on objective, impartial judgments, served the true interests of the nation, while every policy of the parliamentary-democratic Weimar regime was *ipso facto* "political," unobjective, and as such without real legitimacy. Thus although not merely their international relations but every aspect of the scientific-professional life of the Weimar academics was riddled with politics,[62] it was impossible for them to admit this fact.

This categorical rejection of politics was accompanied of course by rejection of those classical utilitarian conceptions of the purpose of the state in which political parties serve as the principal instrumentalities. Not the greatest good for the greatest number, but some higher good is the object of the state. And that higher good? Scholars and scientists generally, but the early-twentieth-century Germans especially, have been inclined toward that "mandarin ideology" which Fritz Ringer has identified and analyzed: "To put it perhaps a little polemically, the state lives neither for the ruler nor for the ruled as a whole; it lives for and through the 'men of culture' and their learning (*die Wissenschaft*)."[63] And while the German natural scientists retained the option of justifying their activities in utilitarian terms—an option which, after November 1918, they chose to exercise far less frequently in the face of a revulsion throughout the educated middle class against the very notion of knowledge for use[64]—implicitly they were at one with the *Geisteswissenschaftler* in feeling that the social, economic, and political order exists for the sake of the science and culture which is created within and supported by the social framework.[65] Although seldom articulated, it was nonetheless this mandarin ideology which made it possible for the German *Gelehrten* to arrogate to themselves the role of representatives of the genuine interests of the German nation and thus to attribute such inordinate importance to their international scientific relations.

---

[62] I have explored this circumstance with special reference to the physicists in "The Environment and Practice of Atomic Physics in Weimar Germany" (Ph.D. Dissertation, University of California, Berkeley, 1967; Ann Arbor: University Microfilms, 1968), and in "Financial Supports and Political Alignments of the German Physicists in the Weimar Republic," *Minerva*, 1973, *11* (in press).

[63] Ringer, *Decline of the German Mandarins*, p. 11, and, in general, pp. 1–13, 113–127. *Cf.* n. 34, and Gerald Feldman, "A German Scientist between Illusion and Reality: Emil Fischer, 1909–1919," in *Deutschland und die Weltpolitik im 19. und 20. Jahrhundert. Festschrift für Fritz Fischer*, I. Geiss and B.-J. Wendt, eds. (Düsseldorf: Bertelsmann, 1973).

[64] See my "Weimar Culture, Causality . . . ."

[65] Consider the following exchange between Sommerfeld and Einstein in Dec. 1918 (*Briefwechsel*, pp. 54–55) regarding the November revolution which overthrew the several monarchies and brought the Social Democrats to power. Sommerfeld writes that he hears that Einstein "believes in the new age and wants to collaborate with it—may God preserve your faith! I find it all unspeakably miserable and stupid. Our enemies are the greatest liars and scoundrels, we the greatest softheads. Not God, but money rules the world." That's *all* that Sommerfeld says, but Einstein knows implicitly what Sommerfeld is *really* worried about, and replies: "I am however firmly convinced that culture-loving Germans will soon again be able to be as proud of their fatherland as ever—with better reason than *before* 1914. I do not believe that the present disorganisation will leave behind permanent damage." Thus it is not the political or economic order, as such, which Einstein supposes to be uppermost in Sommerfeld's mind, but rather the damage to German science consequent upon the disruption of that order, and the belief that a democratic-socialist state will not support and encourage science and culture so liberally or effectively as an imperial one.

VII. THE SUBORDINATION OF THE INTERESTS OF SCIENCE TO THE INTERESTS OF THE NATION

Our brief excursus into the anti-political mandarin ideology of the Weimar academics has not deflected us from our main line of inquiry—the ideology of scientific internationalism considered in relation to the attempts to wield German science as a political instrument. Rather it is an essential preliminary to a discussion of a most characteristic feature of those attempts—the willingness to sacrifice the interests of science, indeed of *German* science, to the interests of the nation. At first sight such abnegation is surprising, for it appears to be inconsistent with an ideology which takes culture as the *raison d'être* of the state and which thus seems to place the interests of science above those of the nation. It becomes intelligible, however, if on the one hand we bear in mind that this anti-political mandarin ideology encouraged an identification of the interests of German science with the "true" interests of the German nation, and on the other hand recall how that element of scientific nationalism which lies behind the ideology of scientific internationalism was reinforced by specifically German notions of nations developing their highest spiritual powers through mutual competition. Under these circumstances the rhetoric which represented German science as one of the few "power factors" remaining to German foreign policy persuaded the German scientists that at least their *formal* international scientific relations were subordinate to, in the service of, and to be dictated by political considerations, even if that should mean denying themselves certain personal and scientific advantages and satisfactions.[66]

To be sure, only the most intransigent nationalists were prepared to enunciate and urge this course explicitly. The farthest I have seen a *physical* scientist go is Georg Struve's proposition that "German science, to whose account we must credit a good bit of the prestige which today [1926] we still enjoy abroad, ought to regard as its most important obligation and task the maintenance of German dignity in the world." And it is most interesting to note that the premises which lead this scion of astronomy's most international family to that conclusion do not contain even the faintest hint of a repudiation of the ideology of scientific internationalism. On the contrary, his memoir on "Astronomie, Völkerhass, Deutsche Würde" opens with a declaration that

> It was certainly one of the most deplorable concomitants of the World War that the national hatred which blazed up everywhere did not stop short of science, and was responsible for a division of the scholarly world into two camps which had no further connection with each other. No discipline felt the damage thereby inflicted upon science

---

[66] It goes without saying that precisely because they identified their own international prestige with the "true" interests of the German nation, the German academics saw that national interest quite differently from Germany's political leaders. They took positions on international scientific relations which were not only detrimental to the progress of their own sciences but objectively disadvantageous to the German state as well. Thus, for example, Peter Polis, Director of the Aachen Meteorological Observatory, pointed out to Reichskanzler Fehrenbach, Sept. 1, 1920, that through the Occupation authorities he had been in correspondence with the meteorological institutes of the Entente, and that these contacts had proved "von besonderem Vorteil" in the adjustment of the German-Belgian boundary. In the light of this circumstance he asked Fehrenbach to judge whether his participation in the forthcoming international meteorological congress in Venice would "dem Reichsinteressen entsprechen, da die wissenschaftlichen Herren von ganz anderen Voraussetzungen ausgehen." (Bundesarchiv, Koblenz; R 43 I/814. Further R 43 I/817, especially for the correspondence between Fritz Haber and Chancellor Luther in 1926–1927.)

more acutely than astronomy. By the nature of the subject, astronomy more than any other science is dependent upon cooperation with professional colleagues in other countries, whether it be a question of coordinated projects which require participation of several observatories, or discoveries which need to be checked and pursued further by other investigators. The transmission of reports in order that newly discovered celestial phenomena may quickly be generally known, the exchange of scientific ideas and opinions, the wide distribution of the astronomical literature is PRACTICALLY A VITAL NECESSITY FOR THIS SCIENCE.[67]

Nonetheless, Struve maintained, for the sake of the dignity of their nation, German astronomers should forgo the advantages of membership in the Astronomical Union of the International Research Council and reject the anticipated invitation to join.

Here we must distinguish two political tendencies. The spokesmen for the great majority of the German academics, unsympathetic toward the Republic and specifically antagonistic toward its policy of "fulfillment" of the Treaty of Versailles, pictured themselves as the legitimate representatives of a scientific great power and sought to use German science as the instrument of an independent, anti-governmental, truly national foreign policy. In all their white papers on international scientific relations, reeking with politics as they were, the principal theme was that science and politics must be kept absolutely separate and therefore Germany can have nothing to do with the international organizations established by the Allies.[68] In particular, the "political

---

[67] A typewritten copy of the 4-page memoir is in the Lorentz Papers, microfilm 9. It is undated, and the internal evidence is contradictory, implying that it was written after Aug. 1926 *and* before July 1926; it was thus evidently written and rewritten in the summer of 1926. The concluding paragraph reads: "Den Gedanken an eine Verständigungs- und Versöhnungspolitik um jeden Preis müssen die deutschen Astronomen entschieden von sich weisen, mag er auch vereinzelte Anhänger in ihren Reihen haben. Der Anschluss an die Union ist für sie keine Lebensfrage, da die astronomische Gesellschaft ihren Zweck völlig erfüllt. Da sollte die deutsche Wissenschaft, zu deren Gunsten ein gutes Stück des Ansehens zu buchen ist, das wir heute noch im Auslande besitzen, es als ihre vornehmste Pflicht und Aufgabe ansehen, die deutsche Würde in der Welt zu wahren." See also Schröder-Gudehus, *Deutsche Wissenschaft*, p. 190, on the distrust of rapprochement as a kind of *Erfüllungspolitik*.

[68] A notable document is the 22-page memoir of Max von Gruber, President of the Bayerische Akademie der Wissenschaften, dated Dec. 5, 1925 (Akademie der Wissenschaften zu Berlin, Archiv, VI a 17, Bd. 7, fol. 124). Claiming to speak for the entire Bavarian Academy, Gruber asserts (p. 8) that the position of the German *Gelehrtenschaft* cannot be otherwise than that they remain ready at any time to participate gladly in those international organizations which offer guarantees that they serve purely scientific goals and whose constitutions enable them to do so, but that they refuse to join organizations which overtly or covertly pursue primarily political goals. Having established this criterion, Gruber then proves that the Germans can have nothing to do with the IRC and especially with the Committee of Intellectual Cooperation of the League of Nations. The IRC is suspect "vor allem" on account of its "demokratische Verfassung," for "wie jede Demokratie wird auch diese die Macht einigen wenigen rücksichtslosen und geschickten Politikern in die Hand spielen" (p. 6). (Characteristically, Gruber did not hesitate to send a copy of his memoir to the Weimar government.) The CCI, being subordinate to the League, is *ipso facto* political (p. 9). Further examples of this fundamental position: Otto Francke, "Bericht des Auslandsausschusses," *Mitt. V. D. H.*, Feb. 1925, 5:29–34, maintains that the IRC is nothing but an organ of the Committee of Intellectual Cooperation of the League, and both pursue only political ends; Ludwig Bernhard, "Locarno und die deutsche Wissenschaft," *Mitt. V. D. H.*, Dec. 1925, 5: 213–216, maintains these are "politicized organizations" and therefore must be replaced by "international associations and institutes which serve pure science"; Heinrich Karo, *Mitt. V. D. H.*, Feb. 1926, 6:25–29, maintains that the IRC represents "a tyrannical domination of science by a few great powers according to purely political postulates," while we Germans "act unceasingly as representatives of pure unpolitical science." Reporting for the *Auslandsausschuss* a year after Germany was invited to join the IRC, Franke, *Mitt. V. D. H.*, Oct. 1927, 7:123–125, emphasized that despite all the pressure from

character" of the International Research Council they saw "chiefly" in a constitution which did not leave national representation solely to academies and equivalent scientific bodies but admitted the possibility that a country's delegates could be named by its government.[69] Here of course is the epitome of the anti-political ideology, serving on the one hand to exclude the Weimar government from any role in international scientific relations and on the other hand serving to shift to the Allies the full responsibility for the absence of those international scientific relations demanded by the ideology of scientific internationalism—to which the Germans remained, of course, quite loyal.

The other political tendency, characteristic of a minority of the German academics (but a larger minority of the natural scientists, and especially of the pure physicists, than of the professorate as a whole), was to accept the Republic as a *pis aller*: in particular to lend support to its foreign policy of conciliation and cooperation with the Allies.[70] Among this group too the concern with the political role of German science was often so great as to take precedence over the parochial or material in-

---

abroad the Verband is persevering in its refusal "if for no other reason than because the Council is a semi-political body; the German *Gelehrtentum*, however, wants to keep science absolutely free of political influences and purposes." In the late 1920s the Allies and neutrals, finding this continuing refusal to join unintelligible, began charging the *Germans* with mixing politics and science. Franke replied, "Bericht über die internationalen wissenschaftlichen Beziehungen," *Mitt. V. D. H.*, Apr. 1929, 9:67–78, by pointing to the Rhineland occupation, reparations, irredentism: "It is a rather cheap view of the matter to try to push these things aside as 'politics.'"

[69] "Den politischen Charakter erblicken wir hauptsächlich darin, dass für die Länder nicht die Akademien und ihnen gleichwertige wissenschaftliche Körperschaften allein Stimmträger sind, sondern neben ihnen auf gleicher Linie auch von den Regierungen ernannte Vertreter." (Unanimous "Beschlüsse der Kartellversammlung [i.e., the cartel of German academies of science: Berlin, Göttingen, Heidelberg, Leipzig, Munich, Vienna] in München am 7. Mai 1926." Auswärtiges Amt, VI W, No. 42, Bd. 1, and also Lorentz Papers, Microfilm 9.) That we are dealing here not with an exclusively German, but a quasi-universal academic ideology is shown by the fact that the leaders of the IRC could try to defend their earlier exclusion of Germany on precisely these same grounds. Namely, for want of a unique national academy "it could only be the German government that joins, and the representatives at meetings would be Government representatives. This was the usual course before the war and our experience shows that it encourages what you wish to avoid: the mixing of science and politics." Arthur Schuster, General Secretary of the IRC, to F.A.F.C. Went, Voorzitter of the afdeeling Natuurkunde of the Amsterdam Academy, Feo. 13, 1924 (Hale Papers, Microfilm 41, frame 1003).

[70] "The policy hitherto pursued by Prussia and the Reich is that in all cases so far as possible invitations of German savants to former enemy countries ought to be accepted." Prussian Kultusminister to the Ministry of External Affairs, June 22, 1923. Quoted from DZA Merseburg by Grundmann, "Zum Boykott." See also Schröder-Gudehus, *Deutsche Wissenschaft*, p. 222.

Some evidence of the relative cooperativeness of the natural scientists, especially of the pure physicists, may be derived from the minutes of discussions in the Ministry of External Affairs. Thus the *Protokoll* of the meeting of Feb. 6, 1925 (cited in n. 57), allows one to arrange the thirteen participating *Gelehrten* in a spectrum with the sociologist Alfred Weber and the (Roman Catholic) physicist Heinrich Konen the most conciliatory, followed then by Sommerfeld; the center of the spectrum is occupied by the applied scientists L. Brauer, A. Nägel, and A. Penck; while the most rabid and recalcitrant are the *Geisteswissenschaftler* G. A. Deissmann, Eduard Meyer, and R. Seeberg. This is confirmed by a report from the Ministry to the ambassador in Paris, Leopold von Hoesch, June 18, 1926 (Auswärtiges Amt, VI B, No. 607, Bd. 1) on a discussion with representatives of the natural science sections of the six academies. Speaking as *individuals*, the reporter said, the natural scientists were all inclined to give a very liberal interpretation to the *Beschlüsse* of the *Kartellversammlung* of May 7 and expressed views which were moderate compared with those of the total *Professorenschaft* embodied in the Verband der Deutschen Hochschulen.

terests of their science itself. This comes through very clearly, for example, in Fritz Haber's negotiations with the representatives of the (U.S.) General Electric Company in the fall of 1926 over the future of the Elektrophysik-Ausschuss, an organization established at the height of the postwar inflation to distribute funds provided by G.E. (and partially matched by German firms) for the support of academic research in "electrophysics" in Germany. Haber showed not the least concern about the *money*— and indeed did not wish even to discuss the general question of the health and welfare of *Elektrophysik* in Germany—which the representatives of G.E. naturally supposed to be the primary consideration. Haber placed all the emphasis upon maintaining the organization, or better still a reconstituted multilateral organization, *qua* international organization, "to hold fast to the connection between the two countries which originated through the American aid."[71]

This general readiness to give precedence to political considerations in the question of international scientific relations led to yet another striking characteristic of these relations in the Weimar period—the more or less voluntary surrender by the individual scientist of responsibility for his own international relations. This was typically the case when it was a question of repudiating international ties or rejecting overtures, for the cognitive dissonance with the ideology of scientific internationalism could be most easily suppressed by surrendering all initiative and thus all responsibility to some higher social unit. Scientific men in Germany and in the Allied countries were at one in taking their "nation" as this social unit; they differed radically, however, when it came to identifying the legitimate spokesmen for their nation. In the last analysis—and often in the first—the Allied scientists were prepared to accept their respective governments as the legitimate spokesmen and so accept the policies of their national governments as authoritative for their own international scientific relations. A striking example of this tendency is the refusal by Arnaud Denjoy, a French mathematician in Dutch employ, to accept a visit from a German colleague in 1920:

> To conclude: the day when the French government believes it has received from yours tokens of good will and finds them sufficient to justify raising Germany again to the normal rank of nations, on that day I will make no further objection in principle to assuming or renewing relations.... But until then I will follow the instructions which the attitude of the government of my country prescribes for me.[72]

---

[71] "... der durch die amerikanische Hilfe begründete Zusammenhang beider Länder festzuhalten...." Haber to Einstein, von Laue, and Planck, Oct. 23, 1926 (Einstein Papers, Institute for Advanced Study). For an account of the Electrophysik-Ausschuss see Steffen Richter, *Forschungsförderung in Deutschland, 1920–1936. Dargestellt am Beispiel der Notgemeinschaft der Deutschen Wissenschaft und ihrem Wirken für das Fach Physik*, Technikgeschichte in Einzeldarstellungen No. 23 (Düsseldorf: VDI-Verlag, 1972), 69 pp.; Forman, *Physics in Weimar Germany*, pp. 316–318.

[72] "Je conclus. Le jour où le gouvernement français croira avoir reçu de la part du vôtre des gages de bonne volonté, et les trouvera suffisants pour justifier le relèvement de l'Allemagne au rang normal des nations, ce jour-là, je ne verrai plus d'objection de principe à nouer ou à reprendre des relations.... Mais jusque-là, j'observe la consigne que l'attitude du gouvernement de mon pays me dicte." A. Denjoy to Otto Blumenthal, Oct. 4, 1920. (Auswärtiges Amt, VI B, No. 296, Bd. 30, "Unberücksichtigt gebliebenes Schreiben von Dr. L. E. J. Brouwer an Seine Exzellenz den Minister für Unterricht, Kunst und Wissenschaft vom 27. September 1922. Symptomatisches zu einer Gefährdung der niederländischen Staatshoheit... Uebersetzung. Als Manuskript gedruckt.") Brouwer was particularly incensed because Blumenthal was then his guest in Holland, while Denjoy, although a French citizen, was then professor in Utrecht and a member of the Amsterdam Academy. *Cf.* A. Schuster to G. E. Hale, Jan. 30, 1922 (Hale Papers, Microfilm 41, frame 907): "a few years

Most German academics, on the contrary, did not recognize in the Weimar regime the legitimate voice of the German nation. This role they arrogated to themselves. Not the German government, but the academic and professional associations, and especially the scientific academies, became authoritative for most German scientists in the matter of international relations. Thus Hans Ludendorff, Director of the Prussian Astrophysical Observatory (and the general's brother), surrenders all responsibility to the Prussian Academy of Sciences:

> We German astronomers, of whom, I believe, the majority have a strong national feeling, find ourselves in a painful dilemma. The national consciousness speaks absolutely *against* participation in the congress of the International Astronomical Union since this organization is subordinate to the International Research Council. The scientific conscience speaks, on the contrary, *for* participation. The discussion today is supposed to serve to help us out of this dilemma. I need scarcely say that Colleague Guthnick and I have not the faintest notion of doing anything whatever in this matter which does not correspond to the wishes of the Academy. We only ask the Academy to relieve us of the decision, in order that we may then, with a clear conscience, act according to the wishes of the Academy.[73]

Still more painful, however, were the dilemmas of that minority of German scientists who were inclined to support their government's policy of reconciliation; to compensate for the odium of their colleagues they received from officials of the Weimar regime much private encouragement but no effective backing. A conversation with Foreign Minister Rathenau might persuade Einstein to reconsider his original refusal of an invitation from the Collège de France, a refusal in which "I gave as the principal reason considerations of solidarity with my colleagues here."[74] But the Reich govern-

---

ago" the British Foreign Office wrote the Treasury "instructing them not to assist any International Union which included Germany before that country was admitted to the League of Nations; but adding that after Germany had been so admitted, they should not give financial assistance to an international body excluding Germany. Although, as I have said, a copy of this letter was sent to me in confidence, I feel myself bound to act in accordance with it; and it seems to me to embody quite a correct principle, and one that should appeal to the neutral nations."

[73] "Wir deutschen Astronomen, von denen, glaube ich, die Mehrzahl ein starkes nationales Empfinden hat, befinden uns in einem argen Dilemma. Das Nationalbewusstsein spricht durchaus *gegen* eine Beteiligung an der Versammlung der U.A.I., da diese vom Conseil des Recherches abhängt. Das wissenschaftliche Gewissen spricht dagegen *für* eine Teilnahme. Die heutige Aussprache soll dazu dienen, uns aus diesem Dilemma herauszuhelfen. Ich brauche nicht zu erwähnen, dass Kollege Guthnick [Paul Guthnick, Director of the University of Berlin Observatory and the only other astronomer among the ordinary members of the Academy] und ich auch nicht im entferntesten daran denken, irgend etwas in dieser Sache zu tun, was den Wünschen der Akademie nicht entspricht.

Wir bitten die Akademie nur, uns die Entscheidung abzunehmen, damit wir dann reinen Gewissens handeln können, wie es die Akademie beschliesst." (Ludendorff, "Erklärung in der Sitzung v. 2 Feb. [1928]," Akad. d. Wiss. Berlin, Archiv, VIa 17, Bd. 10, fol. 29). But the Academy too was anxious to avoid taking the blame for Germany's nonmembership in the international scientific organizations and eventually threw the entire responsibility back upon the professional associations in the several disciplines. (Planck to Einstein, Mar. 23, 1933. Hale Papers, Microfilm 42, frames 990–993.)

[74] Einstein to the Sekretariat der preussischen Akademie der Wissenschaften, Mar. 3, 1922, published by Grundmann, "Der Deutsche Imperialismus," pp. 272–273. This consideration occurs repeatedly in Einstein's letters to H. A. Lorentz, e.g. Aug. 16, 1923 and Jan. 9, 1925 (*AHQP*, Lorentz Microfilm 7). In general, as scientific activities became political demonstrations the German physicists began to look around to see where their leaders were going: Max Born to Einstein, Aug. 25, 1923, in Albert Einstein, Hedwig and Max Born, *Briefwechsel, 1916–1955* (Munich: Nymphenburger, 1969), pp. 116–117, trans. Irene Born as *The Born-Einstein Letters* (New York: Walker & Co., 1971).

ment, unlike the Allied governments, was never willing to take the next step and put pressure upon, let alone act against the wishes of, the recalcitrant majority. In a memorandum occasioned by Haber's report that the IRC had voted to invite Germany to join, Ministerialdirektor Friedrich Heilbron, the officer in the Ministry of External Affairs responsible for such questions, observed that

> According to these indications, we may expect, initially, a struggle within the German academic world between the official representatives of the academies and the individual *Gelehrten* who feel less bound [to the counter-boycott]; and these latter would very much like the Reich to preempt the decision [on the invitation]. This would obviously result in a severe disagreement between the Reich and the academies, and the academies would, presumably, be supported by the governments of their respective states. I am dubious whether we may allow ourselves to be drawn into such a course.[75]

VIII. UNOFFICIAL INTERNATIONAL SCIENTIFIC RELATIONS — PRECONDITIONS FOR INTRANSIGENCE

Despite the anti-political ideology, and despite the mental relief afforded by the renunciation of all personal responsibility, the readiness to sacrifice the interests of science to the interests of the nation called for some rationalization, at least in the form of a denigration of those subordinate elements of the ideology of scientific internationalism asserting the necessity of transnational personal contacts and international collaboration. Brigitte Schröder-Gudehus has pointed out that during the war, when all such relations were necessarily severed, "it was frequently simply denied that any harm at all would arise from forgoing international associations, congresses, etc." and that "this low estimate of international scholarly congresses is, moreover, to be found way down into the ranks of moderate observers."[76]

So also in the postwar period one can find many examples over the whole of the political spectrum of this same tendency to depreciate the importance of organizations for international scientific collaboration and even to deprecate international congresses. As Einstein wrote Lorentz in the summer of 1919: "To exclude for a period of several years German scholars from international social intercourse within the scientific community might perhaps teach them a useful lesson in modesty. It would do little

---

[75] "Nach diesen Andeutungen wird es innerhalb der deutschen Gelehrtenschaft sich zunächst um einen Kampf zwischen den offiziellen Akademievertretern und den sich weniger gebunden fühlenden Gelehrten-Individualitäten drehen, und die letzteren möchten gern, dass das Reich die Entscheidung vorwegnimmt. Dies würde offenbar eine schwere Differenz zwischen Reich und Akademien, die vermutlich von den Länderregierungen gestützt würden, bedeuten. Ob wir uns auf einen solchen Weg ziehen lassen dürfen, ist mir zweifelhaft." F. Heilbron [to Soehring?], July 2, 1926 (Auswärtiges Amt, VI B, No. 607, Bd. 2). The President of the Vienna Academy was thus sadly mistaken in looking to the Reich to help the Austrian *Gelehrten* out of *their* "böses Dilemma. Unsere Regierung wünscht ein Entgegenkommen gegenüber dem Conseil.... Wenn die österreichische Regierung auf die Wiener Akademie drückt und die deutschen Akademien eine ablehnende Haltung einnehmen, so kann—wenn wir die Solidarität mit den deutschen Akademien höher einschätzen als das gute Verhältnis mit Regierung, was ich für das wahrscheinlichste ansehe—ein Konflikt zwischen unserer Akademie und der Regierung entstehen. Wenn wir uns der Regierung fügen, ein Konflikt im deutschen Kartell.... Von ausschlaggebender Bedeutung für mich ist, *was die Deutsche Reichsregierung will.*" R. v. Wettenstein to F. Schmidt-Ott, Dec. 7, 1926 (*loc. cit.*). *Cf.* Schröder-Gudehus, *Deutsche Wissenschaft*, pp. 203, 248, and "Les professeurs allemand," p. 42.

[76] Schröder-Gudehus, *Deutsche Wissenschaft*, pp. 85–86.

harm and might even do some good, but it would be of little significance."⁷⁷ What Einstein is asserting—and critics of opposite political tendency agreed thoroughly—is that the material and psychological satisfactions of such social intercourse are largely irrelevant to the mechanisms of communication most essential to scientific achievement (and to recognition and prestige at the international level). What one never sees, I think, are derogations of the importance of receiving foreign scientific journals (and of wide circulation of German journals abroad). Indeed the subsidization of the purchase of foreign scientific and scholarly periodicals (as well as the publication of German journals) consumed a substantial share of the funds appropriated by the national government for the support of scientific research.⁷⁸

As for the organizations promoting international scientific collaboration, the disparagement of their substantive achievements could serve as justification for refusing collaboration and was in fact one of the reasons most commonly adduced. In 1927 the members of the natural science section of the Göttingen academy were asked to look through the several volumes of *Reports and Proceedings* of the International Research Council and its unions and give their opinion of the work accomplished in their own fields. By far the least negative evaluation was that by the physicists, R. W. Pohl and James Franck. And while they concluded that "for pure physics the interest in an international organization is at the moment not at all pressing," they also maintained that "in the fields bordering on physics there are, on the contrary, numerous very important tasks which without close international cooperation simply cannot be attacked at all."⁷⁹ Yet among the physical scientists it was precisely those representing the mixed physics fields cited by Pohl—geophysics, astronomy, meteorology—who along with the applied physicists tended to take the hardest line in international scientific relations.⁸⁰ This I think we ought to attribute at least in part to the fact that it was these disciplines which were also the major weapon and bore the heaviest responsibility in the attempts to demonstrate Germany's scientific great-power status by forcing concessions from the IRC.

But what of all these "very important tasks which without close international cooperation simply cannot be attacked at all"; were the German scientists really prepared to deny themselves any share in some very important and thus potentially very prestigious tasks? I do not think so. And much evidence indicates that below the level of formal international relations there was a very extensive development of personal

---

⁷⁷ Published in English translation by Otto Nathan and Heinz Norden, eds., *Einstein on Peace* (New York: Schocken Books, 1968), p. 34, and like remarks in another letter to Lorentz of Sept. 21, 1919, *ibid.*, p. 35.

⁷⁸ Zierold, *Deutsche Forschungsgemeinschaft*, pp. 75–85, 92–102.

⁷⁹ "Franck und ich haben jetzt die drei Bände der Conseil-Berichte durchgesehen, soweit es sich um Physik handelt. Wir haben beide den Eindruck, dass es sich bisher in der Hauptsache um Pläne und organisatorische Fragen handelt. An positivien Leistungen halten wir nur die Herausgabe eines grossen Tabellenwerkes für wichtig. Dazu werden von Zeit zu Zeit die grossen Fragen einer internationalen Regelung der Masssysteme treten. Ist also für die reine Physik im Augenblick das Interesse an einer internationalen Organisation durchaus nicht dringend, so finden sich hingegen auf den Grenzgebieten der Physik zahlreiche sehr wichtige Aufgaben, die ohne enge internationale Zusammenarbeit garnicht angegriffen werden können, (Geophysik, Astronomie, Meteorologie, Ausnutzung der Funkentelegraphie für wissenschaftliche Zwecke). gez. Pohl." ("Einige Äusserungen über die wissenschaftliche Betätigung des Conseil international de recherches bezw. der diesem angegliederten Unionen, veranlasst von der Mathematisch-physikalischen Klasse der Gesellschaft der Wissenschaften zu Göttingen," 4 pp.; Akademie der Wissenschaften zu Berlin, Archiv, VIa 17, Bd. 9, fol. 49.)

⁸⁰ This is also Grundmann's view (see n. 70).

scientific contacts and of recruitment of individual German scientists into international collaboration. Thus, the American astrophysicist Charles St. John, who was very active in the affairs of the IRC and its Astronomical Union, reported in the summer of 1922: "I feel satisfied with my whirlwind trip to Germany. I found them apparently much pleased to have an American visit them. . . . I feel satisfied that it will be possible to work together unofficially so that until full relations are renewed we shall be able to coordinate the wavelength work in Germany and elsewhere."[81] In fact, in the evaluations of the work of the IRC by the Göttingen scientists this circumstance—the existence of effective and sufficient informal arrangements—was commonly adduced as an argument *against* joining the organization. It was employed by the astronomer, the geologist, the biologist, and the physiologist; and while not mentioned by the physicists (precisely because of its status as a *negative* argument) such personal contacts and informal collaborative arrangements were quite extensive in physics as well, perhaps above all.

Indeed, I think one can detect among the German physicists a distinct tendency to minimize the cost to themselves of the effort to play great power with German science by treating as "informal" and "unofficial" all those contacts and collaborative arrangements which were in fact of some real importance to them and their work. This is perhaps clearest in the case of Max Planck, who was both strongly committed to the ideology of scientific internationalism and, as one of the permanent secretaries of the Prussian Academy, to the execution of the policy of intransigence in international scientific relations. Thus Planck, wishing to support a projected merger of the English, French, and German physics abstracting journals—a project originating with the IRC and promoted by the League of Nations Committee on Intellectual Cooperation—tried to represent this as an informal, unofficial matter.[82] He failed, largely because of the opposition of the zealots controlling the Deutsche Gesellschaft für technische Physik.[83] Again, when in the summer of 1926 he and a number of other Germans were invited for the first time since the war to a Solvay Congress in Brussels, Planck shut his eyes to the fact that a political test had been employed in determining which Germans were to be invited.[84] Had the fact been explicitly admitted, it would

---

[81] Charles St. John to Lorentz, June 14, 1922. (Lorentz Papers, Microfilm 6). The most often recurring theme in the reports on international scientific relations in the *Mitteilungen* of the Verband der Deutschen Hochschulen is the reprehensibility of "desertion from the German front." This circumstance ought to be taken as evidence of extensive informal relations. Indeed, the 1926 British Association meeting at Oxford, notable for the substantial contingent of German physicists and mathematicians (Iris Runge, *Carl Runge und sein wissenschaftliches Werk*, Göttingen:Vanderhoeck & Ruprecht, 1949, pp. 196-197), had been specifically proscribed by the V. D. H. in a resolution which confidently affirmed that "it is therefore a matter of course that no German *Gelehrter* and no German scientific society will participate in the meeting." *Mitt. V. D. H.*, Dec. 1925, 5:185.

[82] Planck to Lorentz, June 30, 1925 (Lorentz Papers, Microfilm 9).

[83] Georg Gehlhoff to Lorentz, Nov. 16, 1925 (Lorentz Papers, Microfilm 10).

[84] Planck to Lorentz, June 13, 1926 (Lorentz Papers, Microfilm 8), where following the disclaimer "dass es selbstverständlich allein Sache des einladenden Comités ist, hier die Entscheidung zu treffen, nach Gesichtspunkten, die jedenfalls viel zu vielseitig sind, als dass ich sie zu übersehen vermöchte," Planck emphasizes how "ungemein bedrückendes" it is for him that certain participants in the 1911 Congress who have since contributed far more intensively than he to the further development of the quantum theory—specifically, Sommerfeld—will not be present at the 1927 Congress. Sommerfeld's absence from the list of invitees was indeed most conspicuous, and Max Born felt it necessary to apologize to Sommerfeld for accepting an invitation despite Sommerfeld's exclusion. (Born to Sommerfeld, June 15, 1926, *SHQP*, Microfilm 29, Sec. 11.) It is more than likely that Planck's

have precluded his participation and obliged him to urge a boycott of the congress upon the other German invitees.

Throughout the foregoing discussion I have taken for granted the striking circumstance which, in the first instance, draws Weimar Germany to our attention—the vitality of its *wissenschaftlich* culture, especially of that portion centering on the problems of atomic and theoretical physics. The bearers of that culture were indeed professed internationalists, but that "internationalism" was largely subordinate to and in the service of their nationalism. By contrast, the internationalism to be found among leaders of Weimar's literary and artistic culture, especially among the intellectuals associated with the *Weltbühne*, subordinated the nation to the party of humanity.[85] Their response to the anti-semitic political agitation against Einstein was to assure him "in a genuinely international attitude of the sympathy of all free men"[86]— an expression and sentiment which scarcely a single one of Einstein's fellow physicists could have used or shared.[87] Thus if finally we ask what part of the *specific* vitality of Weimar scientific culture derived from a genuinely internationalist impulse, I think the answer must be: none at all.

---

inclusion and Sommerfeld's exclusion were decisively influenced by their respective involvements in Belgian affairs during World War I. Planck had used his good offices to aid Belgian scholars and Ernst Solvay himself, in obtaining passes and other dispensations from the German occupation authorities (Lorentz Papers, Microfilm No. 6). Sommerfeld, on the other hand, had contributed to the efforts of those authorities to promote Flemish cultural nationalism and political separatism; he came away from "Ein Besuch in Gent," *Süddeutsche Monatschefte*, 1918, *15*: 44–46, looking forward to "a voluntary and therefore much more intimate annexation (*Angliederung*) of that part of Belgium which is most valuable to us."

[85] Istvan Deak, *Weimar Germany's Left-Wing Intellectuals. A Political History of the "Weltbühne" and Its Circle* (Berkeley: University of California Press, 1968).

[86] Telegram sent to Einstein from Salzburg signed by Oskar Bie, Joseph Chapiro, Werner Krauss, Andreas Latzko, Alexander Moissi, Max Reinhardt, Johanna Terwin, Helene Thimig, Stefan Zweig, *Berliner Tageblatt*, Aug. 31, 1920, No. 409, p. 3.

[87] Sommerfeld's reaction to the *Weltbühne*: "I have finally gotten hold of this noble little sheet and am revolted (*angewidert*) by its spirit. In our present situation I find this sort of internationalism really abominable (*scheusslich*)..." Einstein and Sommerfeld, *Briefwechsel*, p. 95. Commenting upon the first postwar national scientific congress (Naturforscherversammlung), the reporter for *Vorwärts*, the official newspaper of the German Social Democratic Party, observed (No. 478, 3rd Beilage, p. 1, Sept. 26, 1920): "Science is absolutely (*nun einmal*) not German, but international. And precisely because of its internationality it, just like art, is capable (*berufen*) of uniting and reconciling peoples. But not to a single speaker did it seem even to have occured to allude to this aspect." To my knowledge, among Einstein's fellow physicists only Wilhelm Westphal expressed himself publicly in such terms. See, e.g., Schützler, *Forschen und Wirken*, Vol. I, p. 534.

The Financial Support and Political Alignment of Physicists in Weimar Germany

PAUL FORMAN

THE Weimar republic originated in military defeat and political and social revolution. These political events and the attitudes of German academics toward them were important factors in the scientific life of the period. The economic situation in the aftermath of war and revolution had, however, both directly and indirectly, a greater impact on the personal welfare and the research of German scientists. In the year immediately following the armistice, the mark sank to less than 10 per cent. of its pre-war value and remained there until the summer of 1921. Then began the runaway inflation which the government brought to an abrupt end late in 1923, after the economy and society had been brought to the verge of chaos. Professorial salaries in terms of their purchasing power lagged far behind their pre-war levels, institute budgets still further behind, and savings, whether personal wealth or institutional endowment, were wiped out.[1]

In these circumstances, and in large measure in response to them, a variety of new institutions were created to help support German science and scholarship. Two of the most important of these foundations were the Notgemeinschaft der deutschen Wissenschaft (NGW) and the Helmholtz-Gesellschaft zur Förderung der physikalisch-technischen Forschung. The Notgemeinschaft was the dominant source of funds of the Weimar period. Constitutionally, it was a self-governing union of institutions of research and higher education. Most of the funds which it distributed in support of research and publication in all fields of science and scholarship came however from the central government.[2] The Helmholtz-Gesellschaft, established concurrently and in competition with the Notgemeinschaft, was, legally, an association of the large industrial firms which supplied its funds.

---

[1] Bresciani-Turroni, C., *The Economics of Inflation: A Study of Currency Depreciation in Post-War Germany* (London: Allen and Unwin, 1937); Forman, Paul, *The Environment and Practice of Atomic Physics in Weimar Germany*. Ph.D. Dissertation, University of California, Berkeley, 1967 (Ann Arbor: University Microfilms, 1968), pp. 206–288, cited hereafter as Forman, P., ... *Physics in Weimar Germany*.

[2] Schroeder-Gudehus, Brigitte, "The Argument for the Self-Government and Public Support of Science in Weimar Germany", *Minerva*, X, 4 (October 1972), pp. 537–570; Richter, Steffen, *Forschungsförderung in Deutschland 1920–1936: Dargestellt am Beispiel der Notgemeinschaft der Deutschen Wissenschaft und ihrem Wirken für das Fach Physik*. *Technikgeschichte in Einzeldarstellungen*, Nr. 23 (Düsseldorf: VDI-Verlag, 1972); Zierold, Kurt, *Forschungsförderung in drei Epochen: Deutsche Forschungsgemeinschaft. Geschichte, Arbeitsweise, Kommentar* (Wiesbaden: Franz Steiner Verlag, 1968). I am indebted to Professor Schroeder-Gudehus and Dr. Richter for communicating the results of their research to me prior to publication.

But in this foundation too the intended beneficiaries—academic physicists and engineers—participated very fully in the machinery by which the funds were distributed.³

It is the purpose of this essay, first, to point out that the activities of these two foundations brought substantial increases in the amount, and significant innovations in the mechanisms, of financial support of physical research in Germany in the decade immediately following the First World War; second, to analyse the considerations prompting this amplification of research support and, especially, the distribution of the bulk of that increased support through "project-grants" awarded on the basis of detailed applications reviewed by panels elected from the discipline; and third, to argue that the patterns of distribution of these "project-grants" among German physicists—in particular, differences between the Notgemeinschaft and the Helmholtz-Gesellschaft—were strongly influenced by interrelations between scientific-political alignments within that discipline and political divisions in German society at large.

The Scale of Support for Research in Physics in the Weimar Republic

In 1920–21, it was thought that the scale of support to be provided for physical and technological research by the Helmholtz-Gesellschaft would far exceed that which the Notgemeinschaft could offer.⁴ As it turned out the Helmholtz-Gesellschaft, which lost 80 per cent. of its capital of some 4 million gold marks ($1 million) in the inflation, provided substantially less support, and the Notgemeinschaft a great deal more, than anticipated (Table I). During the years of inflation from 1921, the Notgemeinschaft provided about 50,000 gold marks per year in project-grants to physics, and spent perhaps an additional 100,000 gold marks in 1921–22 to provide machine tools for physics institutes to use in constructing apparatus. In its initial year, 1922, the Helmholtz-Gesellschaft awarded about 25,000 gold marks in project-grants for physical research, but

---

³ There is no published account of the Helmholtz-Gesellschaft. Some information about its organisation and activity is contained in *Die Helmholtz-Gesellschaft . . . in sieben Jahren ihres Wirkens* (n.p., 1928), 49 pp., edited and privately distributed by the secretariat of the Helmholtz-Gesellschaft at the Verein deutscher Eisenhüttenleute, Düsseldorf. I hope to publish shortly a monograph on "The Helmholtz-Gesellschaft: Support of Academic Physical Research by German Industry after the First World War." A preliminary draft of this monograph, referred to hereafter as Forman, P., *The Helmholtz-Ges.* . . . is based principally on the mimeographed minutes of the meetings of the *Verwaltungsrat* and of the *Hauptversammlungen*, surviving records of the secretariat, and correspondence between the principal organisers of the Helmholtz-Gesellschaft, found in the Bibliothek des Vereins deutscher Eisenhüttenleute, in the Historisches Archiv, Gutehoffnungshütte-Sterkrade AG, Oberhausen, and in the Werksarchiv, Farbenfabriken-Bayer, Leverkusen. I am grateful to the directors of the GHH and Bayer archives, Messrs. Bodo Herzog and Peter Göb, and especially to the present general secretary of the Helmholtz-Gesellschaft, Dr. Karl-Heinz Treptow, who recently made available to me the annual lists of applications to the Helmholtz-Gesellschaft for research grants with the ratings assigned them by the *Fachausschüsse*.

⁴ A. Vögler to F. Schmidt-Ott, 21.1.21, and Wilhelm Wien to Schmidt-Ott, 25.7.21 (Geheimes Staatsarchiv, Berlin-Dahlem, Nachlass Schmidt-Ott); Schmidt-Ott to Vögler, 12.9.21 (Bayer, Werksarchiv, 46/11.1); Friedrich Paschen to Arnold Sommerfeld, 15.9.21 (Archive for History of Quantum Physics, Berkeley, Copenhagen, and Philadelphia).

dropped to 5,000 gold marks at the height of the inflation in 1923. From 1924 to 1928–29, the level of project-grant support provided annually by each foundation rose steadily, to about 220,000 marks for the Notgemeinschaft and 55,000 marks for the Helmholtz-Gesellschaft. Moreover, in the late 1920s and early 1930s, the Notgemeinschaft spent another 100–150 thousand marks annually on stipends to support young physicists carrying out particular research projects.[5]

The significance of the project-grants provided by these two foundations may be best appreciated by noting that no more than 100–200,000 marks were even potentially divertible annually to the cost of research as such from the budgets of the physics institutes of the universities and technological institutes in the late 1920s.[6] The contributions of the two major

---

[5] The other new foundations of comparable importance were: (1) those established by the German chemical industry to support chemical education, research and publication—Justus-Liebig-Gesellschaft zur Förderung des chemischen Unterrichts; Emil-Fischer-Gesellschaft zur Förderung der chemischen Forschung; Adolf-Baeyer-Gesellschaft zur Förderung der chemischen Literatur. These foundations contributed about five times as much to academic chemistry as the Helmholtz-Gesellschaft provided for physics and engineering; (2) the *Gesellschaften der Freunde* at each of the German universities and *technische Hochschulen*, which collectively contributed perhaps four times as much as the Helmholtz-Gesellschaft to these institutions, but probably only half as much for research. In both cases, and also in the case of the Helmholtz-Gesellschaft, the initiatives came from Carl Duisberg (1861–1935), who trained as a chemist under Adolf Baeyer at Strassburg, was head of the Bayer chemical firm, a principal architect of the *I. G. Farbenindustrie*, and president of the Reichsverband der Deutschen Industrie. Duisberg maintained a vital interest in the organisation and activity of German chemists, and German scientific and academic life generally. Flechtner, H. J., *Carl Duisberg: Vom Chemiker zum Wirtschaftsführer* (Düsseldorf: Econ-Verlag, 1960) gives some account of his role in institutions which gave patronage to science. Further details, derived chiefly from the Bayer archives, are given in Forman, P., *The Helmholtz-Ges.* . . . . A listing of the " Gesellschaften der Freunde " is included in the *Kalender der deutschen Universitäten und Hochschulen 112*, 464–469 (1932); details of one of the more active may be found in *Universitätsbund Göttingen, Mitteilungen 1*, 14 (1918–1932).

Throughout the Weimar period the Rockefeller Foundation was an important source of medical research funds, indirectly through the NGW, and directly as well. Only in 1929, however, did it initiate a programme of research grants in the physical sciences, from which in the four years 1929–32 one German physicist—Arnold Sommerfeld—and three physical chemists received a total of $50,000. Far more important was the $700,000 appropriated by the Rockefeller Foundation and the International Education Board for the construction and equipment of mathematical and physical institutes in Göttingen and a Kaiser-Wilhelm-Institut für Physik in Dahlem. See Rockefeller Foundation, *Annual Report* (1930), pp. 204, 278, 279, 285; *ibid.*, (1931), pp. 218, 314; *ibid.*, (1932), pp. 247, 247, 326, 339.

[6] The budgets of the institutes of experimental physics of the German universities and *technische Hochschulen* did not as a rule include any funds specifically designated for research. Total budgets including: (i) heat, water, gas, electricity; (ii) salaries of assistants, technicians, custodians; (iii) apparatus, books, and consumable supplies, ranged between 10,000 marks and 40,000 marks per annum in the years immediately before the First World War, with the average for the 21 universities being about 16,000 marks $\pm$ 10 per cent. If one adds the physical institutes of the 16 colleges of technology, for which I have exceedingly fragmentary figures, the average is certainly lowered somewhat. About one third of the budget was appropriated under each of these three headings, but apparatus and supplies received a relatively smaller share in the larger institutes. This heading, under which research equipment falls, seems to have ranged between 2,000 and 8,000 marks, averaging perhaps 4,000 marks $\pm$ 30 per cent. As the primary function of these institutes, and the principal justification of all their expenditures, was instruction, it is undoubtedly an overestimate to assert that one half this amount was available for research.

During the inflation years from 1919 to 1923 the institute budgets are not reliable guides to expenditures for they were commonly exceeded in unspecified ways and amounts. Although the budgets which followed the inflation budgets were generally limited to, or even

foundations thus led to at least a doubling of the real expenditures for the direct support of research in physics compared with the pre-war period.[7]

reduced from, their pre-war levels, by the end of the decade, total budgets seem to have risen 50–150 per cent. at nominal prices (0–70 per cent. at real prices) above their pre-war levels. The average increase was almost certainly less than 80 per cent. in nominal prices or 20 per cent. in real prices. Moreover, as one might expect, these budget increases appear to have taken place largely, if not entirely, under the first two headings; the funds for the purchase of apparatus and supplies were cut somewhat in most states in 1926, and by the end of the decade had apparently, in the majority of cases, not risen much above the nominal pre-war level. If, therefore, we assign an average of 2,000 marks per annum of research funds to each of the 16 physical institutes at the colleges of technology and 24 institutes at the universities, 1,200 marks to each of the 12 institutes for theoretical physics, and 2,000 marks to each of the three X-ray laboratories, we find 100,000 marks available in the annual budgets of institutions of higher education in Germany for application to direct costs of physical research. Extraordinary appropriations are unlikely to have increased that sum by as much as one half. See, Forman, Paul, Heilbron, John L. and Weart, Spencer, " Personnel, Funding, and Productivity in Physics *ca* 1900: A Multinational Statistical Study ", *Historical Studies in the Physical Sciences*, 4 (1973, in press).

For postwar budgets, see Boeck, Carl (ed.), *Deutsche technisch-wissenschaftliche Forschungsstätten*. Teil II: *Die technisch-wissenschaftlichen Forschungsanstalten* (Berlin: VDI-Verlag, 1931), passim; Griewank, Karl, *Staat und Wissenschaft im Deutschen Reich. Schriften zur deutschen Politik*, Heft 17/18 (Freiburg i.B.: Herder, 1927), p. 94. For translation into real terms, see Hoffmann, Walther G. and Müller, H. J., *Das deutsche Volkseinkommen, 1851–1957* (Tübingen: J. C. B. Mohr, 1959), p. 14, where the *per capita* real income in Germany 1925–1929 is found to be 3–5 per cent. lower than in the five years before the war. Thus, by this measure the physical institutes were treated generously by their governments; it was not, however, a measure which the physicists, or the German academics generally, could accept.

[7] Richter, S., *op. cit.*, p. 56, counted 8,800 publications in the three principal German physics journals (*Annalen der Physik, Physikalische Zeitschrift*, and *Zeitschrift für Physik*) 1923–1938, of which 14 per cent. acknowledged support from the *Notgemeinschaft*. Allowing for omissions of acknowledgement, Richter concludes that the *Notgemeinschaft* participated in at least one quarter of the output of physics in Germany in this period. The level of support of research was in fact so high that there was a distinct tendency for the funds to overflow into other channels, *i.e.*, for institute directors to misappropriate Notgemeinschaft and Helmholtz-Gesellschaft funds to more general and instructional purposes: see Walther Gerlach to Württemberg Kultusministerium, 17.12.26 (Universitäts-Archiv, Tübingen, 117/904); A. Vögler and O. Petersen to Reichsministerium des Innern, 27.6.27 (Bundesarchiv, R431/813, Bl. 69–72). A further factor not considered here, but probably of considerable significance, is that the post-war decline in personal income of the German university professor represented a loss of potential research funds. One certainly finds much incidental evidence before the First World War of apparatus purchased and research supported out of personal income, but very little such evidence afterwards.

Notes to TABLE I on p. 43.

a The Notgemeinschaft and Helmholtz-Gesellschaft data are not entirely comparable, as the former include astronomy, astrophysics, geophysics, and meteorology under " physics ". Judging from the titles of the projects supported, as published in annual reports of the NGW, roughly one third of the " physics " grants fall in these fields, and probably a somewhat larger proportion of the funds. Moreover the Notgemeinschaft's support of theoretical physics is included here in " experimental research "; the Helmholtz-Gesellschaft did not support theoretical work as such.

b The Notgemeinschaft's fiscal year began on 1 April; thus " 1922 " is its 1922–23 fiscal year. The Helmholtz-Gesellschaft's fiscal year began late in the autumn (December deadline for grants awarded following spring); thus " 1922 " is its 1921–22 grant year.

c Number of *grants*; the number of grantees was 20–30 per cent. less.

d Amounts in 1000 marks; figures for the inflation years 1921–23 have been reduced to gold value of the mark (gold mark) averaged over the year in the case of the NGW; in December of the previous year in the case of the Helmholtz-Gesellschaft. A tilde indicates my own estimate, good, I judge, to within 10 per cent. in the case of the Helmholtz-Gesellschaft; 20 per cent. in the case of the NGW.

e Elektrophysik-Ausschuss (EPA, Electrophysics committee); the figure for 1922 is the support provided for physics by the Japan Committee (Hoshi Fund).

TABLE I

Support of Research in Physics through Project-Grants[a]

	Notgemeinschaft						Helmholtz-Gesellschaft			
	Grants for "experimental research" in "physics" (incl. theoretical physics, astronomy, geophysics, etc.)				Research fellowships in "physics"		Grants for experimental research in physics		Total grants (physics and engineering)	
Year[b]	Number[c]		Amount[d]							
	EPA[e]	Total	EPA[e]	Total	Number[c]	Amount[d]	Number[c]	Amount[d]	Number[c]	Amount[d]
1921		~200		~100			—	—	—	—
1922	20	54	~15	~30			~75	~27	124	48
1923	62	104	60	~90	4		53	5	110	15
1924	27	57	60	~100	3		41	22	80	56
1925	38	~90	60	~150	4		29	21	71	90
1926		84	60	?	5		28	33		128
1927		154	60	?	29		42	54	97	173
1928		112	60	365	52	217	24	39	70	119
1929		102	50	346	74	236	31	55	77	180
1930				341	72	166	29	37	70	144
1931				232	68	207	28	28	74	97
1932				213	62	144	28	27	66	72
1933				?	69		38	32	86	88

SOURCES: Notgemeinschaft: *Bericht*, I–XII (1922–33); Richter, *Forschungsförderung* ... (1972), pp. 15, 31, 36, 39; Forman, P., ... *Physics in Weimar Germany*, p. 315. Helmholtz-Gesellschaft: Minutes of the meetings of the Administrative Council and the "Minutes" of the meetings of the Administrative Council and the Central Assembly, Central Assembly, 1922–33; annual *Zusammenstellung der Unterstützungsanträge*, 1922–33 (Bibliothek, Verein deutscher Eisenhüttenleute, Düsseldorf).

For notes, see p. 42.

Furthermore, the predominance of the foundations was associated with a change in the mechanisms for the distribution of research support. Funds were no longer supplied "on faith" to the *Ordinarius*-director of the institute to be disbursed entirely at his discretion; they were now granted specifically for research, and only for such projects as could claim priority within the discipline.

The Government of the Reich and the Support of Science

The establishment of the Notgemeinschaft and the Helmholtz-Gesellschaft late in 1920 was the financial and institutional expression of a vivid concern on the part of the central government (Reich) for the plight of science and scholarship (*Wissenschaft*), and of German industry for "physical-technological research". The considerations underlying these concerns were important determinants of the new mechanisms for the support of research and of the criteria applied in the making of grants. At the outset these considerations were on the whole not utilitarian, but were rather political and ideological. Indeed, in the case of public support in the early years of the new regime the most persuasive political arguments were explicitly anti-utilitarian. However surprising this may seem in view of the ubiquitous and emphatic utilitarian justifications for the support of physics before the war, it does indeed correspond to the anti-utilitarian, anti-technological, anti-rationalistic, anti-materialistic, and, fundamentally, anti-scientific attitudes widespread among the educated middle classes—and even shared to some extent by socialists—during the Weimar period. Such attitudes were epitomised in the Prussian Ministry of Education project for reform of the secondary school curriculum in 1924. Taking it for granted that "the economic-political, technical, and positivistic age ... now lies behind us", the Ministry refused to justify any part of the curriculum on utilitarian grounds.[8]

Through the Weimar constitution the central government had acquired a stronger fiscal position vis-à-vis the constituent states (Länder) of the federal system, and moreover had acquired for the first time an explicit mandate to set educational policy and cultivate science and scholarship.[9] The formation of the Notgemeinschaft offered the central government an

[8] Forman, Paul, "Weimar Culture, Causality, and Quantum Theory, 1918–1927: Adaptation by German Physicists and Mathematicians to a Hostile Intellectual Environment", *Historical Studies in the Physical Sciences*, III (Philadelphia: University of Pennsylvania Press, 1971), pp. 1–115, esp. p. 25. See also Sauer, Wolfgang, "Weimar Culture: Experiments in Modernism", *Social Research*, XXXIX (1972), pp. 254–284. For pre-war utilitarianism, see Manegold, Karl-Heinz, *Universität, Technische Hochschule und Industrie: Ein Beitrag zur Emanzipation der Technik im 19. Jahrhundert unter besonderer Berücksichtigung der Bestresbungen Felix Kleins. Schriften zur Wirtschafts- und Sozialgeschichte*, Bd. 16 (Berlin: Duncker & Humblot, 1970).

[9] Gerloff, Wilhelm (ed.), *Handbuch der Finanzwissenschaft*, Band 2: *Besondere Steuerlehre* (Tübingen: J. C. B. Mohr, 1927); Abelein, Manfred, *Die Kulturpolitik des Deutschen Reiches und der Bundesrepublik Deutschland: Ihre verfassungsgeschichtliche Entwicklung und ihre verfassungsrechtlichen Probleme* (Köln, Opladen: Westdeutscher Verlag, 1968).

immediate opportunity to exercise the least controversial part of that mandate, which was the support of research. It welcomed that opportunity because it was anxious to demonstrate that a parliamentary democratic regime was not, as the academics believed, necessarily hostile to higher culture.

German academics had been generally enthusiastic in support of the imperial regime, and the natural scientists and technologists particularly had applauded Wilhelm II, because they regarded it and him as the guarantors of their special position in state and society, and of their special claims—including financial claims—upon them.[10] Under the new regime they saw, as Ernst Troeltsch noted, "the danger of a proletarisation of society . . . the threat of educational reforms which would destroy higher culture, eliminate the leading position of the academic classes, and make the elementary school teacher the spiritual and political ruler of Germany."[11]

Although it could not win over such opponents, the central government, through its generous support of scientific research, did manage largely to deflect such criticisms onto the governments of the states. These became, in the post-inflation period, the objects of exceedingly bitter feeling on the part of the physicists. Even the political liberals among them attributed their stringent budgets to the absence of the influence of the former princely rulers.[12]

The generosity of the central government was not, of course, a simple attempt to purchase the loyalty of the academics. It also expressed the widespread acceptance in German society of the basic tenets of the German academic outlook. This appears especially clearly in the very rapid and extensive adoption of the academics' newly evolved conception of science as a surrogate for power, *i.e.*, science not as ancillary to, but as

---

[10] The most enthusiastic expressions are generally to be found in the orations delivered at the academic festivities held annually in January at most universities in celebration of Wilhelm II's birthday. See also Manegold, Karl-Heinz, *op. cit.*, p. 302. For the basic commitment of even the most outspoken academic critics, see Thimme, Annelise, *Hans Delbrück als Kritiker der Wilhelminischen Epoche. Beiträge zur Geschichte des Parlamentarismus und der politischen Parteien*, Heft 6 (Düsseldorf: Droste, 1955).

[11] Troeltsch, Ernst, *Spektator-Briefe* (Tübingen: J. C. B. Mohr, 1924), quoted by Ringer, Fritz K., *The Decline of the German Mandarins. The German Academic Community, 1890–1933* (Cambridge, Mass.: Harvard University Press, 1969), pp. 206–207. For a striking expression by natural scientists of this same pervasive fear, see the account of H. Timerding's address at the *Naturforscherversammlung* in 1920 reported in the *Berliner Tageblatt*, 448 (23 September, 1920), p. 4.

[12] Haber, Fritz, "Über Staat und Wissenschaft", *Aus Leben und Beruf* (Berlin: Springer, 1927), pp. 158–166, esp. p. 164. The bitterness of the more conservative academics was very evident in a discussion, in the presence of the head of the Bavarian government, of the desirability and feasibility of establishing a large X-ray research laboratory, a Forschungsanstalt für die gesamte Röntgenwissenschaft, in Munich: *Niederschrift über die Sitzung . . . am 14. März 1927*, typescript, 34 pp. (Geheimes Staatsarchiv, München, MA/100/130). Wilhelm Wien, professor of experimental physics of the University of Munich, had actually threatened to sue the Bavarian Ministry of Education for breach of contract were it, in the course of the post-inflation economy drive, to deprive him of one of his assistants: Wien to Verwaltung der wissenschaftlichen Sammlungen des Staates, 9.1.25 (Universitäts-Archiv München, OC-N14).

a substitute for, the military and economic power which had been lost through military defeat. And as Dr. Brigitte Schroeder-Gudehus has pointed out, this anti-utilitarian conception of science as a "substitute for power" received financial expression in the large proportion of funds provided by the central government through the Notgemeinschaft, which were allotted to the humanistic disciplines.[13]

German Industrialists and the Support of Science

These overtly anti-utilitarian appeals could have had only a very limited effect upon the prospective contributors to the Helmholtz-Gesellschaft; more utilitarian arguments were to their taste. Nonetheless, the motives which led to the establishment of this foundation, and the participation of wealthier German businessmen in it, were at least as political as utilitarian. Late in 1920, German industrialists could well afford to give generously. They had enjoyed enormous profits during the war, and now, while the victors were struggling with severe industrial depression, they benefited greatly from the inflation.[14] The tax laws enacted in 1920 were evidently not less but more favourable to large industrial firms, and they encouraged industrial philanthropy by permitting such contributions to be deducted from income in computing income tax.[15] The steady stream of criticism of German industry by parties of the left, and even centre, for its alleged failure to contribute significantly to cultural activities, constituted a further political encouragement to come to the financial aid of science.[16]

Rhenish-Westphalian coal, iron and steel firms, and the banks associated with them, were the principal contributors to the Helmholtz-Gesellschaft, providing two thirds of its initial capital. Less than one fifth came—and that only indirectly—from the still relatively competitive, and largely Berlin-based, electrical engineering and machine tool industries, which were far more likely to reap immediate benefits from "physical-technical" research. By contrast, the nearly completely cartelised German chemical industry provided about one ninth of the capital, even though it could expect relatively little for itself from the research it supported.

[13] See Schroeder-Gudehus, B., *op. cit.* Forman, P., "Scientific Internationalism and the Weimar Physicists: The Ideology and its Manipulation in Germany after the First World War", *Isis*, LXIV (June 1973), pp. 150–180, referred to hereafter as Foreman, P., "Scientific Internationalism . . .".

[14] Forman, P., . . . *Physics in Weimar Germany*, pp. 206–225; Feldman, Gerald D., "German Business between War and Revolution: The Origins of the Stinnes-Legien Agreement", in Ritter, Gerhard A. (ed.), *Entstehung und Wandel der modernen Gesellschaft: Festshrift für Hans Rosenberg* (Berlin: Walter de Gruyter, 1970), pp. 312–341.

[15] Duisberg to F. Klein, 3.9.20 (Bayer, 46/7.2); Reichsminister der Finanzen, Erlass S2209–8111 of 20.1.31, par. B.V.1 (Max-Planck-Gesellschaft, Archiv, I.L.G. Bd II). Such deductions were subsequently disallowed by the 1926 income tax law.

[16] Duisberg to Reichsverband der Deutschen Industrie, 8.4.24 (Bayer, 46/7.2), concerning the criticism by Reich Chancellor Wilhelm Marx of the Catholic Centre Party. From the left: Julius Moses (SPD), *Reichstag, Verhandlungen*, 268. Sitzung, 16.11.22, p. 9005.

Such a distribution of supporting industries is generally consistent with a purely economic explanation for readiness to support openly published research, namely, cartelisation of the industry in question, or some approximation to it through a strong trade association.[17] Yet however illuminating that explanation may be, it is surely not sufficient. To explain why the generous industries were based largely in the northern Rhineland, why heavy industry was relatively so generous, and why the Helmholtz-Gesellschaft took the particular form it did, one must look beyond purely economic considerations to the political and ideological considerations, to a community of social and political interests and beliefs shared by a particular industrial sector and a particular academic faction.

Scientists and Industrialists

Among academics, and not merely among those of the humanistic disciplines, there was a traditional distaste for businessmen, and an abhorrence of the "Americanisation" of institutions of research or higher education.[18] In the Wilhelmian period, even though academics had often implicitly been close political allies of industrial groups—especially in matters of foreign policy—they generally did not like to admit to the taint of such an association.[19] Moreover, however much they sought to influence their governments by propaganda and public pressure—perhaps most regularly and extensively in the area of science and higher educational policy [20]—they never doubted the adherence of the princely rulers and the governmental bureaucracies to their own basic values. In the Weimar period, on the contrary, precisely that confidence was absent, and the need for political allies urgent. Finally, "financial hardship overcame what resistance still remained".[21]

Equally important was the change in the attitude of the industrialists.

[17] Davis, Lance and Kevles, Daniel J., "The National Research Fund: A Case Study in the Theory of Economic Institutions and Industrial Financing of Academic Science", California Institute of Technology, Division of Humanities and Social Sciences, Social Science Working Paper No. 31, October 1972. The German iron and steel industry had a very strong trade association, the Verein Deutscher Eisenhüttenleute.

[18] Manegold, K.-H., *op. cit.*; Flechtner, H. J., *op. cit.*, p. 288. The anti-business attitudes of Adolf von Harnack, the president of the business-supported Kaiser-Wilhelm-Gesellschaft, were very deep. See Schroeder-Gudehus, Brigitte, *op. cit.*; Feldman, Gerald, "A German Scientist between Illusion and Reality: Emil Fischer, 1909–1919", in Geiss, I. and Wendt, B.-J. (eds.), *Deutschland in der Weltpolitik des 19. und 20. Jahrhunderts: Fritz Fischer zum 65. Geburtstag* (Düsseldorf: Bertelsmann, 1973), pp. 341–362.

[19] Schwabe, Klaus, *Wissenschaft und Kriegsmoral: Die deutschen Hochschullehrer und die politischen Grundfragen des ersten Weltkrieges* (Göttingen: Musterschmidt, 1969), pp. 70, 193, 222; Stegmann, Dirk, *Die Erben Bismarcks: Parteien und Verbände in der Spätphase des Wilhelminischen Deutschlands* (Cologne and Berlin: Kiepenheuer & Witsch, 1970).

[20] Burchardt, Lothar, "Wissenschaftspolitik und Reformdiskussion im Wilhelminischen Deutschland", *Konstanzer Blätter für Hochschulfragen*, VIII, 2 (May 1970), pp. 71–84; Manegold, K.-H., *op. cit.*, passim.

[21] Etzold, H., "Zür Wissenschaftspolitik der deutschen chemischen Industrie während der Weimarer Republik", *Wissenschaftliche Zeitschrift der Technischen Hochschule Otto von Guericke, Magdeburg*, XIII (1969), pp. 495–505.

Before the war it was common for the chemists alone among the German university professors to maintain personal relations with industrialists, namely, with the leaders of the German chemical industry.[22] One of the important objectives of the Göttinger Vereinigung zur Förderung der angewandten Physik und Mathematik, founded by the mathematician, Felix Klein, and the chemical industrialist, Henry Böttinger, at the end of the century, had been to extend this " friendly personal contact " beyond the boundaries of academic and industrial chemistry.[23] In fact, however, it was only after the war and the dissolution of the Göttinger Vereinigung that such close relations developed between Göttingen professors of mathematics and physics and at least one of its industrial sponsors—largely on the initiative of the latter.[24] What happened in Göttingen happened in most other German universities. Although the greatest increase in such contacts probably occurred among physicists, the establishment at every university of an " association of friends of the university " and the extensive participation in the same political parties by both academics and industrialists, provided bases for a substantial increase in such contacts among the humanistic academics as well.[25]

This rapprochement between industrialists and academics was part of a tacit social and political alliance. Both groups felt threatened by the new political order, both recognised how much their outlooks had in common. The two groups shared analogous conceptions of *die Wirtschaft* — the economy — and *die Wissenschaft* — learning — as autonomous realms to be governed by strong personalities whose authority within their own factories and institutes must not be limited, and in whose hands lay the true interests of the German people. Each group offered the other moral as well as political support for their common resistance to interference from the parliamentary-democratic—and above all, social-democratic—governments.[26]

---

[22] Such relations between academic chemists and chemical industrialists pervaded scientific life in this field in Germany: Ruske, Walter, *100 Jahre Deutsche Chemische Gesellschaft* (Weinheim-Bergstrasse: Verlag Chemie, 1967); Willstätter, Richard, *From My Life*, Hornig, Lilli S. (trans.) (New York: Benjamin, 1965); Flechtner, H. J., *op. cit.*

[23] Manegold, K.-H., *op. cit.*, p. 225.

[24] Runge, Iris, *Carl Runge und sein wissenschaftliches Werk* (Göttingen, 1949), pp. 182, 189. Einstein, Albert and Born, Max and Hedwig, *Briefwechsel, 1916–1955* (Munchen: Nymphenburger, 1969), pp. 84–85. One can adduce much evidence for the novelty of such relations: Wien, Wilhelm, *Aus der Welt der Wissenschaft* (Leipzig, 1921), p. 235 and *Aus dem Leben und Wirken eines Physikers* (Leipzig: Barth, 1930), p. 150; Fritz Haber to Albert Einstein, 28.6.21 (Einstein Collection, Institute for Advanced Study, Princeton).

[25] A good example is the historian Karl Brandi, who was president of the Universitätsbund of Göttingen, which was founded in 1918, and an active member of the Deutsche Volkspartei; in these capacities he came into personal contact with Rhenish-Westphalian industrial circles.

[26] For the industrialists' attitudes, see Feldman, Gerald D., " The Social and Economic Policies of German Big Business, 1918–1929 ", *American Historical Review*, LXXV (1969), pp. 47–55; Flechtner, H. J., *op. cit.*, pp. 304, 339, 382. For the academics' attitudes: Bleuel, Hans Peter, *Deutschlands Bekenner: Professoren zwischen Kaiserreich und Diktatur* (Bern: Scherz, 1968), pp. 170–178; Ringer, Fritz, *op. cit.*, passim; Forman, P., " Scientific Internationalism. . .". For the industrialists' search for allies, see Böhret,

A small minority of the academics and industrialists committed themselves to a non-socialist but democratic republic and associated themselves with the Deutsche Demokratische Partei (DDP). To this party, so closely associated in its origins and membership with Berlin, belonged the heirs of Germany's two principal electrical engineering firms, Walther Rathenau (AEG) and Carl Friedrich von Siemens.[27] About a third of the academics in the smaller, more conservative universities, along with many small businessmen and commercial employees, supported the Deutsch-Nationale Volkspartei (DNVP).[28] This distinctly nationalistic and extremely conservative party included many of *völkisch* outlook, which comprised anti-parliamentarian, anti-liberal, anti-cosmopolitan and anti-semitic attitudes.[29] Between these two parties, but closer to the DNVP than to the DDP, lay the Deutsche Volkspartei (DVP), the attachment of which to the political system of the republic in its first years was at best equivocal. This was the party preferred by the largest number of academics as well as industrialists; it was the formal expression of the alliance of *Besitz und Bildung*— property and education.

Berlin

But even within this basic pattern of political sympathies and affiliations there was ample room for further political antagonisms. One of the most important and pervasive of these was the opposition between Berlin and the rest of the country. This long-standing tension between metropolis and province had been greatly aggravated by the revolution and the Weimar constitution, so that " Berlin " came to signify all that

---

Carl, *Aktionen gegen die " kalte Sozialisierung," 1926–1930 . . . Schriften zur Wirtschafts- und Sozialgeschichte*, Bd. 3 (Berlin: Duncker & Humblot, 1966); Vögler, Albert, " Wissenschaft, Technik und Wirtschaft. Vortrag in der allgemeinen Sitzungen, Versammlung der Gesellschaft Leutscher Naturforscher und Ärzte, Düsseldorf ", *Naturwissenschaften*, 14 (1926), pp. 44–49. An implicit appeal to these shared attitudes is contained in Wien, Wilhelm, " Die Helmholtz-Gesellschaft und ihre Bedeutung für die deutsche Physik und Wissenschaftliche Technik ", in *Die Helmholtz-Gesellschaft . . . in Sieben Jahren ihres Wirkens* (n.p., 1928). For an explicit offer of alliance, see Friedrich Glum to Krupp von Bohlen, 4.6.29, quoted by Schroeder-Gudehus, Brigitte, *op. cit.*, p. 565.

[27] Hartenstein, Wolfgang, *Die Anfänge der Deutschen Volkspartei, 1918–1920. Beiträge zur Geschichte des Parlamentarismus und der politischen Parteien*, Bd. 22 (Düsseldorf: Droste, 1962); Neumann, Sigmund, *Die Parteien der Weimarer Republik*, 2nd ed. (Stuttgart, 1965); Kessler, Harry Graf, *Walther Rathenau: His Life and Work* (New York, 1930); Siemens, Georg, *Carl Friedrich von Siemens* (Freiburg, Munich: Alber, 1962).

[28] In the 1920 elections in Prussia the university town of Marburg cast 31 per cent. of its votes for the Deutsch-Nationale Volkspartei, 30 per cent. for the Deutsche Volkspartei; the university town of Greifswald cast 27 per cent. for the DNVP, 39 per cent. for the DVP; the town of Göttingen, whose university was relatively liberal, cast only 14 per cent. for the DNVP, 33 per cent. for the DVP (Hartenstein, Wolfgang, *op. cit.*, pp. 246–247).

[29] Jochmann, Werner, " Die Ausbreitung des Antisemitismus ", in Mosse, W. E. (ed.), *Deutsches Judentum in Krieg und Revolution, 1916–1923. Schriftenreihe wissenschaftlicher Abhandlungen des Leo Baeck Instituts*, 25 (Tübingen: J. C. B. Mohr, 1971), pp. 409–510; Mosse, George L., *The Crisis of German Ideology* (New York: Grosset & Dunlap, 1964).

was distasteful in the new social and political order.[30] Among big businessmen, apprehension of the "terror" in Berlin fused with a conception of their Berlin colleagues—among whom the proportion of Jews was very high—as dangerously liberal as well as high-handed and overbearing. "The repeated complaint . . . of arbitrary action by a supposed Berlin tendency" was the plaint of the Berlin industrialist, Kurt Sorge, president of the recently formed Reichsverband der deutschen Industrie.[31] The coal, iron and steel magnates of the northern Rhineland and Westphalia, whose political and social illiberalism was notorious, were especially hostile to Berlin.[32] It was from this group that the Helmholtz-Gesellschaft drew first and most heavily. These two circumstances were intimately connected: industrialists of the Ruhr region, led by Hugo Stinnes' lieutenant, Albert Vögler, an imposing figure on the right wing of the DVP, seized the initiative in order to "shunt aside Berlin", and obtain an organisation "free of the odour of the much disapproved-of capital city".[33]

"Project-Grants" and Review by Peers: Precedents and Rationale

An important phase in the evolution of institutional arrangements for the support of scientific research opened in Germany in the 1920s. There for the first time the bulk of the research funds available to the members of a particular discipline within a national scientific community were distributed to individual investigators for specific investigations, the relative merit of which was decided by referees chosen from and by active scientists themselves. If one accepts the view that scientific merit is an attribute which can be assessed only by persons working within the scientific discipline in question, it follows that such a system of research support ought to be ideally efficacious. Regardless, however, of one's criteria for the efficacy of a system of research support, such a system at least has the virtue of tending to rectify the fundamental inequity of the Western European and American academic systems, built upon the principal "unto him who hath shall be given".

---

[30] The *Berliner Tageblatt* and the *Vossische Zeitung*, the two more or less intellectual, liberal Berlin newspapers—both owned and edited by Jews, and both closely associated with the Deutsche Demokratische Partei—became in the early Weimar period symbols of Berlin and of the republic, and were attacked as such by anti-republicans. See Schwarz, Gotthart, *Theodor Wolff und des " Berliner Tageblatt." Eine liberale Stimme in der deutschen Politik, 1906–1933. Tübinger Studien zur Geschichte und Politik*, Bd. 25 (Tübingen: J. C. B. Mohr, 1968), pp. 124–142.

[31] Sorge in 1920, quoted by Feldman, Gerald, "Big Business and the Kapp Putsch", *Central European History*, IV (1971), pp. 99–130, esp. p. 106. Feldman goes on to say: "Such simplistic attacks on the 'Berliners' are revealing of provincial attitudes toward that city and toward the process of centralisation in the Weimar Republic." They were all the more revealing because, in many cases, they were not at all justified by the facts.

[32] *Ibid.*; Glum, Friedrich, *Zwischen Wissenschaft, Wirtschaft und Politik* (Bonn: Bouvier, 1964), p. 275; Böhme, Helmut, "Emil Kirdorf, Überlegungen zu einer Unternehmerbiographie", *Tradition* 13 (1968), pp. 282–300 and 14 (1969), pp. 21–48.

[33] A. Vögler to C. Duisberg, 27.7.20 (Bayer, 46/7.2); Fritz Haber to Friedrich Schmidt-Ott, 7.8.20, quoted by Zierold, K., *op. cit.*, p. 29.

This system of giving the power of decision in the allocation of funds for research to panels of practising scientists came to the United States only after the Second World War. In the 1920s and 1930s, the allocation of resources was not in the hands of the scientists but rather in large measure in the hands of private foundations " directed by men who," in Robert A. Millikan's view, " have made a life-long study of the problem of disbursing the funds wisely ".[34] In 1922, Millikan, who profited uniquely from the favour of these foundation officials, anticipated that the American precedent would, in time, be imitated throughout the world. In fact, however, it was in the Weimar republic that the precedents for the future were being set—by the Notgemeinschaft and Helmholtz-Gesellschaft.

The prosperous two decades before the First World War had seen a very marked increase in the funds available from traditional sources of support through project-grants in Europe and America, namely, the endowments administered by the academies of science.[35] This period also witnessed a multiplication of the number and types of institutions distributing support for research in this manner. Especially notable were the Jubiläumsstiftung der deutschen Industrie zur Förderung der technischen Wissenschaften, established in 1899 with a capital of 1·5 million marks ; the Caisse de recherches scientifiques which the French government established in 1902 and into which it paid 125,000 francs each year from the profits on pari mutuel betting; the Carnegie Institute of Washington, endowed by Andrew Carnegie with $10 million in 1902, which distributed about $100,000 per year in project-grants; and, finally, the Institut international de physique Solvay, which was established by the Belgian industrialist, Ernest Solvay, just two years before the war and which was expected to do a great deal for physics with about 20,000 francs per year for project-grants, to be awarded by an international committee of nine distinguished physicists.[36]

[34] R. A. Millikan to Henri Bergson, 2.11.22 (California Institute of Technology Archives, Pasadena ; Millikan Papers, Box 11). I do not accept Professor Karl's view that a "takeover by the academics of the management of their own resources was the heart of the intellectual revolution which took place [in the United States] in the 1920s." See Karl, Barry D., " The Power of the Intellect and the Politics of Ideas ", *Daedalus*, 99 (1968), pp. 1002–1035, who is followed by Coben, Stanley, " The Scientific Establishment and the Transmission of Quantum Mechanics to the United States, 1919–32 ", *American Historical Review*, LXXVI (1971), pp. 442–466.

[35] The Paris Académie des Sciences, which until 1904 had only about 10,000 francs per year for research grants (in distinction to prizes, for which it had perhaps 10 times as much to give), received three bequests in the following six years yielding annually 3,000, 50,000, and 124,000 francs respectively. Only somewhat less dramatic rises in endowment for research support through project-grants were enjoyed by the Vienna Academy, the Berlin Academy and the Munich Academy. The Royal Society of London, by contrast, received very little for such purposes in this period, nor was any important institution which made project-grants established in Britain. See Forman, P., Heilbron, J. L. and Weart, S., *op. cit.*

[36] For the Jubiläumstiftung, see Forman, P., " The Helmholtz-Ges . . ." drawing upon the Historisches Archiv, Gutehoffnungshütte. For the Caisse de Recherches Scientifique, see *Rapports Scientifiques* . . . (1904), (Melun, 1905), pp. 1–17, and Crawley, A. E., " France and

Despite this pre-war multiplication of amounts and sources of funds for project-grants, the movement was not on the whole in this direction. In Great Britain where the Royal Society administered the earliest and most perfect precedent for the system of grants made for individual projects on the basis of " peer-review ", the sum—£4,000—provided annually by the government for this programme did not increase in the 40 years before the war and consequently supported a smaller and smaller proportion of British research.[37] The researches supported by the Carnegie Institution's programme of external grants were designated " projects of minor scope " and their small share of the total funds was progressively reduced in favour of the Institution's internal " projects of broad scope ".[38] When plans were being laid for the Kaiser-Wilhelm-Gesellschaft, the most significant institution supporting research in Germany in the pre-war decade, proposals for a programme of external project-grants were turned aside.[39] The shift to the system of project-grant support in Germany after 1918 was thus a substantial reorientation, for which the rationale must be sought.

The Notgemeinschaft, the Helmholtz-Gesellschaft, and the several foundations of the German chemical industry had been created in order to sustain the entire national research enterprise, in a period of political and economic distress. The chemical industry provided such large sums to sustain academic chemistry that during the inflation it could distribute much of its support simply in the form of supplements to the budgets of the university institutes. But in every other field, including physics and engineering, the available funds were a small fraction of total institute budgets, and would have been soaked up without noticeable effect had

---

the Endowment of Research ", *Nature*, LXXXVIII (1912), pp. 317–318. For the Carnegie Institution, see Miller, Howard S., *Dollars for Research. Science and its Patrons in Nineteenth Century America* (Seattle: University of Washington Press, 1970), pp. 166–181. For the Sovay Institute, see Pelseneer, Jean, " Historique des instituts internationaux de physique et de chimie Solvay depuis leur fondation jusqu'à la deuxième guerre mondiale " unpublished typescript in the Archive for History of Quantum Physics, Berkeley, Copenhagen, and Philadelphia, 1962, pp. 20–30.

[37] MacLeod, R. M., " The Royal Society and the Government Grant: Notes on the Administration of Scientific Research, 1849–1914 ", *The Historical Journal*, XIV (1971), pp. 323–358, and " The Support of Victorian Science: The Endowment of Research Movement in Great Britain, 1868–1900 ", *Minerva*, IX, 2 (April 1971), pp. 197–230.

[38] Carnegie Institution of Washington, *Yearbook*, I (1902), pp. xxx–xxxvii; XIII (1914), pp. 33–34. " The Carnegie Institution is not going to do the work in this direction that some of us expected ", Edward B. Rosa of the National Bureau of Standards wrote to Arthur Gordon Webster, 12.3.06. " It is the policy of the institution to undertake a few large projects and to draw out from the smaller enterprises; to give very few grants to individual workers. The President declares that he thinks giving small grants to scientific workers in the universities tends to pauperize rather than enrich " (American Institute of Physics, New York).

[39] Theodor Lewald, a leading civil servant in the Reich Interior Ministry, advocated such a *wissenschaftlicher Kriegsschatz* on the grounds that " one of the chief mistakes of Germany's present scientific organisation is that research has often become virtually impossible for younger investigators, lecturers and extraordinary professors, who do not have the funds of the university institutes at their disposal " (*50 Jahre Kaiser-Wilhelm-Gesellschaft und Max Planck Gesellschaft zur Förderung der Wissenschaften, 1911–1961* [Göttingen: Max-Planck-Gesellschaft, 1961], p. 102).

they been distributed in the traditional way. As its vice-president Fritz Haber explained early in 1921:

> The Notgemeinschaft can in general neither erect buildings, nor bear the costs of heating them, nor pay salaries to individual investigators. It must, except in very special cases, presuppose that the expense of keeping the machinery of scientific research idle is defrayed by some other agency, primarily by the governments of the states.[40]

It could hope only to "raise productivity" by making grants to particular research projects. These grants had in principle to be available to all research workers; consequently the referees who made the decisions had also to be in some way representative of the discipline as a whole.

Partly for these financial reasons, but also partly for constitutional reasons, the intention was to support scientists in universities while avoiding general support for the universities as educational institutions. For the same reasons it was necessary to avoid any grounds for claims by the states for a "geographical" distribution of funds.[41] The solution was to support individual research projects on the basis of merit. The distribution of the funds directly to the research worker, without even informing the administrative officers of his university or the ministry of education of his state that a grant had been made—much less requested—effectively hindered the states from shifting their financial burdens to the Notgemeinschaft or the Helmholtz-Gesellschaft.[42]

Finally, the contributors to these new supporting organisations, the central government on the one hand and industry on the other—unlike the states with their education ministries staffed by the German counterparts of those "men who have made a life-long study of the problem of disbursing the funds wisely"—had no staff of their own with experience and self-confidence in this area. Consequently—and in accordance with the slogan of "self-government" which was current

---

[40] Haber, F., "Die Notgemeinschaft der deutschen Wissenschaft", *Chemiker-Zeitung*, 8.1.21, reprinted in *Aus Leben und Beruf* (Berlin: Springer, 1927), p. 171.

[41] In Germany in the mid-1920s, as in the United States in the mid-1960s, there was a gradual increase of pressure for such "geographical" distributions. See Schroeder-Gudehus, B., *op. cit.*, pp. 556–560; and Zierold, K., *op. cit.*, pp. 25–26.

[42] See Schroeder-Gudehus, B., *op. cit.*; Zierold, K., *op. cit.*, p. 27; Richter, S., *op. cit.*, p. 36. In 1910 when Robert Bosch established a one million mark endowment for the engineering institutes at the Technische Hochschule of Stuttgart, he specified that the contribution from the state of Württemberg must not be diminished on account of his gift, and that any increases which would have been required in the absence of his gift should be made. Heuss, Theodore, *Robert Bosch: Leben und Leistung* (Stuttgart: Rainer Wunderlich, 1946), pp. 193–194. In the Weimar period, the industrial contributors were caught between the desire for public recognition of their contributions and the wish to prevent the governments from evading their own responsibilities. Initially, when industrialists saw themselves as shouldering burdens which governments were financially incapable of carrying, the tendency toward keeping their donations secret seems to have been dominant—except, interestingly, in the case of the chemical industry. After 1924, as industrialists came to regard the governments as capable of greater contributions, they began to recognise that secrecy deprived them of influence; they then sought more publicity for their support of research.

in the first years of the Weimar republic [43]—the funds were placed largely in the hands of the scholars and scientists themselves, for whom the grants for individual projects based on review by peers was the only defensible mechanism of support.

Carl Duisberg, head of the Bayer chemical firm, who planned the Helmholtz-Gesellschaft on the model of his industrial foundations for the support of academic chemistry, had in fact assumed that the Helmholtz-Gesellschaft funds would be distributed as supplements to the budgets of the physical and technological institutes, in the same way as his Liebig-Gesellschaft supported the chemical institutes. But the academic representatives of physics on the executive council of the Helmholtz-Gesellschaft were opposed to such a schematic distribution of even one third of the available funds. Duisberg, although of the opinion that a strict project-grant system would prove unfair and inequitable, acceded to the physicists' wishes.[44]

Inherent Conflicts in the System

Still, important problems remained unsolved. For one thing, the control exercised by the scientists was not complete; the industrial contributors, like the government ministers, had their own views of the value of physical research. Their views were initially influenced by the physicists' view of the nature and significance of their scientific activity—and the differences between them were obscured by the physicists' rhetoric. But these divergences became more apparent as time went on. The particular difficulties in striking a balance between autonomy and control within a system of support by project-grants arose on a national scale for the first time in Weimar Germany; it was only then that scientists achieved sufficient control over sufficiently large resources for the problem to arise.

Equally important difficulties arise when the scientific community or one of its constituent disciplines lacks—as it inevitably must—a unanimous consensus about scientific merit and importance. Although physics is commonly regarded as the discipline least susceptible to this debilitating condition, in Weimar Germany it was most severely afflicted—more so perhaps than any other scientific or scholarly discipline.[45] The affliction was all the more serious because a physicist's scientific views often paralleled closely his political views. This affinity appeared in all areas of scientific life, from university structure and

---

[43] Notgemeinschaft, *Bericht*, V (1926), p. 5; Düwell, Kurt, " Staat und Wissenschaft in der Weimarer Epoche: Zur Kulturpolitik des Ministers C. H. Becker ", *Historische Zeitschrift, Beiheft*, I (1971), pp. 31–74.

[44] Duisberg to F. Klein, 10.10.21 (Bayer, 46/7.2); Forman, P., *The Helmholtz-Ges.* . . .

[45] This is at least suggested by the surviving correspondence regarding formation of *Fachausschüsse* by the Notgemeinschaft in some 20 scientific and scholarly disciplines in 1920–21 (Deutsche Forschungsgemeinschaft, Bad-Godesberg; Akten 86 and 87).

academic appointments, through the organisation of physical associations and journals, to the operations of the institutions for the support of research. " But as it was an axiom among the German men of learning that one must not mix politics with science, it was considered improper to mention the entire subject at all." [46]

Political Antagonism and the Recognition of Scientific Merit

Factions were endemic in the German scientific world in the late nineteenth and early twentieth century. This was commonly attributed to competition for a place in the German academic system. Friedrich Paulsen wrote in 1902:

The anger of the defeated candidates, the resentment felt against the favored ones, the envy of those who have succeeded, the distrust of the influential, all these feelings pass from the university sphere into the scientific literature and give the controversies and polemical discussions the venomous character which they so often reveal in Germany.[47]

The "successful", "favored", and "influential" academics were, above all, those in Berlin. Thus the scientists and scientific institutions in the capital—"the inflated *Berlinerei*", "the great Berlin sausage factory"[48]—became the principal focus of this animosity. Moreover, among scientists as among businessmen, such feelings were compounded with social and political sentiments, in particular a distaste for the liberal, Jewish personalities and attitudes associated with the metropolis.

This association was particularly striking in physics. In the 90 years prior to 1922, three of the six professors of experimental physics at the University had been Jewish in origin: Gustav Magnus, Emil Warburg and Heinrich Rubens. Helmholtz, the only non-Jew to occupy the chair for a significant length of time (1872–88) was himself attacked for his alleged anglo- and judœphilia.[49] All six professors were surrounded by

---

[46] Frank, Philipp, *Einstein: Sein Leben und seine Zeit* (Munich: P. List, 1949), p. 268.

[47] Paulsen, Friedrich, *Die deutschen Universitäten und das Universitätsstudium* (Berlin, 1902), translated as *The German Universities and University Study* (London: Longmans Green, 1906), p. 174.

[48] These expressions, "aufgeblasene Berlinerei", "die grosse Berliner Wurstfabrik", and others to the same effect, occur in the correspondence of the Munich botanist Karl von Goebel around 1900. See Bergdolt, E. (ed.), *Ein deutsches Forscherleben in Briefen aus sechs Jahrzehnten, 1870–1932* (Berlin: Ahnenerbe-Stiftung, 1940), pp. 117–124. The Viennese theoretical physicist, Ludwig Boltzmann, who had also held chairs in Munich and Leipzig, complained that the Berlin Academy boycotted any scientific venture which it did not initiate, much to the detriment of Germany's position in the scholarly world; see *Populäre Schriften* (Leipzig: Barth, 1919), p. 407. For evidence of such among mathematicians, see Grattan-Guinness, I., "An Unpublished Paper by Georg Cantor . . .", *Acta Mathematica*, CXXIV (1970), pp. 65–107, and Biermann, Kurt R., "David Hilbert und die Berliner Akademie", *Mathematische Nachrichten*, XXVII (1963), pp. 377–384.

[49] Helmholtz and his wife were leading figures in the von Schleinitz circle around the Crown Prince Friedrich, whose court was widely regarded as "Jewish-democratic" in outlook. See Paulsen, Friedrich, *An Autobiography* (ed. and trans. by Lorenz, T.) (New York: Columbia University Press, 1938), p. 279; Zöllner, J. C. F., *Das deutsche Volk und seine Professoren. Eine Sammlung von Citaten ohne Commentar* (Leipzig: Staackmann, 1880), pp. 35–36; Zöllner, J. C. F., *Über die Natur der Cometen. Beiträge zur Geschichte*

numerous Jewish assistants and *Privatdozenten*. One of these, Leo Arons, who was an active member of the Social-Democratic Party, had been barred by the Prussian Government from the University in a protracted and notorious affair in the years between 1894 and 1900.[50] In the two decades before the war, moreover, Max Planck's Berlin and Felix Klein's Göttingen were emerging as the centres of a new sort of physics: highly abstract and mathematical, cultivated by an elite which did not soil its hands in experiments, and in which the proportion of Jews was particularly high.[51] Many of the accusations made against Berlin—" Jewish intellectualism ", haughtiness, arrogance and monopoly of the limelight—were also made against the upstart discipline of theoretical physics, so that the tendency to associate the one with the other was unavoidable.

Prior to the First World War, although there was much resentment and many personal antagonisms, the widely shared liberal-nationalist political outlook, the general avoidance of involvement with party politics and the nearly universal commitment to the imperial regime hindered the development of scientific factions along political lines. The outbreak of war aroused a nearly unanimous enthusiasm, but by the summer of 1915 a split had developed within the German academic world between the annexationist majority who wanted " peace through victory " and a minority of moderates who urged a negotiated peace without territorial aggrandisement. Fundamentally the same division persisted in the immediate aftermath of the war in the form of an opposition between the minority of moderates supporting the new regime and the majority of conservatives, shading over to reactionaries and extreme nationalists, who were unsympathetic or openly antagonistic towards the Weimar republic, with its liberal domestic policies and its conciliatory foreign policy.[52]

The leading Berlin physicists and chemists were prominent in the liberal and moderate minority, both as signatories or sympathisers of anti-annexationist manifestos during the war and as advocates of cooperation and compromise in international scientific relations after it.

und Theorie der Erkenntnis, 3rd ed. (Leipzig: Staackmann, 1883), pp. lxv–lxvi, lxxxviii, 77, 85, 363–364, 373–374, 380.
[50] Fricke, Dieter, " Der Fall Arons ", *Zeitschrift für Geschichtswissenschaft*, VIII, part 2 (1960), pp. 1068–1107.
[51] On the " overproduction " of theoretical physicists by the early 1920s, see Einstein, Albert, and Born, M. and H., *op. cit.*, pp. 55, 103. But, on the other hand, there was " an astonishing shortage of able young theorists . . . at least if one demands that they be of Christian origin ", as one did in Greifswald: Clemens Schaefer to Johannes Stark, 14.5.18 (Staatsbibliothek Preussischer Kulturbesitz, Berlin-Dahlem; Nachlass Stark).
[52] Schroeder-Gudehus, B., *Deutsche Wissenschaft und internationale Zusammenarbeit 1914–1928*, Thèse, Université de Genève (Geneva: Imprimerie Dumaret et Golay, 1966); idem, " Les professeurs allemands et la politique du rapprochement ", *Annales d'études internationales* (Geneva), I (1970), pp. 23–44; Schwabe, Klaus, *op. cit.*, p. 185; Bleuel, Hans Peter, *op. cit.*, p. 91; Thimme, Annalise, *op. cit.*; Töpner, Kurt, *Gelehrte Politiker und politisierende Gelehrte. Die Revolution von 1918 im Urteil deutscher Hochschullehrer* (Göttingen: Musterschmidt, 1970); Weisz, Christoph, *Geschichtsauffassung und politisches Denken Münchener Historiker der Weimarer Zeit. Beiträge zu einer historischen Strukturanalyse Bayerns im Industriezeitalter*, Bd. 5 (Berlin: Duncker & Humblot, 1970).

This group included prominent democrats like Albert Einstein and Walter Nernst, and moderates like Fritz Haber, Heinrich Rubens and Emil Warburg. But it also included men like Emil Fischer, Max von Laue and Max Planck, whose basic political inclinations were towards the old regime.[53]

There was also a substantial minority of physicists—almost all outside Berlin—of pronounced reactionary and nationalistic views. The most prominent were Philipp Lenard, professor at the University of Heidelberg, and Johannes Stark, professor at the University of Greifswald, later at Würzburg, both Nobel laureates, both distinguished by their aggressive chauvinism during the war and by their public support of Adolf Hitler during the Weimar period.[54] Wilhelm Wien, professor at the University of Würzburg, and later at Munich, also a Nobel laureate and involved in anglophobic manifestations during the war, was a man of more insight and independence and though sympathetic to Hitler's movement was not directly associated with it.[55] There were also other physicists of more or less *völkisch* persuasion, notably Clemens Schaefer, professor at the University of Marburg, then at Breslau, and Christian Füchtbauer, professor at the University of Rostock.[56] Not very far from this region of the political spectrum, in, say, the right wing of the DVP, stood a considerable fraction of the German physicists. And though they varied greatly on the crucial issue of anti-semitism, as a group they are distinguished by their nationalistic intransigence in international relations.

Thus from the first years of the Weimar period, there were two roughly congruent political divisions among the German physicists. The one, between the moderates and the right; the other, "between Berlin

---

[53] I hope in the future to publish a detailed analysis of the political attitudes and affiliations of the physicists of the Weimar period. See, however, Forman, P., . . . *Physics in Weimar Germany*; "Weimar Culture, Causality . . .", pp. 113–114, and Feldman, G. D., "A German Scientist . . ."

[54] Lenard, P. and Stark, J., "Hitlergeist und Wissenschaft", *Grossdeutsche Zeitung* (Munich), 8.5.24, reprinted in Frey, Fritz, *et al.*, *Philipp Lenard der deutsche Naturforscher: Sein Kampf um nordische Forschung* (Munich and Berlin: Lehmann, 1937), pp. 23–24. Mr. Alan D. Beyerchen of the Department of History of the University of California (Santa Barbara) is preparing a doctoral dissertation on "deutsche Physik" in the Third Reich which will also deal with the earlier political development of Lenard and Stark.

[55] W. Wien to Stark, 21.12.14, 26.1.15, 26.2.15 (Staatsbibliothek Preussischer Kulturbesitz, Nachlass Stark). Lenard, P., "Wilhelm Wien (obit.)", *Volkischer Beobachter*, 12.9.28, Nr. 212, p. 2. The salient features of Wien's political thought and activity, so well known to his colleagues, were largely suppressed in his memoirs and the tendentiously selected extracts from his correspondence published in *Aus dem Leben und Wirken eines Physikers* (Leipzig: Barth, 1930). They may be seen in the *Chronik* of the Wien family in 4 volumes, each 300–500 pages of typescript, compiled by Wien's wife from diaries and correspondence. I am very grateful to Dr. Karl Siebertz of Munich for permission to examine and make extracts from the third volume of this *Chronik*, covering the period 1914–1928.

[56] Schaefer's views appear explicitly in his correspondence with Stark (Staatsbibliothek Preussischer Kulturbesitz, Berlin-Dahlem). For Füchtbauer's views the evidence is more circumstantial: Forman, P., . . . *Physics in Weimar Germany*, p. 470; [Heidorn, Günter. *et al.*], *Geschichte der Universität Rostock*, Band 1: *Die Universität von 1419 bis 1945* (Berlin: Deutscher Verlag der Wissenschaften, 1969), p. 240.

and the Reich".[57] During this period new organisations were founded and existing organisations were reconstructed in virtually all fields of learning, and not least in physics. Despite the political intransigence of the majority of academics, the changes generally followed democratic, federalist lines. The Berlin-based Deutsche Physikalische Gesellschaft was confronted with a secession by the applied physicists, dissatisfied with the predominance of pure physicists and theoretical questions in the society; there was also a movement to deprive the Berlin physicists of their hegemony over the society.[58] Hostility towards Berlin had risen sharply in the last two years of the war. In the spring of 1918, as a gesture of conciliation a non-Berlin physicist was elected for the first time to the society's presidency; nonetheless the tide continued to rise in 1919 and it overflowed in 1920. Einstein, recognising that "from a distance everything appears askew and suspect, especially if it comes from the damned Berliners", assured the physicists in Bavaria that "nonetheless (almost) all of us are gentle as lambs and frightened by our evil reputations".[59]

Drawing on this hostility to Berlin, Stark organised a Fachgemeinschaft Deutscher Hochschullehrer der Physik early in 1920 in opposition to the Berlin-dominated Deutsche Physikalische Gesellschaft, and sought to use his organisation to gain influential positions in the Notgemeinschaft and the Helmholtz-Gesellschaft.[60] Stark lost to his somewhat more moderate political ally, Wilhelm Wien, who emerged as president of a reconstituted Deutsche Physikalische Gesellschaft and as vice-president of the Helmholtz-Gesellschaft. The political overtones to these organisational struggles are clearly audible in Wien's remark to a lay friend that "now, as far as our endeavours in physics are concerned, all anti-semitic tendencies are alien to us—unless one simply identifies Berlin with the Jews".[61]

This political and professional partnership penetrated, moreover, into the realm of scientific ideas and methods. "It is remarkable", Einstein observed in October 1920, "how in these times every evaluation is made

[57] Stark to Sommerfeld, 23.7.20, looking forward to the "*Auseinandersetzung zwischen Berlin und dem Reich*" at the Naturforscherversammlung (Lehrstuhl für Geschichte der Naturwissenschaften und Technik, Universität Stuttgart; Nachlass Sommerfeld).

[58] Riewe, K.-H., "120 Jahre Deutsche Physikalische Gesellschaft", *Physikalische Blatter*, XXI (1965), pp. 3–11; Forman, P., . . . *Physics in Weimar Germany*, pp. 142–170. Both Stark and Emil Warburg attributed the formation of the Deutsche Gesellschaft für technische Physik (1919) to dissatisfaction with "*der in Berlin herrschenden theoretischen Richtung*": Stark, J., *Die gegenwärtige Krisis in der deutschen Physik* (Leipzig: Barth, 1922), p. 24; Warburg, E., "Zur Geschichte der Physikalischen Gesellschaft", *Naturwissenschaften*, XIII (1925), pp. 35–45, esp. p. 36.

[59] Einstein to Arnold Sommerfeld, 18.12.19. Einstein, Albert and Sommerfeld, A., *Briefwechsel*, edited and annotated by Armin Hermann (Basel and Stuttgart: Schwabe, 1968), p. 61.

[60] Richter, S., *op. cit.*, pp. 13–14, Forman, P., "The Helmholtz-Ges. . . ."

[61] Wien to Hans Beggerow, 17.7.21: Wien family, *Chronik, 1914–1928* (typescript), p. 198.

according to political criteria ".[62] Einstein had in mind, in good part, the assessment of his theory of relativity, and that not merely by the general public but by German physicists as well. Although he no longer believed, as he had two months earlier, " that a great part of our [German] physicists were participating " in the public campaign against him and his theory,[63] he nonetheless remained convinced that the opposition of physicists like Lenard, Stark, and Ernst Gehrcke, and the equivocation of Wilhelm Wien, had non-scientific motives :

> I have good reason to believe that motives other than the striving for truth are at the bottom of this enterprise. (Were I a member of the Deutsch-Nationale Volkspartei, with or without swastika, instead of a Jew of libertarian, international convictions, then . . .)[64]

The protagonists and antagonists of the quantum theory of the atom tended to be the same as those who favoured and opposed relativity. Thus one may speak of " reactionaries " who were opposed in principle to an increasingly independent, abstract, and mathematical theoretical physics, in which Jews played so large a role, and of " progressives " who either developed these theories or employed them in their experimental work. Each group believed that the scientific judgement of the members of the opposing faction was strongly influenced by political considerations and discounted it accordingly. Friedrich Paschen, seeking with much difficulty to persuade his non-physicist colleagues at the University of Tübingen to accept a Jewish progressive theorist for the associate professorship vacated by Christian Füchtbauer, interpreted Wilhelm Wien's evaluation of the candidate: " In view of Wien's well-known extra-scientific orientation, his acceptance of Landé's work on the [quantum theory of the] Zeeman effect without qualifications is in my opinion very high praise of Landé by Wien too."[65]

This opposition between "progressives" and "reactionaries" was reinforced and aggravated by the structure of the German university

---

[62] Quoted in English translation by Nathan, Otto and Norden, Heinz (eds.), *Einstein on Peace* (New York: Schocken, 1968), p. 43.

[63] Einstein to Sommerfeld, 6.9.20. Einstein, Albert and Sommerfeld, A., *op. cit.*, p. 69. On this campaign, see Grundmann, Siegfried, " Der deutsche Imperialismus, Einstein und die Relativitätstheorie (1914–1933) ", in *Relativitätstheorie und Weltanschauung* (Berlin: VEB Deutscher Verlag der Wissenschaften, 1967), pp. 155–285, esp. pp. 197–207; idem, " Das moralische Antlitz der Anti-Einstein-Liga ", *Wissenschaftliche Zeitschrift der Technischen Universität Dresden*, XVI (1967), pp. 1623–1626; Clark, Ronald W., *Einstein: The Life and Times* (New York: World, 1971), pp. 255–266.

[64] Einstein, A., " Meine Antwort über die antirelativitätstheoretische G.m.b.H. ", *Berliner Tageblatt*, 27.8.20, Nr. 402, pp. 1–2. (Clark, R. W., *op. cit.*, p. 257, mistranslates *Deutsch-nationaler* as " German national " and so loses much of the point of Einstein's observation.) This is also the one point with which Ernst Gehrcke in replying to Einstein took issue—not with the general rule but with its specific application to him; see *Berliner Tageblatt*, 31.8.20, Nr. 409, p. 3.

[65] Paschen, F., " Bemerkungen zum Schreiben der Fakultäts-Mehrheit ", 11.4.22., (Universitäts-Archiv, Tübingen; 128/Landé). Imputation of political motives to judgements about scientific matters were made more often and more openly by the " reactionary " physicists, for whom it was indeed a central tenet. Lenard, P., " Ein Mahnwort an deutsche Naturforscher ", *Über Äther und Uräther*, 2nd ed. (Leipzig: Barth, 1922), pp. 5–10.

system. Scientific work being largely done, and careers pursued, within hierarchical university institutes which were the fiefs of the single professor,[66] there was inevitably a tendency for entire institutes to assume not merely a uniform scientific character, but a uniform political tone as well. The most striking instance was Philipp Lenard's institute in Heidelberg; even Wilhelm Wien's son, Karl, was surprised by the situation he found there in the summer of 1925. He wrote to his parents:

I haven't yet quite figured out whether one first becomes *völkisch* and then a doctoral candidate, or the reverse. In any case the institute appears to be rather homogeneous in this respect, and in quarrels with the university, the rector, or other officials it is energetically supported by Lenard.[67]

Evidence of a similar, though less extreme, situation in other institutes is not wanting. Thus, for example, Clemens Schaefer, professor and director of the institute at the University of Breslau, congratulating Stark in October 1930 on his recent " slaughter " of Arnold Sommerfeld, the " ring leader " of modern theorists in Germany, could assure Stark that " all of us in the institute were enormously pleased by it ".[68]

The use of political labels—" reactionary ", " conservative ", " progressive "—in describing scientific alignments, is justified both by the content of the scientific positions adopted and by the close correlations between those positions and political orientations. But I do not mean the former terms to be taken purely pejoratively, or the latter as a term of unequivocal praise. Thus, in this period of deep public hostility toward the physicists' efforts to comprehend nature casually, it was by and large the " reactionaries " and " conservatives " who refused to bend with the anti-rational *Zeitgeist*, who continued to insist upon the epistemological ideal of deterministic description and the capacity of the human mind to attain it. Many " progressives ", on the contrary, bowed with that *Zeitgeist*, almost to the point of repudiating the scientific enterprise. A similar circumstance held for another less crucial but more relevant tenet of the physicists' ideology, namely that technology is applied physics and that between the science of physics and scientific technology there is no essential difference. The " progressives ", adapting to the reaction against utilitarianism in post-war Germany—and adopting the popular existentialist *Lebensphilosophie*—repudiated this conception. Physics, the " progressives " maintained, aims only at cognition, at satisfying

---

[66] Busch, Alexander, "The Vicissitudes of the 'Privatdozent': Breakdown and Adaptation in the Recruitment of German University Teachers", *Minerva*, I, 3 (Spring 1963), pp. 319–344; Forman, P., . . . *Physics in Weimar Germany*, pp. 92–100.

[67] Karl Wien from Heidelberg, 26.5.25: Wien Family, *Chronik, 1914–1928* (typescript), p. 286.

[68] Schaefer to J. Stark, 20.10.30 (Staatsbibliothek Preussischer Kulturbesitz; Nachlass Stark). The publications in question are reprinted in Stark, J., *Fortschritte und Probleme der Atomforschung* (Leipzig, 1931), pp. 69–112. Regarding Sommerfeld's role as the " energetic executive secretary " of the " Jewish and philosemitic circle " of mathematicians and theoretical physicists, see Hermann, Armin, " Albert Einstein und Johannes Stark ", *Sudhoffs Archiv*, L (1966), pp. 267–285.

certain innate psychological needs, and thus is essentially different from technology, which seeks knowledge for the sake of power. Although the "conservative" physicists often availed themselves of this anti-utilitarian argument when addressing academic audiences, in general they adhered to their pre-war and war-time outlook, while the "reactionaries" openly advocated the identity of physics and technology, blaming the "progressives'" abstract theoretical schemes for the growing aversion from the application of science.

The Composition of the "Peer-Review" Panels and the Consequences Thereof

Although the central government provided the bulk of the funds dispensed by the Notgemeinschaft, it was content until the late 1920s—for much the same reasons that first led it to supply those funds—to allow the foundation almost complete autonomy. The general policies which emerged thus represented roughly the balance of interests and attitudes within the German academic world generally, although much of the increased support for physical research and for research fellowships in physics after 1925 was obtained by persuading the central government to support programmes of applied research in particular areas—notably the structure and properties of metals.[69] For each discipline, the Notgemeinschaft established a *Fachausschuss*, which was a panel of expert referees, drawn from the relevant field of learning. These panels passed judgement—generally lenient, but almost invariably decisive—upon all applications for support. "These specialised selection panels ought to be so constituted that all the active research workers in the discipline regard the appropriate panel as competent to represent them."[70] Although from 1922 onward the panels were formally elected by these constituencies of active research workers on nomination by the relevant scientific and professional associations, initially the right of selection was conferred upon some academic body. For the panel for physics, this was the mathematical-physical section of the Göttingen Academy of Sciences. This body, dominated, in Stark's parlance, by the "Jewish, mathematically oriented group", had close personal, scientific, and political ties with the Berlin physicists.[71] Although in fact none of the Göttingen Academy's nominees to this panel was a Jew, four of the five physicists named were associated with the Deutsche Demokratische Partei.

---

[69] Schroeder-Gudehus, Brigitte, "The Argument for Self-Government and Public Support of Science in Weimar Germany", *Minerva*, X, 4 (October 1972), pp. 546–550; Richter, S., *op. cit.*, p. 40; Zierold, K., *op. cit.*, pp. 92–93.

[70] Haber, F., *Aus Leben und Beruf* (Berlin: Springer, 1927), p. 169.

[71] Stark to Schmidt-Ott, 15.12.20, quoted by Richter, S., *op. cit.*, p. 14. The Göttingen Academy's nominations to the panel were Carl Runge as chairman, Wilhelm Westphal (Berlin), Arnold Sommerfeld, Wolfgang Gaede and Otto Wiener, an astrophysicist and a geophysicist. The first four named were all liberals associated with the Deutsche Demokratische Partei; Wiener, however, was a political and scientific reactionary.

By the summer of 1920 the organisation of the Notgemeinschaft had come into the hands of Friedrich Schmidt-Ott, who embodied the Prussian monarchy's devotion to culture and who was thus *persona gratissima* in the German academic world. Stark and his allies did not however have much difficulty in arousing the suspicion that the Notgemeinschaft, the Berlin institution contrived chiefly by Fritz Haber, was a "philosemitic foundation".[72] The response—and it has until recently been the typical response of "apolitical" scientific associations— was to co-opt the "radicals", to include first Stark and then Wilhelm Wien in the panel for physics. Nonetheless, the Berlin physicists never lost the inside track in the Notgemeinschaft. When the panel was elected in 1922, Stark's Fachgemeinschaft der Deutschen Hochschullehrer der Physik was excluded from the list of associations to be asked to submit nominations,[73] and neither Stark nor Wilhelm Wien was among those

TABLE II

Grants to, and Research Publications of, Physicists in Berlin

		1922-23	1924-26	1927-29	1930-32
Percentage of research grants made to Berlin physicists by:	Notgemeinschaft Fachaussch. f. Phys. Elektrophysikaussch.	8% 27%	23% 29%	} 23%	?
	Helmholtz-Gesellschaft	4%	1%	1%	2%
Percentage of papers published by Berlin physicists in:	Zeitschrift für Physik	40%	35%	39%	38%
	Annalen der Physik	10%	8%	10%	27%
	Both journals	33%	28%	30%	35%

SOURCES: Notgemeinschaft: *Bericht*, II–IX (1923–30); annual listing of grant applications to the Helmholtz-Gesellschaft (Verein Deutscher Eisenhüttenleute, Bibliothek). The statistics in this table are biased against the Helmholtz-Gesellschaft. Astrophysical and geophysical research were supported as "physics" by the Notgemeinschaft, but were not supported by the Helmholtz-Gesellschaft; these two fields of research were more heavily concentrated in the vicinity of Berlin than was physical research generally. Furthermore, the Notgemeinschaft, but not the Helmholtz-Gesellschaft, supported research in institutes unconnected with institutions of higher education, and such institutes were heavily concentrated in the vicinity of Berlin. Still it is unlikely that these biases could account for as much as half of the Notgemeinschaft percentages.

The *Zeitschrift für Physik* and the *Annalen der Physik* were the two principal vehicles for the publication of physical research in Germany. The former, edited in Berlin, functioned to a certain extent as an organ of the left wing of the German physicists—the latter, edited by Wilhelm Wien until his death in 1928, as an organ of the right wing; see Forman, P. ... *Physics in Weimar Germany*, pp. 183–205. The compilation of the statistics underlying this table was performed by John May.

[72] Schmidt-Ott, F., *Erlebtes und Erstrebtes, 1860–1950* (Wiesbaden: Steiner, 1952), p. 180. Forman, P., ... *Physics in Weimar Germany*, pp. 164, 193, 310.

[73] Richter, S., *op. cit.*, p. 16. The elected committee included only three physicists: Max von Laue, Max Wien, and Jonathan Zenneck—all three nominees of the Deutsche Physikalische Gesellschaft.

elected. Under the chairmanship of Max von Laue from 1922 until 1934, the elected panel seems to have been quite sympathetic—though not unduly partial—to the interests of " Berlin " (Table II).

There were two further sources of support for physics research administered through the Notgemeinschaft. These originated as foreign benefactions during the inflation, and were controlled by autonomous committees well shielded from pressure by " conservatives " and " reactionaries ". The first of these two funds was provided by the Japanese industrialist, Hajami Hoshi, to aid German chemistry, but Haber managed to divert a substantial fraction for the support of atomic physics in 1922 and 1923 by designating Planck, Otto Hahn, and himself as half the disbursing committee. More important still was the $12,500 provided annually from 1923 to 1929 by the General Electric Company, through Willis R. Whitney, on condition that matching grants equal to 10 per cent. were made by the Berlin electrical engineering firms, Siemens and Halske, and AEG. Though intended for the support of experimental research in " electrophysics ", the fund was administered by a committee dominated by " progressives ": James Franck, Fritz Haber, Max von Laue, Walther Nernst, Max Planck, and Max Wien, and, from 1925, Friedrich Paschen. This fund was consequently largely allocated to the support of atomic physics, including some of the most " progressive " work in quantum theory.[74]

The Helmholtz-Gesellschaft was in several respects the antithesis of the Notgemeinschaft ; not merely was it seen as an anti-Berlin foundation, but also as one which would guarantee the industrialists some control over the distribution of their funds. Its constitution placed ultimate authority in a general assembly of the contributors—with votes weighted according to the size of the contribution ; in fact authority resided entirely in the executive council where the great academic figures of the disciplines being supported sat alongside the leaders of Germany's largest industrial firms. Indeed, one of the purposes of the Helmholtz-Gesellschaft was to provide a basis for such commonly desired personal contacts.[75]

The industrialists' conception of the chair-holding institute director as the head of his own little " research factory " (*Wissenschaftsbetrieb*), and as such their social peer, is reflected in the provisions for selecting the representatives of science on the executive council, and also in the provisions for submitting applications: the directors of the " physical-technical " institutes at all German universities and colleges of technology were to form an association, the members of which were empowered to

---

[74] Richter, S., *op. cit.*, pp. 35–37; Forman, P., . . . *Physics in Weimar Germany*, pp. 309–321.
[75] O. Petersen to W. Petersen and E. Lemcke, 5.4.32 (Verein deutscher Eisenhüttenleute; HG 82). For details of the structure and personnel, see Forman, P., *The Helmholtz-Ges.* . . .

elect from among themselves equal numbers of physicists and engineers to three-year terms on the executive council of the Helmholtz-Gesellschaft. All applications for grants had to be submitted through a member of this association. Where the Notgemeinschaft, at least in principle, was concerned to aid younger scholars without institute budgets at their disposal, the Helmholtz-Gesellschaft was rather an organisation of and for the directors of the institutes.

An initial group of six representatives was selected by Wien, Duisberg, and Vögler in October 1920, after Wien himself had been chosen by the two industrialists. Only two of these six were engineers; the rest were physicists: Wilhelm Hallwachs, Planck, Stark, and Wien himself. Planck, though a theorist, was the leading statesman of the discipline and one of the few leading Berlin physicists who, in political attitudes or ethnic origins, would not have been out of place in the executive council of the Helmholtz-Gesellschaft. Duisberg explained apologetically to Felix Klein:

In any case, the representatives of heavy industry would have objected very strenuously if we had taken the physicists in the executive council from the same circle which has hitherto dominated the Deutsche Physikalische Gesellschaft.[76]

Klein, the advocate of "applied physics and mathematics", was dismayed that pure physicists, regardless of their political sympathies, had obtained a dominant position in the Helmholtz-Gesellschaft. This was possible only because the industrialists still accepted the physicists' pre-war and war-time view that pure physics provided the "scientific leadership" for technology, while engineering was merely applied physics.[77] Yet in the Weimar period, it was only the politically extreme conservatives among the pure physicists who still adhered to this belief in the close connection between science and technology; this formed an additional bond between them and the industrialists. On the whole, the anti-Berlin, anti-theoretical faction was in its general outlook as well as politically and "racially" so congenial to the industrial sponsors, and had so firm a hold on the Helmholtz-Gesellschaft from the outset, that its control of the seats on the executive council was almost complete. As a conciliatory gesture, Heinrich Rubens was given Planck's seat when the first elections were held in September 1921. But after Rubens' death the following summer, no Berlin physicist and only one Berlin engineer sat on the executive council until September 1929 when Peter Pringsheim and a Berlin engineer, Georg Schlesinger—both of Jewish origin—were elected.

Although in the spring of 1921 there was little that could still be done about the executive council, the Berlin physicists, sharing the general

---

[76] Duisberg to F. Klein, 22.1.21. (Bayer, 46/7.2.)
[77] [Wien, W.], "Aufruf zum Beitritt in die Helmholtz-Gesellschaft . . ." January 1921.

assumption that the Helmholtz-Gesellschaft was to be the principal source of research funds for their discipline, did concern themselves seriously about the composition of its panel of referees for physics. Wien, though suspicious of his ambitious ally, Stark, had readily agreed that Stark's anti-Berlin Fachgemeinschaft der deutschen Hochschullehrer der Physik, in which the lower academic ranks were included, would be the appropriate body to elect the panel for physics in the Helmholtz-Gesellschaft.[78] The Berlin physicists—including Planck—acting with Schmidt-Ott contended that Stark's Fachgemeinschaft did not meet the Notgemeinschaft's requirement of " objective representation of all scientific interests ", and sought to persuade the industrial sponsors of the Helmholtz-Gesellschaft to accept the Deutsche Physikalische Gesellschaft " as fully representative of the pure physicists ".[79] Wien stamped Schmidt-Ott's efforts " a product of the bureaucratic inclination to centralise everything in Berlin according to a single pattern ".[80] The industrialists regarded them in much the same way, but were more inclined to compromise. The physics panel of the Helmholtz-Gesellschaft was finally composed of three of Schmidt-Ott's nominees—Laue, Gustave Mie, and Wolfgang Gaede—and three physicists elected by the Fachgemeinschaft—Lenard, Jonathan Zenneck (chairman), and Christian Füchtbauer.[81] It too thus remained very largely in the hands of physicists with a definite inclination towards the right and away from Berlin.

And this did have consequences for the distribution of support. Berlin physicists got next to nothing from the Helmholtz-Gesellschaft (see Table II). They submitted few applications and those few were handled very roughly: the rejection rate for Berlin applications was about 40 per cent., compared to about 15 per cent. for all applications. It is my impression that the conservatives applying to the Helmholtz-Gesellschaft requested the largest sums and received benevolent treatment by the panel of referees. Thus, for example, in the 12 years from 1922 to 1933, Clemens Schaefer submitted nine grant applications in his own name totalling 10,100 marks; all but one of these were granted in full, and one in part. His 9,600 marks represented 2 per cent. of all funds provided for physics by the Helmholtz-Gesellschaft in this period. Even more striking is the case of Christian Füchtbauer who submitted 11 applications totalling 27,300 marks. Again, 10 of these were granted in full, and one in part. Füchtbauer's 25,800 marks were 6·6 per cent. of all the funds the Helmholtz-Gesellschaft appropriated to physics in these 12 years.

Still the industrialists on the council were not happy. They were astonished to find that the majority of physicists were immersed in an

---

[78] W. Wien to O. Petersen, 31.3.21 (Verein deutscher Eisenhüttenleute HG82).
[79] Schmidt-Ott to Duisberg, 3.5.21 (Bayer, 46/11.1); Schmidt-Ott to Vögler, 11.4.21 (Bayer, 46/8).
[80] W. Wien to Duisberg, 13.5.21 (Bayer, 46/8).
[81] Helmholtz-Ges., " Niederschrift über die Hauptversammlung . . . 17. September 1921 ", p. 12. (Historisches Archiv, Gutehoffnungshütte, Nr. 400106/49).

atomic world which had no apparent connection with their technological concerns, and they were quite unable to see any significant differences in this regard between the researches of a "reactionary" like Lenard and a "progressive" like James Franck. They showed their dissatisfaction by allowing the executive secretary, Otto Petersen, to make some "general remarks" at the meeting in March 1923 of the executive council:

> ... on the object of the research for which support is requested. He pointed particularly to the fact that among the applications, research projects in pure physics—especially spectral analysis, spectroscopy, radiations of various kinds, and the like—are the most numerous by far. He raised the question whether German *science* ought not also to take greater account of the situation in which the unfortunate conclusion of the war has placed Germany, and to turn more toward such questions as lie in the direction of applied physics.[82]

In the first four years of disbursements by the Helmholtz-Gesellschaft the physicists' share of the total sum available for distribution fell from more than half to less than one quarter, becoming stable in the later 1920s at around 30 per cent. Concomitantly, the type of physics for which support was requested seems to have shifted decidedly away from atomic phenomena toward technological applications and the bulk properties of matter.

Conclusion

In the first 10 years of the republican regime, physical research—above all the fundamental fields of atomic physics and modern theoretical physics—flourished in Germany. This blossoming was due in considerable part to the funds provided by the central government—through the Notgemeinschaft—and industry—through the Helmholtz-Gesellschaft—and especially to the distinctive mode of allocating it: as grants awarded for specific research projects on the basis of recommendations by panels of highly qualified referees.

If the Helmholtz-Gesellschaft contributed very much less than the Notgemeinschaft to this outstanding cultural achievement of the Weimar republic, this was not solely due to the disparity between the financial resources of the two agencies. Rather, in post-war Germany physics had become so deeply and thoroughly affected by "politics" that every social institution of, or closely associated with, this discipline necessarily also had a more or less strongly marked political character, and thus was inevitably partisan.

---

[82] Helmholtz-Ges., "Niederschrift über die Sitzung des Verwaltungsrates ... 21. März 1923", pp. 5-6 (Historisches Archiv, Gutehoffnungshütte, Nr. 30019326/15); *cf.* "Niederschrift ... Verwaltungsrates ... 16. Mai 1924", p. 6 (Verein deutscher Eisenhüttenleute).

1. For science there is no longer any money; the observatory has gone to complete rack and ruin. 2. The astronomy professor begs the government for funds, but in vain. 3. In uttermost desperation he resolves to take up astrology.

4. As astrologer he casts horoscopes for war profiteers. 5. His prophesies are soon in much demand; he makes lots of money. 6. Now the professor has the funds to renovate himself and the observatory; the newest and best instruments are procured.

Th. Th. Heine, "Neue Wege der Wissenschaft," *Simplicissimus*. 25 (26 January 1921), 595.

Edgar F. Smith Memorial Collection, University of Pennsylvania

Historical Studies *in the* Physical Sciences

RUSSELL MCCORMMACH, *Editor*

Third Annual Volume 1971

UNIVERSITY OF PENNSYLVANIA PRESS · PHILADELPHIA

Weimar Culture, Causality, and Quantum Theory, 1918–1927: Adaptation by German Physicists and Mathematicians to a Hostile Intellectual Environment

BY PAUL FORMAN*

> "It is interesting to observe that even physics, a discipline rigorously bound to the results of experiment, is led into paths which run perfectly parallel to the paths of the intellectual movements in other areas [of modern life]." Gustav Mie, inaugural lecture as Professor of Physics, University of Freiburg i.B., 26 January 1925.

I. Weimar Culture as a Hostile Intellectual Environment
   1. As Perceived by the Physicists and Mathematicians 8
   2. As Confirmed by Other Observers 15
   3. Intellectual Allies: Vienna Circle and Bauhaus 19
   4. Educational Ideals and Reforms 23
   5. The Crisis of *Wissenschaft* 26
   6. Spengler's *Decline of the West* 30

II. Adaptation of Ideology to the Intellectual Environment
   1. Introduction 38
   2. From Positivism to *Lebensphilosophie* 40
   3. Capitulation to Spenglerism 48
   4. A Craving for Crises 58

* Department of History, University of Rochester, Rochester, New York 14627.

III. "Dispensing with Causality": Adaptation of Knowledge to the Intellectual Environment

1. Introduction: The Concept of Causality 63
2. The First Intimations of an Issue, 1919–1920 70
3. Conversions to Acausality, 1919–1925
   a. The Earliest Converts: Exner and Weyl 74
   b. 1921, Summer and Fall: von Mises, Schottky, Nernst, et al. 80
   c. Later Notable Conversions: Schrödinger and Reichenbach 87
4. Unregenerates against the Tide, 1922–1923 91
5. The Situation circa 1924 96
6. Causality's Last Stand, 1925–1926 100
7. Conclusion 108

In perhaps the most original and suggestive section of his book on *The Conceptual Development of Quantum Mechanics* Max Jammer contended "that certain philosophical ideas of the late nineteenth century not only prepared the intellectual climate for, but contributed decisively to, the formation of the new conceptions of the modern quantum theory"[1]; specifically, "contingentism, existentialism, pragmatism, and logical empiricism, rose in reaction to traditional rationalism and conventional metaphysics. . . . Their affirmation of a concrete conception of life and their rejection of an abstract intellectualism culminated in their doctrine of free will, their denial of mechanical determinism or of metaphysical causality. United in rejecting causality though on different grounds, these currents of thought prepared, so to speak, the philosophical background for modern quantum mechanics. They contributed with suggestions to the formative stage of the new conceptual scheme and subsequently promoted its acceptance."[2]

These are far-reaching propositions. Properly construed they are,

---

1. M. Jammer, *The Conceptual Development of Quantum Mechanics* (New York: McGraw-Hill, 1966), section 4.2, "The Philosophical Background of Non-classical Interpretations"; on pp. 166-167.
2. *Ibid.*, p. 180. The search for philosophic precedents and influences has otherwise focused almost exclusively upon Bohr's doctrine of complementarity. This issue, which I am not directly concerned with here, has been recently examined once again and the literature reviewed by Gerald Holton, "The Roots of Complementarity," *Daedalus, 99* (Fall, 1970), 1015-1055.

WEIMAR CULTURE, CAUSALITY, AND QUANTUM THEORY, 1918–1927

I think, essentially correct. But it must be said that Jammer did not go very far toward demonstrating them. He displayed such anticausal sentiments among a variety of late nineteenth-century philosophers—French, Danish, and American—but adduced scarcely any evidence to bridge the wide gaps of a quarter century of time, a cultural tradition, and the disciplines of philosophy and physics, which separated their philosophical theses from the development of quantum mechanics by German-speaking Central-European physicists circa 1925. It is not my aim to fill in these gaps, but rather to examine closely the lay of the land on the far side of them. The result is, on the one hand, overwhelming evidence that in the years after the end of the First World War but before the development of an acausal quantum mechanics, under the influence of "currents of thought," large numbers of German physicists, for reasons only incidentally related to developments in their own discipline, distanced themselves from, or explicitly repudiated, causality in physics.

Thus the most important of Jammer's theses—that extrinsic influences led physicists to ardently hope for, actively search for, and willingly embrace an acausal quantum mechanics—is here demonstrated for, but only for, the German cultural sphere. This cultural qualification is essential; it forms the basis of my attempt to provide, on the other hand, an answer to the question—in its general form crucial to all intellectual history—why and how these "currents of thought," evidently of negligible effect upon physicists at the turn of the century, came to exert so strong an influence upon German physicists after 1918. For it seems to me that the historian cannot rest content with vague and equivocal expressions like "prepared the intellectual climate for," or "prepared, so to speak, the philosophical background for," but must insist upon a causal analysis, showing the circumstances under which, and the interactions through which, scientific men are swept up by intellectual currents.

Such an analysis may be either "psychological" or "sociological." That is, it may either consider the mental makeup of the individual scientists concerned, stressing previous intellectual environments and conditioning experiences as determinative of present attitudes, or, on the contrary, it may ignore these factors, treating present mental posture as socially determined response to the immediate intellectual environment and current experiences. I have chosen the latter course, and sought a model in which certain "field variables" and their derivatives at a given place and time are regarded as evoking corresponding attitudes. Though it may seem harsh to stress the social pressure and ignore the emotional pain, though it may seem unsatisfactory to break off our explanatory endeavors at the level of the individual decision, nonetheless I do think the "sociological" the more general and fruitful approach.

The inquiry must begin, then, by characterizing the intellectual milieu in which the German physicists were working and quantum mechanics was developed. This is a formidable problem, above all on account of methodologic difficulties. And the task is especially unattractive to the historian of science, for it obliges him to deal with the "expressions" of nonscientists as well as those of scientists, thus forcing the abandonment of the demarcation criterion by which he seeks to identify and delimit his subject. Nevertheless, with aid and guidance from previous studies by general intellectual historians, especially the work of Fritz K. Ringer, I have addressed this problem in Part I. I show that in the aftermath of Germany's defeat the dominant intellectual tendency in the Weimar academic world was a neo-romantic, existentialist "philosophy of life," reveling in crises and characterized by antagonism toward analytical rationality generally and toward the exact sciences and their technical applications particularly. Implicitly or explicitly, the scientist was the whipping boy of the incessant exhortations to spiritual renewal, while the concept—or the mere word—"causality" symbolized all that was odious in the scientific enterprise.

Now if, as is largely the case even at this late date, the interest of the historian of science is held exclusively by the substantive scientific achievements, he will immediately be struck by a remarkable paradox: this place and period of deep hostility to physics and mathematics was also one of the most creative in the entire history of these enterprises. Faced with this paradox many of us would be tempted to rub our hands with satisfaction, to regard it a welcome refutation of any attempt to impugn the autonomy of these sciences and the sufficiency of intellectualist-internalist history of them. But such an inference would be too hasty. Presupposing the hostility of the intellectual environment, the crucial question is the *nature* of the response of the exact scientists to this circumstance. I had myself previously assumed that in the face of antiscientific currents the pre-

WEIMAR CULTURE, CAUSALITY, AND QUANTUM THEORY, 1918–1927

dominant response in these highly professionalized sciences would be retrenchment, withdrawal into the science and the community of its practitioners, reaffirmation of the discipline's traditional ideology—i.e., its notion of the value, function, motive, goal, and future of scientific activity.[3] *Were* that the case, then, a fortiori, any attempt to attribute a strong and direct influence of that same intellectual environment upon the scientific discourse and dispositions of these same men would appear implausible.

Yet the historian who takes even the most casual notice of the valuations of physical science in contemporary American society, on the one hand, and the present ideological tendencies in these sciences, on the other hand, could scarcely maintain that the predominant response to a hostile intellectual environment is retrenchment. On the contrary, as sentiments of resentment and antagonism toward the scientific enterprise—coupled with a revival of existentialist *Lebensphilosophie*—have become prominent in the last few years, so also have the expressions of and concessions to these same sentiments within the sciences themselves. We are indeed witnessing in America today a widespread and far-reaching accommodation of scientific ideology to a hostile intellectual milieu. As the distinguished physical chemist Franklin A. Long recently stated in both explanation and advocacy of this development: "Faculty, and especially students, are sensitive to social problems, are eager to work on them, and are often prepared to change their previous ways of life to do so. The pressures of discipline orientation and the tradition of individual scholarship are strong among faculty members, but not strong enough to counter the pressures of social concern." And in all of this "responsiveness" there is an astonishing sincerity, a striking absence of cynical, calculated image projection, testifying to a surprising participation of the physical scientists themselves in those fundamentally, often manifestly, antiscientific sentiments.[4]

3. P. Forman, *The Environment and Practice of Atomic Physics in Weimar Germany* (Ph.D. dissertation, Berkeley, 1967; Ann Arbor: University Microfilms, 1968), pp. 11-24.

4. F. A. Long, "Interdisciplinary Problem-Oriented Research in the University [editorial]," *Science, 171* (12 March 1971), 961. Marvin L. Goldberger, "Physics and Environment: How Physicists Can Contribute," *Physics Today* (December 1970), 26-30, and the reply by John Boardman, *ibid.* (February 1971), 9. The new mood, especially the neo-Spenglerianism, in the scientific community is discussed by Bentley Glass in his presidential address to the AAAS, 28 December 1970, "Science: Endless Horizons or Golden Age?" *Science, 171* (8 January 1971), 23-29.

5

But our contemporary experience does not merely lead us to anticipate an ideological accommodation by the Weimar physicists and mathematicians; it also suggests a simple model for the circumstances under which such accommodation is likely to occur. We may suppose that when scientists and their enterprise are enjoying high prestige in their immediate (or otherwise most important) social environment, they are also relatively free to ignore the specific doctrines, sympathies, and antipathies which constitute the corresponding intellectual milieu. With approbation assured, they are free of external pressure, free to follow the internal pressure of the discipline—which usually means free to hold fast to traditional ideology and conceptual predispositions. When, however, scientists and their enterprise are experiencing a loss of prestige, they are impelled to take measures to counter that decline. Drawing upon Karl Hufbauer's factorization of prestige into image and values, one sees that such countermeasures will in general be attempts to alter the public image of science so as to bring that image back into consonance with the public's altered values. But if this is not mere image projection, then such alterations of the image of the scientist and his activity will also involve an alteration of the values and ideology of the science, and may even affect the doctrinal foundations of the discipline—as Theodore Brown has shown of the beleaguered College of Physicians in the latter seventeenth century.[5]

In Parts II and III, I apply this model to the German-speaking exact scientists working in academic environments in the Weimar period. Bearing in mind the radically rearranged scale of values ascendant in the aftermath of Germany's defeat, I explore in Part II the response of these scientific men at the ideological level. This response I have sought primarily in addresses by exact scientists to academically educated general audiences, and especially in their addresses to their assembled universities. The historian is fortunate that the institutions of German academic life provided frequent occasions for addresses before university convocations, and doubly fortunate that it was customary to publish such *Reden*. Conversely,

---

5. K. Hufbauer, "Social Support for Chemistry in Germany During the Eighteenth Century: How and Why Did It Change?" in this volume, pp. 205-231; T. M. Brown, "The College of Physicians and the Acceptance of Iatromechanism in England, 1665-1695," *Bulletin of the History of Medicine,* 44 (1970), 12-30.

the existence of these institutions is both an index and an instrument of the extraordinarily heavy social pressure which the German academic environment could and did exert upon the individual scholar or scientist placed within it. As I illustrate in Part II, there was in fact a strong tendency among German physicists and mathematicians to reshape their own ideology toward congruence with the values and mood of that environment—a repudiation of positivist conceptions of the nature of science, of utilitarian justifications of the pursuit of science, and, in some cases, of the very possibility and value of the scientific enterprise.

Was the tendency toward accommodation, which predominated in the response of this highly professionalized scientific community to its hostile intellectual environment, confined to the ideological level, or did it extend beyond it into the substantive doctrinal content of the science itself? Specifically, are there indications that German physicists and mathematicians were anxious to, and deliberately tried to, alter the character of their disciplines as cognitive enterprises and to alter specific concepts employed within them in order to bring their sciences in closer conformity with the values of the Weimar intellectual milieu? I strongly suspect that the intuitionist movement in mathematics, which won so many adherents and created so much furor in Germany in this period, was primarily an expression of just such inclinations and aims. I am convinced, and in Part III endeavor to demonstrate, that the movement to dispense with causality in physics, which sprang up so suddenly and blossomed so luxuriantly in Germany after 1918, was primarily an effort by German physicists to adapt the content of their science to the values of their intellectual environment.

The explanation of the creativity of this place and period must therefore be sought, in part at least, in the very hostility of the Weimar intellectual milieu. The readiness, the anxiousness of the German physicists to reconstruct the foundations of their science is thus to be construed as a reaction to their negative prestige. Moreover the nature of that reconstruction was itself virtually dictated by the general intellectual environment: if the physicist were to improve his public image he had first and foremost to dispense with causality, with rigorous determinism, that most universally abhorred feature of the physical world picture. And this, of course, turned out to be precisely what was required for the solution of those problems in atomic physics which were then at the focus of the physicists' interest.

ACKNOWLEDGMENTS: I am indebted to Stephen G. Brush, Stanley Goldberg, John L. Heilbron, Karl Hufbauer, Hans Kangro, Fritz K. Ringer, Donald E. Strebel, and the editor of this journal, Russell McCormmach, for their close, critical readings of the typescript and their numerous queries, suggestions, objections, and corrections. I also sincerely thank Ann Schertz and the other staff of the Interlibrary Loan Department of Rush Rhees Library, without whose constant aid it would not have been possible to prosecute this inquiry in Rochester, New York.

I. WEIMAR CULTURE AS A HOSTILE INTELLECTUAL ENVIRONMENT

I.1. As Perceived by the Physicists and Mathematicians

Through the summer of 1918 the German physical scientists, like the rest of the German public, continued to look forward with confidence and satisfaction to a victorious conclusion of the war in which they had been engaged four years. They, perhaps more than any other segment of the German academic world, also felt *self*-confidence and *self*-satisfaction due to their contributions to Germany's military success and to their anticipation of a postwar political and intellectual environment highly favorable to the prosperity and progress of their disciplines. The botanist looking about his institute, bleak and vacant, had to conclude that "probably it will also remain so after the war, for youth will turn to technology and leave so 'unpractical' a discipline as botany lying by the wayside."[6] The chemist, the physicist, the mathematician, however, emphasizing the great practical importance of their subjects during the war and the desirability and inevitability of still closer collaboration with technology in the future, looked forward to yet more, larger, and better stocked institutes and to substantially increased public

6. Karl v. Goebel to Th. Herzog, Munich 19 July 1917, in Goebel, *Ein deutsches Forscherleben in Briefen aus sechs Jahrzehnten, 1870-1932*, ed. Ernst Bergdolt, 2nd ed. (Berlin, 1940), p. 170.

WEIMAR CULTURE, CAUSALITY, AND QUANTUM THEORY, 1918-1927

esteem and academic prestige. "The closer we appear to approach the victorious conclusion of the war," Felix Klein observed in June 1918 before an audience including leaders of German industry and the Prussian government, "the more our thoughts are dominated by the question what, after peace is successfully won, ought then to come." Klein's desiderata ranged from a mathematical institute for himself and his university, through a general reorientation of academic research in the exact sciences to achieve a "preestablished harmony" with the requirements of industry and the military, to a corresponding reorientation of German education at all levels.[7] And at least the first of these desiderata seemed assured as the Prussian Minister of Education, Friedrich Schmidt, came forward to announce a grant of 300,000 Mark. Who, participating in these festivities, could have foreseen that the Göttingen mathematical institute would not be built for another ten years, and then only with American money?[8]

When that "victorious end" which seemed imminent in the summer of 1918 turned suddenly to utter defeat in the fall, the exact scientists found themselves confronting a dramatically transformed scale of public values and thus a drastically altered valuation of their field. That, certainly, was their perception of the situation. Had we no explicit testimony to this effect, we could nonetheless infer it from the defensive tone of the talks given by exact scientists before the assembled faculties and students at academic convocations. While during the latter years of the war such speeches convey self-

---

7. F. Klein, "Festrede zum 20. Stiftungstage [22 June 1918] der Göttinger Vereinigung zur Förderung der Angewandten Physik und Mathematik," *Jahresbericht der Deutschen Mathematiker-Vereinigung*, 27 (1918), Part I, pp. 217-228; on pp. 217, 219. As the philologists noted with some bitterness, during the war the scientists and mathematicians had raised substantially their demands upon secondary school curricula: Robert Neumann, "Politik und Schulreform," *Monatschrift für höhere Schulen*, 18 (1919), 93-106, "Vortrag, gehalten im Berliner Philologen-Verein Februar 1918"; Friedrich Poske and R. von Hanstein, *Der naturwissenschaftliche Unterricht an den höheren Schulen*, Schriften des Deutschen Ausschusses für den mathematischen und naturwissenschaftlichen Unterricht, II. Folge, Heft 5 (Leipzig-Berlin, 1918).

8. *Jahresbericht der D. M.-V.*, 27 (1918), Part 2, p. 47. In 1926 the International Education Board of the Rockefeller Foundation appropriated $275,000 for a mathematical institute. (Geo. W. Gray, *Education on an International Scale, A History of the International Education Board, 1923-1938* [New York, 1941], p. 30; Otto Neugebauer, "Über die Einrichtung des Mathematischen Institutes der Universität Göttingen," *Minerva-Zeitschrift*, 4 [1928], 107-111.)

assurance, confidence in the esteem and good will of the audience, in the Weimar period that is seldom the case. And while it is difficult to display this *tone,* one can at least point to passages alluding more or less explicitly to reproaches against exact science which the speaker clearly supposes to be in his audience's mind. Thus in November 1925 Wilhelm Wien described the great scientific discoveries of the early modern period, especially Newton's derivation of the motion of the planets from the laws of mechanics, as "the first convincing demonstration of the causality [n.b.] of natural processes which revealed to man for the first time the possibility of comprehending nature by the logical force of his intellect." But he then immediately conceded that this program, which the natural scientist finds so grand, has its limitations, and he proceeded to quote Schiller: "Without feeling even for its creator's honor/ Like the dead stroke of the pendulum clock/ Nature devoid of God follows knavishly the law of gravity."[9] The quotation is clearly in response to popular demand, as the astrophysicist, Hans Rosenberg, makes still clearer in his academic address on 18 January 1930: " 'Your subject is, to be sure, the most sublime in space/ But, friend, the sublime does not reside in space,' I hear Schiller-Goethe call out to us."[10]

It is, of course, their audience which Wien, Rosenberg, *et al.* hear calling out these sentiments, and they seek to escape half the reproach by showing that they are themselves at least familiar with the classical literary expressions of German idealism. When, however, the physicist or mathematician was in the audience he had to listen to far sharper reproaches. In March 1921, Friedrich Poske came away from the funeral of the poet Carl Hauptmann smarting at the accusations against the exact natural sciences which he encountered there,[11] accusations apparently much like those which poor Max

9. W. Wien, *Universalität und Einzelforschung. Rektorats-Antrittsrede, gehalten am 28. November 1925,* Münchener Universitätsreden, Heft 5 (Munich, 1926), 19 pp., on 14.
10. H. Rosenberg, *Die Entwicklung des räumlichen Weltbildes der Astronomie. Rede zur Reichsgründungsfeier . . . am 18. Januar 1930* (Kiel, 1930), 27 pp., on 26. The same lines are quoted—with a still deeper "bow before the secret which the other side hides from us"—by Hans Kienle, "Vom Wesen astronomischer Forschung. Rede, gehalten bei der Verfassungsfeier der Universität Göttingen am 29. Juli 1932," *Bremer Beiträge zur Naturwissenschaft, 1* (1933), 113-125, on 125.
11. F. Poske at the Hauptversammlung of the Deutscher Verein zur Förderung des mathematischen und naturwissenschaftlichen Unterrichtes, 31 March 1921. (*Unterrichtsblätter für Mathematik und Naturwissenschaft,* 27 [1921], 34.)

WEIMAR CULTURE, CAUSALITY, AND QUANTUM THEORY, 1918–1927

Born had to listen to daily from his wife, a would-be poet and playwright. Hedwig Born derived a masochistic pleasure from "the feeling of being cast upon an icy lunar landscape" which the company of "objective" natural scientists aroused in her.[12] Nor did she hesitate to let her husband's colleagues know that "it is always like a revelation to me whenever behind the *physicist* I suddenly discover the human being; there are, I mean, also inhuman physicists."[13] Certainly there is no reason to think that Einstein's explanation—"what you call 'Max's materialism' is simply the causal [n.b.] mode of considering things"—alleviated Mrs. Born's disquiet.[14]

Painful as it may have been for the theoretical physicist to have to live with such attitudes, the accusation of *Entseelung,* of destruction of the soul, of the world was not the worst he encountered. As Max von Laue saw it in the summer of 1922, the school of Rudolf Steiner "raises the most serious charges against today's natural science. It is represented as bearing the guilt for the world crisis [n.b.] in which we stand at present, and the whole of the intellectual and material misery bound up with that crisis is charged to natural science's account."[15] The counterattack which Laue published was read "with much pleasure" by his mentor and colleague Max Planck, who thought it "will certainly achieve good effects in wider circles."[16]

---

12. H. Born, "Albert Einstein ganz privat," *Helle Zeit—dunkle Zeit. In memoriam Albert Einstein,* ed. C. Seelig (Zurich, 1956), pp. 35-39, on 36.
13. H. Born to H. A. Kramers, 29 September 1925: "Offengestanden hatte ich früher fast etwas Angst vor Ihnen! Aber die ist ganz verschwunden, seit ich hier die Wärme, den Ernst und die ungekünstelte Kraft Ihres Wesens kennen lernen durfte. Es ist mir immer wie eine Offenbarung, wenn ich neben dem *Physiker* plötzlich den Menschen finde; es gibt nämlich auch unmenschliche Physiker!" (Archive for History of Quantum Physics, Sources for History of Quantum Physics Microfilm 8, Section 2; for descriptions and locations of this archive, see Thomas S. Kuhn, et al., *Sources for History of Quantum Physics. An Inventory and Report,* Memoirs of the American Philosophical Society, Vol. 68 [Philadelphia, 1967].)
14. Einstein to H. Born, 1 September 1919, in Albert Einstein, Hedwig and Max Born, *Briefwechsel, 1916-1955,* edited and annotated by M. Born (Munich, 1969), p. 32. An English translation is being published.
15. M. v. Laue, "Steiner und die Naturwissenschaft," *Deutsche Revue, 47* (1922), 41-49; reprinted in Laue's *Aufsätze und Vorträge = Gesammelte Schriften und Vorträge,* Band III (Braunschweig, 1962), pp. 48-56, on 48.
16. Planck to Laue, 8 July 1922: "Ihren Aufsatz über R. Steiner habe ich mit vielem Vergnügen gelesen. Er . . . wird gewiss in weiteren Kreisen gute Wirkung erzielen." (Handschriftensammlung, Bibliothek, Deutsches Museum, Munich.)

Clearly Planck saw Rudolf Steiner as merely providing the occasion and the ostensible target for rebutting a set of attitudes which he and Laue felt to be widespread among the German educated public. Planck himself adverted to these attitudes and to their danger for science in an address in the Prussian Academy of Sciences a few weeks later.[17] Early in the following year he complained bitterly in a public lecture that "precisely in our age, which plumes itself so highly on its progressiveness, the belief in miracles in the most various forms—occultism, spiritualism, theosophy, and all the numerous shadings, however they may be called—penetrates wide circles of the public, educated and uneducated, more mischievously than ever, despite the stubborn defensive efforts directed against it from the scientific side." Compared to this movement, the agitation of Planck's former *bête noir,* the Monist League, has had, he now allows, "only very meagre success."[18]

It is thus not surprising that the remnants of this largely defunct positivist-monist movement thoroughly agreed with Planck that the Weimar intellectual environment was fundamentally and explicitly antagonistic to science. Drawing upon the universally accepted analogy between contemporary Germany and the period following its defeat by Napoleon, Wilhelm Ostwald thought it evident that "In Germany today we suffer again from a rampant mysticism,

---

17. M. Planck, "Ansprache des vorsitzenden Sekretärs, gehalten in der öffentlichen Sitzung zur Feier des Leibnizischen Jahrestages, 29. Juni 1922," *Preuss. Akad. d. Wiss., Sitzungsber.* (1922), pp. lxxv-lxxvii, reprinted in *Max Planck in seinen Akademie-Ansprachen; Erinnerungsschrift der Deutschen Akademie der Wissenschaften zu Berlin* (Berlin, 1948), pp. 41-48. A similar characterization of the intellectual environment had been given in the fall of 1920 by Artur Schoenflies, *Über allgemeine Gesetzmässigkeiten des Geschehens* [Rektoratsantrittsrede], Frankfurter Universitätsreden XI (Frankfurt, 1920), 16 pp., on 4: "In increasing measure in recent [letzten] years there has developed a conscious hostility to the natural-scientific mode of thought. . . . The fact is that the new mode of thought with force and bluster has fought its way through to success in all fields—in *Wissenschaft* and art, literature and politics, in writing and speaking."

18. M. Planck, *Kausalgesetz und Willensfreiheit. Öffentlicher Vortrag gehalten in der Preuss. Akad. d. Wiss. am 17. Februar 1923* (Berlin, 1923), 52 pp.; reprinted in Planck, *Vorträge und Erinnerungen* (Stuttgart, 1949), pp. 139-168; on 162-163. And again, eight years later, "It is astonishing how many people, particularly from educated circles . . . fall under the sway of these new religions, iridescing with every hue from the most confused mysticism on out to the crassest superstition." ("Wissenschaft und Glaube. Weihnachtsartikel vom Jahre 1930," *ibid.,* pp. 246-249; also quoted at length in Hans Hartmann, *Max Planck als Mensch und Denker* [1953; reprinted Frankfurt, 1964], pp. 52-55, on 52-53.)

WEIMAR CULTURE, CAUSALITY, AND QUANTUM THEORY, 1918–1927

which, as at that time, turns against science and reason as its most dangerous enemies."[19] And even where, as with the theory of relativity, there was great public interest in particular results of physical research, that interest was never, to my knowledge, construed by the physicists as evidencing appreciation and approbation of their enterprise. Rather, it struck Einstein as "peculiarly ironical that many people believe that in the theory of relativity one may find support for the anti-rationalistic tendency of our days."[20]

Arnold Sommerfeld was thus clearly speaking for most of his colleagues when, responding to a request from the most prestigious of the South German monthlies for a contribution to a special number on astrology, he asked:

> Doesn't it strike one as a monstrous anachronism that in the twentieth century a respected periodical sees itself compelled to solicit a discussion about astrology? That wide circles of the educated or half-educated public are attracted more by astrology than by astronomy? That in Munich probably more people get their living from astrology than are active in astronomy? Certainly in Germany this anachronism is based in part upon the misery of the present. The belief in a rational [vernünftig] world order was shaken by the way the war ended and the peace dictated; consequently one seeks salvation in an irrational [unvernünftig] world order. But the reason must lie deeper, for astrology, spiritualism, and Christian Science are flourishing among our enemies also. We are thus evidently confronted once again with a wave of irrationality and romanticism like that which a hundred years ago spread over Europe as a reaction against the rationalism of the eighteenth century and its tendency to make the solution of the riddle of the universe a little too easy. Even though I [wir] have no illusions about being able to hold back this wave by means of arguments based upon reason, nonetheless I [wir] want to throw myself decisively against it.[21]

19. W. Ostwald, *Lebenslinien. Eine Selbstbiographie* (Berlin, 1926-1927), *3*, 442. And again, *ibid.*, *2*, 309, "It is at present considered modern to speak all conceivable evil of the intellect."

20. A. Einstein, *Vossische Zeitung*, 10 July 1921, as quoted by Siegfried Grundmann, "Der Deutsche Imperialismus, Einstein und die Relativitätstheorie (1914-1933)," *Relativitätstheorie und Weltanschauung* (Berlin, 1967), pp. 155-285, on 194.

21. A. Sommerfeld, "Über kosmische Strahlung," *Südd. Monatshefte*, *24* (1927), 195-198; reprinted in Sommerfeld's *Gesammelte Schriften* (Braunschweig, 1968), *4*, 580-583. Cf. Lewis M. Branscomb, Director of the U.S. National Bureau of Standards, *Science*, *171* (12 March 1971), 972: "Astrology is booming; there are three professional astrologers in this country for every astronomer."

13

Although the German physical scientists, regardless of their special discipline, agreed that irrationalism and mysticism were characteristic of the postwar mood, altogether it was the mathematicians and the theoretical physicists who, more than the experimental physicists or the chemists, felt themselves to be the particular objects of odium, both public and private. One cannot withhold a certain sympathy for the Nazi Theodor Vahlen as he confesses in 1923 before the assembled members of his university how "a friendly attitude toward mathematics is so rare that, if we run across it, it really strikes us as especially remarkable."[22] This feeling of facing an antagonistic environment, inside and outside the university, was so generally shared among mathematicians that Gerhard Hessenberg could appeal to it in trying to persuade the theoretical physicist Arnold Sommerfeld to take a course of action which would antagonize an experimental physicist (Friedrich Paschen) to whom Sommerfeld looked for much of his raw material: "But we poor scapegoats of mathematicians have gotten to hear so much evil about ourselves these days—behind our backs as well as to our faces—what difference does a little bit more or less make. . . ."[23] Indeed, these "antimathematical currents," "this onslaught against mathematics" which sprang forth after the war seemed so strong and threatening that in 1920 the German mathematicians joined together in a defense organization, the Mathematischer Reichsverband, whose special task was to protect the position of mathematics in the schools.[24]

22. Th. Vahlen, *Wert und Wesen der Mathematik. Festrede . . . am 15. V. 1923*, Greifswalder Universitätsreden 9 (Greifswald, 1923), 32 pp., on 1. And in this, if in nothing else, Konrad Knopp agreed with Vahlen: "We mathematicians . . . have not been able to obtain, or even merely to retain, the position in public life which mathematics merits." ("Mathematik und Kultur, Ein Vortrag," *Preussische Jahrbücher, 211* [1928], 283-300, on 283.)

23. G. Hessenberg to A. Sommerfeld, 16 June 1922: "Wir armen Sündenböcke von Mathematikern aber haben in diesen Tagen so viel schlechtes, hintenherum, wie auch vorneherum, über uns zu hören bekommen, dass es uns auf ein bischen mehr oder weniger nicht ankommt; der Gerechte hat nun einmal viel zu leiden." (Sources for History of Quantum Physics Microfilm 33, Section 1.)

24. Georg Hamel, as president, at the first general assembly of the Mathematischer Reichsverband, Jena, 23 September 1921, *Jahresbericht der Deutschen Mathematiker-Vereinigung, 31* (1922), Part 2, p. 118. And again at the second general assembly, Leipzig, 22 September 1922, the *Arbeitsausschuss* stressed in its report that "With respect to its place and prestige [*Geltung*] in the schools, mathematics finds itself in a defensive position. The contemporary intellectual currents, directed against intellectualism and rationalism, are decidedly unfavorable to mathematics." (*Ibid., 32* [1923], Part 2, pp. 11-12.)

WEIMAR CULTURE, CAUSALITY, AND QUANTUM THEORY, 1918–1927

The result, then, of this first approach to the problem of establishing the tenor of the intellectual environment within which the Weimar physical scientists worked so productively is unambiguous: the environment was *perceived* by the physical scientists to be markedly hostile. Is it therefore necessary to carry our inquiry any further? One might, after all, argue that it is vain to ask whether these perceptions corresponded to "reality" and that moreover the answer would be of no consequence for the behavior of the physical scientists. Nonetheless the accuracy or inaccuracy of these perceptions is certainly an important datum about these men, a datum which is essential for any attempt to infer their perceptions, and the effects upon their science, of a given intellectual environment. For the purposes of this paper, moreover, it is important to go farther afield in exploring the attitudes toward physical science in Weimar Germany; we need a more detailed specification of those attitudes if we are to determine how far and in what sense the ideology and ideas of the physical scientists may be regarded as responses to their intellectual environment.

I.2. As Confirmed by Other Observers

Unequipped and disinclined to undertake an extensive independent exploration and reconstruction of the Weimar intellectual environment, I have turned to other observers—first intellectual historians, then contemporary observers—seeking their conclusions and their guidance.

For our period and theme there are studies by Georg Lukács,[25] Kurt Sontheimer,[26] Peter Gay,[27] and—most recent, detailed, and relevant—by Fritz Ringer.[28] While these intellectual historians are not specifically concerned with the attitudes toward exact science,

---

25. G. Lukács, *Die Zerstörung der Vernunft. Der Weg des Irrationalismus von Schelling zu Hitler* (Berlin, 1954).
26. K. Sontheimer, *Antidemokratisches Denken in der Weimarer Republik* (Munich, 1962).
27. P. Gay, *Weimar Culture: The Outsider as Insider* (New York, 1968). An only slightly abridged version, omitting however the extensive bibliography, appeared under the same title in Donald Fleming and Bernard Bailyn, eds., *The Intellectual Migration* (Cambridge, Mass., 1969), pp. 11-93.
28. F. K. Ringer, *The Decline of the German Mandarins. The German Academic Community, 1890-1933* (Cambridge, Mass.: Harvard University Press, 1969).

their characterizations of the intellectual milieu do in fact bear directly upon this question. Especially Ringer's examination of academic ideology places before us many of the attitudes toward science which pervaded the Weimar academic world and directs us to many important sources. And despite the diversity of the personal-professional backgrounds, research methods, and ethical-political motivations of these intellectual historians, by and large they give us the same general picture of the Weimar intellectual milieu: rejection of reason as an epistemological instrument because inseparable from positivism-mechanism-materialism, and because, as fundamentally disintegrative, incapable of satisfying the "hunger for wholeness";[29] glorification of "life," intuition, unmediated and unanalyzed experience, with the immediate apprehension of values, and not the dissection of causal nexus, as the proper object of scholarly or scientific activity. This "life-philosophy," of which existentialism was but a variety, Lukács sees as "the dominant ideology of the entire imperialistic period in Germany. . . . In the postwar period virtually all of the widely read bourgeois *Weltanschauungsliteratur* is *lebensphilosophisch*."[30]

With these studies by intellectual historians giving us some confidence that we are not going seriously astray, let us look a little more closely at certain of the programmatic slogans of this life philosophy as epitomized by contemporary observers of Weimar intellectual life. Such characterizations of the intellectual environment will, I think, not merely suggest, irresistibly, a valuation of the physical scientist, but will also force us to recognize the crucial role of the concept of causality.

Within a year of the end of the war these intellectual currents, now monopolizing the movement for educational reform, were flowing everywhere. Discussing "the social-pedagogic demand of the present" in 1920, Alfred Vierkandt could see clearly that "We are generally experiencing today a full rejection of positivism; we are experiencing a new need for unity, a synthetic tendency in all the world of learning [Wissenschaft]—a type of thinking [Eindenken] which primarily emphasizes the organic rather than the mechanical, the living instead of the dead, the concepts of value, purpose, and goal,

29. The phrase is Gay's, *op. cit.* (note 27).
30. Lukács, *op. cit.* (note 25), p. 318.

WEIMAR CULTURE, CAUSALITY, AND QUANTUM THEORY, 1918–1927

instead of causality."[31] A sharper and more penetrating analysis of this call for a "revolution in science [Wissenschaft]" in the early Weimar period is that which Ernst Troeltsch published in 1921.[32] Here causality appears over and over again as the pejorative term epitomizing the tendency in *Wissenschaft* which the new movement rejects: "The methods of these specialized scientific disciplines are those of causal explanation, of natural causality, of psychophysical, psychological, and sociological causality. It is the ultimate intellectualization of our attitude toward the world, the disenchantment of the world, and the path toward an unlimited approximation to a totally causal system [Gesamtkausalsystem] of things."[33]

Troeltsch, in common with many other observers, cites Henri Bergson as perhaps the most important—and the only nonGerman—source of the movement against "all suffocating determinism." "A general sigh of relief follows almost audibly the ever stronger establishment of this system."[34]

> If now we draw all that together, the freedom from positivistic causalism and determinism, the overcoming of neo-Kantian formalism, . . . the orientation toward immediate experience of unanalyzable but understandable cultural tendencies, . . . a new phenomenological platonism which through visions beholds and justifies norms and essences, then one has all elements of the *wissenschaftlichen* revolution in one's hands. . . . It is a neoromanticism as formerly in the *Sturm und Drang*.[35]

---

31. A. Vierkandt, *Die sozialpaedagogische Forderung der Gegenwart* (Berlin, 1920), p. 20, as quoted by F. K. Ringer, "The German Universities and the Crisis of Learning, 1918-1932" (Ph.D. diss., Harvard University, 1960), p. 145.

32. E. Troeltsch, "Die Revolution in der Wissenschaft," *Schmoller's Jahrbuch (Jahrbuch für Gesetzgebung, Verwaltung . . .)*, 45 (1921), 1001-1030. Reprinted in Troeltsch's *Gesammelte Schriften*, Vol. 4: *Aufsätze zur Geistesgeschichte und Religionssoziologie*, ed. Hans Baron (Tübingen, 1925; reprinted 1961), pp. 653-677.

33. *Ibid.*, p. 1020. Cf. Max Weber, "Science as a Vocation [1919]," *From Max Weber: Essays in Sociology*, trans. and ed. H. H. Gerth and C. Wright Mills (1946; reprinted New York, 1958), p. 142: "And today? 'Science as the way to nature' would sound like blasphemy to youth. Today, youth proclaims the opposite: redemption from the intellectualism of science in order to return to one's own nature and therewith to nature in general."

34. *Ibid.*, p. 1005.

35. *Ibid.*, p. 1007. Or again, "The peculiarity of German thought, in the form in which it is nowadays so much emphasized, both outside and inside Germany, is primarily derived from the Romantic Movement . . . a revolution, above all, against the whole of the mathematico-mechanical spirit of science in Western Europe . . ." (Troeltsch, "The Idea of Natural Law and Humanity in World

These intellectual currents, whose sources lay in the prewar period, but which welled up immediately following Germany's defeat, continued to dominate the intellectual milieu in the mid-1920's as in the first years of the Weimar Republic. In 1927, Theodor Litt, reviewing contemporary philosophy and its influence upon the ideal of liberal education [Bildung], found *Lebensphilosophie* to be the strongest intellectual current. It was not a system, not a school, but a general tendency which is only to be defined by what it opposes: "On the one hand . . . the mechanism and determinism of a causal explanation which calculates everything in advance, makes everything comparable, dissolves everything into elements—on the other hand . . . the rationalism and formalism of a logical systematization which deduces everything, classifies everything, subjects everything to concepts."[36]

Litt went on to point out again the often noticed parallel between the rise of *Lebensphilosophie* and the "victorious breakthrough of 'wholistic' convictions" in biology (neovitalism) and psychology (Gestaltism, etc.).[37] It is therefore of some interest to ask what impression a biologist-philosopher of these convictions received of the Weimar intellectual milieu. Eloquent in this connection is Hans

---

Politics [1922]," in Otto Gierke, *Natural Law and the Theory of Society, 1500-1800,* trans. E. Barker [Cambridge, 1934; reprinted Boston, 1957], pp. 201-222, on 210.) Troeltsch emphasized (*op. cit.* [note 32], pp. 1003-1004, 1028-1029), that this "Revolution in der Wissenschaft" was confined to the *Geisteswissenschaften,* that the revolutionary innovations in natural science had no clear *weltanschaulich* significance, and he insisted that the close connection of the natural sciences with technology would prevent their sloughing off rigorous methods, or backsliding into "Naturphilosophie" and dilettantism. But, one would ask Troeltsch, what if, under the influence of these same intellectual currents, the exact scientists should repudiate their connection with technology— as indeed they did. Could we then expect some parallel to the romantic physics of the early nineteenth century?

36. Th. Litt, *Die Philosophie der Gegenwart und ihr Einfluss auf das Bildungsideal,* 2nd ed. (Leipzig, 1927), pp. 32-33. Cf. Friedrich Meinecke, "Über Spengler's Geschichtsbetrachtung," *Wissen und Leben, 16* (1923), 549-561, as reprinted in Meinecke's *Werke, 4* (Stuttgart, 1959), 181-195, characterizing the mood of the times: "One is also tired of having only interconnections of cause and effect [Ursache und Wirkung] demonstrated over and over again according to rational methods of cognition, and tired of performing such demonstrations oneself; one is of the opinion that there is a great deal more in life and humanity than an apparatus of mechanical causality [Kausalitäten]. One has become tired of knowing and thirsty for living. . . ."

37. Litt, *loc. cit.*

Driesch's introduction to *Man and the Universe* (1928); for despite his vitalism, wholism, and idealism he too felt the milieu to be hostile to science and reason. Recognizing that it is "unfashionable" to take account of the results of natural science and that he will be put down as betraying his origin as a scientist, he nonetheless accepts the characterization of his method by the opprobrious epithet "rational" and holds that "the modern contempt for [natural] science is due to the fact that its champions take the concept in too narrow a sense, namely, as denoting a mechanistic view of the world."[38]

The historian of science may feel impelled to object that it is a most serious misconception to regard physics after 1900 as "mechanistic," and that it is a complete misunderstanding of positivism to equate it with mechanism, materialism, or even rationalism. Indeed it is difficult to understand how contemporary observers generally failed to recognize in Mach, Ostwald, and their cohorts a quasi-romantic movement parallel in several respects to *Lebensphilosophie*.[39] But all such objections are, of course, entirely beside the point. The relevant question is only what image the educated public held of the physical scientist and his world view. The image of the mechanistic, rationalistic causalist led inevitably to a negative valuation.

I.3. Intellectual Allies: Vienna Circle and Bauhaus

But can this picture of the physical scientist's intellectual milieu be accurate? When we say "Weimar culture" do we not immediately think also of the Vienna Circle and logical positivism, of the Bauhaus and functionalism, as its typical expressions? And were not *these* movements inherently congenial to rational analysis and the achievements of modern physical science and technology?

Assuredly the Vienna Circle, with its goal of a "wissenschaftliche

---

38. H. Driesch, *Der Mensch und die Welt* (Leipzig, 1928), trans. by W. J. Johnson as *Man and the Universe* (London, 1929), pp. 5-8. Cf. Karl Jaspers, *Die geistige Situation der Gegenwart* (Berlin-Leipzig, 1932), trans. by E. and C. Paul as *Man in the Modern Age* (New York, 1933), p. 159: "Anti-science stalks abroad today amid all parties and sects and manifests its influence among persons of the most diversified outlooks, pulverizing the very substance of rational human existence."

39. Stephen G. Brush, "Thermodynamics and History," *The Graduate Journal*, 7 (1967), 477-565, on 530.

Weltauffassung" based upon empiricism and logical atomistic analysis of conceptual structures, had a very "positive" attitude toward the physical sciences and mathematics. But how characteristic was their brand of philosophy? We are sometimes led to believe that logical positivism, which in fact emerged as a coherent program only in 1929/30, was the dominant current in German philosophy throughout the 1920's. Thus H. Stuart Hughes has the movement in full flower by the early 1920's and represents Ludwig Wittgenstein's *Tractatus Logico-Philosophicus,* of which the German edition (1921) lay virtually unread in the final number of Ostwald's defunct *Annalen der Naturphilosophie,* as "the most influential philosophical work of the post-war years. . . . [T]he neo-positivists . . . were able to rehabilitate the scientific method in philosophy . . . and for another two decades Europe was to be without a philosophy that could speak to the ordinary citizen. . . ."[40]

One need, however, only glance at the manifestos of the Vienna Circle in order to recognize that Hughes has utterly misrepresented the case. In *Wissenschaftliche Weltauffassung: Der Wiener Kreis,* the brochure with which in 1929 the circle first came before the public, "their tone," as Ringer rightly points out, "was that of exasperated outsiders."[41] Indeed the opening lines tell us: "Many assert that metaphysical and theologizing thinking, not only in everyday life, but also in science and scholarship [in der Wissenschaft], are today again increasing. . . . The assertion itself is easily confirmed by a glance at the themes of lectures at the universities

---

40. H. S. Hughes, *Consciousness and Society. The Reorientation of European Social Thought, 1890-1930* (New York, 1958; reprinted New York, n.d.), pp. 399-401. Before we accept as fact that the "ordinary citizen" was deserted by philosophy we should hear what Heinrich Rickert had to say on this score in 1920: "The concept which today dominates the general intellectual atmosphere [die Durchschnittsmeinungen] in an especially high degree seems to us to be best designated by the expression *life*. For some time now it has become ever more frequently used, and plays a great role not only among the popular writers, but also among academic philosophers. 'Erlebnis' and 'lebendig' are favorite words, and there is no opinion which is counted so modern as that it is the task of philosophy to give a doctrine of life, which, shaping itself vitally and genuinely out of experience, is capable of being used by the living human being." *(Die Philosophie des Lebens. Darstellung und Kritik der philosophischen Modeströmungen unserer Zeit.* [Tübingen, 1920], p. 4.)

41. Ringer, *op. cit.* (note 28), p. 308.

WEIMAR CULTURE, CAUSALITY, AND QUANTUM THEORY, 1918–1927

and at the titles of philosophical publications."[42] Writing in 1931 for their *Schriften zur Wissenschaftlichen Weltauffassung*, Philipp Frank, the one professional physicist in the group, repeatedly cited and quoted the "Ganzheitsphilosophie" of Othmar Spann's *Kategorienlehre* (1924) as characteristic of the negative valuation of natural science and mathematics in the prevalent "school philosophy."

> To the discipline which depicts things by means of merely external (quantitative) features the essence of things remains eternally foreign. This is the key to why mathematical-causal natural science is not a comprehending, mentally creative discipline as the *Geisteswissenschaften* are. . . . The quantifying, so-called exact, investigation is on the contrary merely measurement and, since it ignores the essence of things and must decompose them into magnitudes in order to inventory them, it does not deserve the name *Wissenschaft* in the same high sense as the *Geisteswissenschaften*. . . . The question of utility and achieved goals is one thing, the worth [Würde] of genuine *Wissenschaften* concerned with totality and essence is another. Such worth modern mathematical natural science does not possess today.[43]

Far from dominating German philosophy in the 1920's, the Vienna Circle and the corresponding group in Berlin—the Gesellschaft für empirische Philosophie around Hans Reichenbach and Richard von Mises—with their high positive valuation of mathematical natural science represented a rather late and distinctly marginal movement. The impression which in 1929 Sidney Hook brought back to the United States from a year of philosophical study in Germany was that almost all the contemporary schools "are amazingly indifferent to the methods and results of modern physical science." Worse, "The attitude of the German Philosopher to science is not always one of indifference. It is often an attitude of open hostility." The writings of Hans Reichenbach would, Hook thought, be of great interest to the American reader, but in Germany Reichenbach is "ignored by academic philosophers as are all of his

---

42. Verein Ernst Mach, *Wissenschaftliche Weltauffassung. Der Wiener Kreis* (Vienna, 1929), 63 pp., on 9.

43. O. Spann, *Kategorienlehre* (1924), as quoted by P. Frank, *Das Kausalgesetz und seine Grenzen*, Schriften zur wissenschaftlichen Weltauffassung, Band 6 (Vienna, 1932), pp. 54-55.

kind...."[44] Two decades later, sketching the history of the Vienna Circle, Victor Kraft described the great resonance the movement found in western Europe and America, adding ruefully, "It was only in Germany that the Vienna Circle's approach was not taken up at all."[45]

With the Bauhaus the case is somewhat different, for the movement of which Walter Gropius was the leading representative was indeed to a degree characteristic of Weimar culture.[46] Thus in this case we must ask, rather, if the new architecture and the associated movement in design were the expression of an impulse inherently congenial to the methods of the exact sciences or the achievements of modern technology. When one looks at the manifestos of this movement, however, one cannot but be struck by their ambivalence. In the first place, the initial conception and artistic direction of the Bauhaus was largely within the William Morris tradition of a return to handcrafts as a *reaction against* modern technology. When Gropius, in good measure out of simple financial necessity, began to reorient the institution toward industrial design, he had to face tenacious internal resistance. "With absolute conviction I reject the slogan 'Art and Technology—A New Unity,'" Lyonel Feininger wrote in a private letter in August 1923, "this misinterpretation of art is, however, a symptom of our times. And the demand for linking it with technology is absurd from every point of view." An antagonism toward science cum technology was even more explicit in the manifesto which Oskar Schlemmer drafted for the publicity pamphlet of the first Bauhaus exhibition in the summer of 1923.

---

44. S. Hook, "A Personal Impression of Contemporary German Philosophy," *Journal of Philosophy*, 27 (1930), 141-160, on 147, 159. The same view is stated less vigorously by Kurt Grelling, "Philosophy of the Exact Sciences [in Germany]," in *Philosophy Today*, ed. E. L. Schaub (Chicago, 1928), pp. 393-415.

45. V. Kraft, *Der Wiener Kreis. Der Ursprung des Neopositivismus. Ein Kapitel der jüngsten Philosophiegeschichte*, 2nd ed. (Vienna, 1968), p. 8. The first edition was published in 1950.

46. On the architectural side, this is well shown by Barbara Miller Lane, *Architecture and Politics in Germany, 1918-1945* (Cambridge, Mass., 1968). Cf. Gropius' speech before the Thuringian Landtag in Weimar on 9 July 1920: "Based on indisputable facts, I am now going to show convincingly that what the Bauhaus has accomplished is an uninterrupted and logical development that must take place, and already is taking place everywhere in the country." (Hans M. Wingler, ed., *The Bauhaus. Weimar, Dessau, Berlin, Chicago*, trans. W. Jabs and B. Gilbert [Cambridge, Mass., 1969], p. 42.)

WEIMAR CULTURE, CAUSALITY, AND QUANTUM THEORY, 1918–1927

"Reason and science, 'man's greatest powers,' are the regents, and the engineer is the sedate executor of unlimited possibilities. Mathematics, structure, and mechanization are the elements, and power and money are the dictators of these modern phenomena of steel, concrete, glass, and electricity . . . calculation seizes the transcendant world: art becomes a logarithm."[47]

Even Gropius himself, moreover, was thoroughly ambivalent on this question. "My primary aim" in planning the curriculum of the Bauhaus was "training the individual's natural capacities to grasp life as a whole, a single cosmic entity. . . . Our guiding principle was that artistic design is neither an intellectual nor a material affair, but simply an integral part of the stuff of life."[48] And so we return once again to *Lebensphilosophie*.

I.4. Educational Ideals and Reforms

Having gotten a clearer picture of the attitudes toward physical science and analytical rationality prevalent among the educated middle classes, and especially strong in the academic world in the Weimar period, we can now better appreciate the great apprehension with which the mathematicians and physicists viewed the movement for educational reform which followed in the wake of the revolution. And it is worthwhile to examine briefly the educational ideals announced by those in the Prussian Ministry of Education with the power to enact and administer such reforms, for one sees thereby both the resonance which these attitudes found across the political spectrum as well as the imminence of the threat to the physical sciences which these attitudes constituted.

The antirationalist theme was sounded at the opening of the Weimar period by Staatssekretär Carl Heinrich Becker. A distinguished Islamicist, Becker had entered the Prussian Education Ministry during the war and, qua Democrat, was elevated to the top civil service position after the Social Democrats threw out the *Kultus-*

---

47. Wingler, pp. 65-66, 69. The pamphlet was suppressed by the Bauhaus after it had been printed—not because of anything Schlemmer said about science, rationality, or technology, but because he had allowed a favorite Bauhaus slogan, "building the Cathedral of Socialism," to slip into the manifesto.

48. W. Gropius, *The New Architecture and the Bauhaus*, trans. by P. M. Shand (London, 1935), pp. 52, 89.

minister Friedrich Schmidt in November 1918.[49] "The basic evil," Becker asserted in 1919 in a widely read essay on university reform, "is the overvaluing of the purely intellectual in our cultural activity, the exclusive predominance of the rationalistic mode of thought, which had to lead, and has led, to egoism and materialism of the crassest form." And again, in another pamphlet written at this same time, Becker maintained that "our entire educational system is too exclusively oriented toward the intellect. We must acquire again reverence for the irrational."[50] This is all perhaps not too surprising in an academic *Geisteswissenschaftler*. It must give one pause, however, when one finds Becker's superior, the Social Democratic *Kultusminister* Konrad Haenisch propagating the same slogans (recall the rationalist-materialist traditions of his party!): "But if . . . the German people, having suffered for decades from the plight of mechanism and materialism, . . . if in our spiritual life not only the intellectual but also the irrational is to receive its due, then the barriers will have to be broken down which presently separate the universities and the people. . . ."[51]

One thus sees that whatever considerations may have led government officials to support and advance academic research in the physical sciences, the attitude of these "progressive" politicians and

49. Erich Wende, *C. H. Becker, Mensch und Politiker* (Stuttgart, 1959). Cf. Friedrich Schmidt-Ott, *Erlebtes und Erstrebtes, 1860-1950* (Wiesbaden, 1952).

50. C. H. Becker, *Gedanken zur Hochschulreform* (Leipzig, 1919), p. ix; *Kulturpolitische Aufgaben des Reiches* (Leipzig, 1919), p. 55: "Wir müssen wieder Ehrfurcht bekommen vor dem Irrationalen." For further examples see Adolf Grimme, ed., *Kulturverwaltung der zwanziger Jahre: Alte Dokumente und neue Beiträge* (Stuttgart, 1961), pp. 78-79; Wende, *op. cit.* (note 49), p. 305. Cf. the remarks of William D. McElroy, Director of the U.S. National Science Foundation, at Indiana University, 12 October 1970: "In my view, the science community generally should consider more carefully . . . 'the new romanticism,' emphasizing man as an emotional and feeling creature as well as a reasoning one. A healthy dose of this view may counterbalance some of the extreme emphasis upon rational thinking I suspect is endemic within the science community." (*Science*, 170 [1970], 517.)

51. K. Haenisch, *Staat und Hochschule* (Berlin, 1920), pp. 110-111, as quoted by Ringer, *op. cit.* (note 28), p. 282. Haenisch's general readiness to adopt the political and social ideologies of the German academics has been stressed by Hans Peter Bleuel, *Deutschlands Bekenner; Professoren zwischen Kaiserreich und Diktatur* (Berne, 1968), pp. 128-129. Ten years later the Social Democrats once again claimed the Prussian *Kultusministerium*. Their man, Adolph Grimme, was soon writing to Martin Heidegger, 14 May 1930, "as admirer and in a modest sense as pupil. . . . I don't have to tell you how very anxious I am [to get Heidegger to Berlin]. With you here a particular type of philosophy, above all metaphysics, could break through in Berlin." (Grimme, *Briefe*, ed. D. Sauberzweig [Heidelberg, 1967], pp. 36-37.)

WEIMAR CULTURE, CAUSALITY, AND QUANTUM THEORY, 1918–1927

bureaucrats toward the "hard" sciences, and particularly toward the intellectual style they associated with these disciplines, was certainly not unambiguously affirmative.[52] "The mood of decisive circles," Wilhelm Hillers warned the Mathematischer Reichsverband in 1921, "is unfavorable to the natural sciences." And when the Prussian education ministry's plan for the reform of the secondary school curricula finally appeared in the spring of 1924, it proved even worse than had been feared. Taking it for granted that "the economic-political, technical, and positivistic age . . . now lies behind us," the ministry refused to justify any part of the curricula on utilitarian grounds. The claims of mathematics and natural science derived from, and only from, the fact that "not only in Kant but also in Goethe [these types of thinking] have co-determined to the very depths the vital features of German idealism"—which, however, was insufficient merit to save these subjects from substantial reductions in the amount of time allotted them. To the older generation of mathematicians and physicists—to Friedrich Poske, Georg Hamel, Felix Klein—who had been struggling since the nineties to make a generous place for their disciplines in the German secondary schools, it seemed that all had been in vain: "This school reform," Klein remarked bitterly, "signifies for our educational system the end of the century of science."[53]

52. A principal consideration underlying the relatively high level of financial support for academic research in the physical sciences was, once again, prestige—in particular the image of science as a *substitute* for political and economic power. See: Brigitte Schröder-Gudehus, *Deutsche Wissenschaft und internationale Zusammenarbeit* (diss. Geneva, 1966), pp. 181-189, 199; P. Forman, "Scientific Internationalism and the Weimar Physicists," *Science, War, and Internationalism, 1900-1939*, ed. Roger Hahn (in press).
53. W. Hillers, *Jahresbericht der Deutschen Mathematiker-Vereinigung, 31* (1922), Part 2, pp. 120-121. *Die Neuordnung des preussischen höheren Schulwesens: Denkschrift des Preussischen Ministeriums für Wissenschaft, Kunst und Volksbildung* (Berlin, 1924), reprinted in Hans Richert, ed., *Richtlinien für die Lehrpläne der höheren Schulen Preussens*, 7th ed. (Berlin, 1927), *1*, 17-77, on 68-70. F. Klein quoted indirectly by F. Poske, "Der naturwissenschaftliche Unterricht und die Neuordnung des preussischen höheren Schulwesens," *Naturwiss., 13* (1925), 73-75. Klein himself, in a letter to G. Hamel, remarked on the "remarkable circumstance that the development of the German school system has taken an entirely different direction" than the one he had foreseen in his address of June 1918. (*Unterrichtsblätter für Mathematik und Naturwissenschaft, 30* [1924], 44-45). Hamel, speaking for the Mathematischer Reichsverband (*Jahresbericht der Deutschen Mathematiker-Vereinigung, 33* [1924], Part 2, p. 63), quite agreed: "In fact the new school reform signifies the complete repudiation of [Abkehr von] the previous development . . . throws us way back before the time of the first school reform of 1892."

I.5. The Crisis of *Wissenschaft*

In the preceding sections I have explored from several directions the attitudes toward science and reason permeating the intellectual environment of the Weimar physicists and mathematicians. But the intellectual milieu is not fully characterized by a specification of such *substantive* constituents, not even when the catalog of valuations is supplemented by a measure of the intensity with which each of these attitudes is held. To fully characterize an intellectual atmosphere one must specify not merely the likes and dislikes, the sympathies and antipathies, but also the mood, the morale, the accepted view of the contemporary cultural situation, and the common notions of what that situation demanded, or where it must lead.

Turning back once again to the general intellectual historians and to the contemporary observers of the intellectual scene, we find as before remarkable unanimity about this essential dimension of the intellectual milieu: widespread among the educated middle classes, but especially oppressive in academia, was a generalized sense of crisis. Included therein was the permanent political and economic crisis, but, far from being limited to this, the fundamental phenomenon was felt to be a moral and intellectual crisis, a crisis of culture, a crisis of science and scholarship. Fritz Ringer, who has given the closest attention to German academic ideology, and, in particular, to this "crisis of learning," found that:

> Throughout the Weimar period, it was often said in academic circles that a crisis was in progress. No one felt the need to define the exact nature of this crisis, to ask where it came from or what it involved. "Sometimes [the educator Aloys Fischer wrote in 1924], the present situation is represented as a crisis of the . . . economic system only, sometimes as one of politics and of the idea of the state, or as a crisis of the social order. At other times it is conceived more deeply and inclusively as a crisis of the entire intellectual and spiritual culture. . . ." In any case, the crisis existed, if only by virtue of the fact that almost every educated German believed in its reality.[54]

---

54. Ringer, *op. cit.* (note 28), p. 245. It would be more accurate, perhaps, to say that although many academics did indeed feel a need to define the exact nature of this crisis, the diagnoses were often diametrically opposed. For example, Arthur Liebert, *Die geistige Krisis der Gegenwart*, 2nd ed. (Berlin, 1923),

WEIMAR CULTURE, CAUSALITY, AND QUANTUM THEORY, 1918–1927

This notion of a crisis in or of learning, although it had roots running back into the previous century, emerged as a universally cogent cliché only in the aftermath of Germany's defeat. "The phrase 'Krisis der Wissenschaft' has already become a popular slogan in everyone's mouth," the political economist Arthur Salz noted in 1921.[55] So it continued throughout the Weimar period: "The idea of such a crisis of culture [Kulturkrise]," Pierre Viénot observed a decade later, "belongs today to the solid stock of the common habit of thought in Germany. It is a part of the German mentality."[56] And at a quarter century's distance it remained clear to Werner Richter, Becker's lieutenant in charge of the University Section of

---

pp. 7-9, expressed this need very forcefully: the purpose of his essay "is not to establish and portray any one arbitrarily selected crisis from contemporary life, no matter how staggering a force it may possess. The intent is much rather to expose *the* crisis of our time and of simply the entire contemporary world view and life mood, i.e. the concept and meaning of all the individual crises and the common spiritual and metaphysical wellspring by which they are all conditioned and from which they are all fed." This *he* found in "the disastrous historical scepticism and relativism nourished by historicism."

55. A. Salz, *Für die Wissenschaft. Gegen die Gebildeten unter ihren Verächtern* (Munich, 1921), p. 10. Typical for this period, and perhaps for this sort of phenomenon, is the circumstance that even the "opponents" shared in large measure the attitudes they were attacking. Thus Troeltsch, *op. cit.* (note 32), p. 1026, found Salz's essay "very instructive and symptomatic, above all in its almost fatalistic surrender to the anti-scientific, and in that sense revolutionary, currents." Another, more pertinent, example of this circumstance is Adolf von Harnack, historian-theologian but President of the Kaiser-Wilhelm-Gesellschaft. Responding to Karl Barth's challenge, hurling his "Fünfzehn Fragen an die Verächter der wissenschaftlichen Theologie unter den Theologen [1923]," Harnack warned against efforts "to revile, indeed eliminate reason [Vernunft]. . . . Does not gnostic occultism even now raise itself upon the ruins?" Yet on another occasion ("Stufen wissenschaftlicher Erkenntnis [1930]") we find this spokesman for the interests of the natural sciences declaring that "our intellect [Verstand] is the born mathematical physicist; like the mathematical physicist it abstracts, it calculates, it weighs." But "this abstracting method which corresponds to mechanism" is incapable of grasping the "life," "forms," "wholes" which surround us. Moreover natural science is only the second step in the cognitive hierarchy; above it stands knowledge of life, followed by knowledge of man, while the fifth, last, and highest step is occupied by philosophy. (Harnack, *Ausgewählte Reden und Aufsätze*, ed. A. von Zahn-Harnack and Axel von Harnack [Berlin, 1951], pp. 132-134, 177-180.)

56. P. Viénot, *Ungewisses Deutschland. Zur Krise seiner bürgerlichen Kultur*, trans. Eva Mertens (Frankfurt a.M., 1931), pp. 24-25. Typical is the observation of the former Prussian Minister of Education Otto Boelitz, in *Grundsätzliches zur Kulturlage der Gegenwart*, Flugschriften der deutschen Volkspartei, Folge 77 (Berlin, 1931), p. 5: "so gehe ich aus von der Kennzeichnung der Kultur der Gegenwart als einer Kulturkrisis."

the Prussian Ministry of Education, that "The self-image [Selbstverständnis] of that period was decisively influenced by the consciousness of a crisis of culture."[57] I will return in Part II to consider how this generalized sense of crisis may have affected the rhetoric and the *Selbstverständnis* of the Weimar physicists and mathematicians; here I would only emphasize that implicit in this sense of crisis was a negative valuation of the traditional scientific disciplines, methods, and practitioners. If the educated public were convinced that "today's *Wissenschaft,* together with its methods, has run up a blind alley,"[58] then, inevitably, the stature of those who had cultivated these sciences by these methods would be considerably diminished. Conversely, if the scholar or scientist was to maintain his prestige in his academic environment and beyond, he too would have to acknowledge and affirm the crisis, would have to repudiate the traditional methods and doctrines of his discipline.[59]

It was, of course, especially the radical *Lebensphilosophen* who pressed this interpretation of the crisis of learning. As a crisis of "causal monism," of positivistic methods in *Wissenschaft,* the crisis of learning must be followed by a revolution which liquidates this barren and intolerable mechanism in favor of a "new *Wissenschaft"* of values, intuition, feeling, of the living, the organic.[60] But the

---

57. W. Richter, *Wissenschaft und Geist in der Weimarer Republik,* Arbeitsgemeinschaft für Forschung des Landes Nordrhein-Westfalen, Geisteswissenschaften, Nr. 80 (Cologne, 1958), 31 pp., on 11.

58. A. Salz, *op. cit.* (note 55), p. 10. Cf. Hermann Weyl, "Felix Kleins Stellung in der mathematischen Gegenwart," *Naturwiss., 18* (1930), 4-11, on 6: "der Typus des Gelehrten und die Wissenschaft in ihrer Geltung und ihrem Wert während der letzten Jahrzehnte in Frage gestellt waren, im Zeichen der Krisis standen." (Reprinted in Weyl's *Gesammelte Abhandlungen* [Berlin, 1968], *3,* 292-299.)

59. The Weimar academics were indeed much agitated by a variety of perceived threats to their social prestige, their intellectual leadership, and their economic situation. Yet, by and large, they saw democracy and republican institutions as the cause of their falling esteem, and rather looked to *Lebensphilosophie* to restore their power and prestige. Cf. note 40, above.

60. E. Troeltsch, *op. cit.* (note 32), p. 1023. To take but one of innumerable examples: Ernst Barthel, Privatdozent U. Köln, in an essay on "Mechanischer und organischer Naturbegriff," *Annalen der Philosophie, 5* (1925), 57-76, argued that the "three principles" of space, time, and causality have been sufficiently analyzed and recognized as the "fundaments of rational thought. . . . It is, however, also generally conceded that the usual rational thinking through of the world totality assuming these principles leads to essential antinomies and incomprehensibilities, which bring thought to the limits of its competence. The organic conception of nature would like to ask itself now, whether it could

WEIMAR CULTURE, CAUSALITY, AND QUANTUM THEORY, 1918-1927

conviction of the reality of this crisis of learning—including the image of science in a cul de sac—was even more widely spread than formal *Lebensphilosophie* itself. Thus "the collapse of science" was also trumpeted vigorously by Hugo Dingler, a prolific and widely read philosopher of physics, whose orientation was strongly rationalistic: "The state where nothing is any longer really certain, everything is possible and at the same time every possible position is also maintained, where there is no longer any basis and any guidelines, nothing, nothing which may be considered certain—in a word, chaos, collapse. In that state we stand; right in the middle of it. The public does not suspect it, and the *Gelehrten*, often frantically, shut their eyes."[61] And in Dingler's view "this new collapse of science, in whose midst we stand . . . consists in the collapse of the belief in the certainty of the experimental principle," i.e., of the possibility of establishing the truth of a theory upon its agreement with experiment.[62]

Here the "crisis of learning" begins to touch the Weimar physicists very nearly. And when we find the aging and conservative professor of Experimental Physics at Privatdozent Dingler's institution taking advantage of his term as rector to contradict and deplore the notion of "the collapse of science," at least in respect of his own science,[63] then I think it fair to infer that the physicists who could not or would not join the revolution saw such doctrines as diminishing their claims and threatening their prestige both in the academic world and among the public at large.

---

possibly be that the one and only way of thinking is to make three abstract principles the basis for explanation of a world full of concrete living contents, or whether, on the contrary, also the opposite way . . ." (p. 71). The exploration of this "opposite way" leads to the conclusion of the article (pp. 75-76) that "the quality of the phenomena and their mutual connection lies in a region of noncausal [nichtkausal] harmony, which is to be grasped only by the intuition" and that, in general, one may distinguish between "mechanical" and "organic" scientific research in that the aim of the former is "practically utilizable, hypothetical causal abstractions, of the latter an intuitive cognition of the immanent essential connections."

61. H. Dingler, *Der Zusammenbruch der Wissenschaft und der Primat der Philosophie*, 2nd ed. (Munich, 1931), p. 10. The second edition differs from the first, 1926, only in the addition of supplementary notes.

62. *Loc. cit.*

63. W. Wien, *Vergangenheit, Gegenwart und Zukunft der Physik. Rede, gehalten beim Stiftungsfest der Universität München am 19. Juni 1926*, Münchener Universitätsreden, Heft 7 (Munich, 1926), 18 pp., on 18.

I.6. Spengler's *Decline of the West*

The crisis of culture, the revolution in *Wissenschaft*, radical *Lebensphilosophie*, all proclaimed and epitomized by a sweeping theory of world history in which—"das ist das Neue"[64]—physics and mathematics are treated alongside art, music, and religion as wholly culturally conditioned. The first volume of Oswald Spengler's *Der Untergang des Abendlandes*,[65] in which the theory is presented and to which the extensive discussions of science are largely confined, appeared in 1918. Although at certain points it bore the stamp of the wartime mind—as, for example, in its positive valuation of technology—on the whole its fatalistic-relativistic pessimism was precisely the right tone for a defeated Germany. In five years the first volume went through thirty printings, and by 1926 the revised edition published in 1923 had gone through a further thirty printings —altogether 100,000 copies in a country with scarcely three times that number of college graduates. Almost universally read in academic circles—by the physicists, too, as we shall see—the typical professorial reaction was: "Of my discipline Spengler understands, of course, not the first thing, but aside from that the book is brilliant."[66]

Ernst Troeltsch saw the first volume of the *Untergang* as the paradigm of the revolution in science: "It is the first decisive public

---

64. E. Troeltsch's review of the first volume of *Der Untergang* in the *Historische Zeitschrift* (1919), reprinted in Troeltsch's *Ges. Schr., 4 (op. cit.,* note 32), 682. Troeltsch found it, inter alia, "ein bedeutsames Kulturdokument aus der Zeit einer geistigen Krisis der deutschen Wissenschaft."

65. O. Spengler, *Der Untergang des Abendlandes. Umrisse einer Morphologie der Weltgeschichte*. Vol. 1: *Gestalt und Wirklichkeit* (Munich, 1918). The first thirty-two editions are unaltered, and of these the third through thirty-second (1920-1922) have essentially the same pagination; I refer to these latter editions as the "orig. ed." Editions 33 through 47, published in 1923, are the revised edition; they all have the same pagination and are referred to as the "rev. ed." The English translation of the revised edition by C. F. Atkinson, *The Decline of the West*, Vol. 1: *Form and Actuality* (New York: Knopf, 1926) is referred to as the "Eng. ed."

66. Or so it appeared to Gerhard Hessenberg, *Vom Sinn der Zahlen. Akademische Antrittsrede, gehalten an der Universität Tübingen am 8. Dezember 1921* (Leipzig, 1922), 56 pp., on 31. So also had it appeared to Hessenberg's friend Leonard Nelson, *Spuk; Einweihung in das Geheimnis der Wahrsagerkunst Oswald Spenglers* . . . (Leipzig, 1921). And in 1923 Friedrich Meinecke noted (*op. cit.,* note 36) that "When the first volume of the *Decline of the West* appeared one frequently heard from the circle of professional scholars [Fachgelehrten] the judgment: 'What he says about my field is, indeed, complete nonsense. But all the rest is very ingenious [geistreich].' "

revelation of the new *Wissenschaft,* and thereupon rests a great part of its captivating effect." For Lukács, Spengler is the characteristic representative of the *Lebensphilosophie* of the postwar years.[67] And for us his book is all the more valuable as an index of the attitudes toward science and reason in the intellectual environment of the early Weimar period because, on the one hand, it gives a prominent place to physics and mathematics, and because, on the other hand, it was the one expression of those attitudes to which the Weimar physicists and mathematicians uniformly exposed themselves.

The Spenglerian account of world history is based on the proposition that the principal cultures are autonomous organisms, each wholly unique apart from a common life cycle. Every cultural manifestation—art, science, or whatever—is simply and solely an expression of the soul of that particular culture and as such is neither "valid" nor even comprehensible outside that culture, i.e., at any other time or place: "Each culture has its own new possibilities of self-expression which arise, ripen, decay, and never return. There is not *one* sculpture, *one* painting, *one* mathematics, *one* physics, but many, each in its deepest essence different from the other, each limited in duration and self-contained."[68]

After sketching his program in the "Introduction," Spengler aims in his first chapter, "The Meaning of Numbers," to establish this thesis of mutual nonintelligibility once and for all by proving (by iteration) that "There is not and cannot be number as such. There are several number-worlds because there are several cultures."[69] Likewise, "There is no mathematic, but only mathematics."[70] In the following chapters this radical relativism is extended to natural science, above all, to physics.

And in fact, in the historian's view there is only a *history of physics*. All its systems now appear to him as neither correct nor incorrect, but

---

67. Troeltsch, *op. cit.* (note 32), p. 1014. Lukács, *Zerstörung der Vernunft (op. cit.,* note 25), pp. 364-378. The same general view of Spengler's role is taken by Helmut Kuhn, "Das geistige Gesicht der Weimarer Zeit," *Zeitschr. f. Politik, 8* (1961), 1-10.
68. Orig. ed., p. 29; rev. ed., p. 29; Eng. ed., p. 21.
69. Orig. ed., p. 85; rev. ed., p. 81; Eng. ed., p. 59.
70. "Es gibt keine Mathematik, es gibt nur Mathematiken." (Orig. ed., p. 88; rev. ed., p. 83; Eng. ed., p. 60.)

as historically, psychologically conditioned by the character of the epoch, and representing that character more or less completely.[71]

... a firstrate scientist of the time of Archimedes would have declared himself, after a thorough study of our modern theoretical physics, quite unable to comprehend how anyone could assert such arbitrary, grotesque, and involved notions to be science, still less how they could be claimed as necessary consequences from actual facts.[72]

Writing off all criteria for the truth of a scientific theory as themselves culture-bound illusions, and dismissing with a wave of his hand the argument from the fact that "the machine works," as Boltzmann put it, Spengler maintained: "There simply are no conceptions other than anthropomorphic conceptions . . . so is it certainly with every physical theory, no matter how well founded it is supposed to be. Every such is itself a myth, and in all its features anthropomorphically preformed. There is no pure natural science, there is not even a natural science which could be designated as common to all men [als allgemein menschlich]."[73] And although Spengler was a bit less categorical in the second edition, altogether his extension of extreme cultural relativism to physics and mathematics was meant and was received as a direct challenge to the ideology of the exact scientists. At first they might refuse to hear anything of it, but, asked repeatedly for their reactions, within a year or two they all had to confront it.

Still more important for our present inquiry than Spengler's notorious theses regarding the nonobjectivity of the exact sciences are Spengler's specific interpretations of post-Renaissance physics, its content and its future. The content of Western physics and mathematics is, of course, but an expression of the soul of Western culture —of the "Faustian" culture, as Spengler calls it. And the essential, determinative characteristic of "Faustian" science is, we are no longer surprised to learn, "the *Kausalitätsprinzip*—the logical form of the Faustian world-feeling:"[74] "We see then that the causality-principle, in the form in which it is selfevidently necessary for us —the agreed basis of truth for our mathematics, physics and philos-

---

71. Orig. ed., p. 167; deleted in rev. ed.
72. Orig. ed., p. 530; rev. ed., p. 491; Eng. ed., p. 380.
73. Orig. ed., p. 533; deleted in rev. ed.
74. Orig. ed., p. 551; deleted in rev. ed.

ophy—is a Western and, more strictly speaking, Baroque phenomenon. . . ."[75] But although causality has reigned supreme in modern exact science, it is nonetheless—here comes the *Lebensphilosophie*—an artificial construction erected as a defense against the more fundamental, and fundamentally irrational, notion of destiny, *Schicksal*. This, indeed, is the "key" to the problem of world history:

> I mean the opposition of the *destiny-idea* and the *causality-principle,* an opposition which, in its deep world-shaping necessity, has never hitherto been recognized as such. . . . Destiny is the word for an indescribable inner certainty. One makes the essence of the causal clear by means of a physical or epistemological system, by means of numbers, by means of conceptual analyses. . . . The one requires us to dismember, the other to create, and therein lies the relation of destiny to life and causality to death.[76]

Thus we have the fundamental *lebensphilosophisch* theme, with which we are already all too familiar, inflated to cosmic proportions. Over and over again Spengler equates causality, conceptual analysis, and physics, and flays them across the stage of world history.

> For the principle of causality is late, unusual, and only for the energetic intellect of higher cultures a secure, somewhat artificial possession. Out of it speaks fear of the world. Into it the intellect banishes the demonical in the form of a continually valid necessity, which rigid [starr] and soul-destroying is spread over the physical world-picture. Causality is coextensive with the concept of law. There are only causal-laws.[77]

> The abstract savant, the natural scientist, the thinker in systems, whose entire mental existence is founded upon the principle of causality, is a "late" manifestation of the hatred of the powers of destiny, of the incomprehensible.[78]

> The words "time" and "destiny," for anyone who uses them instinctively, touch life itself in its deepest depths—life as a whole, which is not to be separated from lived experience. On the other hand, physics,

---

75. Orig. ed., p. 549; rev. ed., pp. 507-508; Eng. ed., p. 392.
76. Orig. ed., pp. 164-165; slightly, but insignificantly, altered in the rev. ed., pp. 154-155; Eng. ed., pp. 117-118.
77. Orig. ed., p. 165; first sentence modified and second and third sentences omitted in rev. ed., p. 155; Eng. ed., p. 118.
78. Orig. ed., pp. 168-169; rev. ed., p. 158; Eng. ed., p. 120.

reason, *must* separate them. The livingly-experienced in itself, detached from the living act of the observer and become an object, dead, inorganic, rigid [starr]—that is now Nature as mechanism, i.e., as something to be exhausted mathematically. . . . This is the eternal embarrassment of all physics as the expression of a soul. All physics is treatment of the problem of motion, in which lies the problem of life itself, not as if it could one day be solved, but even though it is unsolvable.[79]

Striking in this last passage is Spengler's elaboration of a notional complex advanced by Bergson and soon to be codified as existentialism in such works as Heidegger's *Being and Time*.[80] Time, Spengler assures us, is "something intensely personal," indeed, "we ourselves, insofar as we live, are time." It follows, therefore, that physics really "has nothing whatever to do with time," knows no direction of time, eliminates time in favor of a "web of cause and effect . . . of timeless duration."[81] Likewise to be noted for later reference is Spengler's favorite epithet for causality—*starr*, i.e., stiff, rigid; it is intended to evoke and reinforce an antithesis between causality and life, an association of causality with death (cf. *die Totenstarre*, rigor mortis).[82]

Spengler's indictment of physics=causality is all the weightier because he pretends to be a connoisseur of the physical sciences and modern technology, for whom "the depths and refinement of mathematical and physical theories are a joy," who "would sooner have the splendidly clear, highly intellectual forms of a fast steamer, of a steel structure, of a precision lathe, the subtlety and elegance of certain chemical and optical processes, than all the pickings and stealings of present day applied art, architecture and painting included."[83] *He* is not to be dismissed as an aesthete, a romantic; *he* is a hard-headed realist who, with a full appreciation of modern physics, "our ripest and strictest science,"[84] "the masterpiece of the Faustian spirit,"[85]

---

79. Orig. ed., pp. 542-543; rev. ed., pp. 501-502, where the sentence "This is the eternal embarrassment . . ." is deleted; Eng. ed., pp. 388-389. Cf. Harnack (*op. cit.*, note 55): "our intellect is the born mathematical physicist."
80. Published in Edmund Husserl's *Jahrbuch für Philosophie und phänomenologische Forschung*, 8 (1927), 1-438.
81. Orig. ed., pp. 170-172; rev. ed., pp. 158-160; Eng. ed., pp. 120-122.
82. Thus orig. ed., pp. 69, 165, 167, 574, and p. 156 of the rev. ed. where an additional *starr* is added: "die starre Weltmaske der Kausalität."
83. Orig. ed., p. 60; rev. ed., p. 60; Eng. ed., pp. 43-44.
84. Orig. ed., p. 215; rev. ed., p. 205; Eng. ed., p. 156.
85. Orig. ed., p. 608; deleted in rev. ed.

tells us what sort of cultural manifestation it really is, and what, according to the ineluctable cycle of cultural development, its fate must be.

> Before us there stands a last spiritual crisis that will involve all Europe and America. What its course will be late Hellenism tells us. The tyranny of reason—of which we are not conscious, for the present generation is its apex—is in every culture an epoch between man and old-man, and no more. Its most distinct expression is the cult of the exact sciences, of dialectics, of demonstration, of causality.[86]

> Now, the history of the higher cultures shows that "science" is a transitory spectacle, belonging only to the autumn and winter of their life-courses, and that . . . a few centuries suffice for the complete exhaustion of its possibilities. Classical science faded out between the battle of Cannae [216 B.C.] and that of Actium [31 B.C.] and made way for the world outlook of the "second religiousness." And from this it is possible to calculate in advance the end of western natural science.[87]

> It remains now to sketch the last stage of western science. From our standpoint of today the already declining route is clearly visible. . . . In this very century, I prophesy, in the age of scientific-critical Alexandrianism, resignation will overcome the will to victory of science. European science is advancing toward self-destruction through refinement of the intellect. . . . But from skepsis a path leads to the "second religiousness." . . . No one yet believes in the exhaustion of the spirit even though we already feel it acutely in all our limbs. But two hundred years of civilization and orgies of scientificness—then one is fed up. Not the individual, but the soul of the culture itself has had enough, and expresses this by choosing to put into the historical field of the day ever smaller, narrower, and more unfruitful researchers . . . in physics as in chemistry, in biology as in mathematics, the great masters are dead, and we are experiencing today the decrescendo of the stragglers, who arrange, collect, and conclude, like the Alexandrians of the Roman period.[88]

---

86. Orig. ed., pp. 607-608; rev. ed., p. 551; Eng. ed., p. 424.
87. Orig. ed., p. 532; rev. ed., p. 492, where the phrase "and made way for the world-outlook of the 'second religiousness' " was added and the final sentence softened to: "And from this it is possible to foresee a date at which our Western scientific thought shall have reached the limit of its evolution." (Eng. ed., p. 381.)
88. Orig. ed., pp. 607-609; rev. ed., pp. 551-553, where altered slightly but insignificantly; Eng. ed., pp. 424-425.

I would happily let this stand—and it was meant to stand—as the measure of Spengler's world-historical vision, for a more erroneous description and valuation of early twentieth-century physics could scarcely be devised. Yet as perverse and denigrating as this image is, it must still be recognized for what is was—an integral part of an analysis of Western culture, its present state, and its future prospects, which expressed and shaped the notions and inclinations of the educated middle classes in postwar Germany.

In another respect, however, Spengler's analysis of contemporary physics, confused and contradictory like all else in his treatise, shows a flash of prescience. For physics in his generation is not merely plodding forward in a beaten track, tying up loose ends, it is also, according to Spengler, disintegrating *and* metamorphizing, undergoing a transformation of the goals and principles of scientific explanation paralleling the *Zeitgeist,* the "second religiousness."

> Western European physics—let no one deceive himself—has reached the limit of its possibilities. . . . This is the origin of the sudden and annihilating doubt that has arisen about things that even yesterday were the unchallenged foundation of physical theory, about the meaning of the energy principle, the concepts of mass, space, absolute time, and causal natural laws generally [n.b.] . . . this doubt extends to the very possibility of a natural science. What deep and utterly unconscious skepsis lies, for example, in the rapidly increasing use of enumerative and statistical methods, which aim only at the probability of the results, and forgo in advance the absolute exactitude of the laws of nature, as one understood it in hopeful earlier generations.[89]

It is not, of course, to the quantum theory which Spengler refers here in the original edition. In speaking of the concepts of mass, space, time, energy he evidently has above all the theory of relativity in mind, and it is primarily to the atomistic foundation of the second law of thermodynamics that the remarks about statistics and probability refer. But the talk of doubt about the concept of causal natural laws points beyond these theories; I do not myself know just what Spengler, the seer, has in mind,[90] but his images and associa-

---

89. Orig. ed., pp. 596-597; rev. ed., pp. 541-542, where again modified insignificantly; Eng. ed., pp. 417-418.

90. Uncited, like virtually all of Spengler's sources, but almost certainly important was Wilhelm Wien's "Ziele und Methoden der theoretischen Physik.

tions are certainly suggestive of those which, as I show in Part III, soon after begin to appear in the writings of German theoretical physicists.

> Statistics belong, like chronology, to the domain of the organic, to fluctuating life, to destiny and incident and not to the world of exact laws and timeless eternal mechanics. . . .

> The more dynamics exhausts its inner possibilities as it nears the goal . . . the more insistently the organic necessity of destiny asserts itself side by side with the inorganic necessity of causality. . . . The course of this process is marked by the appearance of a whole series of daring hypotheses, all of like sort. . . . Above all, this is manifested in the bizarre hypotheses of atomic disintegration. . . . This destiny strikes only a few individuals in an aggregate of radioactive atoms, the neighbors being entirely unaffected.[91]

Here, then, is the fate and the salvation of physics—a reunification of thought and feeling, a self-discovery of physics as a fundamentally religious-anthropomorphic expression:

> The goal reached, the vast and ever more meaningless and threadbare fabric woven by natural science falls apart. It was, after all, nothing but the inner structure of the mind. . . . But what appears under the fabric is once again the earliest and deepest, the myth, immediate becoming, life itself. . . . Out of the religious soulfulness of the gothic there grew up the urban intellect, the alter ego of irreligious natural science, overshadowing the original world feeling. But today, in the sunset of the scientific epoch, in the stage of victorious skepsis, the clouds dissolve and the quiet landscape of the morning reappears in all distinctness . . . weary after its striving, the Western science returns to its spiritual home.[92]

Festrede . . . Universität zu Würzburg. Gehalten am 11. Mai 1914," printed in several places at that time and reprinted in Wien, *Aus der Welt der Wissenschaft. Vorträge und Aufsätze* (Leipzig, 1921), pp. 150-171. If this is the case, the doubts about causality which Spengler thinks he finds among the physicists, like so much of what Spengler reads into and out of his scientific sources, is simply a confusion. If, however, Spengler is drawing upon Max Planck's *Festrede* of 4 August 1914 (see note 158 below) he has evidently understood his author very well.

91. Orig. ed., pp. 603, 605-606; rev. ed., pp. 547 (where "timeless eternal mechanics" becomes "timeless causality"), 549-550; Eng. ed., pp. 421, 423.

92. Orig. ed., pp. 614-615; rev. ed., pp. 556-557; Eng. ed., pp. 427-428.

37

II. ADAPTATION OF IDEOLOGY TO THE INTELLECTUAL ENVIRONMENT

II.1. Introduction

Spengler epitomizes for us a set of attitudes, widely diffused among educated Germans, explicitly hostile to the ideology of the exact sciences and to particular concepts employed within them. In the remainder of this paper I explore some aspects of the response of the representatives of these sciences in German-speaking Central Europe—in the first instance of the response at the level of ideology; i.e., I explore the effect of this intellectual environment upon the professed justifications of scientific activity, upon the epistemological stance of the exact scientists, and upon their elan, their esprit, their confidence in the future of their discipline.

I do not, however, undertake to construct here a comprehensive typology of ideological responses to the Weimar intellectual milieu, but limit myself to illustrating a few of the more striking ideological adaptations. The resulting picture, emphasizing examples of accommodation, but not examples of resistance, is necessarily one-sided. Nonetheless, the imbalance is not as great as one might expect, both because the instances of a physicist or mathematician forcefully advancing ideals antithetic to those of his milieu are rare indeed, especially before the last years of the Weimar period,[93,94] and

---

93. Apart from the literature discussed below (Part III) in connection with the dispute over the law of causality, examples from the early Weimar period are almost exclusively in the form of, and limited to, rejections and rebuttals of Spengler's book and theses. Such are the tracts by Leonard Nelson and Gerhard Hessenberg cited in note 66; P. Riebesell, "Die Mathematik und die Naturwissenschaften in Spengler's 'Untergang des Abendlandes,'" *Naturwiss.*, 8 (1920), 507-509; and the preface to the second edition of Franz Exner's *Vorlesungen über die physikalischen Grundlagen der Naturwissenschaften* (Leipzig-Vienna, 1922), pp. vi-xiii. Other affirmations of the "scientific approach" are at the very least ambivalent: so, for example, the "Introduction" to Max Born's *Die Relativitätstheorie Einsteins* (1920), trans. H. L. Brose as *Einstein's Theory of Relativity* (London, 1924), pp. 1-6. In reprinting the "Einleitung" as the first selection in Born's *Physik im Wandel meiner Zeit*, 4th enlarged ed. (Braunschweig, 1966) the highly characteristic epigraph from Goethe was omitted: "The most perfect pleasure of the thinking human being is to have successfully researched that which is researchable, and to calmly revere that which is unresearchable."

94. Only two outspoken, unambivalent affirmations of an antithetic ideal before an academic audience are known to me; both are from the *late* Weimar period when, as Sontheimer has pointed out *(op. cit.,* note 26, pp. 43 ff.), there

WEIMAR CULTURE, CAUSALITY, AND QUANTUM THEORY, 1918-1927

because, as we shall see, it is often the case that the same scientist who in one context offers resistance to the antiscientific currents of his milieu, in another context can be found flirting with propositions intimately associated with those same currents. Moreover, this one side of the scientist's response is particularly interesting and instructive as largely contradicting the usual assumption of the intellectual autonomy of modern professionalized scientific disciplines. It thus provides essential motivation and support for my contention in Part III that the German physicists' predisposition toward acausal laws of nature likewise arose as a form of accommodation to their intellectual environment.

---

was a general stiffening of resistance among intellectuals to irrationalism: Richard von Mises, *Über das naturwissenschaftliche Weltbild der Gegenwart. Rede bei der Feier . . . der Berliner Universität . . . am 27. Juli 1930* (Berlin, 1930), 29 pp., reprinted in *Naturwiss., 18* (1930), 885-899; Konrad Knopp, "Der Einfluss der Naturwissenschaft auf das moderne Bildungsideal," in H. Gerber, ed., *Die Universität. Ihre Geschichte, Aufgabe und Bedeutung in der Gegenwart. Öffentliche Vorträge der Universität Tübingen, Wintersemester 1932-33* (Stuttgart, 1933), pp. 189-217. Mention might here also be made of Wilhelm Blaschke, *Leonardo und die Naturwissenschaften. Rede, gehalten am 10. November 1927, zum Antritt des Rektoramts an der Universität Hamburg,* Hamburger mathematische Einzelschriften 4 (Leipzig, 1928), 15 pp.; of Walther Kossel's *Die Einheit der Naturwissenschaft. Rede beim Antritt des Rektorats der . . . Universität [Kiel] am 5. März 1929* (Kiel, 1929), 22 pp.; and certainly also of the closing lines of David Hilbert's public address at the Naturforscherversammlung in Königsberg in September 1930, "Naturerkennen und Logik," *Naturwiss., 18* (1930), 959-963, reprinted in Hilbert's *Gesammelte Abhandlungen, 3* (Berlin, 1935; reprinted New York, 1965), 378-387: "he who is sensible of the truth of the liberal way of thinking and world view which shines forth from these words of Jacobi, he does not succumb to reactionary and unfruitful scepticism; he will not believe those who today with philosophic mien and superior tone prophesy the decline of western culture [den Kulturuntergang] and take pleasure in declaring 'ignorabimus.' For the mathematician there is no 'ignorabimus,' and equally for natural science, in my opinion, there is none whatsoever. . . . The true reason that Comte was not able to find an insoluble problem is, in my opinion, that there simply does not exist an insoluble problem. In place of this foolish 'ignorabimus' let our solution on the contrary be: We must know,/ We will know." Hermann Weyl, "Zu David Hilberts siebzigstem Geburtstag," *Naturwiss., 20* (22 January 1932), 57-58, reprinted in Weyl's *Ges. Abhl., 3,* 346-347, quoted this last sentence, adding that "our contemporaries are not glad to hear that sort of thing; they see in it shallow-minded rationalism or human presumption and with a torrent of confused words appeal to 'life itself' or the deeper 'existential truth' or the 'creatureliness' of man as justifying their repudiation of reason [Ratio]. And granted; a sentence here and there in Hilbert's address sounds suspiciously like the words with which Gottfried Keller ridiculed his natural scientist. . . . Nonetheless, one does Hilbert an injustice if one tosses his rationalism into the same pot with, say, that of a Haeckel."

39

As today with the "ecology" fad, so also in the Weimar period it was the biologist who could most easily adapt his ideology and values to those of his intellectual milieu. Life, that central symbol, was his own subject. Paraphrasing a spokesman for the discipline,[95] of all natural sciences biology merits an especially ample place in the school curriculum for it is least deserving of the reproach of aiming at "knowledge for power"; its mission is to counter the alienation from nature in our technical age; it provides the link between the *Naturwissenschaften* and the *Geisteswissenschaften* because it works in part with the concept of scientific law, but also with the techniques of understanding and imparting of meaning; it brings us to the edge of the irrational and teaches us to respect that which is beyond rational investigation. Nor are these arguments merely for public and governmental consumption; the eminent embryologist Hans Spemann, recommending to his eldest son a recent work on the philosophy of education, praises the author, Eduard Spranger, because he "above all treats the living and the spiritual with reverence and the love of the artist; he lets it live and doesn't pluck it apart into dead little pieces."[96] But although the materials are richer in some respects, and the evidence of the influence of the intellectual milieu more flagrant, biology is not my subject.

II.2. From Positivism to *Lebensphilosophie*

From his term as rector of the University of Würzburg in 1914 until his death in 1928 Wilhelm Wien was—alongside Max Planck—the most prominent spokesman for physics in Germany. His semipopular essays and addresses fall especially densely in the period 1918–1926, and so offer us a most striking example of the very quick change of tune which followed Germany's defeat in the First World War. On 1 May 1918, speaking in Dorpat on "Physics and the

---

95. Philipp Depdolla, "Biologie," *Unterrichtsblätter für Mathematik und Naturwissenschaft,* 37 (1931), 183-190.

96. Letter of 12 October 1928 quoted in H. Spemann, *Forschung und Leben,* ed. Friedrich Wilhelm Spemann (Stuttgart, 1943), p. 229. One cannot but feel some sympathy for Spemann who had to bear through the early twentieth century the cross of "Entwicklungs*mechanik*" which the late nineteenth century had fastened upon his field.

WEIMAR CULTURE, CAUSALITY, AND QUANTUM THEORY, 1918-1927

Theory of Knowledge," Wien emphasized the independence and autonomy of physics—from philosophy, especially; for aid, physics calls only upon mathematics, chemistry, and technology. Helmholtz, he stated flatly and unapologetically, was a "pure empiricist in particular opposition to the German idealist philosophy, above all to Hegel." More important, the bulk of the lecture was given over to a discussion of Ernst Mach's views, which Wien accepted in large measure with some conventionalist and some realist modifications.[97] In his Machianism and, more to the point, in his readiness to advocate it before a general academic audience, Wien was at this moment by no means unusual. That same summer, in his address as rector of the University of Göttingen, Hermann Th. Simon, applied physicist, delivered himself of a thoroughly positivist account of how we obtain knowledge of the world through adaptation of our ideas to our sensations, basing himself upon Mach and Avenarius.[98]

Compare, now, Willy Wien as he appears in September 1919 in an article commemorating the twenty-fifth anniversary of Helmholtz' death.[99] In an apologetic tone Wien explained that while it is true that Helmholtz became an empiricist "through opposition to the Hegelian school"—n.b., not to idealism—and that he was never able to give up this position, nonetheless Helmholtz was always concerned with the "totality of the sciences," and always had "ideal," not "material," goals in view, always aimed at the "dominion of the spirit." Wien thus implicitly concedes the series of equations made repeatedly by the antagonists of modern science—empiricism = positivism = narrow specialization = utilitarianism = materialism—attempting only to make an exception of Helmholtz, who, we are assured, were he alive today would look to "German idealism" to put us on our feet again. In this essay there is, of course, not the faintest whiff of positivism; no mention of Mach at all. By February 1920, Wien's own "idealism" had matured so far that in addressing

---

97. W. Wien, "Physik und Erkenntnistheorie. Vortrag, gehalten in Dorpat am 1. Mai 1918," *Aus der Welt der Wissenschaft. Vorträge und Aufsätze* (Leipzig, 1921), pp. 209-234.

98. H. Th. Simon, *Leben und Wissenschaft, Wissenschaft und Leben. Rektoratsrede zur Jahresfeier der Georgia Augusta [Universität Göttingen] am 26. Juni 1918* (Leipzig, 1918), 32 pp.

99. W. Wien, "Hermann von Helmholtz," reprinted from *Naturwiss.*, 7 (5 September 1919), 645-648, in *Aus der Welt der Wissenschaft*, pp. 86-94.

41

a public session of the Prussian Academy of Sciences on "The Connections Between Physics and Other Disciplines" he represented the "postulate of the cognizability of nature" as "in the final analysis not so very far from the fundamental idea of the Hegelian philosophy of identity,"[100] and in November 1925 he could be found shedding tears before his assembled university over the abandonment, some twenty years earlier, of the requirement that philosophy, the "unifying discipline," be one of the subordinate subjects in every Ph.D. examination at a German university.[101]

Wien may well be the only German physicist to have expressed regret for that requirement, but he is certainly far from unusual in announcing a recent reversal of an earlier, deplorable trend toward fragmentation and isolation of physics from other disciplines,[102] or in thoroughly suppressing his earlier positivism. I know of only one instance during the entire Weimar period of a German physicist venturing, in a general academic address, to mention Mach's name with clear approbation and to associate himself with Mach's epistemological doctrines. Nor was it mere coincidence that in taking this courageous stand at the *end* of the Weimar period Richard von Mises refused to associate himself with the demand for synthesis, "counting it"—as did Mach—"the highest philosophy to tolerate an incomplete world view."[103]

The renunciation of positivism was intimately connected—for

---

100. W. Wien, "Über die Beziehungen der Physik zu andern Wissenschaften. Öffentlicher Vortrag, gehalten in der Preussischen Akademie der Wissenschaften in Berlin am 27. Februar 1920," *Aus der Welt der Wiss.*, pp. 16-40, on 28. And again, eighteen months later, we can find Walther Nernst pursuing the question in his *Rektorats-Antrittsrede* "ob nicht, wie fast stets bei derartigen starken geistigen Strömungen, auch in der Identitätsphilosophie ein gesunder Kern steckt." (*Naturwiss.*, 10 [1922], 489-495, on 490.)

101. W. Wien, *Universalität und Einzelforschung. Rektorats-Antrittsrede, gehalten am 28. November 1925*, Münchener Universitätsreden, Heft 5 (Munich, 1926), 19 pp., on 14-15. All the German universities except Berlin had abandoned this requirement at the beginning of the century.

102. The rebutting of the charge of fragmentation, of disintegration of the world picture by specialized research, is surely the single most common theme of general academic lectures by physicists. So Hermann Weyl, *op. cit.* (note 58): "To be sure one hears over and over complaints about the extent of specialization in the sciences. I believe, however, that on the whole in recent decades the situation has gotten better rather than worse." Likewise, Walther Kossel, *Die Einheit der Naturwissenschaft* (*op. cit.*, note 94).

103. R. von Mises, *Über das naturwissenschaftliche Weltbild* (*op. cit.*, note 94), p. 27. W. Ostwald, *Lebenslinien* (*op. cit.*, note 19), 2, 312.

WEIMAR CULTURE, CAUSALITY, AND QUANTUM THEORY, 1918-1927

Wien as for all his fellow physicists—with a renunciation of "knowledge for power," of the harnessing of nature, of utility as the object, motive, or justification for scientific research. In June 1914, reviewing as rector of the University of Würzburg the development of the German universities in the preceding century, Wien had given only one measure of achievement for the fields of physics and chemistry, viz. that they "have created the solid foundations upon which the pillars of our industry are erected," and he had reproached the universities for failing to incorporate the *Technische Hochschulen* as technical faculties.[104] In May 1918 Wien's lecture on "Physics and Epistemology," rejecting any liaison with philosophy, was followed a few days later by one on "Physics and Technology," whose basic, endlessly exemplified theme was the "support and stimulation" which *these* two fields have received from one another and should continue to receive in the future.[105] In the following years Wien did indeed play a key role in the creation and operation of the Helmholtz-Gesellschaft, which for the first time channeled substantial financial support from German industry into the physical institutes of the German universities. Yet in his academic addresses of the Weimar period he never let slip even a single word about this compromising connection.[106]

The shift in that element of the ideology defining "the significance of physical research" was announced by Wien in his address before the Prussian Academy of Sciences in February 1920. This question, we are told, can be judged from two very different points of view. The first sees physical research as aiming at "human domination over the recalcitrant forces of nature." Wien implicitly recognizes

104. W. Wien, "Die neuere Entwicklung unserer Universitäten und ihre Stellung im deutschen Geistesleben. Rede für den Festakt in der neuen Universität am 29. Juni 1914," *Aus der Welt der Wissenschaft*, pp. 1-15, on 14.
105. W. Wien, "Physik und Technik, Vortrag, gehalten in Reval am 6. Mai 1918," *Aus der Welt der Wiss.*, pp. 235-263.
106. W. Wien, "Die Helmholtz-Gesellschaft und ihre Bedeutung für die deutsche Physik . . .," *Die Helmholtz-Gesellschaft zur Förderung der physikalisch-technischen Forschung in sieben Jahren ihres Wirkens* [privately printed], 1928), pp. 7-11. This renunciation of utility in academia was not wholly uncompensated. The slogan "knowledge for power" was exchanged for "knowledge as a substitute for power," "Wissenschaft als Macht-Ersatz." See B. Schröder, *op. cit.* (note 52). It must be said, moreover, that the academic chemists, unlike the physicists, do not appear to have become at all bashful about discussing technical applications and justifying their science through them.

43

that his audience is very hostile to any such conception, and after pointing out that it need not necessarily proceed from a "purely materialistic mode of thought," passes on to the second and proper point of view, one "free of all striving toward a goal [Zielstrebigen]." Physical research is, in truth, nothing but the expression of "the pure human instinct for inquiry"; "it arises solely from an inner need of the human spirit"—note the implicit *Lebensphilosophie*—"a craving," Wien explains, "to grasp the causality of events."[107] To this new line Wien held fast in all further academic addresses. In November 1925, as rector of the University of Munich, his discussion of "Universality and Specialized Research" touches on technological applications of physical science in only one paragraph, and there only in order to stress that scientific ideas which appear at first to refer to but a narrow field may turn out to have enormous practical consequences. He is very careful, however, not to appear to be praising science on that account. Throughout the lecture Wien totally abstains from any attempt to justify science by utility. On the contrary, the goal of science is *culture*: "The significance of a scientific achievement can ultimately only be measured by the effect which it has upon the intellectual life"; "the results of research are worthless if they are not taken up into the culture."[108]

As Wien's reference to causality suggests, he had very strong views on which results of research ought to be taken up into the culture and which ought not. And although, as will appear in Part III, many of Wien's colleagues took a different view of the concept of causality, they were in complete agreement with him on the motive, goal, and justification for physical research. The "common driving force" of

107. W. Wien, *op. cit.* (note 100), p. 28. Cf. the mineralogist Gottlob Linck, *Über Wesen und Wert der Universität. Rede, gehalten . . . am 19. Juni 1920 . . . zu Jena vom Rektor der Universität* (Jena, 1920), p. 4: "Wie der Hirsch schreiet nach frischem Wasser, so schreiet unsere Seele nach Erkenntnis."
108. W. Wien, *op. cit.* (note 101), pp. 13, 7, and 19, respectively. Cf. the theoretical physicist Erwin Madelung, *Die Bedeutung der Wissenschaft im Rahmen unserer Kultur. Rede anlässlich der Übernahme des Rektorates* [1931], Frankfurter Universitätsreden 39 (Frankfurt a. M., 1932), 16 pp., who explains (pp. 2-4) that by "Kultur" he understands "everything which broadens and enriches our inner life, which is, I believe, in sufficient accord with the customary usage. . . . We want, thus, to put completely aside here the consideration of practical utilizability and of the dead piling up of information, however important it may be. We want only to ask ourselves what spiritual needs [geistige Bedürfnisse] we have and to what extent these are satisfied by science." Naturally, "our needs arise out of the dark spring of our living existence."

WEIMAR CULTURE, CAUSALITY, AND QUANTUM THEORY, 1918–1927

all the research in the university—as the members of the University of Greifswald learned from their new rector, the *applied physicist* Friedrich Krüger—is "the innate human urge for ever new knowledge. . . . All external driving forces originating in considerations of utility and necessity for our physical existence and its advancement are completely without influence." And, after this introduction, Krüger devoted his *Rektoratsrede* to the question of the heat death of the universe![109]

This ideology tinged with *Lebensphilosophie*—that the value of physics lies in and derives from the fact that it is the expression of an unanalyzable and irreducible human drive (for cognition of nature, in this case)—attained by 1929 the status of orthodox physical doctrine, occupying the opening pages of the volume on "General Foundations of Physics" in the new edition of the *Handbuch der Physik*. Firmly rejecting technological application as a measure of the value of physical knowledge, and, less firmly, rejecting the foundation of a *Weltanschauung* as the goal of physical research, Hans Reichenbach explained that "the most important thing that one can say about it [i.e., doing physics] is that it is a need, that it grows up out of the human being just like the wish to live, or to play, or to form a community with others."[110]

At first sight it seems most surprising that even a Reichenbach, i.e., even a representative of "rigorous" logical empiricism, should have taken over the standards of value and the ideology for the physical enterprise from an intellectual milieu specifically hostile to his philosophical position. This circumstance becomes less surprising

109. F. Krüger, *Materie und Energie im Welt-Geschehen. Rektoratsrede,* Greifswalder Universitätsreden 15 (Greifswald, 1928), 29 pp. "Alle äusseren Triebkräfte des Nützlichen und für die äussere Existenz und ihre Förderung Notwendigen kommen nicht in Frage" (p. 3). Only at the end of his address, apropos of artificial disintegration of atoms, did Krüger make any reference at all to the application of scientific knowledge, namely to "extracting the energy of the atom as one of the greatest technical problems worthy of the most strenuous efforts . . . consequently we see at present a mighty contest in the laboratories of the civilized nations . . . to find the methods for extracting this energy" (pp. 28-29).
110. H. Reichenbach, "Ziele und Wege der physikalischen Erkenntnis," *Handbuch der Physik,* Band 4: *Allgemeine Grundlagen der Physik,* ed. H. Thirring (Berlin, 1929), pp. 1-80, on 1-2. As for the new discipline of "technical physics," Reichenbach dismisses it (p. 11) with the remark that even though it has now attained a place in German universities "sie ist doch ihrem Wesen nach eine Technik und keine Wissenschaft," for a *Wissenschaft* aims solely at "Erkenntnis."

45

if one recalls that the positivist tradition itself contained a substantial element of *Lebensphilosophie* and that, moreover, there was a solid Machian precedent for regarding natural science as the outgrowth of a basic human drive. How horrified Hedwig Born would have been had she known that her favorite Einsteinian apothegm—"I feel such solidarity with everything living, that it is all one to me where the individual begins and ends," which she heard at his bedside as he lay critically ill in 1917/18 and which she found so beautiful that she quoted it over and over again to Einstein himself[111]—was also in fact a most genuinely Machian-positivist sentiment.

Yet with such sentiments we have by no means reached the limits of the community of values and attitudes between the physical scientist and his *lebensphilosophisch* intellectual milieu. The "unbroken life force" of mathematics, the "concrete nearness to life" of applied mathematics, the *"lebensvollste* interlacing" of mathematics with natural science and technology, the training of students by "living interaction to spontaneous scientific work," the maintaining of mathematics "in contact with the concrete stuff of life," and avoiding the danger that it will "rigidify as a pure distant-from-life form"—these examples of "life" rhetoric are all drawn from a single address, a bare half dozen pages, by Richard Courant.[112] To be sure Courant's rhetoric of "life," "organism," "spontaneity," "ecstacy," "phantasy," "instinct," "intuition" is considerably more exuberant than that of most of his colleagues. It is indicative, nonetheless, of a substantial participation by the physicist and mathematician in the values of his general cultural milieu.

Another striking example of this participation is the assent of the exact scientists to the proposition that feeling and intellect are antithetic, incapable of coexisting in an exalted state in a single individual, and that feeling is the higher quality.[113] Einstein

111. Born-Einstein, *Briefwechsel (op. cit.,* note 14), p. 113.
112. R. Courant, "Über die allgemeine Bedeutung des mathematischen Denkens," *Naturwiss., 16* (1928), 89-94. "Vortrag, Tagung Deutscher Philologen und Schulmänner, Göttingen, September 1927." Under the cover of this rhetoric, however, Courant basically stands firm upon the traditional intellectualist conception and cognitive claims of mathematics.
113. The "classical" exponent of this thesis was Ludwig Klages, who is still counted a great seer in Germany today. Indeed the press of the Deutsche Physikalische Gesellschaft recently published a collection of essays—*Physik, Gleichung*

WEIMAR CULTURE, CAUSALITY, AND QUANTUM THEORY, 1918-1927

expressed this very well in a letter to H. A. Lorentz: "In your case Nature had the rare impulse to unite with a sharp mind warm feeling. If only this were often so. . . ."[114] Given this ranking, it is not surprising that the physicists and mathematicians responded so ineffectually to the charge of "intellectualism" leveled by Becker *et al.* Nor is it surprising that they themselves sought the motive, goal, and justification for their own scientific activity in feeling and instinct, in the wish for, not in the fact of, cognition. As Wilhelm Ostwald, now in his old age, so succinctly put it, "the reason, and science along with it, is only a servant of the feeling."[115]

Here, moreover, we must add that most curious simile in Reichenbach's explanation of why one does physics—it is "like the wish to form a community with others," the characteristic emotional need of the Weimar period. In the famous "Introduction" to his book on *Einstein's Theory of Relativity,* written in 1920, Max Born gave expression to this same longing for "community," for participation in some whole which transcends the individual, maintaining that "all religions, philosophies, sciences are procedures designed for the purpose of expanding the 'I' to the 'we'." What distinguishes the natural scientist is his resolve—"often shuddering"—to achieve this goal by sacrificing the absolute for the sake of objectivity. And it is in *this* way that, for the physicist, "the pain of spiritual loneliness disappears, the bridge to kindred spirits is formed."[116] And with

---

und Gleichnis. Vorträge und Aufsätze über Physik (Mosbach i. B., 1967)—by Eberhard Buchwald, a mediocre theorist who in the 1920's was struggling to make a career in applied physics. There one can read, pp. 68-80, Buchwald's contribution to *Ludwig Klages, Erforscher und Künder des Lebens* (Linz, 1947), and learn what Klages, "enthroned" alongside Heraclitus, Goethe, and Nietzsche, can mean to a "dankerfüllten Fachphysiker."

114. Einstein to H. A. Lorentz, 3 April 1917. Microfilm of Lorentz papers in the Algemeen Rijksarchief, The Hague, deposited in the Archive for History of Quantum Physics (see note 13); on microfilm nr. 6. For the higher status of feeling: Einstein to Hedwig Born, 1 Sept. 1919, *Briefwechsel* (op. cit., note 14), pp. 32-33, and Einstein to Jacob Laub [1909?], as quoted by Carl Seelig, *Albert Einstein: Eine dokumentarische Biographie* (Zurich, 1954), p. 117.

115. W. Ostwald, "Von der Formel zur Form," *B.Z. am Mittag,* 3 Nov. 1926, as quoted in Grete Ostwald, *Wilhelm Ostwald, mein Vater* (Stuttgart, 1953), pp. 229-230.

116. M. Born, "Introduction" to *Einstein's Theory of Relativity* (op. cit., note 93). Born's inclination toward what one might call "futurist *Lebensphilosophie*"—such as one finds expressed by Walther Rathenau, for example—is revealed in his recommendation of Richard N. Coudenhove-Kalergi's *Apologie*

this concession that science is primarily a means for satisfying certain emotional needs, we have arrived again at the very axiom of the "revolution in science."

II.3. Capitulation to Spenglerism

A remarkable readiness of the Weimar physicists to adapt their ideology to the values of their milieu has, I think, been shown in the preceding section. Yet the adaptations displayed there, while they involve redefinitions of the motivation and justification for doing physics, do not explicitly alter the fundamental conceptions of scientific method, renounce the cognitive claims of the exact sciences, or resign confidence in their future development—as demanded and predicted by Spengler. But if we shift our focus slightly, from the physicists to the theoretical physicists and mathematicians, there appears a distinct tendency to carry the ideological adaptation into these vital regions and to adopt specific propositions which at the time were attributed to, or especially closely associated with, the *Untergang des Abendlandes*.[117]

A most interesting and most suggestive case of a clearly discernible Spenglerian influence is to be found in Richard von Mises' inaugural

---

der Technik (Leipzig, 1922), "dessen Inhalt mir sehr eingeleuchtet hat." (Born to Einstein, 7 April 1923, *Briefwechsel* [*op. cit.*, note 14], p. 110.) On science as primarily a means to satisfy emotional needs, compare E. Madelung (*op. cit.*, note 108), p. 14, replying to those who place a higher value upon the *Geisteswissenschaften* than upon the *Naturwissenschaften*: "As for the value judgement, I want to emphasize once again that it obviously doesn't matter so very much what one does and what methods one employs, but far more how one views one's own activity, whether one feels richer and freer through it, and more secure vis-à-vis life's changing forms."

117. The shift in focus away from the experimental physicists is in good part simply a consequence of their relative inarticulateness. But not entirely. If in general they did not go so far in their ideological accommodation as the theoretical physicists and mathematicians, that was also in part because they had less to recant. In the Weimar period the theoretical physicists seem to have drawn closer to the mathematicians with whom they were lumped for opprobrium in the public's mind, and with whom their relations could in some respects be less constrained than those which they maintained with the experimental physicists. Some twenty notable theoretical physicists held membership in the Deutsche Mathematiker-Vereinigung in 1924, half of whom had joined in or after 1918 (*Jahresbericht*, 34 [1925], Part 2, pp. 49-92), and that despite the fact that the annual meetings of the Deutsche Physikalische Gesellschaft and the Deutsche Mathematiker-Vereinigung were ordinarily held at the same time and place, thus largely obviating the need for formal membership in both organizations.

WEIMAR CULTURE, CAUSALITY, AND QUANTUM THEORY, 1918–1927

(and farewell) lecture as Professor of Mechanics at the *Technische Hochschule* Dresden—delivered in February 1920, after von Mises had accepted a chair of applied mathematics at the University of Berlin. Considerable stir was created by von Mises' contention—or rather concession to the intellectual milieu—that the "age of technology," to which the *Technische Hochschulen* owed their rise, was on its way out. His advice to these institutions was that they do their best to get onto the wave of the future by entering the field which was destined to replace technology in the "culture consciousness," namely speculative natural science, particularly relativity and atomic physics. In these subjects, he asserted, we have had for the past two decades a period like that of Copernicus, Galileo, and Kepler. "It is not a question of new facts of any sort, nor of new theoretical propositions, nor even of new methods of research, but, if I may say it—taking this word in its philosophical sense—of new intuitions [Anschauungen] of the world." Atomic physics has taken up again "the question of the old alchemists"; "numerical harmonies, even numerical mysteries play a role, reminding one no less of the ideas of the pythagoreans than of some of the cabbalists."[118]

Astonishing as these remarks are from the mouth of a convinced positivist, impossible as it may be to find their like two years earlier, much as they remind us of Spengler's prediction that a new mysticism was the fate and salvation of natural science, still *prima facie* evidence of a connection with the *Decline of the West* is wanting. In fact, the immediate precedent and probable inspiration for von Mises' reference to numerical harmonies and mysteries is an article on "A Number Mystery in the Theory of the Zeeman Effect" by Arnold Sommerfeld which appeared in *Die Naturwissenschaften* a few weeks earlier, as well as the preface to Sommerfeld's *Atomic Structure and Spectral Lines*, which had appeared late in 1919.[119]

118. R. v. Mises, *Naturwissenschaft und Technik der Gegenwart. Eine akademische Rede mit Zusätzen*, Abhandlungen und Vorträge aus dem Gebiete der Mathematik, Naturwissenschaft und Technik, Heft 8 (Leipzig, 1922), 32 pp., on 2, 5, 16, respectively. The initial publication without the *Zusätze* had been in the *Zeitschr. des Vereins deutscher Ingenieure, 64* (1920), 687-690, 717-719.

119. A. Sommerfeld, "Ein Zahlenmysterium in der Theorie des Zeemaneffektes," *Naturwiss., 8* (23 Jan. 1920), 61-64, on which see my "Alfred Landé and the Anomalous Zeeman Effect, 1919-1921," *Historical Studies in the Physical Sciences, 2* (1970), 153-261; Sommerfeld, *Atombau und Spektrallinien*, 1st ed. (Braunschweig, 1919), p. viii. Sommerfeld was sufficiently pleased with this *Sphärenmusik* passage to reprint it in the 2nd, 3rd, and 4th editions, 1920-1924.

There Sommerfeld had spoken of "the mysterious organ upon which nature plays the spectral music" of the atomic spheres. In the future Sommerfeld was to go considerably further in this direction. A ceremonial address at a public session of the Bavarian Academy of Sciences in July 1925 offered Sommerfeld the opportunity to stress that "hand in hand with this turn toward the arithmetical goes a certain inclination of modern physics toward pythagorean number mysticism. Precisely the most successful researchers in the field of theoretical spectral analysis—Balmer, Rydberg, Ritz—were pronounced number mystics. . . . If only Kepler could have experienced today's quantum theory! He would have seen the most daring dreams of his youth realized. . . ."[120]

It is true that, having indulged himself in such rhetoric at some length, Sommerfeld concluded with the hope that he will "not be suspected of speaking in favor of mysticism in the ordinary sense, as it comes out in the astrological, metaphysical, and spiritualistic impulses of our time." Nothing is farther from his intent, he insisted; he was not speaking of human things, but only of laws of nature, and he meant rather to be attacking "conventionalism," "positivism," and "Machian philosophy."[121] Yet it is perfectly clear that, despite the disclaimers, Sommerfeld was indeed catering to the antirational as well as the antipositivist inclinations of his audience, that he was trying to project an image of physics that would find favor with his audi-

---

120. A. Sommerfeld, *Die Bedeutung der Röntgenstrahlen für die heutige Physik. Festrede, gehalten in der öffentlichen Sitzung der B. Akademie der Wissenschaften . . . am 15. Juli 1925* (Munich, 1925), 17 pp., reprinted in Sommerfeld's *Ges. Schr.* (*op. cit.*, note 21), *4*, 564-579, on 573-574.

121. *Ibid.*, pp. 575-576. Cf. Georg Hamel, President of the Mathematischer Reichsverband, whom we have also previously seen throwing himself against the wave of irrationalism and anti-intellectualism, summarizing his *Rektoratsrede* delivered 30 June 1928 at the Technische Hochschule Berlin: "Mathematics customarily appears as the rational science per se; to the layman the mathematician is a calculator. In opposition thereto I maintain the thesis that mathematics is an art, and that, in the last analysis, it is conditioned not logically but transcendentally. . . . The mathematician is a poet. Like the dramatist he creates a form. . . . The problem of the irrational numbers leads mathematics into metaphysics. . . . The genuine foundation for all of mathematics I see in Kant's pure *Anschauung*. . . . In conclusion I take issue with the misconception that my remarks represent a repudiation of intellectualism. Although the irrational basis of mathematics has been clearly recognized, that does not alter in the least the mathematician's obligation to proceed purely logically within his science, with the greatest care, with precise modes of inference." (Hamel, "Ueber die philosophische Stellung der Mathematik," *Forschungen und Fortschritte, 4* [1928], 267.)

WEIMAR CULTURE, CAUSALITY, AND QUANTUM THEORY, 1918-1927

tors and raise the prestige of the discipline in their eyes. And one cannot help but be struck at the close correspondence between this image and that which Spengler sketched in the final pages of the *Decline of the West*.

But let us return to von Mises—upon whom a direct influence of Spengler can in fact be established by September 1921. When at this time von Mises added an appendix to the republication of his lecture of February 1920, his tone had changed entirely, his optimism and enthusiasm had disappeared. Von Mises had largely, and *explicitly*, adopted Spengler's perspective and assumptions. It is "at least highly probable that the towering structure, under construction for the past five centuries, of a Western culture oriented entirely toward cognition and performance will collapse in the following centuries. From this standpoint one must count the theory of relativity and modern atomic physics as among the last building stones destined to crown the structure." Accepting Spengler's doctrine that cultures, as "living organisms," are fundamentally incommensurable, von Mises declared it "entirely out of the question" that the culture which succeeds ours will "continue the exact sciences in our sense." Nor can such views be dismissed as pessimism—"as if the man, who conscious of his old age and the inevitability of his death, is a pessimist because he faces the fact and acts accordingly."[122] What, one wonders, would it mean for a physicist or mathematician to "act accordingly"? Could it possibly mean that he strives to alter the content of his science and the very nature of the scientific enterprise, in order to fulfill Spengler's prophesies?

What is perhaps most striking and appalling about the von Mises of September 1921 is the failure of nerve, the complete loss—just as Spengler predicted—of the esprit, the self-confidence which we expect from the mathematical physicist. And in this von Mises was by no means unique. One can find, on the contrary, many examples—most

---

122. R. v. Mises, *op. cit.* (note 118), p. 32. Cf. note 103 and text thereto. In general the shifts in von Mises' mental posture correspond quite closely to those which Georg Steinhausen, *Deutsche Geistes- und Kulturgeschichte von 1870 bis zur Gegenwart* (Halle a. d. Saale, 1931), p. 4, found to be typical for the Weimar period. Namely, initially, despite the political and military collapse, a certain euphoria over the wholly new epoch thus begun; this mood was, however, very soon displaced by disillusionment and an "Untergangsstimmung," which, again, had entirely disappeared by the late Weimar period when there was a strong tendency "to fall back into the old mental grooves."

often in addresses before general academic audiences—of theoretical physicists and applied mathematicians denigrating the capacity of their discipline to attain true, or even valuable, knowledge. The earliest such is, perhaps, the passage which in the spring of 1918 Hermann Weyl placed as a conclusion to the first edition of *Space-Time-Matter*.[123] Theoretical physics is, Weyl maintained, entirely analogous to formal logic. "True" propositions must conform to logic, but logic is incapable of judging the "truth" of the propositions it manipulates; so also reality conforms to the laws of physics, but physics is incapable of informing us about the reality which its laws govern. Is, perhaps, this reality—these "darker depths" than the mathematician can grasp with his methods[124,125]—Spengler's "immediate becoming, life itself"? Indeed, as we shall see in Part III, it is.

A still more striking example of these same "annihilating doubts" is offered us by Gustav Doetsch in his inaugural lecture as Privatdozent for applied mathematics at the University of Halle, 27 January 1922. There, in conclusion, pointing back to his exposition of the "Meaning of Applied Mathematics," Doetsch burst forth:

> Such *rationalistic dogmatism* is the characteristic expression of *that* intellectual epoch which is at this moment perishing [im Untergehen]. It is the spirit, one could say, of the *age of natural science,* which, essentially, coincided with the 19th century, and which in our days is sinking with violent convulsions into its grave in order to make room

---

123. H. Weyl, *Raum-Zeit-Materie. Vorlesungen über allgemeine Relativitätstheorie,* 1st ed. (Berlin, 1918), pp. 226-227; again, somewhat more fully, in the 3rd ed. (Berlin, 1919), pp. 262-263. In the fall of 1920, when preparing the fourth edition (Berlin, 1921), Weyl struck this conclusion, replacing it by an attack on causality.

124. *Ibid.,* 1st ed., p. 9; 3rd ed., p. 9; 4th ed., p. 9.

125. Slightly later, and perhaps owing something to Weyl, is Paul Gruner, *Die Neuorientierung der Physik. Rektoratsrede, gehalten an . . . der Universität Bern den 26. November 1921* (Bern, 1922), 23 pp. In conclusion, theoretical physicist Gruner conceded that although "es mag dem Naturforscher schwer fallen," his field cannot satisfy "das Sehnen nach absoluter Wahrheit" which fills contemporary academic youth. "Dem blossen Denken und Beobachten der Naturwissenschaften ist dieses intuitive Schauen versagt." Natural science can tell us nothing of the meaning of the world and our life; the disaster we are experiencing today is due to this intellectual advance without ethical-religious foundation. These are, of course, precisely the charges which theoretical physicist von Laue, taking Rudolf Steiner as his ostensible target, was trying to rebut a few months later. (See note 15 and text thereto.)

for a new spirit, a new life-feeling . . . this epoch, at whose beginning we unquestionably find ourselves today, is fed up with this rationalistic attitude. Whether we direct our attention toward expressionism in art, or to more recent philosophical tendencies, which in many ways have not yet emerged entirely distinctly, or to any other area of life and thought whatsoever, we find everywhere an ever stronger *aversion* for *that* spirit which believed that it had to express, and that it could express, everything whatsoever in dry words, in one formula—an aversion deriving from the unconscious feeling: *this* path has never and will never lead us to the *essence* of things, we must try to get "nearer" to the object, to transfer ourselves inside of it itself. Whether the new path leads to the goal, or whether it can only get us closer, may be left undecided. Here my intent was only to point out in the *domain of natural science itself,* which has served as a model for so many others, that the *mathematical* treatment of the material of experience does not begin to impart information about the essence of the world, that is, to yield true cognition.[126]

And still Doetsch was not quite finished. After this tirade against his discipline he quoted Hegel's dictum that mathematics is "kein Begreifen," and clinched his case by observing that if Hegel "should not be regarded as the proper person to bring applied mathematics to a correct estimation of itself, then I refer to the words of our most brilliant contemporary mathematician, Hermann Weyl, in whose famous work, *Raum-Zeit-Materie.* . . ." It was, of course, rather daring of Doetsch to speak his mind *so* freely—although his general academic audience must have been very happy indeed to hear their views confirmed by a mathematician. It was, however, foolhardy of Privatdozent Doetsch to publish such sentiments in the journal of the German Society of Mathematicians, where it was read by, and had necessarily to offend, senior and influential colleagues. Indeed it may well have cost him a chair.[127]

126. G. Doetsch, "Der Sinn der angewandten Mathematik," *Jahresbericht der Deutschen Mathematiker-Vereinigung, 31* (1922), 222-233, on 231-232. "Antrittsvorlesung gelegentlich der Umhabilitierung von der T. H. Hannover an die Universität Halle a/S. am 27. Januar 1922."
127. *Ibid.,* p. 233. Otto Blumenthal to Theodor von Karman, 8 July 1923: "Das Ergebnis Ihrer Anfragen betreffs papabler Mathematiker finde ich im höchsten Masse betrübend. Ich habe von Doetsch nicht die Ansicht, dass er für uns brauchbar ist. C. Müller hat uns seiner Zeit ein sehr vernünftiges und abfälliges Gutachten über ihn gegeben. Auch habe ich von einem Vortrag, den er in Halle über das Wesen der angewandten Mathematik (D.M.V.) gehalten

And finally, to place in evidence an example from the latter part of our period, consider the picture of "The Peculiar Nature of the Mathematician's Mind" which Max Dehn, Professor of Pure and Applied Mathematics, held up before his assembled university in January 1928. Painted in pure Spenglerian style, the characteristic mental tone of the contemporary [German] mathematician is skepsis, mistrust of reason, self-inculpation, pessimism, and resignation:

> This somewhat sceptical attitude of many a contemporary mathematician is reinforced by what is going on in the neighboring field of physics. Here it appears to be the case that the physical phenomena no longer admit of being construed consistently [widerspruchslos] in a mathematical four-dimensional space-time manifold. Up to now we were able to provide physics with sufficiently freely built scaffolding for its ever bolder constructions. Now, however, in certain reflections arising from important investigations of the finest structure of matter, physics is perhaps in the process of cutting itself loose from mathematics. [Dehn is, *inter alia*, two years behind the times—see sections III.4–6.]
>
> All this has impelled many of us to be somewhat sceptical in more general questions as well. The fundamental conviction of every philosopher that the world can be comprehended consistently [widerspruchslos] by the human reason is, for the mathematician, no longer certain. ... This attitude is, to be sure, not entirely original; it is reminiscent of the thought of the later Eleatics at the time of the foundation crisis in ancient Greece.
>
> Out of this skepsis there develops a certain resignation, a kind of mistrust for the power of the human mind in general.
>
> ... because of the boundedness of human intellectual power a limit is set to abstraction, to the departure from the intuition. Beyond this limit no further development is possible. But contemporary mathe-

---

hat, einen ungünstigen Eindruck erhalten, müsste mir allerdings den Artikel noch mal durchsehen" (Karman Papers, California Institute of Technology Archives, Box 13). By 1924 Doetsch had become reconciled to his discipline, and had even persuaded himself that although mathematics and natural science—"immediately casting everything into *rational* schemata"—"can never uncover the real meaning of the world and its interconnections, ... nonetheless the mind [Geist] at least comes ever nearer to that which it would really like to grasp." ("Sinn der reinen Mathematik und ihrer Anwendungen," *Kant-Studien, 29* (1924), 439-459, on 458-459; "Vorträge ... Januar 1924.") And later that same year Doetsch received his first chair.

WEIMAR CULTURE, CAUSALITY, AND QUANTUM THEORY, 1918–1927

matics is by no means dead, and naturally, even in topology, for example, a man can and hopefully will come who simplifies the processes so much . . . that a new development sets in. . . . Such achievements will, however, scarcely arise in the course of an organized routine. But if the mathematician is already complaining for this very reason—that in consequence of the modern development finally even the pursuit of his science has become organized—then he must properly say to himself: *mea maxima culpa*. For through mathematics the constructive power of the human being first unfolded, and thus brought forth the age of technology. And if, confronted by this disaster which he has brought about, the mathematician is seized with despair, then, for the third time, resignation saves him.[128]

The foregoing examples—especially the cases of von Mises and Doetsch—demonstrate most clearly that there were mathematical physicists who went so far in assimilating the values and mood of their intellectual milieu as to effectively repudiate their own discipline. They show, moreover, that this process of ideological adaptation to the intellectual environment was, either explicitly or implicitly, in large measure a capitulation to Spenglerism. These cases are extreme, of course, and as such atypical. Yet the stages by which von Mises advanced to this extreme, and the readiness of even a Sommerfeld to flirt with the very antiscientific tendencies he deplored, makes it difficult to avoid the conclusion that most German mathematicians and physicists largely participated in, or accommodated their persona to, a generally Spenglerian point of view.

This conclusion is supported by the combination of ample evidence that Spengler's book was read by many, if not most, German physicists and mathematicians and the remarkable paucity of public criticism by representatives of these disciplines. In reviewing the

---

128. M. Dehn, *Über die geistige Eigenart des Mathematikers. Rede anlässlich der Gründungsfeier des Deutschen Reiches am 18. Januar 1928*, Frankfurter Universitätsreden 28 (Frankfurt, 1928), 20 pp., on 15, 18. This address by one of Hilbert's oldest students (Ph.D. Göttingen, 1899) seems to have drawn considerable attention. Otto Neugebauer quoted it, without citing it, in concluding his exposition of the elaborate installations of the new Göttingen mathematical institute (*op. cit.*, note 8). Although Neugebauer's final lines were in defense of "organisation," he did not fail to concede to Dehn that "Gewiss lässt sich auch für eine solche Auffassung viel ins Feld führen." One may therefore suppose that Hilbert's remarks in September 1930 (*op. cit.*, note 94) were meant as much for his colleagues—and former students—as for his lay audience.

Untergang in 1919 Troeltsch had emphasized the desirability of such criticism, but reported that "to be sure, when I asked one of our most eminent mathematicians and physicists [Planck?] to give his opinion of the book, and briefly described Spengler's principal theses, he refused to read any part of it."[129] But that reaction was either untypical or changed very quickly, for I have seen explicit references to Spengler, either suggesting or demonstrating acquaintance with his book, by Max Born, Albert Einstein, Franz Exner, Philipp Frank, Gerhard Hessenberg, Pascual Jordan, Konrad Knopp, Richard von Mises, Friedrich Poske, Hermann Weyl, and Wilhelm Wien.[130] This list, which I expect could be substantially lengthened, is already of such extraordinary length as to make it virtually certain that Spengler's theses, and not merely the public enthusiasm for them, were generally known to the Weimar physicists and mathematicians. And yet they did or said remarkably little to oppose them. Reviewing the literature of the "controversy over Spengler" in 1922 Manfred Schroeter found that "both the cornerposts of the book, the first and sixth chapters, mathematics and physics, have remained almost unanswered." Indeed, Schroeter was able to find very few criticisms by mathematicians, and only one by a physicist—Wilhelm Wien.[131]

Where and when a physicist or mathematician came forward to attack Spengler it was almost invariably in defense of that most basic tenet of the scientific ideology, the autonomy, objectivity, and

---

129. E. Troeltsch, *Ges. Schr.* (*op. cit.*, note 32), *4*, 682.

130. Born-Einstein, *Briefwechsel* (*op. cit.*, note 14), p. 44; F. Exner, *op. cit.* (note 93), preface; P. Frank, *op. cit.* (note 43), p. 54; G. Hessenberg, *op. cit.* (note 66); P. Jordan, *Anschauliche Quantentheorie* (Berlin, 1936), p. 279; K. Knopp, *op. cit.* (note 94), p. 208, cf. note 131; R. v. Mises, *op. cit.* (note 117), p. 32; F. Poske, "Anschauliche und abstrakte Begriffsdefinitionen im physikalischen Unterricht. Vortrag, Naturforscherversammlung, Nauheim, September 1920," *Zeitschr. f. den physikal. u. chemisch. Unterricht, 34* (1921), 97-103 ("Den immerhin unvermeidlich intellektualistischen Charakter der theoretischen Physik hat neuerdings Oswald Spengler in seinem Buche, *Der Übergang* [sic] *des Abendlandes,* hervorgehoben," and Poske reluctantly agreed, without an inkling of criticism); H. Weyl, "Das Raumproblem," *Jahresber. d. Dtsch. Mathematiker-Vereinigung, 31* (1922), 205-221, reprinted in Weyl's *Gesammelte Abhandlungen,* ed. K. Chandrasekharan, 4 vols. (Berlin, 1968), *2*, 332; W. Wien, *op. cit.* (note 100), pp. 36-39.

131. M. Schroeter, *Der Streit um Spengler. Kritik seiner Kritiker* (Munich, 1922), pp. 56-57, 70. The only scientist critics he knows are W. Wien, O. Neurath [sociologist], Leonard Nelson [philosopher], G. Hessenberg, and K. Knopp, who published a "sehr absprechend" review.

universality of scientific knowledge.¹³² This notion Spengler claimed to have exploded by demonstrating that there are no immanent, invariant criteria of knowledge, that the science of a period is dependent in toto upon its *Lebensgefühl*. Yet for every opponent of Spengler's thesis one can cite another exact scientist who, more or less explicitly and more or less fully, identified himself with this doctrinal touchstone of Spenglerism.¹³³ And once again von Mises provides evidence

---

132. Thus the rebuttals cited in note 93.

133. Thus, for example, Gustav Mie, *Das Problem der Materie. Öffentliche Antrittsrede* [as Professor of Physics at the University of Freiburg i. Br.], *gehalten am 26. Januar 1925* (Freiburg in Baden, 1925), pp. 23-24: "Ich bin der Überzeugung, dass in der Geschichte der geistigen Bewegungen überall ein Zusammenhang der verschiedensten Gebiete des Geisteslebens zu beobachten ist. Der Atomismus ist ein Kind des 18. Jahrhunderts und des Rationalismus. . . . Deswegen sind wir im Begriff, uns einem andern Weltbild zuzuwenden. Ich glaube, dass dieses neue Weltbild gewisse charakteristische Züge trägt, die auch sonst im Bilde des modernen Geisteslebens auffallen, ich meine, das Suchen nach einer grossen Einheit und nach einem allgemeinen Zusammenhang im physikalischen Geschehen, der sehr gegen das Auseinanderfallen in die einzelnen Atome kontrastiert. Es ist eine interessante Beobachtung, dass auch die streng an experimentelle Erfahrungen gebundene Physik auf Bahnen geführt wird, die zu den Bahnen der geistigen Bewegungen auf anderen Gebieten durchaus parallel verlaufen."

Again R. Courant, Sept. 1927 (*op. cit.*, note 112), p. 90: "It is surely no accident that the *Umschwung* in the orientation of mathematics from naive productivity to rigorous scientificness is temporally parallel with the great intellectual and social transformations which the European world has undergone since the beginning of the French revolution."

Again E. Madelung, 1931 (*op. cit.*, note 108), pp. 4-6: "I can say without exaggerating that I have not the least interest in the world, but only in the picture [Bild] that I possess of it. Out of this picture I draw my joys and sufferings, my fear and hope, my feeling of comfort and of sorrow. . . . By means of language a communal world-picture is created as a convention. . . . We designate today as '*die Wissenschaft*' our stock of knowledge [Besitz an Wissen] which is codified by the written word and sanctified by convention."

And once again, E. Schrödinger, *Über Indeterminismus in der Physik. Ist die Naturwissenschaft milieubedingt? Zwei Vorträge zur Kritik der naturwissenschaftlichen Erkenntnis* (Leipzig, Barth, 1932), pp. 38-39: "Unsere Kultur bildet ein Ganzes. Auch wer das Glück hatte, die Forschung zu seinem Hauptberuf zu machen—ganz abgesehen davon, dass dies nicht die einzigen sind, die sie fördern,—ist doch nicht nur Botaniker, nur Physiker, nur Chemiker. Vormittags auf dem Katheder spricht er wohl der Hauptsache nach bloss von seinem Fach. An demselben Abend sitzt er in einer politischen Versammlung, hört und spricht ganz andere Dinge,—steht ein andermal im Kreis einer Weltanschauungsgemeinde, wo wieder von anderem die Rede ist. Man liest Romane und Gedichte, geht ins Theater, treibt Musik, macht Reisen, sieht Bilder, Skulpturen, Architektur— und vor allem, man liest und spricht viel über diese und andere Dinge. Kurz wir alle sind Mitglieder unseres Kulturmilieus. Sobald bei einer Sache die Einstellung unseres Interesses überhaupt eine Rolle spielt, muss das Milieu, der Kulturkreis, der Zeitgeist oder wie man es sonst nennen will, seinen Einfluss

that the repudiation of this tenet of the scientific ideology was, in some cases and to some extent, a capitulation to the *Untergang, per se*. Thus in February 1920 von Mises had still been a good enough positivist to deny the influence of political and social conditions or the associated *Lebensgefühl* on the quantity, vitality, direction, or content of the higher intellectual productions. By September 1921, however, he had, as we have seen, gone over to Spengler in this respect as well.[134]

II.4. A Craving for Crises

The exploration of the forms and extent of the ideological adaptation by the physical scientist to his environment must not stop at that unmarked and undefinable frontier where motivation and metaphysics end and the scientific activity itself begins. For to the ideology belongs not merely the general conceptions of the nature and goals of scientific activity, not merely the morale and esprit of the scientist, but also the scientist's perception of the state of his discipline, his hopes, fears, and expectations for its future development. Here, then, we return to the notion and mood of crisis, the conviction of a crisis of culture and of science, which was an essential component of the persona of the Weimar academics.

But before inquiring how far the German mathematical-physical community was likewise infected by this mood, how far a craving for crises affected the exact scientist's perception of the significance and bearing of specific scientific problems, it is worthwhile to emphasize how ready the mathematicians and physicists were to serve themselves with the crisis rhetoric when addressing a general academic audience. For as the notion of crisis became a cliché, it also became an entrée, a ploy to achieve instant "relevance," to establish rapport between the scientist and his auditors. By applying the word "crisis"

üben. Es werden sich auf allen Gebieten einer Kultur gemeinsame weltanschauliche Züge und, noch sehr viel zahlreicher, gemeinsame stilistische Züge vorfinden, in der Politik, in der Kunst, in der Wissenschaft. Wenn es gelingt, sie auch in der exakten Naturwissenschaft aufzuweisen, wird eine Art Indizienbeweis für Subjektivität and Milieubedingtheit erbracht sein." Translated very freely by James Murphy in Schrödinger, *Science, Theory, and Man* (New York, 1957), pp. 98-100.

134. R. v. Mises, *op. cit.* (note 117), pp. 3, 32.

WEIMAR CULTURE, CAUSALITY, AND QUANTUM THEORY, 1918-1927

to his own discipline the scientist has not only made contact with his audience, but has *ipso facto* shown that his field—and he himself—is "with it," sharing the spirit of the times. A presumption is thus insinuated, and often explicitly stated, that in the course of this crisis his science will shed all those characteristics which the academic audience finds most objectionable.[135,136]

But now, unless we are willing to charge duplicity and suppose that the physicists and mathematicians were engaged in a cynical manipulation of their image, I think we must allow that their accommodation to the intellectual environment penetrated deeper

---

135. For example: Walter Schottky, "Das Kausalproblem der Quantentheorie als eine Grundfrage der modernen Naturforschung überhaupt. Versuch einer gemeinverständlichen Darstellung," *Naturwiss., 9* (1921), 492-496, 506-511, opening paragraph: "Darstellung der Krisis in der sich die heutige Physik befindet"; M. H., "Ein Vortrag Einsteins: 'Neue Ergebnisse über die Eigenschaften des Lichtes'," *Neue Züricher Zeitung* (20 June 1922), nr. 808, p. 2: "The reason for the choice of precisely this theme was offered by the circumstance that in respect to the problem of the nature of light physics finds itself today in a severe crisis [sich heute . . . in einer schweren Krise befindet]"; Leo Graetz, *Alte Vorstellungen und neue Tatsachen der Physk. Drei Vorlesungen* (Leipzig, 1925), p. 1; Wilhelm Wien, "Kausalität und Statistik," *Illustrierte Zeitung* (Leipzig), Nr. 4169 (February 1925), pp. 192-196: "Das Aufwerfen dieser Fragen hat die ganze theoretische Physik in eine Krisis gebracht"; Erhard Schmidt, *Über Gewissheit in der Mathematik. Rede zum Antritt des Rektorats der . . . Universität zu Berlin am 15. Oktober 1929* (Berlin, 1930), p. 12: "so steht in der Tat die mathematische Gewissheit in einer Krise"; Richard Gans, *Die Physik der letzten dreissig Jahre. Rede, gehalten bei der Reichsgründungsfeier am 18. Januar 1930*, Königsberger Universitätsreden 7 (Königsberg, 1930), p. 1: ". . . eine Krisis in der Physik heraufbeschworen hat, wie sie in unserer Wissenschaft vorher nicht bekannt war"; once again, Gans, "Der Zufall in der Physik," *Schriften der Königsberger Gelehrten Gesellschaft, Naturwissenschaftliche Klasse, 4* (1927), 113-125, opens: "I believe, however, that I may be permitted to claim your interest insofar as I want to try to show you what things excite the physicist today, and how he exerts himself to wind his way out of the most serious crisis in which our science has ever found itself"; Hans Hahn, ed., *Krise und Neuaufbau in den exakten Wissenschaften: Fünf Wiener Vorträge* (Leipzig-Vienna, 1933), Vorwort: "The growing interest of ever wider circles for the exact sciences is surely above all a seeking after one of the regions which are removed from the world of crises. . . . In truth the exact sciences are by no means secure from crises and precisely in recent decades, from theoretical physics on out into logic, they have been shaken by severe crises."

136. One can find prewar precedents for this "crisis talk" in popular essays and academic addresses. So, for example, Paul Ehrenfest's inaugural lecture as Professor of Theoretical Physics in Leiden, *Zur Krise der Lichtaether-Hypothese* (Leiden-Berlin, 1913), reprinted in Ehrenfest's *Collected Scientific Papers*, ed. M. J. Klein (Amsterdam, 1959), pp. 306-327. Just how widespread this was, and just what its significance was, I am not able to say, but it is my impression that it was then more common in France (Poincaré, Abel Rey) than in Germany.

than the rhetoric. Indeed, the rhetoric itself reacts back upon the persona of the scientist, upon his view of the conceptual situation in his science, of the extent and character of the reconstruction necessary or desirable. In fact, in this period, both mathematics and physics—but above all *German* mathematicians and physicists—went through deep and far-reaching crises, whose very definitions showed the most intimate relation with the principal currents of the Weimar intellectual milieu.

"The New Crisis in the Foundations of Mathematics" proclaimed by Hermann Weyl was precipitated virtually out of thin air in the two or three years following Germany's defeat. With extraordinary suddenness the German mathematical community began to feel how insecure were the foundations upon which the entire structure of mathematical analysis rested, how dubious the methods by which that edifice had been erected. Now, with quasi-religious enthusiasm, considerable numbers of German mathematicians rallied to L. E. J. Brouwer's standard calling for a complete reconstruction of mathematics, a redefinition of the enterprise, which, appropriately enough, went under the name "intuitionism."[137,138] The seriousness of this movement and its consequences may be judged by the vehemence of David Hilbert's counterattack in the spring of 1922. "If Weyl notices an 'inner untenability of the foundations upon which the construction of the empire [Reich] rests,' and worries himself over 'the threatening dissolution of the polity [Staatswesen] of analysis,' then he is seeing ghosts." Weyl and Brouwer are trying "to erect a repressive dictatorship [Verbotsdiktatur]"; to follow "such reformers" is to risk losing the most valuable treasures of mathematics. "No, Brouwer is not, as Weyl believes, the revolution, but only the repetition, with old means, of an attempted putsch . . . and now

---

137. H. Weyl, *Das Kontinuum* (Berlin, 1918; reprinted New York, 1962); "Der circulus vitiosus in der heutigen Begründung der Analysis," *Jahresber. d. Dtsch. Mathematiker-Vereinigung*, 28 (1919), 85-92, reprinted in K. Chandrasekharan, ed., *Gesammelte Abhandlungen von Hermann Weyl* (Berlin, 1968), 2, 43-50; "Über die neue Grundlagenkrise der Mathematik," *Math. Zeitschr.*, 10 (1921), 39-79, reprinted in *Ges. Abhl.*, 2, 143-180.

138. For description, bibliography, and documents of intuitionism: Abraham A. Fraenkel and Y. Bar-Hillel, *Foundations of Set Theory* (Amsterdam, 1958), especially pp. 203-204; Jean van Heijenoort, *From Frege to Gödel. A Source Book in Mathematical Logic, 1879-1931* (Cambridge, Mass., 1967); Constance Reid, *Hilbert* (New York, 1969).

WEIMAR CULTURE, CAUSALITY, AND QUANTUM THEORY, 1918-1927

especially, with the government [Staatsmacht] so well armed and secured by Frege, Dedekind, and Cantor, condemned to failure from the start."[139]

Can one read this rhetoric and not suppose that both Weyl and Hilbert at the very least saw close parallels between the crisis in mathematics and the political crises then wracking Germany, that their sense of the significance of the mathematical issues was colored by their perceptions of the political issues, that perhaps this crisis in mathematics depended for its very existence upon the social-intellectual atmosphere in the aftermath of Germany's defeat? Looking back thirty years afterward, Weyl almost conceded as much, and in fact the "crisis" itself, never resolved, eventually simply ceased to be felt.[140]

Turning to physics one finds once again a notable internal crisis. This is the "crisis of the old quantum theory" which gripped atomic physicists—first and foremost the Germans—in the years before the introduction of the quantum mechanics in 1925/26.[141] I have myself devoted some effort to the intriguing problem of isolating the particular difficulties and frustrations which led at a particular moment to a conviction that "the whole system of concepts

---

139. D. Hilbert, "Neubegründung der Mathematik. Erste Mitteilung," *Abhandlungen aus dem Math. Seminar der Hamburgischen Universität, 1* (1922), 157-177, reprinted in Hilbert's *Gesammelte Abhandlungen, 3* (Berlin, 1935; reprinted New York, 1965), 157-177, on 159-160. Hilbert had delivered this tirade against his most brilliant pupil as a lecture at a number of universities before printing it. And it is important to note that the adherents of the intuitionist movement not only admitted its destructive impact, but seem almost to have welcomed that consequence in a spirit of abnegation and resignation: "That proceeding from this standpoint only a part, perhaps only a wretched [kümmerlich] part, of classical mathematics is tenable is a bitter but inevitable fact." (Weyl, "Diskussionsbemerkung zu dem zweiten Hilbertschen Vortrag über die Grundlagen der Mathematik," *ibid., 6* [1928], 86-88, reprinted in Weyl, *Ges. Abhl., 3*, 147-149, and translated in Heijenoort, *Source Book*, pp. 482-484.)

140. H. Weyl, "Nachtrag, Juni 1955" to "Über die neue Grundlagenkrise...," *Ges. Abhl., 2*, 179: "Only with some reluctance do I acknowledge [bekenne ich mich zu] these lectures, whose occasionally quite bombastic style reflects the mood of an agitated period—the period directly after the First World War." One might even hear in Hilbert's rhetoric a quite literal warning to Weyl and his other friends against the political attitudes of the leader whom they had chosen to follow. See Schröder, *op. cit.* (note 52), pp. 219-220; Reid, *Hilbert*, p. 188.

141. Thomas S. Kuhn, "The Crisis of the Old Quantum Theory, 1922-1925," address delivered at the American Philosophical Society, April 1966. Friedrich Hund, *Geschichte der Quantentheorie* (Mannheim, 1967), p. 103.

of physics must be reconstructed from the ground up," as Max Born asserted in the summer of 1923.[142] And while it is undoubtedly true that the internal developments in atomic physics were important in precipitating this widespread sense of crisis among German-speaking Central European physicists, and that these internal developments were necessary to give the crisis a sharp focus, nonetheless it now seems evident to me that these internal developments were not in themselves sufficient conditions. The *possibility* of the crisis of the old quantum theory was, I think, dependent upon the physicists' own craving for crises, arising from participation in, and adaptation to, the Weimar intellectual milieu.

Of this predisposition to perceive the state of physics as critical, we have many examples between the summer of 1921 and the summer of 1922, which is to say in the year immediately preceding that in which the crisis of the old quantum theory was precipitated. Taking only those cases in which the crisis is proclaimed in the title itself, there is Richard von Mises's lecture "On the Present Crisis in Mechanics" of September 1921, Johannes Stark's pamphlet on *The Present Crisis in German Physics* of June 1922, Joseph Petzoldt's remarks "Concerning the Crisis of the Causality Concept" of July 1922, and Albert Einstein's popular article "On the Present Crisis in Theoretical Physics," dated August 1922.[143] Very roughly speaking each of these physicists is pointing in the same direction, viz. toward the quantum theory. There, of course, the agreement ends; each is putting his finger upon a largely, or completely, different "problem." But that very circumstance—the widespread but initially poorly focused application of the word and notion of a crisis —suggests most strongly that the crisis of the old quantum theory,

---

142. M. Born, "Quantentheorie und Störungsrechnung," *Naturwiss., 11* (1923), 537-542, quoted in my essay on "The Doublet Riddle and Atomic Physics *circa* 1924," *Isis, 59* (1968), 156-174.

143. R. v. Mises, "Über die gegenwärtige Krise der Mechanik," *Zeitschr. f. angewandte Math. u. Mech., 1* (1921), 425-431, reprinted in *Naturwiss., 10* (1922), 25-29, and reprinted once again in v. Mises' *Selected Papers* (Providence, R.I., 1964), *2,* 478-487, lecture, Math.-Phys. Congress, Jena, September 1921; J. Stark, *Die gegenwärtige Krisis in der Deutschen Physik* (Leipzig, 1922), 32 pp., preface dated "Anfang Juni 1922"; J. Petzoldt, "Zur Krisis des Kausalitätsbegriffs," *Naturwiss., 10* (1922), 693-695, dated 2 July 1922, to which Walter Schottky replied 6 October 1922 under the same title, *ibid.,* p. 982; A. Einstein, "Über die gegenwärtige Krise der theoretischen Physik," *Kaizo* (Tokyo), *4* (Dec. 1922), 1-8, dated August 1922.

WEIMAR CULTURE, CAUSALITY, AND QUANTUM THEORY, 1918-1927

far from being forced upon the German physicists, was more than welcome to them.

And here again, as with "intuitionism" in mathematics, one cannot help but be struck by the extraordinary convenience of the chief slogan of this crisis: the failure of mechanics. However appropriate this slogan may have been as a diagnosis of the internal difficulties in theoretical atomic physics, it certainly was *most* appropriate as a code word signaling the physicists' intent to rid their discipline of its most obnoxious elements. Conversely, the almost universal conviction among the German atomic physicists that this crisis was going to last a long, long time—although in fact it was "resolved" within two or three years by the discovery of the quantum mechanics—can be understood in part as a reluctance to contemplate giving up their fashionable and praiseworthy plight, but also in part as an expression of a Spenglerian pessimism: "in my heart I am once again convinced that this quantum mechanics"—which I, Werner Heisenberg, have just discovered—"is the answer, for which reason Kramers accuses me of optimism."[144]

III. "DISPENSING WITH CAUSALITY":[145,146] ADAPTATION OF KNOWLEDGE TO THE INTELLECTUAL ENVIRONMENT

III.1. Introduction. The Concept of Causality

Composing his article "On the Present Crisis in Theoretical Physics" for a popular audience in August 1922, Einstein began with a definition of the aim and structure of physical theory. "It is

---

144. ". . . mich des Optimismus anklagt. . . ." W. Heisenberg to Wolfgang Pauli, 29 June 1925, as quoted by B. L. van der Waerden, *Sources of Quantum Mechanics* (Amsterdam, 1967; reprinted New York, 1969), p. 27. Heisenberg's paper propounding this quantum mechanics, which was indeed "schon richtig," appeared as "Über quantentheoretische Umdeutung kinematischer und mechanischer Beziehungen," *Zeitschr. f. Phys.*, *33* (18 Sept. 1925), 879-893, received 29 July 1925; it is translated in v. d. Waerden, *op. cit.*, pp. 261-276, and reprinted in M. Born, W. Heisenberg, P. Jordan, *Zur Begründung der Matrizenmechanik*, Dokumente der Naturwissenschaft, Abteilung Physik, Band 2, ed. Armin Hermann (Stuttgart, 1962), pp. 31-45.

145. This is the title of the opening chapter of Albrecht Mendelssohn-Bartholdy, *The War and German Society* (New Haven, 1937). The eminent emigré legal scholar there asserted that "War canceled causality. It seemed to do so, at least, to the German people . . . the people as a whole, regardless of their interest in politics, their state of education, or their profession and walk in life,

the goal of theoretical physics," Einstein maintained, "to create a logical [!] conceptual system, resting upon the smallest possible number of mutually independent hypotheses, which allows one to comprehend causally [!] the entire complex of physical processes."[147] On the whole the conception of theoretical physics expressed by this definition is neither unfamiliar nor surprising; it was probably shared by most physicists of the period.[148] Yet two of Einstein's restrictions upon a physical theory seem either superfluous (of course a *logical* conceptual system, what else?) or gratuitous (why must one comprehend physical processes *causally*?). It is precisely these seemingly superfluous and gratuitous additions to what is otherwise the common creed of his colleagues which signal issues Einstein evidently regarded as crucial to theoretical physics as an enterprise. The first of these issues—a logical, as opposed to, say, an intuitive structure of physical theory—I will leave largely aside.[149]

---

realize the change quite clearly, long before it could be measured by historians or sociologists" (p. 20).

146. Earlier studies touching upon the pre-1926 movement to dispense with causality: Victor F. Lenzen, "The Philosophy of Nature in the Light of Contemporary Physics," *University of California Publications in Philosophy*, 5 (1924), 27-48; Alois Gatterer, S.J., *Das Problem des statistischen Naturgesetzes*, Philosophie und Grenzwissenschaften, herausgegeben vom Innsbrucker Institut für scholastische Philosophie, 1. Band, 1. Heft (Innsbruck, 1924), 70 pp.; Stefan Kis, *Das Kausalitätsprinzip in der Physik*, doctoral diss., U. Greifswald (Greifswald, 1925), 35 pp.; Hugo Bergmann, *Der Kampf um das Kausalgesetz in der jüngsten Physik*, Sammlung Vieweg, Heft 98 (Braunschweig, 1929), 78 pp.; Philipp Frank, *Das Causalgesetz und seine Grenzen*, Schriften zur wissenschaftlichen Weltauffassung, Band 6 (Vienna, 1932); Ernst Cassirer, *Determinism and Indeterminism in Modern Physics. Historical and Systematic Studies of the Problem of Causality*, trans. O. T. Benfey (New Haven, 1956); Max Jammer, *The Conceptual Development of Quantum Mechanics* (New York, 1966).

147. A. Einstein, *op. cit.* (note 143), p. 1.

148. The conception is essentially that stated most fully and forcefully by Pierre Duhem, *La Théorie physique: son objet, sa structure*, 2nd ed. (Paris, 1914), trans. by P. P. Wiener as *The Aim and Structure of Physical Theory* (Princeton, 1954), pp. 19, 52, 107, *et passim*.

149. There are two aspects to this issue: 1) a rationalist-irrationalist opposition which is reflected in attitudes toward causality, and which I will therefore touch upon in passing; 2) an opposition, cutting across this first alignment, between intuitiveness and abstractness, *Anschaulichkeit* and *Unanschaulichkeit*, as desirable or necessary in a physical theory. The demand for *Anschaulichkeit* was, once again, very closely connected with the predilections and antipathies characteristic of the Weimar intellectual environment. This opposition played an important role in physics and mathematics in Germany in the Weimar period (see notes 235 and 237), and especially in the Nazi period. I do not, however, attempt here to deal with this difficult problem.

WEIMAR CULTURE, CAUSALITY, AND QUANTUM THEORY, 1918–1927

It is the second issue, the "violent dispute over the significance of the law of causality"—that is the way Max Planck saw the situation in February 1923[150]—which I will describe and analyze.

But what is meant by "causality," or, rather, what notion did the physicists and the philosophers who stood closest to them associate with this term at this time? That notion was, in a word, lawfulness. "The principle of causality," Moritz Schlick explained in 1920, "is . . . the general expression of the fact that everything which happens in nature is subjected to laws which hold without exception."[151] (Cf. Spengler's "Causality is coextensive with the concept of law. There are only causal laws."[152]) And even ten years later Heisenberg, in his *Physical Principles of the Quantum Theory*, could still broach this question by declaring that "the resolution of the paradoxes of atomic physics can be accomplished only by further renunciation of old and cherished ideas. Most important of these is the idea that natural phenomena obey exact laws—the principle of causality."[153] Of course Heisenberg immediately improved upon this formulation by introducing distinctions which were made and employed only after the development of quantum mechanics. In the period 1919–1925, however, both physicists and philosophers held this essentially Kantian notion of causality as conformity to law, so that, as Hans Reichenbach put it in 1920, "if there is cognition of nature, then the principle of causality is valid, for, without this principle, cognition, by its very meaning, is impossible."[154]

---

150. M. Planck, *op. cit.* (note 18), p. 140: "Seit langem ist über die Bedeutung des Kausalgesetzes in der Natur- und Geisteswelt . . . nicht so heftig gestritten worden wie in unseren Tagen. . . . Fast hat es den Anschein, als ob die denkende Menschheit bezüglich dieser Fragen in zwei getrennte Lager gespalten ist."

151. M. Schlick, "Naturphilosophische Betrachtungen über das Kausalprinzip," *Naturwiss.*, 8 (11 June 1920), 461-474; opening paragraph. This was still the general view five years later when *Kis* (*op. cit.*, note 146), p. 3, maintained that "Das Causalprinzip ist das Allgemeinste unter den Prinzipien der Naturwissenschaften. Es besagt, dass die Naturvorgänge nach strengen Gesetzen ablaufen, es ist mit den Worten Machs unsere Zuversicht in die Gesetzmässigkeit der Natur."

152. Note 77. Cf. Walter Rauschenberger, reviewing Ernst Berg, *Das Problem der Kausalität* (Berlin, 1920) in *Kant-Studien*, 26 (1921), 174: "Verfasser ist also strenger Determinist. Sein Kampf gegen das Kausalitätsgesetz erscheint deshalb nicht recht verständlich. Kausalität und Gesetzmässigkeit wurden bisher stets als gleichbedeutend angesehen." I.e., determinism = causality = lawfulness.

153. W. Heisenberg, *The Physical Principles of the Quantum Theory* (Chicago, 1930), p. 62.

154. H. Reichenbach, "Philosophische Kritik der Wahrscheinlichkeitsrech-

But now, if such was the physicist's notion of causality, how indeed could he even contemplate dispensing with this principle? The point was made in 1924 by Alois Gatterer, philosopher and Jesuit, as "a sort of *argumentum ad hominem* which ought to give pause to especially those physicists who, like [Franz] Exner, take pleasure in trying to degrade all physical and chemical laws to statistical laws, and nonetheless at the same time, full of confidence and pride, pursue magnificent researches on the constitution of the chemical atom. How, I ask, can one approach this research with hope of success, and devote oneself actively to it, if one secretly nurtures the conviction that the elementary processes proceed, at least in part, lawlessly at random . . .?"[155] Gatterer's question is a good one, and his appeal—given the physicists' own notion of causality—ought to have been a strong one; and yet, evidently, it simply was not cogent. Thus if we find physicists repudiating causality—and taking pleasure in doing so—without making any attempt to critically analyze and revise the notion itself, then I think we must construe such repudiations as directed against the sort of cognitive enterprise in which physicists theretofore had understood themselves to be engaged.

Precedents for such a reaction against the cognitive enterprise of physics do exist, and, as Stephen Brush has made us recognize, such sentiments provided much of the steam behind the positivist movement at the turn of the century.[156] Where Mach was content to challenge the universal validity of the laws of mechanics, the radical fringe around the positivist-monist standard advanced to the denial of *any* exact laws for atomic processes with the intent of making room for "an element of indeterminacy, spontaneity, or

---

nung," *Naturwiss,* 8 (20 February 1920), 146-153: "Dass eine funktionelle Form existiert, garantiert die Kausalität. . . . Allerdings ist es *denkbar,* dass das Naturgeschehen ohne funktionelle Abhängigkeiten verliefe; aber wenn es eine *Erkenntnis der Natur* gibt, dann gilt das Kausalprinzip, denn ohne dieses ist Erkenntnis ihrem Sinne nach nicht möglich. . . . Wir könnten durchaus zu einer physikalischen Erkenntnis kommen, wenn das Energiegesetz nicht gilt; die Gleichungen würden dann eben anders lauten; aber ohne Geltung des Kausalgesetzes wäre Erkenntnis unmöglich, weil wir überhaupt keine quantitativen Funktionalbeziehungen aufstellen könnten."

155. A. Gatterer, *Das Problem des statistischen Naturgesetzes* (*op. cit.,* note 146), pp. 45-46.

156. S. G. Brush, "Thermodynamics and History," *The Graduate Journal,* 7 (1967), 477-565.

absolute chance in nature," as C. S. Peirce urged in *The Monist* in 1892.[157] Taking acausality in Peirce's sense—and that is, fundamentally, the sense in which it was understood in the early 1920's—I do not know of any other notable physicist who publically advocated this doctrine in the following quarter century, i.e., before the end of the First World War.[158]

157. Quoted by Brush, *ibid.*, p. 531. In this, Comte's latter day disciples were but accepting with relish what Comte believed but dreaded, namely that "the natural laws . . . could not remain rigorously compatible . . . with a too detailed investigation." Quoted by Emile Meyerson, *Identity and Reality*, trans. Kate Loewenberg (1930; reprinted New York, 1962), p. 20. It should be noted, however, that the editor of *The Monist*, Paul Carus, was scandalized: "Mr. Charles S. Peirce's Onslaught on the Doctrine of Necessity," *Monist*, 2 (1892), 560-582.

158. The persistence of a subterranean anticausality current is however suggested by some public refutations of the notion in the intervening period. On 3 August 1914, addressing a founder's day convocation at the close of his term as rector of the University of Berlin, Max Planck acknowledged that "this dualism" between causal and statistical laws which has arisen as a result of the introduction of the statistical point of view into physics "is regarded by many as unsatisfying." Consequently attempts have been made to deny that there are any dynamical (= causal) laws whatsoever and to regard all regularity as statistical: "the concept of an absolute necessity would be lifted from physics entirely. Such a view must, however, very quickly show itself to be an error as disastrous as it is shortsighted." (Planck, "Dynamische und statistische Gesetzmässigkeit," *Physikalische Abhandlungen und Vorträge* [Braunschweig, 1958], 3, 77-90, on 86.) In a posthumously published contribution to the issue of *Die Naturwissenschaften* dedicated to Planck on his sixtieth birthday (23 April 1918), Marian von Smoluchowski pointed to "the tendency which holds sway today to reduce the totality of physical laws . . . to the statistics of hidden elementary events," and he regarded it as "perfectly possible" that in time Lorentz' theory of electrons, relativity, and the law of conservation of energy will also be subjected to this program. Yet Smoluchowski's aim was to show that "chance—in the sense of the word ordinarily used in physics—can perfectly well result from exactly defined lawbound [gesetzmässig] causes," and he emphasized that the calculus of probabilities is thus not to be regarded as a new principle of research but "merely a simplifying statistical schematization of certain functional interconnections which arise very frequently." (Smoluchowski, "Über den Begriff des Zufalls und den Ursprung der Wahrscheinlichkeitsgesetze in der Physik," *Naturwiss.*, 6 [1918], 253-263.)

Although neither Planck nor Smoluchowski identifies the physicists they mean to be refuting, it may well be that both are thinking especially of Franz Exner, whose views Smoluchowski surely knew from his years in Vienna, and Planck may have learned from Exner's *Rektoratsrede* at the University of Vienna: *Über Gesetze in Naturwissenschaft und Humanistik. Inaugurationsrede, gehalten am 15. X. 1908* (Vienna, 1909), 45 pp. This publication, which came to my attention only after the final draft of the present essay was completed, does indeed maintain that "Alles Geschehen in der Natur ist das Resultat zufälliger Ereignisse" (p. 42), and that "if we were capable of slowing down the molecular motion so enormously that we could follow the individual molecular processes, then we would perceive nothing but a chaos of chance events, in which we would seek in vain for any regularity" (p. 13). Exner then went on to sketch a "unified

It must, of course, be acknowledged that in precisely this period Mach himself, the positivist movement in general, and even neo-Kantians like Cassirer were waging a campaign against quite a different concept of causality, so that by 1918 Friedrich Poske could well observe that "it has recently become the fashion among those concerned with the theory of scientific method to throw the causal concept onto the scrap heap."[159] At issue here was not, however, the notion of conformity to law, but rather the "metaphysical," "animistic," "fetishistic" doctrine of cause and effect (*Ursache und Wirkung*) as an ontological assumption, which Mach and his allies wished to replace by the mathematical conception of function.[160] "This putting together of functional interconnections is what theoretical physics is really all about," Wilhelm Wien explained to a lay audience in 1914; "causality—i.e., *der Satz von Ursache und Wirkung*—has nothing to do with the business."[161] And by 1918 this point of view had become almost a matter of course among physicists and the philosophers closely associated with them—

---

and comprehensive world picture" in which all law is but the expression of the law of large numbers, and the want of laws in the *Geisteswissenschaften* is due neither to the peculiar nature of their subject matter, nor to "Das Lebendige," nor to free will, but results simply from the relatively small number of equally chance events underlying the phenomena they study.

It is noteworthy, moreover, that P. Frank, *Kausalgesetz* (*op. cit.*, note 146), pp. 56-58, treated "Die energetische Naturauffassung" as one of the "Kausalitätsfeindliche Strömungen," and asserted that "at the time when the ideas of the Ostwaldian *Naturphilosophie* appeared to be dominant among those active in natural science and even among the majority of the laity interested in natural science, one could regard the notion of mechanical causality in Laplace's sense as having been disposed of, and the introduction of soul-like factors seemed necessary." This, like so much of Frank's book, is largely blather; nonetheless, it is suggestive.

159. F. Poske, *Zeitschr. f. den physikal. u. chem. Unterricht*, 31 (March 1918), 39.

160. E. Mach, *The Analysis of the Sensations*, trans. from the 5th edition, 1906, by S. Waterlow (reprinted New York, 1959), p. 81; Joseph Petzoldt, "Naturwissenschaft," *Handwörterbuch der Naturwissenschaften* (Jena, 1912), 7, 50-94, on 79-80; Paul Volkmann, *Einführung in das Studium der theoretischen Physik . . . mit einer Einleitung in der Theorie der physikalischen Erkenntnis*, 2nd enlarged ed. (Leipzig-Berlin, 1913), pp. 385-398; Ernst Cassirer, *Substance and Function*, trans. from the German edition, 1910, by W. C. and M. C. Swabey (Chicago, 1923; reprinted New York, 1953), Ch. 5; Hans Kelsen, *Society and Nature* [German title: *Kausalität und Vergeltung*]: *A Sociological Inquiry* (Chicago, 1943), p. 381.

161. W. Wien, "Ziele und Methoden der theoretischen Physik [1914]" (*op. cit.*, note 90), p. 156.

resisted only by a few arch-reactionaries like Ernst Gehrcke[162]—so that "causality," stripped of all ontologic overtones, was taken as equivalent to functional determination. Often, but not always, causality was further specified as the "Laplacian" conception of the necessary and sufficient conditions for such complete determination, viz. a cross section of the "world" at a given moment in time;[163] such a conception followed not merely from classical dynamics but equally from the very notion of a field theory.

Beginning with causality as the postulate of the lawfulness of natural processes, we have ended with causality as rigorous determinism. One might object that there is room for several distinct positions between these two conceptions. The possibility of satisfying a (weaker) postulate of lawfulness without demanding that every detail of every natural process be unambiguously determined did not entirely escape physicists in the years before the discovery of a quantum mechanics having this general character. Nonetheless, the essential point is that in the period treated in this paper every such suggestion of a relaxation of complete determinism was advanced as, and regarded as, a failure or abandonment of *causality*. In fact, as we proceed we will occasionally find the word "causality" being used in several senses narrower than, not wider than, "determinism"—as equivalent to the laws of classical mechanics, to the conservation of energy and momentum, to visualization in space and time, to the absence of action at a distance, to action by contact, or to description by differential equations. And again, in many instances these special definitions of causality were advanced in conjunction with, and as the justification for, an assertion of the invalidity of the law of causality. In every instance, however, such special definitions of causality, and a fortiori the general requirement of unambiguous

162. E. Gehrcke, *Physik und Erkenntnistheorie*, Wissenschaft und Hypothese 12 (Leipzig-Berlin, 1921), pp. 43-51.
163. E.g., Rudolf Carnap, "Dreidimensionalität des Raumes und Kausalität," *Annalen der Philosophie*, 4 (1924), 105-130, who, laying the groundwork preliminary to demonstrating his principal thesis that "die Dreidimensionalität des Raumes (gleichbedeutend mit der Vierdimensionalität des Weltgeschehens) ist die logische Folge der Gesetzmässigkeit des Geschehens," notes: "Die Geltung der Kausalität im Sinne der Physik besagt: in der physikalischen Welt herrschen determinierende Gesetze, und zwar sind alle Vorgänge eindeutig bestimmt, wenn die Gesamtheit der Vorgänge eines beliebig kleinen Zeitabschnittes bestimmt ist. Die Begriffe 'bewirken,' 'Ursache' u. dgl. haben also mit dem physikalischen Begriff der Kausalität nichts zu tun."

determination, were held to be equivalent to the assumption of the comprehensibility of nature, and repudiated or defended as such.

III.2. The First Intimations of an Issue, 1919-1920

If one examines the annual indices to German books and periodicals in the first decades of the century, one finds a remarkable number of articles and tracts with the word "causality" in their title. Most striking, however, is the spate of such tracts in the five years 1918-1922.[164] Typically, these are short answers to the riddle of the universe, the revelations of enthusiasts rather than the ruminations of academics. (They show, *inter alia*, that Spengler was not alone in seeing causality as the key to that riddle.) Yet the German academics too were anxious not to be left out of this discussion; in 1915 the Prussian Academy of Sciences had offered a prize for the best history of the causal problem since Descartes, awarding it in 1919 to a devoted student of the noted determinist Benno Erdmann.[165]

It is also at just this time—I know of no example earlier than 1919—that intimations of this issue appear in the private correspondence and public addresses of German physicists. In June 1919, replying to a lost letter from Max Born, Einstein asked ironically, "Is a hardboiled x-brother and determinist allowed to say with tears in his eyes that he has lost faith in humanity. Precisely the instinctive behavior of our contemporaries in political matters is suited to maintain a vivid belief in determinism."[166] Here Einstein is, on the one hand, gently ridiculing Born for feeling sorry for himself and his country by reminding Born of the public image of the theoretical physicist—hardboiled determinist—and, on the other hand, Einstein is making a small joke which can only be to some point if it were a recognized fact that the law of causality was under attack in the social sphere and under discussion among physicists.

Einstein could still joke about the matter with Born in early

---

164. *Deutsches Bücherverzeichnis*, 6 (1915-1920), 770; *10* (1921-1925), 1298, lists ten such.
165. Else Wentscher, *Geschichte des Kausalproblems in der neueren Philosophie* (Leipzig, 1921), preface.
166. Einstein-Born, *Briefwechsel* (*op. cit.*, note 14), pp. 29-30.

WEIMAR CULTURE, CAUSALITY, AND QUANTUM THEORY, 1918–1927

December,[167] but at the end of January 1920 his tone had become most serious, for in the meantime Born, in a long letter also lost, had evidently confessed to Einstein that he was willing to entertain the idea of acausality, supporting himself upon arguments of his subordinate, Einstein's former student, Otto Stern.[168] "That business of causality plagues me a great deal too," Einstein conceded, shaken by Born's defection but also anxious not to give offense by too categorical an assertion of his own "very very great reluctance to forgo *complete* causality." Yet even more interesting than these remarks themselves is the association of ideas which their precise location in Einstein's long letter—clearly a point by point reply to Born's—reveals. They occur toward the end, immediately following not unsympathetic remarks on Oswald Spengler, whose *Decline of the West*, published the year before, included barbs directed at both Einstein and Born. "Sometimes in the evening," Einstein allowed, "one likes to entertain one of his propositions, and in the morning smiles about it."[169]

167. *Ibid.*, p. 38. Born had published a short popular exposition of general relativity, "Raum, Zeit und Schwerkraft," *Frankfurter Zeitung*, 23 Nov. 1919, Nr. 876, pp. 1-3, in which, among other jabs at Kant, he remarked that "Wer diese Entwicklung [of relativity theory] miterlebt hat, der wird sich des Zweifels am a priorischen Character auch anderer Kategorien des Denkens nicht erwehren können." Born was then singled out for attack by Robert Drill, "Ordnung und Chaos. Beiträge zur Geschichte von der Erhaltung der Kraft," *Frankfurter Zeitung*, 30 Nov. 1919, Nr. 895, pp. 1-2; 1 Dec. 1919, Nr. 899, p. 1, a wild, fanatical Kantian who sought to prove the a priori character of the [metaphysical, ontological] concept of causality using the very concrete example of our anticipation of the taste of a piece of *Wurst*. This tickled Einstein: "Sein Nachweis der Causalität a priori ist wahrhaft erhebend."

168. Einstein to Born, 27 Jan. 1920, *Briefwechsel*, pp. 42-45. It would be most interesting to know just what Born's and Stern's views were. Some indication of them was probably contained in a talk on "Wahrscheinlichkeit und Kausalität in der Physik" which Born delivered to the Physikalischer Verein in Frankfurt on 27 July 1920 (*Jahresbericht*, 1919-1925, p. 107), but which appears to have remained unprinted.

169. Einstein, 27 Jan. 1920, *loc. cit.*: "Der Spengler hat auch mich nicht verschont. Man lässt sich gern manchmal am Abend von ihm etwas suggerieren und lächelt am Morgen darüber. Man sieht, dass die ganze Monomanie aus der Schullehrer-Mathematik kommt. Euklid-Cartesius ist sein Gegensatz, den er nun in alles hinarbeitet, aber—wie man gern zugibt—mit Geist. Solche Dinge sind amüsant, und, wenn morgen einer mit dem nötigen Geist das Gegenteil sagt, so ist es wieder amüsant, und was *wahr* ist, weiss der Teufel!"
"Spengler didn't spare me either" is a puzzle, for, so far as I can see, in the original edition (cf. note 65) relativity is handled rather unpejoratively as "die letzte Form der faustichen Natur," pp. 599-601, while Born is not mentioned

71

What proposition might Einstein have had in mind? Might he not have been thinking of Spengler's most fundamental proposition, the axis of the system, "the opposition of the destiny-idea and the causality-principle"? Is the juxtaposition in Einstein's letter of Spengler's *Untergang* and the issue of causality in atomic physics pure chance? Is it not more likely that in Born's mind and/or in Einstein's mind there is an intimate, although perhaps not fully conscious, association between the physicists' new and sudden inclination to forgo causality and Spengler's enormously popular culture-criticism in which the physicist "whose entire mental existence is founded upon the principle of causality" symbolizes the late and decadent fear of the irrational, the incomprehensible?[170]

Such an association between Spengler and the issue of acausality is certainly explicit in Wilhelm Wien's public lecture on "The Connections of Physics with Other Disciplines," which he delivered just one month later, at the end of February 1920, in the Prussian Academy of Sciences. Previously I used this lecture to illustrate a chameleonlike adaptation of the physicists' ideology to changes in the intellectual environment. Here, however, it stands as the first of a series of attempts to draw a clear line between that environment and physics as a cognitive enterprise. The apparent contradiction reflects rather the distinction I made in Part II between the peripheral and the central features of scientific ideology. Although Wien was ready to advance a new conception of the wellsprings and social-intellectual function of scientific activity designed to make the enterprise seem worthwhile in the public's eye, he was unwilling to compromise those ideological tenets which he regarded as essential to the scientific method and its cognitive goals. The true source and value of natural science lies, to be sure, in "an inner need of the human mind," but that need is for a particular kind of knowledge;

---

at all. It is then only in the revised edition (1923) that relativity is described as "a ruthlessly cynical working hypothesis," and the space given it much reduced, pp. 544-545 (Eng. ed., p. 419), while Born, mistaken for a chemist ignorant of mathematics, receives his just deserts in a footnote on pp. 205-206 (Eng. ed., p. 156).

170. See notes 77 and 78 and texts thereto. P. Frank, *op. cit.* (note 146), devoted a chapter to "Kausalitätsfeindliche Strömungen." He suggested that such currents were widespread in the general intellectual milieu of the Weimar period, but gave only one example: Spengler.

it is a "longing to comprehend the causality of the course of phenomena [die Kausalität des Geschehens]."

Wien's motivation for incorporating causality in the very definition of natural science becomes quite evident when, at the end of his lecture, he comes to the *Decline of the West*. While conceding that there is ample evidence of the accuracy of Spengler's characterization of our present cultural situation, Wien rejects in principle the notion of historical laws, of any necessary course of history. All such laws can be and are repeatedly violated "by irrational expressions of the human spirit." Turning Spengler against Spengler, Wien emphasizes that if we suppose a generally valid law of aging of cultures and use it to predict the future of our own, "then we reintroduce a covert causality into history." But to adapt Kant's well-known epigram, "I would like to assert that there is so much the more true historical science in history the less it contains of physics. ... Causality is the foundation of the physical world picture, but it is a category [Denkform] of our mind [Geist] and cannot be employed again for the analysis of the same spirit [Geist], whose effects it is the task of history to portray."[171]

Embracing causality in order to effect a separation between his discipline and his milieu, Wien has then the task of repulsing Spengler's attempt to make physics culture-bound, and especially Spengler's contention that contemporary physics, as its nerve failed, was renouncing causality. It is nature which has compelled the physicists to resort increasingly to the use of statistics; it is neither a sign of "decadence" nor of any renunciation of causality. On the contrary, every utilization of statistics "postulates causality," but because of the great complexity the causal interconnections cannot be traced in detail. Where in 1914 Wien stressed that of all the natural sciences it is theoretical physics in which the personality has the greatest scope and importance, both in constructing theories and influencing the course of scientific development, now, contra Spengler, he is at pains to emphasize that "however strongly the shaping of physical modes of thought depends upon the constitution of the physicist, it is nonetheless decisively determined by the nature of the things themselves." Archimedes' results accord entirely with

171. W. Wien (*op. cit.*, note 100); the quotations are from pp. 20, 35, and 38-39, respectively.

our own, and our results will in all probability be utilizable by physicists of a later culture.[172]

The attention which Wien gives to Spengler, his focus upon the issue of causality, and his consistent effort to isolate physics—as a cognitive enterprise for which causality is the defining characteristic—from its acausal, irrational historical milieu, all suggest that he sensed an intimate connection between the treasonable murmurings against causality among his colleagues and Spengler's brilliant expression of certain powerful currents in the contemporary milieu.

III.3. Conversions to Acausality, 1919–1925

a. The Earliest Converts: Exner and Weyl

It was not until 1919, when Franz Exner was a full seventy years old, that his *Lectures on the Physical Foundations of the Natural Sciences* were printed.[173] And although the crucial concluding lectures on laws of nature may have been worked out long before in the mind and the conversation of the distinguished Viennese spectroscopist, they were probably not even included in the course as delivered to the public before the war.[174] The argument begins with the assertion that *none* of our laws of nature is exact. From this postulate—and here perhaps is the link with the late nineteenth-century positivist-monist repudiations of causality—Exner jumped to the conclusion that "causality" does not obtain, that if examined sufficiently closely during sufficiently short time intervals the motion of a falling body would be found to be perfectly random, directed up as often as down.[175] The apparent lawfulness which we discover at

---

172. *Ibid.*, p. 37; W. Wien, "Ziele und Methoden der theoretischen Physik. Festrede . . . 1914" (*op. cit.*, note 90), pp. 152-154.

173. F. Exner, *Vorlesungen über die physikalischen Grundlagen der Naturwissenschaften* (1st ed., Vienna, 1919; 2nd enlarged ed., Leipzig and Vienna, 1922). The crucial 86th-94th *Vorlesungen* are in all essential respects identical in both editions. The following quotations are from the preface to the first edition and from the 93rd and 94th *Vorlesungen*.

174. Internal evidence in these concluding lectures on laws of nature and Exner's reference to them in the preface as an "Anhang" argue that they were never delivered as such, but synthesized during the war when the book was written. See, however, note 158.

175. Exner claimed (86th *Vorlesung*, p. 658 in 2nd ed.) to have obtained Ludwig Boltzmann's assent to this proposition.

the macroscopic level is then "explained" by Exner's second thesis that all macroscopic natural laws are essentially statistical in character, the regularity arising in some unspecified way out of the collaboration of the random motions. The speculation that *all* macroscopic laws are essentially statistical, that none is exact, was by no means unprecedented. What is novel is the leap from that supposition to the conclusion that causality fails. For this leap no justification is offered, and the problem of how perfectly acausal microscopic motions result in statistical regularities is not even raised by Exner.

Exner the experimentalist takes a radical nominalist-empiricist stance: the absolutely rigorous laws "are a creation of man and not a piece of nature." Nor have we the right to postulate even "the existence of an absolute causality," least of all on the grounds that it is necessary in order for us to understand nature. "Nature does not inquire at all whether men understand her or not, and we are not to construct a nature adequate to our understanding, but our task is simply to come to terms with that which is given as best we can." Although Exner cannot consistently maintain his empiricist posture and also categorically deny the existence of causality at the microscopic level, he wants very much to do so in order "to arrive at a unified world picture" in which *all* law is purely statistical, a world of pure chance. He therefore does his best to convince his (lay) readers of the implausibility of the existence of such a causal substratum, switching back and forth between, and largely confounding, the question of the validity of the laws of classical mechanics in the atomic domain and the validity of the principle of causality in the same domain.

Influential as Exner's lectures indeed were, they have in many respects an archaic air. Exner is a curious mixture of the philosophical currents of the two preceding generations, a self-confessed mechanist-materialist yet clearly also a positivist in his view of scientific constructs. Of the *Lebensphilosophie* and existentialism which will figure so prominently in most of the following conversions to acausality there is scarcely a hint. Radioactivity and Brownian motion are the most recent developments in physics to which he refers; in his efforts to cast doubt upon the causal character of atomic processes he omits to serve himself with a "quantum" of any sort. Thus the first of the calls for a renunciation of causality was clearly independent of the problems raised by the quantum theory of the atom or radiation.

Apart from Exner, the earliest to speak out against causality was Hermann Weyl. Weyl was a phenomenologist of quite a different sort from his Machian ex-brothers. As Privatdozent at Göttingen shortly before the war Weyl had fallen under the influence of Edmund Husserl's program of "pure phenomenology." This Platonizing phenomenology of the mind, based upon intense introspection, had originated in epistemological concerns but in this period was degenerating into existentialism. Dating from 1917 is the first avowed intrusion of Weyl's philosophical outlook into his scientific work—his own attempt to place the continuum on an intuitionist foundation.[176,177] But as I indicated in Section II.4, Weyl soon

---

176. H. Weyl, "Erkenntnis und Besinnung (Ein Lebensrückblick)," *Studia Philosophia* (1954), as reprinted in Weyl's *Ges. Abhl.*, *4*, 631-649, recalled that in his student years at Göttingen, 1906-1910, his adolescent Kantianism was converted to positivism—he read Mach, Poincaré, and F. A. Lange—and only shortly before his departure for Zurich in 1913 "was it Husserl, then, who led me out of positivism to a freer view of the world once again." The contact with his colleague was mediated by one of Husserl's numerous enthusiastic students, Helene Joseph, whom Weyl married in 1913. Explicit citations of Husserl first appeared in *Das Kontinuum* (Berlin, 1918), written in 1917, and in the introduction to *Raum-Zeit-Materie*, 1st ed. (Berlin, 1918), preface dated Easter 1918. Weyl's extraordinary deference to Husserl is evident in the repeated quotations, always with full approbation, in his *Philosophy of Mathematics and Natural Science* (Princeton: Princeton University Press, 1949), translated from the German edition of 1927. In return the Husserl school was happy to lean upon Weyl and claim him for one of their own: Oskar Becker, "Beiträge zur phänomenologischen Begründung der Geometrie und ihrer physikalischen Anwendungen," *Jahrbuch für Philosophie und phänomenologische Forschung*, 6 (1923), 385-560, on 387-388. By 1928, however, the success of formalist metamathematics had shaken Weyl's allegiance: "If Hilbert's view prevails over intuitionism, as appears to be the case, *then I see in this a decisive defeat of the philosophical attitude of pure phenomenology*, which thus proves to be insufficient for the understanding of creative science even in the area of cognition that is most primal and most readily open to evidence—mathematics." (Weyl, *op. cit.*, note 139). See also Peter Beisswanger, "Hermann Weyl and Mathematical Texts," *Ratio*, *8* (1966), 25-45.

177. An admirable and uniquely intelligible account of this remarkable intellectual phenomenon is given by Herbert Spiegelberg, *The Phenomenological Movement: A Historical Introduction*, 2nd ed., 2 vols. (The Hague, 1965), which may be supplemented by Joseph J. Kockelmans and Theodore J. Kisiel, eds., *Phenomenology and the Natural Sciences: Essays and Translations* (Evanston, Ill., 1970).

WEIMAR CULTURE, CAUSALITY, AND QUANTUM THEORY, 1918–1927

became the principal champion of Brouwerian intuitionism in Germany. That Weyl saw an intimate connection between intuitionism in mathematics and acausality in physics emerges quite clearly from his initial manifesto against causality, "The Relation of the Causal to the Statistical Approach in Physics," printed in August 1920.[178,179] "Are statistics merely a shortcut to certain consequences of causal laws," Weyl asks, "or do they imply that no rigorous causal interconnection governs the world and that, instead, 'chance' is to be recognized alongside law as an independent power restricting the validity of the law? The physicists are today entirely of the first opinion." And yesterday, in the spring of 1918, Weyl had been too, having in his proposed extension of general relativity made an "attempt," as he admits, "at carrying through the idea of a pure

---

178. H. Weyl, "Das Verhältnis der kausalen zur statistischen Betrachtungsweise in der Physik," *Schweizerische Medizinische Wochenschrift*, 50 (19 August 1920), 737-741, reprinted in Weyl's *Ges. Abhl.*, 2, 113-122. Weyl had prepared an address with this title for a symposium on "the Significance of Probability for Natural Science and Medicine" which had been organized by Heinrich Zangger, professor of forensic medicine in Zurich, for the annual congress of the Schweizerische Naturforschende Gesellschaft in Lugano in September 1918. The congress was canceled, however, due to the grippe epidemic, so that the earliest statement we have of Weyl's position is the 500-word abstract of the address Weyl delivered the following year, 8 September 1919; *Schweizerische Naturforschende Gesellschaft, Verhandlungen* (1919), 2. Teil, pp. 152-153. This abstract has the same general structure as the printed paper, includes Husserlian phenomenological-existentialist jargon ("das nur im Willen erlebte 'Grund-sein,'" etc.), and concludes with an affirmation of Weyl's belief "that at the basis of statistics there lies an independent principle which is not to be reduced to causality." Nonetheless, it suggests very strongly that in the fall of 1919 Weyl had not yet advanced as far as his position of August 1920, and that, in particular, he had not yet made the connection with intuitionism in mathematics, the repudiation of causality and "der reinen Gesetzesphysik" being based solely upon their incompatibility with the "für unser ganzes Erleben fundamentale Einsinnigkeit der Zeit."

179. Although he failed to mention Hermann Weyl in this connection, A. d'Abro, *Decline of Mechanism* (New York, 1939), reprinted as *The Rise of the New Physics*, 1 vol. in 2 (New York, 1952), pp. vii, 212, justified inclusion of a chapter treating "The Controversies on the Nature of Mathematics" in his historical exposition of the quantum theory on the grounds that "In our opinion these controversies originate from the same psychological differences which appear to be responsible for the current controversy concerning the principle of causality in physics." In concluding that chapter d'Abro suggested "It might even be said that modern physics is witnessing the same crisis that we have been discussing in mathematics: the quantum theorists occupy the position of the intuitionists while Einstein and Planck occupy that of the formalists." I think d'Abro's conjecture is essentially correct.

physics of law for the entirety of the world."[180] But now Weyl has changed his mind and is placing himself in opposition to the prevailing opinion. Why? He has certain dissatisfactions with classical statistical mechanics and the treatment of fluctuation phenomena, but the real issue, he admits is that

> finally and above all, it is the essence of the continuum that it cannot be grasped as a rigid [starr] existing thing, but only as *something which is in the act of an inwardly directed unending process of becoming.* . . . In a *given* continuum, of course, this process of becoming can have reached only a certain point, i.e. the quantitative relations in an intuitively given piece S of the world [regarded as a four-dimensional continuum of events] are merely approximate, determinable only with a certain latitude, not merely in consequence of the limited precision of my sense organs and measuring instruments, but because *they are in themselves afflicted with a sort of vagueness.* . . . And only "at the end of all time," so to speak, . . . would the unending process of becoming S be completed, and S sustain in itself that degree of definiteness which mathematical physics postulates as its ideal. . . . Thus the rigid [starr] pressure of natural causality relaxes, and there remains, without prejudice to the validity of natural laws, *room for autonomous decisions* [Entscheidungen], *causally absolutely independent of one another,* whose locus I consider to be the elementary quanta of matter. These "decisions" are what is *actually real* in the world.[181]

I have quoted Weyl at some length, both because he goes on at some length and because a mere ascription of such radically existentialist views and motives would very likely be dismissed as incredible. Yet, clearly, these motives are primary. Weyl has resolved to abandon the ideal of a pure field physics—for which he had labored so hard and achieved such striking success—and adopted matter, or rather its free will, as the ultimate reality. The field and its laws, like geometry before Einstein, were now a mere backdrop. Why? Because it seemed necessary in order to escape the determinism which the field conception involved. Here, in the fall of 1919 and the summer of 1920 Weyl says not a word about Planck's quantum

---

180. H. Weyl, *Ges. Abhl.*, 2, 116-117; "Gravitation und Elektrizität, *Preuss. Akad. der Wiss., Berlin, Sitzungsber.* (30 May 1918), 465-480, reprinted *Ges. Abhl.*, 2, 29-42, and trans., with additional notes, in H. A. Lorentz, et al., *The Principle of Relativity* (London, 1923; reprinted New York, 1952), pp. 201-216.
181. H. Weyl, *Ges. Abhl.*, 2, 121-122.

of action. It has evidently not yet occurred to him that the quantum theory could be dragged in to provide an ostensible physical basis for his existentialist repudiation of causality. It was only in the fall of 1920, when preparing the fourth edition of *Space-Time-Matter,* that Weyl seized upon the quantum theory as compelling him to say "clearly and distinctly that physics in its present state is simply no longer capable of supporting the belief in a closed causality of material nature resting upon rigorously exact laws." There Weyl also added that crucial existentialist consideration which had been with him for some time—the repudiation of determinism restores the unidirectionality of time, "the most fundamental fact of our experience of time," which field physics denied us a priori.[182] Thus, "not only is matter restored to its old claim to reality, but also the genuine idea of causality, of *Verursachung,* as we experience it most immediately in our will, awakes to new life. Branded as fetishism by Mach . . ." etc., etc.[183]

It seems pretty clear—and indeed it is characteristic of the acausalists—that the sort of primary reality which Weyl would have matter enjoy is simply not a sort of reality which is accessible to physical cognition. Thus by the summer of 1924, in carrying his "Leibnizian agent-theory of matter" to its logical conclusion, Weyl was led back to the field as the primary *physical* reality:

> the material particle itself is not even a point in space, but is something entirely outside the category of extension. . . . It is analogous to the Ego, whose actions, despite the fact that it is itself nonextensional, always have their origin, through its body, at a definite place in the world continuum. Yet whatever this field exciting agent may be in its inner essence—perhaps life and will—in physics we consider it only in terms of the field actions which are excited by it and we are able to characterize it numerically (charge, mass) only by virtue of these field actions.[184]

---

182. H. Weyl, *Raum-Zeit-Materie,* 4th ed. (Berlin, 1921), pp. 283-284; *Space-Time-Matter,* trans. from 4th ed. by H. L. Brose (London, 1922; reprinted New York, 1952), pp. 310-312. The preface to this edition is dated November 1920. Again, with a more precise statement of "causality," in the 5th ed. (Berlin, 1923), pp. 286-287.
183. H. Weyl, "Feld and Materie," *Annalen der Physik,* 65 (1921), 541-563, received 28 May 1921; reprinted in Weyl's *Ges. Abhl.,* 2, 237-259, on 255.
184. H. Weyl, "Was ist Materie?" *Naturwiss.,* 12 (11, 18, 25 July 1924), 561-569, 585-593, 604-611; *Ges. Abhl.,* 2, 486-510, on 510.

Weyl was now able to reconcile himself to this resurrection of the field because he thought he had finally found an escape from the proposition that the classical field theories embody and impose the Laplacian conception of causality. In a semipopular article in the form of a dialogue, Weyl argued that "according to the general theory of relativity the concept of the relative motion of several bodies with respect to one another is just as little tenable as that of the absolute motion of a single body." Consequently, the principle of causality cannot involve these untenable states of motion, and so reduces to the assertion that "the world of events only depends upon, and must be unambiguously determined by, the charge and mass of all material particles. Since this is obviously absurd . . . that principle of causality must be abandoned."[185]

b. 1921, Summer and Fall: von Mises, Schottky, Nernst, *et al.*

The quasi-religious conversions to acausality, of which Weyl's is the earliest example, became a common phenomenon in the German physical community during the summer and fall of 1921. As if swept up in a great awakening, one physicist after the other strode before a general academic audience to renounce the satanic doctrine of causality and to proclaim the glad tidings that the physicists are about to release the world from bondage to it. The cases known to me are: Walter Schottky in June, Richard von Mises in September, Walther Nernst in October.[186]

The conversion of von Mises to acausality is particularly interesting not only because it shows the suddenness with which this regeneration could take place and its essential independence of the difficulties encountered in atomic physics, but also because it provides *prima facie* evidence of a direct connection between the repudiation of causality by a loyal scion of Austrian positivism and

---

185. H. Weyl, "Massenträgheit und Kosmos. Ein Dialog," *Naturwiss.*, *12* (14 March 1924), 197-204; *Ges. Abhl.*, *2*, 478-485.
186. W. Schottky, "Das Kausalproblem der Quantentheorie als eine Grundfrage der modernen Naturforschung überhaupt. Versuch einer gemeinverständlichen Darstellung," *Naturwiss.*, *9* (24 and 30 June 1921), 492-496, 506-511; R. von Mises, "Über die gegenwärtige Krise der Mechanik" (*op. cit.*, note 143); W. Nernst, *Zum Gültigkeitsbereich der Naturgesetze* (Berlin, 1921), 26 pp., reprinted in *Naturwiss.*, *10* (26 May 1922), 489-495. This is Nernst's inaugural lecture as rector of the University of Berlin, 15 October 1921.

WEIMAR CULTURE, CAUSALITY, AND QUANTUM THEORY, 1918–1927

his capitulation to the *Weltschmerz* of Spengler's *Decline of the West*. In von Mises' inaugural (and farewell) address as delivered in February 1920 at the Technische Hochschule Dresden, and as printed in August 1920, causality was still handled unself-consciously and unpejoratively as equivalent to physical explanation. "We see now in our time, how a new and simply enormous field of phenomena, the multiplicity of the chemical elements, is drawn into the realm of causal explanation." And von Mises takes it for granted that the goal of atomic physics, as of all natural science, is and must be "to explain all these phenomena on the basis of a very few principles, to reveal their causality."[187] But when one turns to the thoroughly Spenglerian appendix which von Mises added in September 1921 to the republication of this lecture, one finds his attitude toward causality—as toward so much else—entirely transformed. Every electrical, every thermal, every optical process is a statistical phenomenon and as such fundamentally incompatible with the concept of causality. So long as we base ourselves upon that concept "the quantum theory and everything connected therewith must appear as an insoluble riddle. Whoever traces back the history of physical cognition cannot help but recognize that here an essential *alteration of our mode of thinking*, of the entire scheme of 'physical explanation,' is inexorably demanded and is gradually being prepared."[188]

Admittedly, von Mises has invoked the quantum theory as the occasion for the repudiation of causality. But he was not willing that it be *more* than the occasion, that, in particular, his own discipline of applied classical mechanics remain saddled with the stigma of causality. In this same month, September 1921, at the first of the annual German physics-mathematics congresses, von Mises read his colleagues a lecture—or better, made a public confession before an assembly of his peers—regarding "The Present Crisis in Mechanics."

> Stated in the briefest form, this question—in whose negative answer I discern the crisis in the present state of mechanics—runs thus: can we still assume that all phenomena of motion and equilibrium which we observe in visible bodies are explicable within the framework of the

187. R. von Mises, *Naturwissenschaft und Technik der Gegenwart* (*op. cit.*, note 118), p. 19.
188. *Ibid.*, p. 30.

81

Newtonian axioms and their extensions. In other words, can the temporal course of every motion of an arbitrarily delimited portion of mass be unambiguously determined by specifying the initial state and assuming some appropriate force law to be acting? . . . All that I want to try to show here is that the accumulated facts which we possess today make it evident that it is highly improbable that this goal of classical mechanics could ever be attained, and that other, perfectly definite and no longer unfamiliar, considerations are destined to relieve or to supplement the rigid causal structure [den starren Kausalaufbau] of the classical theory . . . whether the sacrifice be great or small, whether we find it difficult or easy, it seemed to me unavoidable for once clearly and frankly to state that within the purely empirical mechanics there are phenomena of motion and equilibrium which will forever escape an explanation on the basis of the differential equations of mechanics. . . .[189]

One cannot help but be struck by the "me too" tone of von Mises' repudiation of "the stiff causal structure" of classical mechanics and his representation of that renunciation as an act of moral virtue. Yet it is also precisely this tone which suggests that a conversion to acausality carried with it significant social approbation, social rewards so substantial that von Mises could not bear to let the atomic physicists monopolize them.

Although Weyl had already turned to the quantum theory in seeking support and ammunition for his attack on causality, Walter Schottky seems to have been the first atomic physicist to publish an acausal manifesto treating "The Problem of Causality in the Quantum Theory as a Basic Question for Modern Natural Science as a Whole."[190] Schottky's article of June 1921, subtitled "Attempt at a Popular Exposition," is clearly an expanded version of a lecture—very likely an inaugural lecture as Privatdozent for theoretical physics at the University of Würzburg, where he had recently habilitated after several years at the research laboratories of Siemens and Halske in Berlin. Schottky feels sure that inasmuch as one is accustomed to regard the rigorous laws of physics as a model and ideal for "all analytical contemplation of nature," a general and historical

189. R. v. Mises, *op. cit.* (note 143), *Selected Papers*, 2, 482, 487.
190. W. Schottky, *op. cit.* (note 186).

WEIMAR CULTURE, CAUSALITY, AND QUANTUM THEORY, 1918-1927

exposition of the "crisis," the "revolution in the basic conception of the form and range of physical laws" which is in preparation, will be welcome to his audience.[191]

In the first installment of the article Schottky builds up to the proposition that the electromagnetic field and its variables are finished, done. For, he argues with impeccable logic, if we don't know the laws of the interaction of atoms with radiation, but yet can only observe the electromagnetic field quantities through their interactions with matter, then these "state variables of the field theory . . . no longer possess any significance whatsoever for scientific research." Allowing that "that is a consequence which to be sure thus far only very few physicists have accepted," Schottky proceeds immediately to ask what sort of observable quantities, and what sort of connections between them, are to be put in place of the electromagnetic field. And the answer: "The law of causality itself, with its complete conditioning of the coming phenomena by the present and past phenomena, appears . . . to be placed in doubt."[192]

So much in the first installment. In the second installment we discover the conventional electromagnetic field variables and equations back at work, and all that remains of the earlier "analysis" is the insistance that any solution to the problem of the interaction of atoms and radiation must cancel causality. Schottky's first proposal is the oft-recurring conjecture that the field equations determine merely the *rate* at which the quantal elementary processes take place. But this "at first sight very attractive way is impassable," for Einstein has told him that because of the inexact fulfillment of the conservation laws, in the course of sufficiently long times a motion with arbitrarily large velocity could arise out of nothing—a point which escaped Bohr, Kramers, and Slater three years later.

Schottky now turns to his own pet idea that there is a direct connection of the emitting with the absorbing atom by retarded action at a distance, so that at the moment when a quantum is emitted it is already predetermined where, when, and by what atom it will be absorbed. But is this not causality with a vengence, a physics à la Calvin? That is certainly how Tetrode, an under-appreciated Dutch theoretical physicist, represented the case when, exactly one year

191. *Ibid.*, p. 492.
192. *Ibid.*, pp. 495-496.

later, he published the outlines of a theory based upon this same conception.[193] Yet such a thought never enters Schottky's mind; all that he sees is a failure of causality arising from the fact that it is no longer possible "to conceive the course of events like a continually and uniformly flowing stream," that because the "unbreakable threads" connecting emission and absorption extend infinitely far towards past and future, it is no longer possible in principle to predict the future from a cross section of the world at a given moment in time. And finally, to make the acausality doubly sure, Schottky asserts categorically—but inexplicably—that these elementary acts of emission and absorption, the precise positions of the beginnings and ends of these unbreakable threads, are indeterminate, "without direct cause and without direct effect," "outside the relation of cause and effect."[194]

Thus Privatdozent Schottky. Is perhaps the demonstration of the failure of causality which Geheimrat Professor Walther Nernst offered four months later in his inaugural lecture as rector of the University of Berlin—and which produced a correspondingly greater stir—less tendentious, less shallow and fallacious? Scarcely. Here again, what is most striking is the author's resolve to sink the law of causality by hook or by crook. And his motive for doing so is clear enough: "But, now, can philosophy and natural science really assert with certainty that, for example, every human action is the unambiguous result of the circumstances prevailing at the moment? If absolutely rigorous laws of nature controlled the course of all events, one would in fact scarcely be able to escape from this conclusion." But philosophy has adopted this position only because it has been tyrannized by the exact natural sciences, whose "conception of the principle of causality as an absolutely rigorous law of nature laced

---

193. H. Tetrode, "Über den Wirkungszusammenhang der Welt. Eine Erweiterung der klassischen Dynamik," *Zeitschr. f. Phys., 10* (1922), 317-328, received 14 June 1922. Tetrode is generally critical of "der einseitig gerichteten [!], zum Teil zufallsmässig bedingten Kausalität" to which the modern development of physics, above all the theory of field action, has led. But this conception of causality is not aboriginal in the human mind; therefore why not consider another conception. The result is remarkably like the Bohr-Kramers-Slater theory —the electro-magnetic field becomes unreal, the conservation of energy and momentum is statistical—but with exactly the opposite intent, namely to strengthen rather than relax determinism.

194. W. Schottky, *op. cit.* (note 186), pp. 509-511.

WEIMAR CULTURE, CAUSALITY, AND QUANTUM THEORY, 1918-1927

the mind [Geist] in Spanish boots, and it is therefore at present the obligation of research in natural science to loosen these fetters sufficiently so that the free stride of philosophical thought is no longer hindered."[195]

In outline Nernst's argument is that, first, the principle of causality implies the existence of exact natural laws, but none of the natural laws with which we are acquainted is exact, ergo it is possible, even likely, that causality does not obtain. (A debt to Exner is *not* acknowledged.) Second, even if it should be the case that the motions of individual molecules follow exact laws, we may postulate that the fluctuations in the zero-point energy of the æther disrupt these motions. As there are no experimental means for isolating a portion of the æther, the ideal of identically prepared, isolated systems is in principle unrealizable. "The law of causality demands that in the case of identical initial conditions, two different systems will follow identical courses in their changes; now, however, we conclude that two systems of this sort do not admit of being realized at all."[196] Nernst is not, of course, prefiguring a quantum field theory in which the fluctuations of the æther are themselves in principle indeterminate, but rather he implicitly assumes that, as with any classical field, the time, place, and manner of such fluctuations would be completely determined if the state of the *entire* æther could be specified. This possibility Nernst can exclude only on the grounds that "then we come to an infinitely extended system, in the face of which our laws of thought fail."[197]

Thus it is clear that although Nernst wishes with all his heart and soul to renounce causality, he is simply unable to free himself from the implicit assumption that the world *really is* causal. Nernst himself had begun to perceive this by the spring of 1922 when his lecture was republished in *Die Naturwissenschaften*. He then added a postscript pointing out that "most religions maintain that all events

195. W. Nernst, *op. cit.* (note 186), pp. 492, 495. The following quotations are from pp. 494-495.
196. Cf. Werner Heisenberg, "Über den anschaulichen Inhalt der quantentheoretischen Kinematik und Mechanik," *Zeitschr. f. Phys.*, 43 (1927), 172-198, received 23 March 1927: "But in the sharp formulation of the law of causality: 'If we know the present exactly, we can calculate the future' it is not the conclusion but the presupposition which is false. We are unable in principle to get to know the present in all of its determinative elements" (p. 197).
197. In the original: "unsere Denkgesetze versagen."

occur according to the will of a most high intelligence, and thus with complete logic, which is identical to the requirement of the principle of causality." Therefore "it is obviously less a question of whether or not one regards the principle of causality as rigorously valid, but much more a question of whether one conceives the natural processes to be comprehensible or, on the contrary, holds that the human mind is incapable of following these processes down to their last details." This latter is, now, Nernst's position—"only statistical mean values of the course of events are accessible to our natural-scientific cogniton"—and so we see once again that the repudiation of causality is in fact a repudiation of both reason itself and the cognitive enterprise in which physicists had theretofore been engaged.[198]

Apart from their common theme of *ignorabimus*, the three cases just examined—von Mises, Schottky, Nernst—show a remarkable temporal coincidence, suggesting a wave of conversions to acausality. And if one recalls that there were at just this moment no specific developments in physics which could plausibly be regarded as the source of such acausal convictions, then one can scarcely escape the conclusion that what we are dealing with is, essentially, a capitulation to those intellectual currents in the German academic world which we charted in Part I. Moreover, I am inclined to regard this capitulation as a very widespread phenomenon precisely because of the lack of negative evidence. The only other general academic lecture by a theoretical physicist at this moment with which I am acquainted contains, to be sure, no explicit renunciation of the principle of causality, but the clearest indications that it is a controversial issue: upon assuming the rectorate of the University of Berne, Paul Gruner made the most strenuous efforts to hang the opprobri-

---

198. Nernst is really quite old fashioned in his physical conceptions, and tries only to draw modish conclusions from them. His postulate that the motions and interactions of a *sub*-atomic mechanical system (the æther) perturb those of atomic-molecular mechanical systems, so that the laws of motion of a single gas molecule would express only mean values, had been entertained by Ludwig Boltzmann a quarter century earlier—without, of course, any failure of causality having been seen therein. (Boltzmann, *Vorlesungen über Gastheorie* [Leipzig, 1896-1898], trans. by Stephen G. Brush as *Lectures on Gas Theory* [Berkeley, 1964], p. 449.) In proposing that the fluctuations of the zero-point energy of the æther are responsible for triggering the decay of radioactive atoms Nernst is, in truth, adopting a causal explanation and mechanism for this prime example of an apparently acausal natural process.

WEIMAR CULTURE, CAUSALITY, AND QUANTUM THEORY, 1918–1927

ous epithet "causal" upon the mechanistic-materialistic world view and to sink the two together.[199]

c. Later Notable Conversions: Schrödinger and Reichenbach

In the fall of 1921, Erwin Schrödinger came to Zurich as professor of theoretical physics at the University, and so also came into contact with Hermann Weyl. Schrödinger had earlier been in close personal contact with Franz Exner as student, assistant, and Privatdozent in Vienna before the war. And when, one year later, he delivered his public inaugural lecture, he too delivered himself of a manifesto against causality bearing much resemblance to those issued on like occasions a year earlier. Schrödinger's manifesto, however, is distinguished not merely by its tight exposition and fine literary form, but also by its stress upon Exner's priority and importance.[200]

The principle of causality is the postulate "that every natural process or event is absolutely and quantitatively determined at least through the totality of circumstances or physical conditions that accompany its appearance." But "in the past four or five decades physical research has demonstrated perfectly clearly that for at least the overwhelming majority of phenomena, the regularity and invariability of whose courses has led to the postulation of general causality, the common root of the observed rigorous lawfulness is— *chance*." Now, insofar as the physical laws are statistical, they do not *require* that the individual molecular events be rigorously causally determined. (It was "Exner who in 1919, for the first time, with complete philosophical clarity" pointed out the groundlessness of the common assumption that molecular processes are in fact causal.) Moreover, Schrödinger finds most unsatisfying the duality in the laws of nature implied by the assumption of rigorous causality in the microcosm. "In the world of visible phenomena"—governed as it is by statistics, and thus by the concept of pure number—"we have

---

199. P. Gruner, *Die Neuorientierung der Physik. Rektoratsrede* (*op. cit.*, note 125), pp. 5, 11.
200. E. Schrödinger, "Was ist ein Naturgesetz?" *Naturwiss.*, 17 (4 Jan. 1929), 9-11; trans. as "What is a Law of Nature?" in Schrödinger, *Science, Theory, and Man* (New York, 1957), pp. 133-147. This was Schrödinger's inaugural lecture as professor of theoretical physics at the University of Zurich, 9 December 1922, which remained unprinted at the time.

clear intelligibility, but behind this a dark, eternally unintelligible imperative, an enigmatic 'must.'" (Compare Spengler: "Out of the principle of causality speaks fear of the world. Into it the intellect banishes the demonic in the form of a continually valid necessity, which rigid and soul-destroying is spread over the physical world picture."[201]) "This duplication of the laws of nature," Schrödinger continues, "reminds one too much of the animistic duplication of natural *objects* for me to believe in its tenability." And he concludes his lecture by asserting that the solution to our difficulties in atomic physics will depend upon "liberation from the rooted prejudice of absolute causality."

But here again the most striking features of the manifesto are, on the one hand, the quasi-moral terms in which causality—"ein dunkles, ewig unverstandenes Machtgebot"—is repudiated and, on the other hand, the frivolousness with which the objections to dispensing with causality are dismissed. And so once again there seems good reason to regard the conversion as a form of accommodation to the intellectual environment—especially good reason inasmuch as Schrödinger himself was prepared to admit the Spenglerian thesis that physical theory is an expression of, and thus conforms itself to, the *Zeitgeist*.[202]

I am acquainted with one further clear and dramatic example of a quasi-religious repudiation of causality in the years before quantum mechanics: Hans Reichenbach's conversion in the fall of 1925. In 1924, when he wrote his *Axiomatization of the Relativistic Theory of Space and Time,* Reichenbach still adhered firmly to the ideal of causality.[203] And even as late as August, or possibly September, 1925, Reichenbach could open a popular article on "Probability Laws and Causal Laws" by asserting that the law of causality, "this supreme law," is the precondition for the application of mathematics to physics and thus for physics to be an exact science.[204] But the

---

201. See note 77 and note 158.
202. Schrödinger, *op. cit.* (notes 133, 228, and 235).
203. H. Reichenbach, *Axiomatik der relativistischen Raum-Zeit-Lehre* (Braunschweig, 1924), trans. by Maria Reichenbach as *Axiomatization of the Theory of Relativity* (Berkeley-Los Angeles, 1969), p. 15.
204. H. Reichenbach, "Wahrscheinlichkeitsgesetze und Kausalgesetze," *Die Umschau,* 29 (3 October 1925), 789-792.

further one reads in this article the clearer it becomes that Reichenbach's allegiance to causality is beginning to waver. "Will we one day see the old ideal of physics realized, and comprehend the atomic world perfectly rigorously? Many researchers, including the most significant, believe this. . . . Others, on the contrary, and also among them significant researchers, are of the opinion that here perhaps there is an intrinsic limit to all explanation whatsoever." And Reichenbach concludes: "One is not permitted to say that under any circumstances it must be possible to find a causal explanation at the atomic level. Rather, the decision on this question must be reserved to physics itself, and cannot be made by philosophy."

Thus far our logical empiricist Reichenbach, in August or September 1925. Consider now the paper on "The Causal Structure of the World" which the notorious existentialist Reichenbach wrote in the following month or two.[205] The opening section—which carries a most curious subtitle: "Determinism and the Problem of the 'Now'"—begins: "It has become usual to regard the hypothesis of causality in physics as so self-evidently necessary that one no longer even thinks of subjecting it to criticism. And for the most part one does not notice at all to what a high degree this hypothesis is an extrapolation above and beyond the facts of experience. The assertion that without the hypothesis of causality no exact knowledge of nature is possible exhausts the customary defense of this standpoint." Here one searches in vain for the anticipated citation of Reichenbach's own earlier publications. "In what follows it will be shown that even without the hypothesis of rigorous causality it is possible to give a quantitative description of the course of nature which does everything that physics can possibly do. . . ." From a brief analysis of the concept of causality there then emerges very quickly, and essentially without argument, the "conclusion" that causality in the sense of determinism is an unjustified and needless extrapolation: "For physics the hypothesis of determinism is completely empty." It is therefore to be discarded, and in its place is set the concept of probability, taken as fundamental and irreducible.[206]

205. H. Reichenbach, "Die Kausalstruktur der Welt und der Unterschied von Vergangenheit und Zukunft," *Bayerische Akad. d. Wiss., München, math.-naturwiss. Abteilung, Sitzungsber.* (1925), pp. 133-175. Presented by C. Carathéodory in the session of 7 November 1925.
206. *Ibid.*, pp. 133, 136.

What is the occasion, the motive, the driving force behind this revolution? Is it perhaps that the decision which Reichenbach the logical empiricist had reserved for physics has suddenly fallen? Of any such developments we hear not a word. Rather we are assured that "It is the demand for a minimum of assumptions which compels us to renounce rigorous causality." Which is to say that existentialist philosophy, disguised as logical empiricism, has preempted the decision. But at this point Reichenbach the existentialist strips off his disguise: various investigations, notably those of Reichenbach the logical empiricist, have shown that the idea of a causal chain is closely connected with the topology of time, that is, with the fundamental concepts "earlier," "later," and "simultaneously."

> But [our existentialist Reichenbach stresses] what these investigations could not resolve is the problem of the "now" . . . the "now-point" as experience [Erlebnis] of the boundary between the past and the future. . . . An "earlier" and "later" exist also for determinism, but there is no "now"; there is no distinguishable point in time. And the feeling that my own existence is a reality, whereas Plato's life only throws its shadow into reality, must be an error. That, however, contradicts the entire orientation of our existence; we have a completely different attitude towards the future than towards the past. And unless one wants to regard every single one of our actions, every thought which accompanies us in the ordering of our daily life, as a single huge error, then determinism must be false. . . . If one renounces it, the contradiction with our elemental life-feeling can be avoided. Of course such a feeling must not be decisive if reason speaks cogently against it—let one therefore first analyze reason to see if the maintenance of determinism is necessary. And that it is not.[207]

In the suddenness of the conversion to acausality, in its explicit independence of recent developments in atomic physics, and in its perfectly manifest connection with a capitulation to existentialist *Lebensphilosophie*, Reichenbach's case is certainly extreme. Yet every one of the cases I have examined—and most especially those of Weyl, von Mises, and Schrödinger—share these characteristics to some extent. Excepting Exner, all of them have the qualities of a quasi-religious experience, of a rebirth, of contrition for past sins— in a word, of a conversion. When our converts attempted to demon-

207. *Ibid.*, pp. 138-141.

WEIMAR CULTURE, CAUSALITY, AND QUANTUM THEORY, 1918–1927

strate the necessity for this renunciation of causality, their arguments, as often as not, ought logically to have led to the opposite conclusion. From this I think one must infer that they fully anticipated that *any* argument advanced by a physicist as a demonstration of the failure of causality would be received by their audience with uncritical applause. And when one recalls that the audiences for most of these renunciations of causality were, in the first instance, the whole body of a university assembled on a ceremonial occasion, then I think it reasonable to construe such renunciations as attempts to alter, or at least receive a special dispensation from, an unbearably opprobrious public image of the theoretical physicist as a "hard-boiled determinist."

III.4. Unregenerates against the Tide, 1922–1923

The wave of conversions to acausality in the latter part of 1921 prompted a series of public demonstrations in support of causality by "the most significant" theoretical physicists. Planck and Einstein—Nernst's and von Mises' colleagues at the University of Berlin—were quite disturbed; they felt that their colleagues were (unwittingly) betraying their calling, and carrying fuel to the antiscientific fires then raging in Germany. In 1922 and 1923 they both came forward to rebuke such rashness and to defend the principle of causality in physics and beyond.

Among the first, however, to raise his voice was Mach's old bulldog, Joseph Petzoldt, who in a long letter to the editor of *Die Naturwissenschaften* "Concerning the Crisis of the Concept of Causality" lectured Schottkey and Nernst like schoolboys.[208] The questions which they have dragged up were thoroughly considered and disposed of more than two decades ago. To Schottky he pointed out that temporal action at a distance is quite as compatible with the Machian concept of causality as is spacial action at a distance. To Nernst he declared firmly that while it is *conceivable* the regularity of nature could fail, "there is no limit to the 'understanding' [des 'Begreifens']." Only Schottky replied to Petzoldt, and his rebuttal was weak, vague, and disingenuous—"it goes without saying that the

208. J. Petzoldt, "Zur Krisis des Kausalitätsbegriffs," *Naturwiss.*, 10 (11 August 1922), 693-695, dated 2 July 1922.

physicists too are not glad . . . to renounce the assumption that *all* events are tied together by laws."²⁰⁹ One thus sees how little prepared the converts were to meet criticism, how disconcerting they found it, and how readily they could be silenced by it.

On 29 June 1922, some weeks before the publication of Petzoldt's letter, Max Planck, as secretary of the Prussian Academy, took the occasion of the annual public session in honor of their spiritual founder Leibniz to affirm the transcendental character of the law of causality and to reprimand academician Nernst—naturally, without naming him—for his irresponsible talk.²¹⁰ When the quantum hypothesis shall have been developed sufficiently so that one can properly speak of a quantum theory, that will be the proper moment to consider its consequences for our scientific-causal thought. "Meanwhile groping speculation offers itself the most various possibilities, whose rich profusion admonishes critical caution all the more as precisely at the present time not inconsiderable dangers to the sure advance of scientific work have arisen from various sides." Chief among these dangers is penetration by a "lively, but basically unfruitful dilettantism," confusing and fusing science and religion, seeking "directly and relatively effortlessly to pluck the golden fruits of knowledge and bliss from the rich tree of life, in contrast to the so-called school or guild science, which only in hard, protracted, specialized studies is able to gather one tiny little grain after another into its barn. Today it cannot yet be foreseen when and where these colorfully iridescing foam bubbles will finally burst. . . . Vis-à-vis such intellectual currents the academies find themselves in a substantially better protected situation than their sister institutions the universities, which have to stand far more directly against the shifting surge of the waves of public life."²¹¹ Evidently then, Planck, too, saw, or at least sensed, an intimate connection between an anticausal manifesto by a rector of the University of Berlin and that constellation of attitudes which made the Weimar intellectual milieu seem to the theoretical physicist so hostile to his enterprise.

Early in the following year on 17 February 1923, Planck devoted

---

209. W. Schottky, "Zur Krisis des Kausalitätsbegriffes," *Naturwiss.*, 10 (1922), 982, dated 6 October 1922.
210. M. Planck, "Ansprache des vorsitzenden Sekretärs" (*op. cit.*, note 17).
211. *Ibid.*, pp. 46-48.

WEIMAR CULTURE, CAUSALITY, AND QUANTUM THEORY, 1918–1927

an entire public lecture, again in the Prussian Academy, to a most uncompromising and courageous reaffirmation of allegiance to the principle of causality—not merely in the natural sciences, but in the *Geisteswissenschaften* too.[212] Planck knew full well that in this "violent dispute" over causality, "splitting the intellectuals into two camps," one of reason and one of feeling, the bulk of his audience lay within the latter camp, that much of what he said would "provoke" them and might even appear "a blasphemy, as cheap as it is intolerable."[213] Nonetheless he proceeded to tell his audience that "the assumption of a causality without exception, of a complete determinism, forms the presupposition and the precondition for scientific [*wissenschaftlich*] cognition." And anticipating precisely the issues which the uncertainty principle and complementarity were to raise, Planck knew well in advance what position he would adopt: "But has it then—one could now certainly ask—any sense whatsoever to continue speaking of a definite causal interconnection when no one in the world is capable of actually comprehending that causal interconnection as such? . . . Absolutely. . . . For causality is . . . transcendental, it is entirely independent of the constitution of the inquiring intellect, indeed it would retain its significance even in the complete absence of a knowing subject."[214]

Again in the summer of 1923 Planck took the opportunity offered by his contribution to the issue of *Die Naturwissenschaften* commemorating the tenth anniversary of the Bohr atom in order to warn his colleagues against those "eminent physicists"—unnamed, of course, but evidently Exner, Nernst, Schrödinger, and, yes, Bohr himself—"who want to allow the principles of the classical theory basically only a statistical significance. . . . Such a conception seems to me, however, to shoot far over and beyond the target, if only because with the abandonment of classical dynamics they simultaneously pull out the foundations of every rational statistics."[215]

212. M. Planck, *Kausalgesetz und Willensfreiheit* (*op. cit.*, note 18).
213. *Ibid.*, pp. 140, 160. See note 150.
214. *Ibid.*, p. 161. In his address of 3 August 1914 (*op. cit.*, note 158), pp. 78, 88-89, Planck had asserted these same propositions equally categorically but without any suggestion that his views were unwelcome to his audience.
215. M. Planck, "Die Bohrsche Atomtheorie," *Naturwiss.*, *11* (6 July 1923, Bohr Heft), 535-537; reprinted in Planck's *Physikalische Abhandlungen und Vorträge* (Braunschweig, 1958), 2, 543-545. Note that, like Exner, Planck, too, confuses the validity of classical dynamics in the atomic domain with the validity

93

On this issue of causality Planck and Einstein were in complete agreement, and their stand together against the rising tide of acausal sentiment contributed to the preservation of a close personal bond between these two men despite the wide divergence in their political and social views. Writing to Einstein on 22 October 1921, a week after Nernst's *Rektoratsrede,* Planck, as president of the Gesellschaft deutscher Naturforscher und Ärzte for 1922, appealed to Einstein's "fine feeling for causal interconnections," and so succeeded in overcoming Einstein's resolve to boycott an organization which, he felt, had treated him meanly the previous year.[216]

Apart from insisting that the adjective "causal" occupy a conspicuous place in his definitions of the goal and function of scientific activity, Einstein was not given to dogmatizing publicly and popularly on this issue.[217] His own efforts were devoted to searching in the

---

of causality. Bohr had recently associated himself with "the view, which has been advocated from various sides, that, in contrast to the description of natural phenomena in classical physics in which it is always a question only of statistical results of a great number of individual processes, a description of atomic processes in terms of space and time cannot be carried through in a manner free from contradiction by the use of conceptions borrowed from classical electrodynamics. . . ." ("Über die Anwendung der Quantentheorie auf den Atombau. I. Die Grundpostulate der Quantentheorie," *Zeitschr. f. Phys., 13* [ca. 1 Feb. 1923], 117-165, on 157; English translation in *Cambridge Philosophical Society, Proceedings* [1924], supplement, 42 pp., on 35.) It was, however, only in 1924 that Bohr spoke of "a causal description in space and time."(*Op. cit.,* note 226, p. 790.) By 1927 Bohr had ceased to regard "causal" descriptions and "space-time" descriptions as equivalent, and saw them, rather, as "complementary." (*Op. cit.,* note 241.)

216. "Es ist doch sonst bei Ihrem feinen Gefühl für Kausalzusammenhänge nicht Ihre Art, bei sachlichen Überlegungen allgemeinen Gefühlsstimmungen den entscheidenden Einfluss zu gewähren." Planck to Einstein, 22 October 1921, Einstein Collection, Institute for Advanced Study, Princeton. The experience at Nauheim, September 1920, had left a very bad taste in Einstein's mouth; nonetheless he agreed to deliver a major lecture at the following, hundredth anniversary congress in Leipzig. That summer, however, in the aftermath of Rathenau's assassination, Einstein felt compelled to withdraw from public life, and from Germany, for a time.

217. See note 147 and text thereto; cf., Einstein, "Das Kompton'sche Experiment. Ist die Wissenschaft um ihrer selbst willen da?" *Berliner Tageblatt,* 20 April 1924, Nr. 189, I. Beiblatt (Readex Microprint edition of the publications of A. Einstein, Nr. 147), where Einstein maintains that the great educational task of science "darin besteht, das Streben nach kausalem Erkennen in der Gesamtheit zu wecken und wach zu erhalten." Addressing a popular audience in June 1922 on "New Results Regarding the Nature of Light," *op. cit.* (note 135), "Einstein in conclusion gave expression to his opinion that considering the great advances in our knowledge of nature one can count upon a future solution of this problem also, and that the human consciousness possesses the necessary capabilities [Voraussetzungen] for the comprehension of the natural processes."

WEIMAR CULTURE, CAUSALITY, AND QUANTUM THEORY, 1918–1927

field-theoretic apparatus of general relativity for a super-causal solution to the quantum problem by means of over-determined systems of differential equations.[218] Any program—e.g., Tetrode's—to solve the problem by tightening rather than loosening the causal interconnections he greeted most enthusiastically, while efforts in the opposite direction—e.g., the Bohr-Kramers-Slater theory—he received most cooly and critically.[219] Einstein was convinced, and rightly so, that his fellow physicists were rushing to embrace a failure of causality without having made any serious attempt to explore the possibilities for a causal solution. In order to advertise this point,

---

218. How long had Einstein been pursuing this program? Russell McCormmach, "Einstein, Lorentz, and the Electron Theory," *Historical Studies in the Physical Sciences*, 2 (1970), 41-88, especially 83-84, raising the general problem of Einstein's conversion to a field approach, locates that reorientation in the years 1907-1909, and sees Einstein as aiming thenceforth at "a field theory with quantum solutions, not a quantum 'mechanics.'" Einstein's own statements in the 1920's of his quantum-theoretical program are quite consistent with this early date. Thus in January 1920 he wrote Born: "I believe now as before [nach wie vor] that one must seek an overdetermination by differential equations in such a way that the *solutions* no longer have the character of a continuum. But how??" (*Briefwechsel* [*op. cit.*, note 14], p. 43.) And again on 28 June 1929, receiving from Planck's hands the second Planck Medal of the German Physical Society—the first had gone to Planck himself—Einstein implied that it had *always* been and always would be his program: "There were two ideas, especially, around which my ardent exertions grouped themselves. The evolution of the world [das Naturgeschehen] seems to be so largely determined that not only the temporal course, but also even the initial state is largely bound by law. To this idea I believed I had to give expression by finding overdetermined systems of differential equations. The postulate of general relativity as well as the hypothesis of the unified structure of physical space, or the field, were supposed to serve as guideposts in this search. There the goal stands, unattained. And there was scarcely a fellow physicist to be found who shared my hope of arriving by this route at a deeper understanding of reality. What I found on the subject of quanta are only chance insights [Gelegenheitseinsichten] or, to a certain extent, fragments, which broke off in the course of my fruitless exertions upon the great problem. I am ashamed now to receive for this so high an honor.

"Despite the fact that I believe strongly that we will not remain stuck at a sub-causality, but rather, ultimately, we will even arrive at a super-causality in the sense indicated, nonetheless I most highly admire the contributions of the younger generation of physicists which are comprised under the name "quantum mechanics," and I believe in the deep truth-content of this theory; only I believe that the restriction to statistical laws will be only temporary." (Einstein, "Ansprache . . . an Prof. Planck," *Forschungen und Fortschritte*, 5 [1929], 248.)

219. Writing to Paul Ehrenfest late in August 1922 (SHQP Microfilm Nr. 1; the letter is undated), Einstein recommended "eine sehr geistvolle Arbeit von Tetrode über das Quantenproblem. Vielleicht hat er Recht; jedenfalls zeigt er sich durch diese Arbeit als Kopf ersten Ranges. Schon lange hat mich nichts mehr so elementar gepackt." For his reactions to the Bohr-Kramers-Slater theory: Martin J. Klein, "The First Phase of the Bohr-Einstein Dialogue," *Historical Studies in the Physical Sciences*, 2 (1971), 1-39, on 32-33.

Einstein published in December 1923 a sketch of his own program, despite the fact that he had made essentially no progress with it.[220]

Altogether, were one unacquainted with the overwhelming anti-causal sentiment in the Weimar intellectual environment and the social pressures to which a physicist stepping before a general academic audience was exposed, one would have to be surprised at just how few physicists came forward to defend causality, and take issue with their colleagues who were, in fact, repudiating physics as a cognitive enterprise. It seems reasonable to suppose, however, that although few had the courage to brand themselves publicly as determinists, many a senior and influential colleague let it be known with what displeasure he viewed these capitulations to anti-scientific currents.[221] And such intimidation may well have been responsible for the decline in the number of full-scale manifestoes against causality by physicists after the end of 1921.

III.5. The Situation circa 1924

Although public silence seems to have been imposed fairly effectively—only in 1929 did Schrödinger allow his manifesto to be printed, while Reichenbach's was interred in the proceedings of the Munich academy—the tide against causality was not stemmed. There are numerous indications that privately the question continued to be "much discussed,"[222] and it was the impression of a contemporary

---

220. A. Einstein, "Bietet die Feldtheorie Möglichkeiten für die Lösung des Quantenproblems?" *Preuss. Akad. d. Wiss., phys.-math. Kl., Sitzungsber.* (13 Dec. 1923), pp. 359-364, published 15 January 1924. It must be said, however, that Einstein gave a rather different impression of the origins of his paper and his intentions in publishing it when mentioning it to H. A. Lorentz, 25 December 1923 (for source see note 114): "Ich sehe eine Möglichkeit den Quantentatsachen von der Feldtheorie aus beizukommen unter Preisgabe der mechanischen Gleichungen. Das mechanische Verhalten der Elektronen (Singularitäten) soll durch überbestimmte Feldgleichungen mitbestimmt werden. Leider sind die mathematischen Schwierigkeiten für meine Kräfte zu gross. Ich habe deshalb durch eine kurze Abhandlung das Interesse der Fachgenossen auf die Methode zu lenken versucht."

221. For intimations of such a distaste for "polemics" see my "Doublet Riddle" (*op. cit.*, note 142), p. 171.

222. Wolfgang Pauli, "Quantentheorie," *Handbuch der Physik*, Band 23: *Quanten* (Berlin, 1926), p. 11: the moment of transition of a single excited atom "appears, according to the present state of our knowledge, to be determined solely by chance. It is a much discussed but still undecided question whether we have to regard this as a fundamental failure of the causal description

WEIMAR CULTURE, CAUSALITY, AND QUANTUM THEORY, 1918–1927

observer that considerable sympathy for, and more or less explicit avowals of belief in, acausality were to be met with "ever more frequently."[223] Where in 1922 Friedrich Poske was simply shocked by Nernst's renunciation of causality, a year later he "warmly recommended" the second edition of Exner's lectures.[224]

And in seeking the grounds for this relatively undramatic but quite definite drift away from causality circa 1924 one can finally point to specific recent developments in atomic physics. For, as I discussed in Section II.4, in 1923 and 1924 the atomic physicists were becoming convinced of the fundamental inadequacy of the extant quantum theory of the atom—which supposed classical mechanics to be valid for motions within the stationary states—and were beginning to doubt the reality of the visualizable atomic models to which that theory had been applied. I argued there that the Weimar intellectual milieu at the very least facilitated the precipitation of a generalized conviction of a crisis of the old quantum theory, and

---

of nature, or only as a temporary incompleteness of the theoretical formulation." (The article was written in 1924-1925.) Compare the remarks which H. A. Kramers added in this same connection to the German translation of his popular account of the Bohr theory, originally written in Danish jointly with Helge Holst: Kramers posed for the first time the question whether the probability laws have an underlying causal mechanism or "das physikalische Kausalitätsgesetz in Wirklichkeit nicht gilt." He then warned against stamping this latter conception as an epistemological impossibility, and added that "for the moment it is certainly rather a matter of taste which alternative one prefers, and perhaps will remain so forever. The actual choice affects the methods of physical research far less than one would at first perhaps like to believe." (*Das Atom und die Bohr'sche Theorie seines Baues* [Berlin, 1925], p. 139. The preface is dated March 1925.)

223. A Gatterer, *op. cit.* (note 146), p. 47; also p. 36. Although Nernst seems to have refrained from printing anything further on the question, he was not entirely silent. On 11 February 1925 he delivered one of those popular lectures to the lay members of the Kaiser Wilhelm Gesellschaft which were "supposed to serve to give them an insight into the scientific work of the institutes" under the title "Causalgesetz und neuere Naturforschung." It was noticed in the *Mitteilungen der Gesellschaft Deutscher Naturforscher und Ärzte*, 2 (April 1925), 10.

224. F. Poske, *Zeitschr. f. den physikal. u. chem. Unterricht*, 35 (July 1922), 188-189, emphasized that Nernst's parallel between loosening of the causal principle and "certain theological doctrines . . . makes clear how earth-shattering his conception, if it were accepted, would have to be for the entire *Weltanschauung.*" In March of 1923 Poske, *ibid.*, 36, 133-134, merely described the position taken in Exner's "especially noteworthy" final chapter, observing that "this conception is closely related to other recently expressed views according to which the role of the law of causality has been played out, and causeless chance governs."

I emphasized how very apt, from the point of view of an adaptation to the intellectual environment, the principal diagnosis—"the failure of mechanics"—indeed was. Yet however much this crisis and rallying cry themselves owed to precisely the same intellectual currents which were driving the reaction against causality, in the present connection the important fact is that at this moment the antimechanical and anticausal movements coalesced, reinforcing one another. The confluence and synergy of these movements appears all the more intelligible if one recalls the persistent tendency, evidenced by such diverse figures as Exner and Planck, to confuse and confound the validity of the laws of classical mechanics and the validity of the law of causality.

Now, finally, after all the posturing before popular audiences, we find the first attempts to *do* a little acausal physics. The earliest of these, appropriately enough, we owe to a quasi-crank, Hans Albrecht Senftleben. The program advanced in his paper "On the Foundation of the 'Quantum Theory'" of November 1923 included such prescient postulates as that "natural phenomena generally are to be regarded as statistical effects of totalities of elementary molecular processes which are themselves not subject to the requirement of causality," and that "Planck's constant h limits in principle the possibility of describing a process in space and time with arbitrary accuracy." Moreover, Senftleben was not entirely ignored.[225]

But when it comes to attention, few papers could compare with that published by Bohr, Kramers, and Slater in the spring of 1924. In January John Clarke Slater, fresh from the two Cambridges, had carried to Copenhagen a semi-deterministic space-time picture of light quanta traveling along the Poynting vector of a virtual radiation field which—and this was novel—Slater assumed to be continually emitted by atoms throughout their existence in stationary states. "When this view was presented to Professor Bohr and Dr. Kramers," Slater recalled not long afterward, "they pointed out that

---

225. H. A. Senftleben, "Zur Grundlegung der 'Quantentheorie,'" *Zeitschr. f. Phys.*, 22 (March 1924), 127-156, received 13 November 1923. Quotations from pp. 129-131.

Kis, *op. cit.* (note 146), discussed Senftleben quite seriously. In the summer of 1924 Kramers visited him in a sanatorium in Denmark. (Letters to Bohr and Kramers of 23 August and 8 October 1924 in the Archive for History of Quantum Physics.)

WEIMAR CULTURE, CAUSALITY, AND QUANTUM THEORY, 1918–1927

the advantages of this essential feature would be kept, although rejecting the corpuscular theory, by using the field to induce a probability of transition rather than by guiding corpuscular quanta. ... Under their suggestion, I became persuaded that the simplicity of mechanism obtained by rejecting a corpuscular theory more than made up for the loss involved in discarding conservation of energy and rational causation [n.b.], and the paper ... was written."[226] And it is, I think, only by reference to the widespread acausal sentiment that one can understand the immediate and widespread assent which the theory received in Germany, even though it was in fact hardly a theory at all but rather a vague suggestion of how, renouncing causality, one might try to give a "formal" account of the interaction between atoms and radiation.[227]

226. N. Bohr, H. A. Kramers, and J. C. Slater, "Über die Quantentheorie der Strahlung," *Zeitschr. f. Phys.*, 24 (ca. 20 May 1924), 69-87, received 22 February 1924, dated January 1924. The publication of the paper was probably delayed in order that it not appear earlier than the English version in the May issue of the *Philosophical Magazine*, 47 (1924), 785-802. J. C. Slater, "The Nature of Radiation," *Nature*, 116 (1925), 278, dated 25 July 1925; quoted by van der Waerden *op. cit.* (note 144), pp. 13-14, who also reprints "On the Quantum Theory of Radiation," pp. 159-176. The very real difference between Slater's original notion and the view to which Bohr and Kramers persuaded him suggests a distinction between *probabilistic* and *acausal* approaches. Thus the guiding field approaches to the problem of light quanta, which had long been commonplace, and de Broglie's suggestion of a wave as a guiding field for material particles, were probabilistic, but only by anachronistically imposing Heisenberg's uncertainty principle can one say that they abandoned causality. Their proponents did *not* suppose it impossible to get behind these probabilities to the determinants of the individual events. An *acausal* theory, on the contrary, is one which excludes this possibility in advance. Thus the Bohr-Kramers-Slater interpretation was formed from Slater's original proposal by precluding in principle "rational causation" in the interaction of atoms and radiation. This feature was then made more palatable by stressing the "formal character" of their description of the interaction, in contrast, one might add, with Slater's "physical" picture.

227. Interesting for its testimony to the extent and strength of belief which the "theory" received in Germany—as for much else—is W. Pauli to H. A. Kramers, 27 July 1925 (Archive for History of Quantum Physics, SHQP Microfilm Nr. 8, Section 9): emphasizing that he does not want to be mistaken for one of the "true believers," "Ich halte es überhaupt für ein ungeheures Glück, dass die Auffassung von Bohr, Kramers und Slater durch die schönen Experimente von Geiger u. Bothe sowie durch die kürzlich erschienenen von Compton so schnell widerlegt worden ist. Es ist zwar natürlich richtig, dass Bohr selbst, auch wenn diese Experimente nicht gemacht worden wären, nicht mehr an dieser Auffassung festgehalten hätte. Aber viele ausgezeichnete Physiker (wie z.B. Ladenburg, Mie, Born) hätten daran festgehalten und diese unglückselige Abhandlung von Bohr, Kramers und Slater wäre vielleicht für lange ein Hemmnis des Fortschrittes der theoretischen Physik geworden!"

Certainly that same essentially moral feeling underlying Schrödinger's repudiation of causality predominated in his response to Bohr, Kramers, Slater. Having demonstrated that the "Exner-Bohr" conception of statistical conservation of energy involves an unbounded random walk of the energy content of a closed system, Schrödinger did *not* conclude that the theory is impossible, but rather, "clutching at it with both hands," he saw in it a demonstration that "a certain stability in the course of the world sub specie aeternitatis can only subsist through an *interconnection* of every individual system with the entire rest of the world. . . . Is it idle play with ideas," Schrödinger asked rhetorically, "if one is, in this connection, struck by the similarity with social, ethical, cultural phenomena?" Clearly Schrödinger thought one ought to be, and that that recognition should be decisive.[228]

III.6. Causality's Last Stand, 1925–1926

We are now approaching the end of the development which I have been trying to trace, that is of the rise of a will to believe that causality does not obtain at the atomic level *before* the invention of an acausal quantum mechanics. With the introduction of Heisenberg's matrix mechanics in the fall of 1925 and of Schrödinger's wave mechanics in the spring of 1926, physicists realized relatively quickly that that belief no longer had to rest primarily upon ethical considerations or to involve a purely gratuitous renunciation of the possibility of exact knowledge of atomic processes. The grounds of argument and belief were thereby substantially altered. I will not attempt here to treat the growing realization of this new situation

---

228. E. Schrödinger, "Bohrs neue Strahlungshypothese und der Energiesatz," *Naturwiss.*, 12 (5 September 1924), 720-724. Schrödinger, peculiarly, seems to have seen in the Bohr-Kramers-Slater proposal an attempt to rid the quantum theory of discontinuities, a goal he then pursued in and through the wave mechanics which he began to develop late in 1925 on the basis of de Broglie's ideas. Writing to Wilhelm Wien on 18 June 1926, Schrödinger observed: "Es scheint ja, dass zur Zeit nicht auf allen Seiten die Ueberzeugung besteht, dass eine Abkehr von den grundsätzlichen Diskontinuitäten unbedingt zu begrüssen ist, *wenn* es damit geht. Ich aber habe immer mit Inbrunst gehofft, dass das möglich sein wird und würde mit beiden Händen zugegriffen haben—wie ich bei Bohr-Kramers-Slater mit beiden Händen zugriff—auch wenn der Zufall nicht gerade mir selbst den ersten (mit Rücksicht auf de Broglie muss ich richtiger sagen: den zweiten) Zipfel in die Hände gespielt hätte." (Archive for History of Quantum Physics.)

WEIMAR CULTURE, CAUSALITY, AND QUANTUM THEORY, 1918–1927

in any detail, but only emphasize once again how conscious the physicists were of the fact that they were playing before an audience hostile to causality=mechanism=rationalism, and how anxious many were to play up to that audience.

Not all, however, did so. During this period it was Wilhelm Wien who assumed again the role of champion of causality. In January 1925 he had taken his case to the general public through the pages of the Leipzig *Illustrierte Zeitung* where his denial that the quantum theory has, will, or could lead to an abandonment of the law of causality threaded its way among pictures of cabinet meetings and catastrophes, opera balls and carnival costumes. "The notion that nature is comprehensible . . . is identical with the conviction that all natural processes can be reduced to causality, to invariably valid natural laws." Of all purely philosophical notions the concept of causality has had the greatest impact on the development of humanity. It is responsible for the suppression of superstition, for modern natural science, and for the revolutions in technology and industry (n.b., the audience was nonacademic). Although the problem of the interaction of atoms and radiation "has brought all of theoretical physics into a crisis which will occupy it for a long time," the present form of the quantum theory can only be transitional, for "a statistics without a causal foundation will never be recognized by physics as something final."[229]

During the academic year 1925–1926 Wien fully exploited the platform available to him as rector of the University of Munich, speaking out in defense of causality in both his official addresses.[230]

---

229. W. Wien, "Kausalität und Statistik," *Illustrierte Zeitung* (Leipzig), Nr. 4169 (Feb. 1925), pp. 192, 194, 196. Max Planck was by no means silent: "Physikalische Gesetzlichkeit im Lichte neuerer Forschung," *Vorträge und Erinnerungen* (Stuttgart, 1949), pp. 183-205, especially pp. 184, 194-196; also reprinted in Planck, *Physikalische Abhandlungen* (Braunschweig, 1958), 3, 159-171. This lecture was delivered on 14 February 1926 in Düsseldorf, and again on 17 February in the Auditorium Maximum of the University of Berlin. (*Forschungen und Fortschritte*, 2 [15 March 1926], 50.)

230. W. Wien, *Universalität und Einzelforschung. Rektorats-Antrittsrede gehalten am 28. November 1925*, Münchener Universitätsreden, Heft 5 (Munich, 1926), 19 pp.; *Vergangenheit, Gegenwart und Zukunft der Physik. Rede gehalten beim Stiftungsfest der Universität München am 19. Juni 1926*, Münchener Universitätsreden, Heft 7 (Munich, 1926), 18 pp. In his one other published academic address, *Goethe und die Physik. Vortrag gehalten in der Münchener Universität am 9. Mai 1923* (Leipzig, 1923), 39 pp., on p. 5, Wien had made a point of owning his allegiance to causality: "Accustomed to seek the law of causality everywhere, the physicists ever and again give themselves great pains to uncover the reasons which led Goethe to his unfavorable attitude toward physics."

Although his inaugural lecture of November 1925 contained no reference to the current situation in physics, Wien nonetheless took the opportunity, as we saw in Section I.1, to stress the historical importance of causality, equating it once again with the conviction that nature can be comprehended by the logical force of the human intellect, and then went on to criticize Langbehn, Chamberlain, and Spengler for their antirationalism and pessimism. The slightly equivocal tone of this lecture had, however, disappeared entirely in June 1926 when, towards the end of his term as rector, Wien spoke at the annual founder's day ceremonies on "The Past, Present, and Future of Physics," or, more accurately, on causality in the past, present, and future of physics. The theme first appears on page 4 of the printed text as the capacity of the human intellect to grasp the causality of natural processes, continues on pages 6–8 where it is emphasized that, even when the laws are statistical, causality must reign at the level of the elementary processes, and reaches a climax on pages 10 and 11 where Bohr is attacked directly and by name.

Here one must recall that, supporting himself in part upon Heisenberg's discovery of a way to do atomic physics while renouncing the goal of a detailed picture of intra-atomic motions and mechanisms, Bohr had recently been expressing far more openly and categorically his hope and belief that such pictures were impossible in principle, that physics was faced "with an essential failure of the pictures in space and time on which the description of natural phenomena has hitherto been based."[231] Quoting these words, Wien then sought to reprimand and silence Bohr and all others of like convictions with that same demand for self-censorship which Planck had advanced so successfully in 1922: "The physicists have always openly displayed before all the world the difficulties with which they have to contend. . . . But we must be very careful with pronouncements whose significance extends far beyond the limits of the field of physics." And Wien then went on to assert in the strongest terms that there is no physical field which is closed

---

231. W. Heisenberg, *op. cit.* (note 144). N. Bohr, "Atomic Theory and Mechanics," lecture at the Sixth Congress of Scandinavian Mathematicians, 31 August 1925, and revised before publication in *Nature, 116* (5 December 1925), 845-852; reprinted in Bohr, *Atomic Theory and the Description of Nature* (Cambridge, 1934), pp. 25-51; quotation from 34-35. The German text, "Atomtheorie und Mechanik," appeared in *Naturwiss., 14* (January 1926), 1-10.

WEIMAR CULTURE, CAUSALITY, AND QUANTUM THEORY, 1918–1927

to our understanding, and that physicists will not rest until they have subjected atomic processes to the law of causality.[232]

At this point, having dealt with Bohr and causality, Wien turned upon his colleague, the Professor of Theoretical Physics, Arnold Sommerfeld—without, of course, naming *him*. Although Wien had readily adapted his justifications for doing physics to the changing public values, he had nonetheless been concerned to shield the enterprise itself from the influence of the Wiemar cultural milieu. Sommerfeld's "Atomystik," on the contrary, dressed up for the public with pythagorean numerical harmonies and number mysteries, was not merely an attempt to use the quantum theory to play up to the ambient antirationalism, but represented an actual research program. "The number mysticism," Wien hoped and expected, "would be supplanted by the cool logic of physical thought; not perhaps to everyone's joy. For mysticism often exerts upon many minds a greater force of attraction than the cold and sober physical mode of thought. It is far from my intent to attack mysticism as such. There are many areas of the life of the soul from which mysticism cannot be excluded; but in physics it does not belong. A physics in which mysticism governs, or even collaborates, relinquishes the ground from which it draws its strength, and ceases to deserve its name." Wien then concluded his lecture by reaffirming once again his confidence that "insight into the causal interconnections of natural processes will continue to be possible," suggesting that those who express doubts on this score are just suffering from mental exhaustion, and perhaps also on that account are inclined to harken to pessimistic words about the *Untergang des Abendlandes* or the *Zusammenbruch der Naturwissenschaft*.[233]

The confidence and corresponding aggressiveness which Wien manifested on the issue of causality in the spring of 1926 derived

---

232. W. Wien, *Vergangenheit, Gegenwart und Zukunft der Physik*, (*op. cit.*, note 230), p. 10. Cf. note 221.

233. *Ibid.*, pp. 15, 18. We may perhaps read this as a veiled allusion to the breakdown which Bohr suffered in 1921 and which often threatened to recur. Wien's hostility toward Sommerfeld's "Atomystik" and his agressiveness due to confidence in Schrödinger's wave mechanics is corroborated by Werner Heisenberg's recent memoirs, *Der Teil und das Ganze: Gespräche im Umkreis der Atomphysik* (Münich, 1969), trans. [often quite inaccurately] as *Physics and Beyond: Encounters and Conversations* (New York, 1971), pp. 104-105, and 72-73, respectively.

chiefly from Erwin Schrödinger's papers on wave mechanics which Wien was then publishing in his journal, the *Annalen der Physik*. Having repudiated causality for social-ethical reasons in 1922–1924, by the fall of 1925 Schrödinger had converted back to causality for what were most probably personal-political reasons.[234] He now conceived and developed the wave mechanics as a causal space-time description of atomic processes in opposition to the Copenhagen-Göttingen matrix mechanics. To accept their contention that such a description is not possible "would be equivalent to a complete surrender." For, Schrödinger argued in February 1926 in his second paper, "we really cannot change the forms of thought, and what cannot be understood within them cannot be understood at all. There are such things—but I do not believe that the structure of the atom is one of them."[235]

Yet at just that moment in June 1926 when Wien, armed with Schrödinger's theory, was striking out so vigorously, the anticipated victory was being transformed into defeat by Max Born's statistical interpretation of the wave function, building an abandonment of causality right into the foundations of the wave mechanics.[236] "The

234. V. V. Raman and Paul Forman, "Why Was It Schrödinger Who Developed de Broglie's Ideas?" *Historical Studies in the Physical Sciences*, 1 (1969), 291-314. A collection of seventeen letters from Schrödinger to W. Wien, December 1925-November 1927, which has recently come to light, strengthens the case advanced in that publication. Xerox copies of these letters have been deposited in the Archive for History of Quantum Physics.

235. E. Schrödinger, "Quantisierung als Eigenwertproblem (Zweite Mitteilung)," *Ann. d. Phys.*, 79 (April 1926), 489-527, reprinted in Schrödinger, *Die Wellenmechanik*, Dokumente der Naturwissenschaft, Abteilung Physik, Band 3, ed. Armin Hermann (Stuttgart, 1963), pp. 25-63; on 509 and 45, respectively. A partial, and occasionally quite erroneous, translation is included in Gunther Ludwig, *Wave Mechanics, Selected Readings in Physics*, ed. D. ter Haar (Oxford, 1968), pp. 106-126, on 120-121.

On 25 August 1926, Schrödinger wrote W. Wien that: "Ich möchte aber heute nicht mehr gern mit Born annehmen, dass solch ein einzelnes Ereignis [e.g., the interaction of an electron with an atom] 'absolut zufällig' d.h. vollkommen undeterminiert ist. Ich glaube heute nicht mehr, dass man mit dieser Auffassung (für die ich vor vier Jahren sehr lebhaft eingetreten bin) viel gewinnt. . . . Bohrs Standpunkt, eine räumlich-zeitliche Beschreibung sei unmöglich, lehne ich a limine ab. Die Physik besteht nicht nur aus Atomforschung, die Wissenschaft nicht nur aus Physik und das Leben nicht nur aus Wissenschaft. Der Zweck der Atomforschung ist, unsere diesbezüglichen *Erfahrungen* unserem übrigen Denken einzufügen. Dieses ganze übrige Denken bewegt sich, soweit es die Aussenwelt betrifft, in Raum und Zeit." (Archive for History of Quantum Physics.) One thus meets once again (cf. note 228) Schrödinger's insistent demand that scientific views conform with world views.

236. Max Born, *Zur statistischen Deutung der Quantentheorie*, Dok. der Naturwiss., Abt. Physik, Bd. 1, ed. Armin Hermann (Stuttgart, 1962).

true state of affairs," Heisenberg declared in the spring of 1927, "can be characterized thus: Because all experiments are subject to the laws of quantum mechanics, . . . quantum mechanics establishes definitively the fact that the law of causality is not valid."[237] And once again, when one sees how rapidly this failure of causality was accepted by physicists not merely as a definitive feature of the theory, but equally of reality, one can scarcely escape the conclusion that such a result, far from being regretted, was greeted with relief and satisfaction. The atomic physicists had fulfilled the obligation which Nernst—and their social-intellectual milieu—had laid upon them.

That conclusion is surely also suggested by the physicists' general anxiousness to carry the good news to the educated public—Heisenberg published a popular article retailing his conclusions even before his "technical" paper was printed[238]—but also from the terms

---

237. W. Heisenberg, "Über den anschaulichen Inhalt . . .," (*op. cit.*, note 196), p. 197, received 23 March 1927. A full year earlier Senftleben, *Physikal. Berichte, 7* (April 1926), 520, had pointed to Heisenberg's paper initiating the matrix mechanics (*op. cit.*, note 144) as an example of the recent tendency to accept "to a certain degree" the view he had advanced in 1923 (*op. cit.*, note 225). Presumably Senftleben would have regarded the principle of indeterminacy which Heisenberg now propounded as merely the consummation of that process of acceptance of his own views.

238. W. Heisenberg, "Über die Grundprinzipien der 'Quantenmechanik'," *Forschungen und Fortschritte, 3* (10 April 1927), 83: "so scheint durch die neuere Entwicklung der Atomphysik die Ungültigkeit oder jedenfalls die Gegenstandslosigkeit des Kausalgesetzes definitiv festgestellt." Considered biographically, this enthusiasm is not unexpected. Heisenberg stresses repeatedly in his memoirs (*op. cit.*, note 233) that when he entered upon the study of theoretical physics at the University of Munich in the Fall of 1920 he had been active in the German *Jugendbewegung* for some years, and he continued so for some years afterward. Although Heisenberg is studiously vague about the particular organization in the politically variegated youth movement to which he belonged—W. Z. Laquer asserts that Heisenberg was a *Weisser Ritter*, and the following observations are especially applicable to this generally rightist group—the intellectual orientation of the movement as a whole has been well characterized by Theodor Wilhelm: "the *Jugendbewegung* is firmly and deeply embedded in that glorification of undivided life, which was intoned by Nietzsche, systematized in the *Lebensphilosophie* of the beginning of the century, paraphrased by the movements for reform in art and pedagogy, and from which the Hitler movement, too, profited in its own way." In fact it was the most radical antagonists of the exact sciences among the vulgar *Lebensphilosophen*—Ludwig Klages, Hermann Keyserling, Rudolf Steiner—who had the greatest following and exerted the strongest influence within the *Jugendbewegung*. Laquer quotes the leader in charge of the youth movement's career counseling office—never mind that he was a communist—contending in November 1918 that some professions were "without value for our future community and its plans to conquer the world"; heading that list was, naturally, physics, followed by chemistry, medicine, and engineering.

105

in which they presented these glad tidings. In a public lecture at the University of Hamburg early in 1927 Arnold Sommerfeld raised "the question which is discussed so much these days, whether the rigid pattern [starre Form] of causality which we have inherited from the 18th century"—read enlightenment, utilitarianism, materialism, etc.—"and from the rationalistic science of mechanics, is appropriate to our contemporary body of experience."[239] And when the question is posed in this form there is no doubt either about the answer which his audience wished to hear. Or again, consider

---

Although this orientation is never permitted to appear explicitly in Heisenberg's memoirs, it may be read between the lines. Thus Heisenberg represents himself (pp. 19, 27) as being forced to defend his decision to make a career of theoretical physics—which, interestingly, he claims to have done on the grounds that theoretical physics has "thrown up problems that challenge the whole philosophical basis of science, the structure of space and time, and even the validity of causal laws." Refusing to choose between theoretical physics and the *Jugendbewegung*, during his first two years at the University Heisenberg divided himself between "two quite different worlds. . . . Both worlds were so filled with intense activity that I was often in a state of great agitation, the more so as I found it difficult to shuttle between the two." The nature and intensity of that agitation becomes clearer if one recalls, on the one hand, that the youth movement organizations of the Weimar period, the *Bunde*, unlike present or Anglo-Saxon organizations, demanded a total commitment—as Theodor Wilhelm says, "Man verschrieb sich seinem Bund ganz"—and also notes, on the other hand, that Heisenberg's monitor in his second world, Wolfgang Pauli, was the very epitome of all that the youth movement detested: unathletic, hedonistic, indifferent to nature, addicted to urban night life, sarcastic, cynical, incisively critical, and Jewish to boot. (Walter Z. Laquer, *Young Germany: A History of the German Youth Movement* [New York, 1962], pp. 34, 102, 116, 141; Theodor Wilhelm and Wilhelm Ehmer in Werner Kindt, ed., *Grundschriften der deutschen Jugendbewegung* [Düsseldorf-Köln, 1963], pp. 12, 232.)

239. A. Sommerfeld, "Zum gegenwärtigen Stande der Atomphysik. Vortrag, gehalten auf Einladung der naturwissenschaftlichen Fakultät zu Hamburg," *Physikalische Zeitschr.*, 28 (1927), 231-239, received 18 February 1927, reprinted in Sommerfeld's *Gesammelte Schriften* (Braunschweig, 1968), 4, 584-592, on 588. In the pre-quantum mechanical period, too, although never willing to renounce the full and unique determination of physical processes, neither could Sommerfeld resist the temptation to play up to the anticausal sentiments of a popular audience. Thus in addressing a general session at the Innsbruck Naturforscherversammlung, September 1924, he passed over the use of transition probabilities, the Bohr-Kramers-Slater paper, etc., without comment, but took the structure of the semi-empirical formulas for the relative intensities of spectral lines as the occasion for opening the prospect of a "teleologische Umbildung der Kausalität." Sommerfeld, "Grundlagen der Quantentheorie und des Bohrschen Atommodelles," *Naturwiss.*, 12 (21 November 1924), 1047-1049; *Ges. Schr.*, 4, 535-543. Cf. note 31 and text thereto.

WEIMAR CULTURE, CAUSALITY, AND QUANTUM THEORY, 1918–1927

the terms in which Max Born discussed the same question in the *Vossische Zeitung*, Berlin's highbrow liberal newspaper, in the spring of 1928. After defining causality as determinism, and adding that all previous laws of physics had that characteristic, Born observed that "such a conception of nature is deterministic and mechanistic. There is no place in it for freedom of any sort, whether of the will or of a higher power. And it is that which makes this view so highly valued by all 'good rationalists.' " But happily physics has now discovered new laws which give it an entirely different character.[240] That character, Bohr had stressed repeatedly in his lectures at Como and at the Solvay Congress the previous fall, is an "inherent 'irrationality' "; indeed "the inevitability of the feature of irrationality characterizing the quantum postulate" was accepted most willingly by Bohr, who showed no sympathy for

240. M. Born, *Vossische Zeitung*, 12 April 1928, as quoted at length by H. Bergmann, *Der Kampf um das Kausalgesetz* (*op. cit.*, note 146), pp. 34-37. Born is reviewing Emanuel Lasker's *Die Kultur in Gefahr* (Berlin, 1928), 64 pp. Lasker, himself much provoked by the professional physicists over the theory of relativity, adopted a very provocative tone: "The old axiom 'from nothing comes nothing' is refuted by the new discovery that the principle of causality is not valid. It's hard to say from whom the genial notion came. Inspired by the spirit of the age of c [the velocity of light] the prophets of the new doctrine had this bright idea which is destined to make world history. Long it grew in secret, carefully weighed and considered, until it has now celebrated in the *Handbuch der Philosophie* [i.e., Weyl, *op. cit.*, note 177] its entrance into the realm of science. . . . The new result runs: in physics and chemistry the principle of causality holds only probably. The old idea of the necessity, unambiguity, and regularity of the laws of nature is ridiculous. The pattern for a law of nature is the lottery. Until further notice. It depends upon what we decide. We believe in principle in the power of experiment. Our council decides the meaning of the experiment—by majority decision. . . . Unfortunately there are a few experimentalists who don't understand the meaning of their own experiments. They still struggle for the old, outmoded view. Rigid habits of thought! The interpretation of an experiment is reserved solely to those who understand experiments and at the same time have a high flying, world embracing imagination. The opinion of those who do not satisfy both these conditions doesn't count. The physicist who is content to measure remains an artisan. He becomes an artist only when he is also a philosopher. The philosopher in turn is negligible if he isn't stamped as an experimental physicist. The physicist-philosopher alone is permitted to interpret and evaluate experiments. . . . The true instrument of the physicist-philosopher is illumination. . . . We are prepared to debate with anyone who is both physicist and philosopher and accepts our methods. To debate with other people would be a waste of time, and we have quite enough work to do turning science into new pathways. Just at this moment we have our hands full replacing the principle of causality by another which we will postulate, and which we will then impose upon the philosophers" (pp. 20-22).

Schrödinger's attempt "to remove the irrational element expressed in the quantum postulate."[241]

It is true that Sommerfeld himself, even as he raised the question of "the rigid pattern of causality," stressed that it was not his intent to call into question "the lawlike definiteness of the physical processes," and elsewhere, as we saw in Section I.1, was at this time actually writing against the less academic forms of the contemporary romantic reaction. But it seems to me that this circumstance only strengthens the inference that an acausal quantum mechanics was particularly welcome to the German physicists because of the irresistible opportunity it offered of improving their public image. Now they too could polemicize against the rigid, rationalistic concept of causality and hope to recover lost prestige thereby.

III.7. Conclusion

In an interview with Einstein in 1932, James Gardner Murphy, an Irish literary man with wide acquaintance among the German theoretical physicists, remarked that "it is now the fashion in physical science to attribute something like free will even to the routine processes of inorganic nature." "That nonsense," Einstein replied, "is not merely nonsense. It is objectionable nonsense. . . . Quantum physics has presented us with very complex processes and to meet them we must further enlarge and refine our concept of causality." Murphy: "You'll have a hard job of it, because you'll be going out of fashion . . . scientists live in the world just like other people. Some of them go to political meetings and the theater and mostly all that I know, at least here in Germany, are readers of current literature. They cannot escape the influence of the *milieu* in which they live. And that *milieu* at the present time is charac-

---

241. N. Bohr, "The Quantum Postulate and the Recent Development of Atomic Theory," *Nature, 121* (14 April 1928), 580-590, reprinted in Bohr, *Atomic Theory and the Description of Nature* (Cambridge, 1934), 52-91, on 580, 586, 590, and 54, 75, 91, respectively; German translation in *Naturwiss., 16* (1928), 245-257. Cf. Philipp Frank, "Gibt es ein irrationales Moment in den Theorien der modernen Physik?" *Neue Züricher Zeitung* (17 December 1928), Nr. 2355, who is at pains to combat this notion which had already been seized upon gleefully by Adolf Koelsch, "Die Verpersönlichung des Elektrons," *ibid.* (20 October 1928), Nr. 1910.

WEIMAR CULTURE, CAUSALITY, AND QUANTUM THEORY, 1918-1927

terized largely by a struggle to get rid of the causal chain in which the world has entangled itself."[242]

Murphy's assertion of the inescapability of the influence of the milieu is the more worthy of our attention as it is but a paraphrase of a passage from a lecture by Schrödinger, "Is Natural Science Conditioned by the Milieu?," published earlier that year.[243] Murphy's own contribution is the specific identification of hostility toward causality as the dominant characteristic of the contemporary milieu, and the implication that the scientist's attitude toward this particular concept had virtually been determined thereby.

Schrödinger's and Murphy's analysis is, as the foregoing investigation has shown, remarkably accurate, at least for the German-speaking Central European physicists. Their craving for crises, their readiness to adapt their ideology to the values of their social-intellectual environment argue a substantial and largely indiscriminate participation in the attitudes of their academic milieu, a readiness to swim along in the intellectual currents of the day. This circumstance is the more surprising if one bears in mind that the values characteristic of these intellectual currents which set in so strongly after Germany's defeat were fundamentally antithetical to the scientific enterprise. Indeed the mathematical physicist, the personification of analytical rationality, was often singled out as the prime exemplar of a despicable way of grasping the world. Above all, with astonishing unanimity, it was the physicist's attempt to subject the world to the rigid, dead hand of the law of causality—to use the rhetoric Spengler made so popular—which was taken to epitomize all that was most detestable in the scientific enterprise. These two circumstances—hostile environment and accommodation to its values—were then found to be linked by much direct and indirect evidence suggesting that the accommodation was in response to the hostility. Stated in terms of Karl Hufbauer's distinctions: suddenly deprived by a change in public values of the approbation and prestige which they had enjoyed before and during World War I, the German physicists were impelled to alter their ideology

---

242. "Epilogue: A Socratic Dialogue. Planck-Einstein-Murphy," in Max Planck, *Where Is Science Going?* trans. James Murphy (New York, Norton, 1932), pp. 201-221, on 201-205.
243. Schrödinger, *op. cit.* (note 143).

109

and even the content of their science in order to recover a favorable public image. In particular, many resolved that one way or another, they must rid themselves of the albatross of causality.

In support of this general interpretation I illustrated and emphasized the fact that the program of dispensing with causality in physics was, on the one hand, advanced quite suddenly *after* 1918 and, on the other hand, that it achieved a very substantial following among German physicists *before* it was "justified" by the advent of a fundamentally acausal quantum mechanics. I contended, moreover, that the scientific context and content, the form and level of exposition, the social occasions and the chosen vehicles for publication of manifestoes against causality, all point inescapably to the conclusion that substantive problems in atomic physics played only a secondary role in the genesis of this acausal persuasion, that the most important factor was the social-intellectual pressure exerted upon the physicists as members of the German academic community.

And here, saving perhaps the case of Hermann Weyl, it was not a question of "philosophical" influences in any serious intellectual sense. By far the single most influential "thinker" was Spengler, and that only because the *Untergang des Abendlandes,* the concentrated expression of the existentialist *Lebensphilosophie* that was diffused through the intellectual atmosphere, was read with attention by most German mathematicians and physicists on account of the prominent role Spengler had given their sciences. Thus, excepting Franz Exner, the philosophical theses of the latter nineteenth century to which Jammer has drawn attention, while they may perfectly well have some ultimate responsibility for the ideational content of the *Lebensphilosophie* of the Weimar period, played, *per se, an sich,* a negligible role in the sudden rise of anticausal sentiment among German physicists after the First World War. Rather, it was only as and when this romantic reaction against exact science had achieved sufficient popularity inside and outside the university to seriously undermine the social standing of the physicists and mathematicians that they were impelled to come to terms with it.

There are, moreover, many indications that this accommodationist strategy met with considerable success. The "objectionable nonsense" about the free will of electrons which philosophers, aided and abetted by physicists, were talking in the late 1920's, constituted in

WEIMAR CULTURE, CAUSALITY, AND QUANTUM THEORY, 1918–1927

fact a very favorable press. Although distasteful to Einstein, this image of modern physics was exactly suited to the taste of the educated public of the Weimar period. And I would emphasize that much of the nonsense announced with great fanfare by philosophers in the late 1920's owed nothing whatsoever to the quantum mechanics discovered in 1925–1926, but was based wholly and solely upon the manifestoes against causality issued by physicists before that date. Such, for example, were the articles which Ludwig von Bertalanffy published in 1927 gloating over the fact that "in physics itself views are coming to be accepted which in biology would be designated as vitalistic.... The causal world picture of the physicist is dissolving—into its place steps one which recognizes individuality, even for the molecular process.... Indeed that allusion of Nernst's to the freedom of will of the theologians can even be employed to support one of Spengler's most controversial ideas: that modern physics, renouncing rigorous causality and exact laws of nature, will give way to a new mysticism."[244]

---

244. L. v. Bertalanffy, "Über die Bedeutung der Umwälzung in der Physik für die Biologie," *Biologisches Zentralblatt, 47* (Nov. 1927), 653-662; on 653-656. Likewise, Bertalanffy, "Über die neue Lebensauffassung," *Annalen der Philosophie, 6* (Sept. 1927), 250-264. A somewhat more delicate picture, again based solely upon pre-1925 sources, was painted by Karl Joël, "Überwindung des 19. Jahrhunderts im Denken der Gegenwart," *Kant-Studien, 32* (1927), 475-518, especially 482-487.
   Not every *Lebenspilosoph* reacted in this way; in fact it appears that many a transcendental idealist, *lebensphilosophisch*-existentialist academic philosopher resented the attempts of the physicists to escape from the stocks of causality and usurp their role of national *Seelsorger*. A very early example is Kurt Riezler who in 1923 noted that recent developments in physics "have induced a number [*Reihe*] of natural scientists to express the hope, or at least to hint, that utilizing these and perhaps other discoveries still to be made the concept and the range of validity of natural laws will be transformed in such a way as would allow the bridge from natural processes to historical processes to be espied, and the gulf which appears to separate necessity and freedom to be closed." Riezler aims, therefore, "to probe the question whether, how far, and by what route that bridge to the world of the spirit and of freedom, which some natural scientists believe they espy, can and may be sought." His conclusion is that the cobbler should stick to his last, that "the second presupposition of natural science, determination, is likewise invariable. The natural scientist—[Nernst cited here]—who wants to see 'the bands of the law of causality' loosened, saws off the branch upon which he sits," while the philosopher on the contrary is not restricted to so narrow a conception of the world. ("Über das Wunder gültiger Naturgesetze. Eine naturphilosophische Studie," *Dioskuren: Jahrbuch für Geisteswissenschaften,* 2 [1923], 238-274, on 238, 257.) By 1925, however, Riezler was no longer quite so categorical: "Die Hypothese der Kausalität," *Die Akademie: eine Sammlung von Aufsätzen aus dem Arbeitskreis Erlangen 4* (1925), 116-146, esp. 143.

One must admit that Bertalanffy's equation of the renunciation of causality with mysticism is not wholly unjustified. For as we saw, the manifestoes by physicists against causality before 1925 were issued not in spite of, but much rather because of, the general belief that "an abandonment of determinism would signify a renunciation of the comprehensibility of nature." Far from engaging in any critical analysis of the concept of causality, directed toward the relaxation of determinism without renouncing *a priori* the comprehensibility of nature, these physicists actually reveled in that consequence, stressed the failure of analytical rationality, implicitly repudiated the cognitive enterprise in which physics had theretofore been engaged.[245]

For this reason the acausality movement could not but arouse opposition within the German physics community. Indeed one has here the most characteristic difference between those physicists who hastened to renounce causality and those who clung to it even after the discovery of quantum mechanics. For Exner, Schottky, Nernst, and Bohr the failure of causality was essentially a failure of the human intellect; Weyl, von Mises, and Reichenbach went even further, expressing an existentialist revulsion against intellectuality. On the other hand those few physicists—strikingly and significantly few—who came forward to publicly oppose dispensing with causality all based their cases upon the value of rationality and their faith in the capacity of the human intellect to comprehend the natural world: so Einstein, Petzoldt, Planck, Schrödinger (after his reconversion), and W. Wien (vis-à-vis inorganic nature). And for this reason also I have not been able to, nor indeed wished to, maintain a perfectly neutral stance in my exposé. Although a readiness to view atomic processes as involving a "failure of causality" proved to be, and remains, a most fruitful approach, before the introduction of a rational acausal quantum mechanics the movement to dispense with

---

245. The quotation is from S. Kis, *op. cit.* (note 146), p. 33. The first serious attempt to reanalyze the concept of causality in order to determine just how much of it is required for the comprehensibility of nature appears to be Eino Kaila, *Der Satz vom Ausgleich des Zufalls und das Kausalprinzip. Erkenntnislogische Studien* (Turku [Åbo], 1924 = *Annales universitatis fennicae Aboensis*, Series B, Tom. 2, No. 2), to which we may perhaps add Reichenbach's, *op. cit.* (note 205).

WEIMAR CULTURE, CAUSALITY, AND QUANTUM THEORY, 1918–1927

causality expressed less a research program than a proposal to sacrifice physics, indeed the scientific enterprise, to the *Zeitgeist*. My sympathies have consequently been with the conservatives in their defense of reason, rather than with the "progressives" in their denigration of it.

But if this social-intellectual phenomenon is to be comprehended, in part at least, by means of a dichotomy between progressives and conservatives, then correlations might be anticipated between a physicist's position on the causality issue and his general intellectual-political orientation. And in fact, paralleling Ringer's observation that early in the Weimar period the "modernist" academics tended to be "methodologically adventurous," one finds that, by and large, those physicists who were readiest and earliest to repudiate causality had either distinctly "progressive" political views by the standards of their social class and the German academic world, and/or had an unusually close interest in, or contact with, contemporary literature. Nernst, who in his youth had wished to become a poet and who retained his interest in literature throughout his life, was also one of the few German physicists who publicly associated themselves with the cause of parliamentary democracy. Von Mises, although politically conservative and nationalistic, was on his way to becoming the foremost authority on the young Rilke. Born and Weyl were both well disposed toward the German republic, at least at its birth —in itself a sufficiently unusual sentiment in the German academic world—and both had literary wives. On the other hand, with the notable exception of Einstein, those who defended causality tended to be highly principled political conservatives and/or interested in classical literature. Such were Planck, Schrödinger, and Max von Laue—who kept their knowledge of Greek well polished. Standing to their right was W. Wien. And finally to the causalist camp one may add the outright reactionaries: Ernst Gehrcke, Erwin Lohr, Philipp Lenard, and Johannes Stark.[246]

---

246. F. A. Lindemann and F. Simon, "Walther Nernst, 1864-1941," *Obituary Notices of Fellows of the Royal Society of London*, 4 (1942), 101-112. Nernst had subscribed to a manifesto in the summer of 1917 calling for democratic political reforms in Prussia, and he had participated in the 1926 meeting of republican-parliamentary university professors: Klaus Schwabe, *Wissenschaft und Kriegsmoral: Die deutschen Hochschullehrer und die politischen Grundfragen des*

This very circumstance—that the alignment within the German physics community over the issue of causality correlates closely with the intellectual and political temper of the individual physicist—reminds us, however, that the "sociological" model employed in this paper cannot be the whole truth. It provides a general framework, and seems to work especially well in certain extreme cases. But in order to account for its special applicability to some physicists and its special inapplicability to others one must invoke precisely those factors which are excluded from the model—individual personality and intellectual biography. The mechanism advanced for the entrainment of the German physicists and mathematicians by the *Zeitgeist* is thus clearly not sufficient. And it may be that

---

Ersten Weltkrieges (Göttingen, 1969), p. 264, note 229; Wilhelm Kahl, *et al., Die deutschen Universitäten und der heutige Staat: Referate erstattet auf der Weimarer Tagung deutscher Hochschullehrer am 23. und 24. April 1926* (Tübingen, 1926), pp. 38-39. Von Mises' nationalistic attitudes are reflected in his repeated promotion of political boycotts by German scientists of international congresses. For example: Th. von Kármán to von Mises, 11 December 1923, and P. Debye to von Kármán, 1 May 1926, in the Kármán Papers, California Institute of Technology Archives, but especially the correspondence between von Mises and L. E. J. Brouwer in 1928 in the von Mises Papers, Niels Bohr Library, American Institute of Physics, New York. There are many indications of Max Born's political attitudes in his correspondence with Einstein (*op. cit.*, note 14), where specimens of Hedwig Born's poetry are also published. Hermann Weyl's initial attitude toward the German republic may be glimpsed in a letter to Einstein of 16 November 1918 in the Einstein Collection, Institute for Advanced Study, Princeton, and in the Weyl Nachlass, Bibliothek der Eidgenössischen Technischen Hochschule, Zurich; his continuing attachment to democracy is suggested in the talk by which he introduced himself to the Göttingen mathematics students in 1930, quoted at length in his "Rückblick auf Zürich aus dem Jahre 1930," *Schweizerische Hochschulzeitung, 28* (1955), 180-189, reprinted in Weyl's *Ges. Abhl., 4*, 650-654. There he also described his decision to accept the call to Göttingen as resulting from discussions, carried on in his imagination, with Jacob Burkhardt and Hermann Hesse. Weyl's wife Helene, the former disciple of Husserl, translated Ortega y Gasset into German. Planck's political conservatism is well known, but not quite so deep as it is often represented: he and von Laue were both members of the Deutsche Volkspartei in the Weimar period. (Wilhelm Westphal, "Der Mensch Max von Laue," *Physikalische Blätter, 16* [1960], 549-551; Friedrich Herneck, *Bahnbrecher des Atomzeitalters* [Berlin, 1969], pp. 303-304). Schrödinger's conservative nationalism is implicit in his correspondence with Wilhelm Wien, cited in note 234, most strongly in a letter of 26 April 1927 describing his emotions at the sight of the German countryside upon returning from the United States. Schrödinger frequently used Greek and Latin titles for his research notebooks (Archive for History of Quantum Physics); von Laue frequently used quotations from these languages in his publications.

WEIMAR CULTURE, CAUSALITY, AND QUANTUM THEORY, 1918–1927

examination of other episodes of entrainment in the late nineteenth and early twentieth centuries will prove that it is also not necessary. But be that as it may, it seems difficult to deny that the shifts in scientific ideology and the anticipated shifts in scientific doctrine exposed in this paper were *in effect* adaptations to the Weimar intellectual environment. Moreover, whatever similarities one may find in the mental posture of non-German exact scientists in this same period, there is one feature which cannot, I think, be found outside the German cultural sphere: a repudiation of "causality."

Kausalität, Anschaulichkeit, and *Individualität,* or, How Cultural Values Prescribed the Character and the Lessons Ascribed to Quantum Mechanics*

Paul Forman

Quantum mechanics is the fundamental theory of atomic and subatomic physics. It is the basic set of rules held by the theorist whether he be calculating the electrical resistance of a new semiconductor to be built in an industrial laboratory or the mass of a new elementary particle to be produced by a high-energy accelerator. These rules were found in 1925-26 and elaborated in the following years in a scientific milieu whose language was German and whose center lay in Germany. Of course, Niels Bohr's institute was in Copenhagen. And P.A.M. Dirac, a young research student at Cambridge University, was among the earliest and most important contributors. Overwhelmingly, however, the creation of quantum mechanics was an enterprise of Germans and Austrians. Consequently, as a product of Germany it may appropriately be considered in relation to German culture.

---

*This paper is reprinted here with correction of a small number of typographical errors in the original publication. This paper was originally presented at a workshop in Lecce, Italy, Sept. 3-6, 1979, organized by the Istituto di Fisica and the Istituto di Matematica of the Università di Lecce and by the Gruppo di Storia della Scienza dell' Istituto di Fisica 'G. Marconi' of the Università di Roma. It has been published in slightly different form in Italian in the proceedings of that workshop: M. De Maria, E. Donini et al., *Fisica e Società negli Anni '20* (Milan: CLUP/CLUED, 1980), pp. 15-34; and in German in Nico Stehr and Volker Meja, eds., *Wissenssoziologie,* Sonderheft 22 of the *Kölner Zeitschrift für Soziologie und Sozialpsychologie* (Opladen: Westdeutscher Verlag, 1981), pp. 393-406. Comparable endeavors to explain the origin and interpretation of quantum mechanics along cultural lines are: Lewis Feuer, *Einstein and the Generations of Science* (New York: Basic Books, 1974); S.G. Brush, "The chimerical cat: the philosophy of quantum mechanics in historical perspective," *Social Studies of Science* 10 (1980):393-447.

The following discussion of the character of quantum mechanics is in the form of three probes. Each opens with a statement of the *true* bearing of the theory relative to one of three concepts or characteristics paraded in German in the title:causality, individuality, and intuitive evidentiality or visualizability. This is followed by a brief exposition of the *alleged* character of the theory regarding that concept - alleged first of all by the creators of the theory. These allegations are found to diverge markedly from what, a moment before, had been posited as warranted regarding these concepts. Each probe then concludes with an explanation of this disparity between the true and alleged character of quantum mechanics. The explanation given makes appeal to the role of that concept in German-speaking Central European culture in the era of the Weimar Republic, the culture of the region and period in which quantum mechanics was created.

I speak of the *true* character of quantum mechanics. I trust the reader will not therefore suppose that I believe in, or am logically obliged to believe in, a *transcendental* truth of any sort. What is at issue here is not whether quantum mechanics or any other scientific theory is True - whether it represents the world as it *really* is. At issue is merely whether the theory has certain specified characteristics: whether it represents the world in an *anschaulich* manner, whether it admits ascription of individuality to the subatomic entities with which it deals, and whether it warrants the denial of causality in the operations of nature. Granted that quantum mechanics, or any other representation of the world, can never be established as True, for we can never make a direct comparison of the representation with that which the representation claims to represent. Our case is essentially different. That which is represented - the theory - is presented to us directly as an intellectually apprehensible entity. It is quite reasonable here to ask for open-minded observers capable of being brought by reason and evidence to recognize "what is the case" (to use the expression of the early Wittgenstein).

I also speak of unwarranted interpretations of quantum mechanics as explicable by reference to the broader cultural milieu in which its creators lived and worked. I trust the reader will not suppose that I am oblivious to the internal culture of physics which creates and maintains an ample inventory of preconceptions about the explanatory models to be employed in the description of nature and the interpretations to be placed upon them. This internal culture, along with individual idiosyncrasy, can be held responsible for the greatest part of the ever-present misunderstanding and misinterpretation of scientific innovations - not least by the innovators themselves. It is characteristic of such scientific conservatism that the innovator himself, together with most of his contemporaries, fails to appreciate how wide and deep is his break with tradition.[1] But our case is just the opposite: all sorts of "lessons" of a thoroughly and intentionally revolutionary character are announced nearly simultaneously with the discovery of a theory which, properly understood, barely supports, or even flatly contradicts, the claims founded upon it. We are dealing here not with the commonplace misconstructions in which new wine is poured into the old bottles of scientific culture. We have, rather, the exceptional, but by no means unprecedented, circumstance that the concocting of a new wine is taken as an opportunity to cast away the old-fashioned bottles and blow new ones to a modish form.

My purpose is not to demonstrate this proposition, but merely to make it plausible. With respect to *Kausalität*, I have argued the point at length in an earlier publication.[2] With respect to *Anschaulichkeit* and *Individualität*, I present a hypothesis and a program of investigation. The apothegmatic form is chosen chiefly for concision and distinctness of expression.

Kausalität

Quantum Mechanics as Statistical Theory of Atomic Processes

An essentially statistical character was given matrix mechanics, the earliest form of the new quantum mechanics, as it developed along the Göttingen-Hamburg-Copenhagen axis in the summer and autumn of 1925. The "quantum mechanical variables" which Werner Heisenberg conceived, and which Max Born recognized as matrices, were probabilities of atomic transitions, or algebraic combinations and derivatives of such probabilities - more precisely, probability amplitudes. To be sure, the wave mechanics conceived by Erwin Schrödinger in the winter of 1925-26, as an antithesis to matrix mechanics, was intended to provide a continuous, space-time description of the behavior of individual particles. But the wave equation and its solution, the wave function, did not admit of such an interpretation compatible with the facts of atomic physics.

Schrödinger himself, by demonstrating the formal equivalence of matrix and wave mechanics in the spring of 1926, opened the way to Born's statistical interpretation of the wave function in the summer of that year. Thus wave mechanics too became what matrix mechanics had been all along, a statistical theory of atomic processes. And this latter expression permits only one minimal construction, which is natural and free of all unnecessary ontology: that the wave function describes an ensemble of identically prepared systems, and thus gives only the distribution of outcomes of a large number of sensibly identical experiments; the wave function does *not* describe an individual particle.[3]

Again from the very outset, the inventors of quantum mechanics were confronted with the puzzling circumstance that the quantum mechanical variables - whether represented by matrices, differential operators, or "q-numbers" - do not always commute with each other. As a consequence of this essential peculiarity of quantum mechanics, the statistical distributions of any pair of variables specifying the dynamical state of a particular degree of freedom of an atomic system are not independent. Rather, they are subject to a condition - the Heisenberg uncertainty principle - which places a lower limit on the extent to which the spreads of the statistical distributions of the values assumed by those two variables may be reduced simultaneously.

Quantum Mechanics Alleged to Demonstrate the Failure of Causality

It was in conjunction with his first resort to wave mechanics - to treat the collision of an electron with an atom - that Born, in the summer of 1926, proposed the probabilistic (statistical) interpretation of Schrödinger's wave function. In his brief initial announcement of the results of this investigation, Born stressed that "from the standpoint of our quantum mechanics there is no quantity which causally determines the effect of a collision in the individual case." He then freely confessed that "I myself am inclined to abandon determinedness in the atomic world." [4] In the following months there were further declarations by German physicists that quantum mechanics shows that causality does not hold.[5] Their culmination was Heisenberg's famous paper setting out the uncertainty principle. "The true state of affairs," Heisenberg there declared, "can be characterized thus: because all experiments are subject to the laws of quantum mechanics . . . quantum mechanics establishes definitively the fact that the law of causality is not valid." [6]

Notice that Heisenberg's statement is categoric. Not only has he dropped Born's restriction, "in the atomic world," but his assertion is so general, so sweeping, that it could be supposed to apply even to the sphere of human will and action. Heisenberg intended nothing less. Clear allusions in that direction were already being made by Sommerfeld and others. Bohr - with whom Heisenberg was then in daily contact - would soon come forward with explicit parallels between the laws of quantum mechanics and the laws of human thought.[7] Derivations of freedom of the will from quantum mechanics became more common and more dogmatic in the following years, and were extended by Bohr in a vitalistic sense to preclude physical description of the processes of life.[8]

There is great disparity between quantum mechanics, per se, and the world-view implications immediately ascribed to it. Quantum mechanics is merely a statistical theory. As Einstein repeatedly but vainly emphasized, it cannot be regarded as a complete description of an independently subsisting microscopic world. Nor can it be regarded as an appropriate conceptual basis for describing our macroscopic world, where, unquestionably, we deal with individual objects and events, not statistical ensembles.[9] Thus even categoric statements about the invalidity of the law of causality in the *physical* world go much too far, not least because they slur over the fact that quantum mechanics is a deterministic theory of probabilities. As for the still farther-reaching world-view implications ascribed to quantum mechanics - that it ensures free will, or the impossibility of a physicochemical explanation of life - one must say that these are completely unwarranted.

Why This Misuse of Quantum Mechanics for Sweeping Epistemic Renunciations?

These immediate anticausal interpretations of quantum mechanics and inflations of its significance for the issue of determinism were limited to German-speaking Central Europe. British physicists, even those intimately involved with atomic physics and the nascent quantum mechanics, remained oblivious to the broader epistemic significance of the theory until their consciousness was raised by Central European colleagues. The same was largely true of American physicists.[10] French physicists were unhappy about the indeterminism of quantum mechanics, and drew few world-view implications from it.[11]

We are therefore led to look for special circumstances, unique, or largely unique, to the environment of German-speaking Central European physicists, which may have called forth their misuses of quantum mechanics. Briefly stated, the critical characteristics of this cultural milieu were:

Antiintellectualistic, romantic irrationalism, celebrating "life" and immediate, unanalyzed experience. "I maintain," said Richard Müller-Freienfels in the introduction to his *Philosophie der Individualität*, "and in this I am at one with many recent thinkers, that rational logic is the basis of only one kind of knowing, not of knowing generally. Life is more than rational *Wissenschaft*, and to me philosophy is not merely *Wissenschaft*, but knowing, knowing even also of that which does not enter into *Wissenschaft*. Indeed, philosophy is more even than knowing; philosophy is itself life." [12]

Antipathy toward causality. As is to be expected with such a widely diffused but highly confused *Lebensphilosophie*, its proponents defined the movement chiefly in negative terms, that is, by what they opposed. Most regularly and insistently that target was causality, "the mechanism and determinism of a causal explanation which calculates everything in advance, makes everything comparable, dissolves everything into elements." [13]

Antagonism toward physical science, and more particularly toward the

theoretical physicist. This arose, first, because the physicist's cognitive goals epitomized that which *Lebensphilosophie* rejected; second, because the physicist was held responsible for modern technology and industry, and thus for our alienation from nature, from our primitive roots, from the simple life, and so on.[14]

The German physicist, especially the theoretical physicist, who before and during World War I enjoyed so much esteem, now, in the Weimar period, found himself despised. Gone were those values and attitudes which had ensured him an honored place in German society and in the eyes of his university, displaced by others distinctly hostile to the theoretical physicist's enterprise. In an effort to realign their discipline with dominant cultural values, individual German physicists had been delivering manifestos against causality even before the discovery of quantum mechanics. Typically, they asserted that through wrestling with the problems of atomic physics their science would find the way to free itself from this odious axiom.[15] Thus we should not be surprised that once a nondeterministic theory of atomic processes was at hand, German physicists were disposed to view it and represent it in public as providing that liberation from causality so generally desired.

Anschaulichkeit

Quantum Mechanics as Abstract, Unpictorial Theory

When in the spring of 1925 Heisenberg began to work along those lines which led him to matrix mechanics, it was with an emphatic commitment to give up pictorial atomic models. Heisenberg sought to give concrete expression to Bohr's and Pauli's contention that a solution to the baffling problems facing atomic theorists would lie in a renunciation, for atomic processes, of causal description in space and time.[16] (Causality and space-time description, which in 1927 Bohr would find to be complementary, were in 1924-25 regarded by him and others as more or less equivalent.) Certainly the last epithet one could apply to the highly abstract theory created by Heisenberg, Born, Jordan, and Dirac is *anschaulich*, intuitive, pictorial.

Wave mechanics, on the contrary, was conceived by its creator as a highly *anschaulich* description of subatomic entities and atomic processes. However, the apparent *Anschaulichkeit* of this version of quantum mechanics proved to be largely illusory. First, the "waves" or "clouds" are not of mass or electric density as Schrödinger pictured them, but of probability density. Second, the waves are not in real space at all but in configuration space, a space with as many dimensions as there are degrees of freedom in the atomic system. Finally, the wave function does not describe the behavior of an individual particle (let alone the particle, per se), but of a statistical ensemble of identical particles or systems.[17]

The Epithet Anschaulich *is Applied to Quantum Mechanics*

The program behind matrix mechanics was to create an atomic dynamic free of any *anschaulich*, pictorial, elements. Yet within months of the discovery of this unpictorial calculus, Heisenberg was complaining to his friend Pauli that "I'm always infuriated whenever I hear the theory referred to only by the name matrix physics." What Heisenberg heard in his colleagues' words and tone, and reacted to so defensively, was an antipathy toward a theory so emphatically *unanschaulich*.[18]

Born, the less resolute character, saw in his probablistic interpretation of Schrödinger's wave function a *via media*. While enjoining renunciation of "the causal definiteness of the individual events," it nonetheless allowed "the retention of the familiar conceptions of space and time, in which the events take place in an entirely normal way."[19] Heisenberg was unhappy about Born's deviation from the quantum-mechanical creed, his compromise with the "populäre Anschaulichkeit" of wave mechanics.[20] At first Heisenberg tried to dismiss the problem by arguing that "the hypothesis of the atomistic structure of matter is, from the outset, *unanschaulich*."[21] He soon realized that the issue could not be ignored: the theory had to be cleared of the stigma of *Unanschaulichkeit*.

Heisenberg's preoccupation with this problem is evident in his essay presenting the program and achievements of *Quantenmechanik* to a wider scientific public in the autumn of 1926. Using *Anschauung, anschaulich*, etc., once in every 250 words, sixteen times in all, Heisenberg concluded that "thus far there is still some essential feature missing in our picture."[22] The result of Heisenberg's continuing struggle with the problem through the winter of 1926-27 was his uncertainty principle paper, which he titled "On the Intuitive [*anschaulich*] Content of the Quantum-Theoretic Kinematics and Mechanics."[23]

There and thereafter Heisenberg sought to remove the stigma of *Unanschaulichkeit* by redefining the "intuitive" quality so as to make it predicable of his irremediably unpictorial quantum mechanics. This redefinition equated "intuitive" to "satisfactory" in a strictly positivist sense: "From an *anschaulich* theory in this sense one ought thus demand only that it be without contradiction with itself, and that it permit prediction, without ambiguity, of the results of all imaginable experiments in its domain."[24] Perhaps the best index of the wantonness of this solution is that in their colleagues' eyes Heisenberg and Born (who joined in this redefinition) had simply inverted the ordinary, accepted signification of the word. Thus we find Arnold Sommerfeld, Heisenberg's teacher, speaking of "the fuzziness or *Unanschaulichkeit* (in Heisenberg's mode of expression the *Anschaulichkeit*) of our contemporary world picture." [25]

Why This Dissimulation of the Unanschaulich Character of Quantum Mechanics?

The factors operating on the originators of quantum mechanics so as effectively to prescribe their representation of their theory as *anschaulich* were largely those presented above in the discussion of *Kausalität*. As regards *Anschaulichkeit*, the true character of the theory ran counter to the *Zeitgeist*, rather than, as with causality, offering physicists an opportunity to sail with the intellectual currents. Where *Kausalität* was the pejorative code word in the antiintellectualist "philosophy of life," *Anschauung, Anschaulichkeit*, etc. were among the principal shibboleths of this *Lebensphilosophie*. It is no mere coincidence that just at this time German mathematicians were enchanted by the program of L.E.J. Brouwer, a Dutch romantic antiintellectualist and anticausalist, who called for an intuitionistic mathematics. As with acausality among physicists, the attraction this program exerted on German mathematicians was not diminished but rather heightened by requiring a severe restriction of their epistemic ambitions. Dissimulation of the *unanschaulich* character of quantum mechanics, the attempt to redefine *Anschaulichkeit* in order that the epithet could be applied to this unintuitive theory, was chiefly a result of the physicists' weak social-ideological position in post-World War I Germany; they simply had not the strength to oppose the *Zeitgeist*. [26]

Individualität

Quantum Mechanics Precludes Individuality

Attacking the "complex of catchwords" that was complementarity, Schrödinger observed in 1949 that "during these twenty years of empty talk, and because of this empty talk, the most important result of the 'new mechanics' has fallen from sight - the physically *and* philosophically most significant finding. Every physicist knows of it, but one does not speak of it, regards it as less important, veils the subject with a convenient but obscuring jargon. The *particle is not an identifiable individual*. There is no longer individuality in the absolute sense." [27]

This, the most direct and least arguable consequence of the new quantum mechanics for our world-view, has two bases. The first is in quantum mechanics per se, the uncertainty principle, which, it is commonly said, by placing limits on our capacity to follow the path of a particle, prevents us from knowing whether we have confused the particle with another identical particle. But this statement clearly begs the question. For if the particles had any individuality from the point of view of the quantum mechanics, there would be no danger of confusing them. Rather, one should simply say that the quantum theory has no place, no means, for ascribing individuality to subatomic particles of a given species, and that experimentation is found fully to confirm the consequences of this logical exigency.

The other basis for asserting that the elements of the physical world picture lack any individuality is quantum statistical mechanics, the so-called Bose-Einstein and Fermi-Dirac statistics. Introduced independently of, but coincidentally with, quantum mechanics, the new statistics result from new rules for determining the relative probability of the various possible states of a complex system. These new rules differ from the rules underlying classical statistical mechanics precisely in that the older rules assumed the distinguishability, hence individuality, of the particles comprised by the system. The new rules, by contrast, give emphatic expression to the assumption of the indistinguishability, the absence of any individuality, of the subatomic particles of a given species.

Although there had been from the middle of the nineteenth century general agreement that the atoms of a given element were "as alike as manufactured articles," that was scarcely ever taken to imply absolute indistinguishability. Nor did the classical physical theories permit any natural expression of such a notion. [28] It was only quantum mechanics *cum* quantum statistics which, *nolens volens*, deprived the elements of our physical world picture of individuality, as a matter of principle.

Quantum Mechanics is Alleged to Demonstrate Individuality

Among the lessons of quantum mechanics none is clearer or surer than the denial of individuality to subatomic particles. Nor was this most striking consequence of the theory completely overlooked. Heisenberg, in his autumn 1926 essay on "*Quantenmechanik*," noted as one of the principal results of the new quantum statistics that "the individuality of a particle can be lost." [29] But never again did this simple truth flow from Heisenberg's pen. Whether Bohr too saw this truth for even a brief moment, I do not know. But when, in the summer of 1927, he first pronounced on the issue, it was to advance the perverse thesis that the main bearing of quantum theory was to demonstrate not merely the individuality of atomic processes, but indeed the "indestructible individuality" of material particles. Then and later Bohr went farther to draw explicit analogies with the individuality of living organisms and human personalities.[30] No other physicists, not even Heisenberg and Rosenfeld who were Bohr's loyal disciples in philosophical matters, were prepared to go nearly so far as he in celebrating individuality as the main lesson of atomic physics.[31] Yet certainly none openly contradicted Bohr's contention, while many found it easy and natural to tip their hats, not only in Germany but throughout Western Europe and in the United States.[32]

Misrepresentation of the Ontologic Bearing of the Quantum Mechanics and Statistics

Individuality is traditionally one of the strongest cultural values among the educated elites of the West, and especially so in Germany, where the ideal of the autonomous individual personality compensated Germans for their tradition of authoritarianism in the social and political spheres.[33] As this ideal was central to romanticism, it had once again great appeal in Germany due to the strong romantic tradition of that linguistic and cultural region.[34] In his perceptive study of developments in German academic ideology around the turn of the century, Fritz Ringer noted "the amazingly frequent reappearance of certain themes and images in modern German academic literature." Ringer found that "the principle of individuality was one such theme," and emphasized that "it is impossible to imagine the mandarin creed without the concept of individuality." [35]

In the Weimar period, commonly seen by contemporary observers as a return to the romanticism of a century earlier, this cultural value was given even freer reign than in the decades before World War I when progressivist ideology was ascendant in Germany as elsewhere in Europe and the United States.[36] Although that progressivist ideology retained an important place for "individualism" - which in any case must be sharply distinguished from "individuality" - it placed the greatest emphasis on rational organization and coordination. The mode of research support characteristic of the prewar period was establishment of large, centralized research institutes to pursue tightly integrated research programs. In the Weimar period, on the contrary, the predominant and characteristic pattern of support was wide distribution of small grants to individuals to carry out their own independently conceived research projects.[37] Although there were many factors behind this striking reorientation, the romantic emphasis on the individuality of the researcher was important.

In Weimar Germany invocations of *Individualität* could be heard everywhere.[38] Books with Individualität in their title were just about as numerous as those with *Kausalität*.[39] The shibboleth answered perfectly to the irrationalist *Lebensphilosophie* of the period. This was stated straight out by Theodor Haering, professor of philosophy at Tübingen, in the opening sentence of one of those numerous books: "From the middle ages stems the expression *individuum est ineffabile*, individuality cannot be comprehended with words and concepts; our own irrationalism-happy age would perhaps best say instead: 'individuality is something irrational.'" [40]

This equation of individuality with irrationalism is fully explicit in Bohr's first presentation of the doctrine of atomic individuality in the summer of 1927: the quantum postulate implies on the one hand the individuality of atomic processes, and on the other expresses an "inherent irrationality." [41] Thus we should not be surprised to find irrationality and individuality linked with failure of causality. A biologist-philosopher, basing himself not on the new quantum mechanics, but on the anticausal manifestos by German physicists in the early 1920s, could write at this same time that "the causal world picture of the physicist is dissolving; into its place steps one which recognizes individuality, even for the molecular process." [42] Nor should we expect Bohr's to be the only attempt to introduce a "principle of individuality" into atomic physics. Arthur Korn, a German physicist who still clung to the mechanical tradition of the late nineteenth century, proposed in 1926 to make the bridge to quantum theory by modifying the deterministic character of classical mechanics through an *Individualitätsprinzip* whose effect makes itself felt in very high-frequency oscillatory processes.[43]

Recapitulation and Conclusion

The quantum mechanics introduced in 1925-26 proved to be an incompletely causal theory. It went some distance toward meeting the obligation which many German-speaking Central European physicists had assumed under pressure from their environment: to eliminate *Kausalität* from their science and world picture. These physicists sought to make the most of acausality in quantum mechanics, exaggerating and trumpeting it. The theory grossly violated the requirements of *Anschaulichkeit*, one of the strongest positive values of the cultural milieu. Far from making this intrinsic feature of quantum mechanics one of the epistemologically pertinent "lessons" of atomic physics, its originators sought to free their theory from the stigma of *Unanschaulichkeit*. This they did by redefining the meaning the word was to have in physics. Finally, the one inarguable "lesson" of quantum mechanics, and the statistics integral to it, is that individuality does not exist in the atomic world. This clear implication, so much at variance with the cultural values of Germany and the West, physicists either suppressed, or, flying unchallenged in the face of the facts, maintained the diametrically opposite proposition.

My conclusion is that there was little connection between quantum mechanics and the philosophic constructions placed on it, or the world-view implications drawn from it. The physicists allowed themselves, and were allowed by others, to make the theory out to be whatever they wanted it to be - better, whatever their cultural milieu obliged them to want it to be.

This conclusion is admittedly radical. But it does not touch the question of the social construction of reality so directly as one might at first be inclined to suppose. It is neither a statement about the physicists' practice in their laboratories nor about the physicists' theories as descriptions of reality. It is rather a meta-meta statement, a statement about the physicists' statements about their descriptions of reality.[44]

Notes

1. Thomas S. Kuhn has been steadily concerned with this social/epistemological phenomenon, most recently and directly in *Black-Body Theory and the Quantum Discontinuity, 1894-1912* (New York: Oxford University Press, 1978).

2. Paul Forman, "Weimar culture, causality, and quantum theory: adaptation by German physicists and mathematicians to a hostile intellectual environment," *Historical Studies in the Physical Sciences* 3(1971):1-115.

3. Albert Einstein, "Elementare Überlegungen zur Interpretation der Grundlagen der Quanten-Mechanik," in *Scientific Papers Presented to Max Born* (New York: Hafner, 1953), pp. 33-40, and reproduced in Einstein, *Collected Writings* (New York: Readex Microprint, 1960), item 266. It appears, however, that my interpretation of Einstein as well as quantum mechanics may be superficial: Arthur Fine "What is Einstein's statistical interpretation, or, is it Einstein for whom Bell's theorem tolls?" Forthcoming in *Topoi* (1984).

4. Max Born, "Zur Quantenmechanik der Stossvorgänge: vorläufige Mitteilung," *Zeitschrift für Physik* 37(1926):863-67.

5. E.g., Arnold Sommerfeld, "Zum gegenwärtigen Stande der Atomphysik," *Physikalische Zeitschrift* 28 (1927):231-39; P. Jordan, "Kausalität und Statistik in der modernen Physik," *Naturwissenschaften* 15(1927):105-10.

6. Werner Heisenberg, "Über den anschaulichen Inhalt der quantentheoretischen Kinematik und Mechanik," *Zeitschrift für Physik* 43 (1927):172-98; also, "Über die

Grundprinzipien der 'Quantenmechanik'," *Forschungen und Fortschritte* 3(1927):83.

7. Niels Bohr, "The quantum postulate and the recent development of atomic theory," *Nature* 121(1928):580-90, being the Como Congress lecture, Sept. 16, 1927.

8. Niels Bohr, *Atomtheorie und Naturbeschreibung* (Berlin: Springer, 1931), pp. 14-15, 66, 76; Forman, op. cit., n. 2, pp. 108-9.

9. The one important exception was A.S. Eddington. See Paul Forman, "Reception of an Acausal Quantum Mechanics in Germany and Britain," in *The Reception of Unconventional Science*, ed. S. Mauskopf, "AAAS selected Symposium, 25" (Washington, D.C.: AAAS, 1979), pp. 11-50.

10. J.R. Oppenheimer in Göttingen in the autumn of 1926 was probably already alert to these issues, for he was duly impressed by the "fantastically impregnable metaphysical disingenuousness" of the German physicists he met there. Quoted by A.K. Smith and C. Weiner, *Robert Oppenheimer: Letters and Recollections* (Cambridge, Mass.: Harvard University Press, 1980), p. 100. However, American physicists do not appear to have begun to draw sweeping indeterministic conclusions from quantum mechanics until 1929-30.

11. Some evidence on this point is given by R. Maiocchi, "Paul Langevin's epistemological considerations on quantum mechanics and the reactions to them in French culture before World War II," *Scientia* 110(1975):493-518. (In Italian with English translation.)

12. Richard Müller-Freienfels, *Philosophie der Individualität*, 2nd ed. (Leipzig: Meiner, 1923), p. 4. The first edition appeared in 1921.

13. Theodor Litt, *Die Philosophie der Gegenwart und ihr Einfluss auf das Bildungsideal* (1925), as quoted in Forman, op. cit., n. 2, p. 18.

14. Forman, op. cit., n. 2, pp. 8-19.

15. E.g., Hermann Weyl, Richard v. Mises, Walter Nernst, ibid., pp. 76-86.

16. Daniel Serwer, "Unmechanischer Zwang: Pauli, Heisenberg, and the Rejection of the Mechanical Atom, 1923-1925," *Historical Studies in the Physical Sciences* 8(1977):189-256; Wolfgang Pauli, *Wissenschaftlicher Briefwechsel*, vol. 1: 1919-

1929, ed. A. Hermann, K.v. Meyenn, V.F. Weisskopf (New York: Springer, 1979), p. 231 et passim.

17. While almost all physicists, of whatever nationality, from 1927 until the present would, if pressed, acknowledge this fact, when left to themselves, and especially when popularizing their theory, they often pretend that the wave function describes not only the behavior of individual particles but indeed the particles themselves.

18. Heisenberg to Wolfgang Pauli, Nov. 16, 1925, in Pauli, *Briefwechsel* (1979), op. cit., n. 16, p. 255; Heisenberg to Einstein, Nov. 16, 1925 [sic] (Einstein Archive, Princeton). For the following see also my "Reception of an Acausal Quantum mechanics," op. cit., n. 9. A different view is taken by Arthur I. Miller, "Visualization lost and Regained: The Genesis of the Quantum Theory in the Period 1913-27," in *On Aesthetics in Science*, ed. J. Wechsler (Cambridge, Mass.: MIT, 1978), pp. 73-102.

19. Max Born, "Quantenmechanik der Stossvorgänge," *Zeitschrift für Physik* 38(1926):803-27, p. 826.

20. Heisenberg, op. cit., n. 6, p. 196, was so wrought up as to use this expression in a scientific publication, in which he claimed to reveal the genuine *Anschaulichkei*t of quantum mechanics.

21. Heisenberg, "Quantentheoretische Mechanik," Deutsche Mathematiker-Vereinigung, *Jahresbericht* 36(1927):24-25; abstract of a lecture at the Naturforscherversammlung, Düsseldorf, Sept. 23, 1926.

22. Heisenberg, "Quantenmechanik," *Naturwissenschaften* 14(1926):989-94. This, the elaborated text of the lecture cited in the previous note, shows considerable difference from the abstract presumably prepared in advance of delivery.

23. Heisenberg, op. cit., n. 6.

24. Max Born and Werner Heisenberg, "La mécanique des quanta," *Electrons et photons: rapports et discussions du cinquième conseil de physique [Instituts Solvay]...1927* (Paris: Gauthier-Villars, 1928), p. 144. The typescript of the original German text of this report is at the Humanities Research Center, University of Texas, Austin. The first page, with the passage in question, is reproduced in Albert C. Lewis, *Albert Einstein, 1879-1955. A Centenary Exhibit...* (Austin, Texas: University of Texas, Humanities Research Center, 1979), unpaginated. Heisenberg's

initial redefinition of *Anschaulichkeit*, given in the opening lines of "Über den anschaulichen Inhalt," op. cit., n. 6, did not yet deviate quite so far from the accepted meaning of the word.

25. Arnold Sommerfeld, "Über Anschaulichkeit in der modernen Physik," *Unterrichtsblätter für Mathematik und Naturwissenschaft* 36(1930):161-67.

26. Forman, op. cit., n. 2, pp. 60-61; L.E.J. Brouwer, *Collected Works*, vol. 1: *Philosophy and Foundations of Mathematics* (Amsterdam: North Holland, 1975), pp. 2-4, 481-87; T. Tonietti, "A research proposal to study the formalist and intuitionist mathematicians of the Weimar Republic," *Historia Matematica* 9(1982):61-64. While none but the proponents of the matrix mechanics had any interest in dissimulating the character of this version of quantum mechanics, *Anschaulichkeit* was so general a cultural value, within physics as well as outside of it, that Schrödinger's wave mechanics had, as a seemingly *anschaulich* theory, a much greater appeal than did matrix mechanics, no less in Britain and the United States than in Germany. Recently M. Beller has used citation data to test, and confirm, the relative appeal and utilization of matrix and wave mechanics.

27. Schrödinger to Sommerfeld, Feb. 13, 1949 (Archive for History of Quantum Physics). Lewis Feuer put his finger on "individuality" in *Einstein and the Generations of Science* (New York: Basic Books, 1964), pp. 116-17, 198-99, in conjunction with a psychological characterization of Niels Bohr which is right on the mark. However, Feuer has not connected *Individualität* in the right way with the other two sides of the triangle - quantum mechanics and the cultural milieu.

28. Wilhelm Ostwald, *Individuality and Immortality* (Boston: Houghton Mifflin, 1906), p. 46, saw, from his monist viewpoint, that "individuality means limitations and unhappiness." Consequently he maintained that individuality was not to be found at the fundament of the scientific world picture, which, for him, was the basis of our personal, social, and ethical world picture. The more pertinent precedent is contained in the final chapter of Josiah Willard Gibbs, *Elementary Principles in Statistical Mechanics* (New York: Scribner's Sons, 1902). But there is no evidence that even the most acute of Gibbs' contemporaries and immediate successors were impressed by the logical lacuna which Gibbs was attempting to fill. See: M.J. Klein, *Paul Ehrenfest*, vol. 1 (Amsterdam: North Holland, 1970), p. 136.

29. Heisenberg, op. cit., n. 22, p.993.

30. Bohr, op. cit., n. 8, pp. 14-15, 37, 43, 65.

31. Carl F. v. Weizsäcker, also a disciple of Bohr in lessons of atomic physics, is a puzzle in re individuality. His collected semipopular writings, *Zum Weltbild der Physik*, 7th ed. (Stuttgart: Hirzel, 1958), is full of complementarity but has not a word on individuality. Nonetheless there are various indirect indications that Weizsäcker promoted *Individualität* as a principal lesson of atomic physics. Thus in 1943 there appeared in Germany, under Weizsäcker's wing and under the title *Elementarteilchen, Individualität und Wechselwirkung*, a translation of Louis de Broglie, *Continu et discontinu en physique moderne* (Paris: Michel, 1941). Again, Weizsäcker's student, Klaus M. Meyer-Abich produced, with repeated citations of his mentor, an exceptionally acute and sympathetic study of Niels Bohr's physical thought, under the title *Korrespondenz, Individualität und Komplementarität* (Wiesbaden: Steiner, 1965), in which Bohr's claims regarding individuality in atomic physics are accepted in full.

32. For example, V.F. Lenzen, "Individuality in Atomism," in *The Problem of the Individual*, "University of California Publications in Philosophy, vol. 20," (Berkeley: Univ. of Calif. Press, 1937), pp. 31-52.

33. E.g. Leonard Krieger, The *German Idea of Freedom* (Boston: Beacon Press, 1957), p. 130 et passim.

34. Peter Kapitza, *Die frühromantische Theorie der Mischung: Über den Zusammenhang von romantischer Dichtungstheorie und zeitgenössischer Chemie* (Munich: Fink, 1968), pp. 158-59, 176-78; Ludwig W. Kahn, *Social Ideals in German Literature, 1770-1830* (New York: Columbia University Press, 1938), pp. 6-7, 14-23, 91-95, et passim. Reprinted: New York: AMS Press, 1969.

35. Fritz Ringer, *Decline of the German Mandarins* (Cambridge, Mass.: Harvard University Press, 1969), p. 108.

36. I know no study of this progressivist phase in Euro-American culture; I have pointed to a few characteristic phenomena in an unpublished (and still incomplete) study of the social institutions of the physicists at the turn of the century.

37. Paul Forman, "Financial support and political alignment of physicists in Weimar Germany," *Minerva* 12 (1974):39-66.

38. A classical statement of the thesis of German individuality vis-á-vis the West is Ernst Troeltsch, "The idea of natural law and humanity in world politics," address to

the Hochschule für Politik, Berlin, 1922, translated as an appendix to Otto von Gierke, *Natural Law and the Theory of Society, 1500 to 1800*, 2 vols. (Cambridge: Cambridge University Press, 1934), and also appended to the condensation published by Cambridge University Press in one volume in 1950. Troeltsch's affirmation of the value of *Individualität* is all the more significant because of his general opposition to the radical *Lebensphilosophie* of the period.

39. As listed in the subject index of *Deutsches Bücherverzeichnis*, vols. 7-16 (Leipzig: Börsenverein der deutschen Buchhändler, 1926-32), covering the years 1921-30.

40. Theodor Haering, Über *Individualität in Natur- und Geisteswelt* (Leipzig & Berlin: Teubner, 1926), p. 1.

41. Niels Bohr, op. cit., n. 7.

42. Ludwig Bertalanffy as quoted in Forman, op. cit., n. 2, p. 111.

43. Arthur Korn, *Die Konstitution der chemischen Atome* (Berlin: Siemens, 1926), pp. 118-19; *Physikalische Zeitschrift* 27(1926):802.

44. Or a meta-meta-meta statement if one chooses to distinguish between the formalism of quantum mechanics and the ordinary language interpretation/explication of that formalism.

The Reception of an Acausal Quantum Mechanics in Germany and Britain

Paul Forman

In considering the reception of a scientific innovation, whether unconventional or otherwise, it is useful to distinguish--as Aristotle would surely have--three phases: short term, midterm, and long term. I will not attempt to characterize any of these phases except the first, for my discussion of the response to the advent of an acausal quantum mechanics will be limited to the short term--and that, moreover, only in Central Europe and the British Isles. In this initial phase, extending roughly two years, from the autumn of 1925 to the autumn of 1927, responses were, as I will argue, conditioned largely by prior expectations, predilections, and prejudices. This, however tautologously, I take as the defining characteristic of the immediate reception of innovations: people, and physicists, not only tend to find what they are looking for, but also fail to recognize what they are not prepared to see. Thus it is necessary to examine the period prior to the introduction of quantum mechanics quite as closely as that immediately following the event.

The issue, or unconventionality, to be examined is "the abandonment of causality" in the description of atomic processes. Causality, for the early twentieth century physicist, meant complete lawfulness of Nature, determinism. Unrealizable in practice, it was nonetheless regarded as the ideal goal of the science, and an assumption essential to its pursuit. While many macroscopic laws were even then being recognized as merely statistical regularities--notably the second law of thermodynamics--it was always supposed that the underlying microscopic events were completely deterministic.[1] The quantum mechanics of atomic structure and processes introduced by Heisenberg and Schrödinger in 1925/26 proved, on the contrary, to be a self-consistent formalism giving descriptions apparently adequate to the test of experiment, yet providing in general not deterministic, but only probablistic predictions about the state or behavior of these most fundamental physical entities.[2]

The reception accorded these unquestionable advances in the description of these basic processes was, as I will evidence, just what one would anticipate from the antecedants. Among German-speaking Central-European physicists (including, in particular, the originators of the theory) there was immediate recognition of, and assumption of a posture toward, the acausal aspect of the theory and its world-view implications--corresponding to the "violent dispute over the significance of the law of causality" which had been raging in Germany for some years.[3] Meanwhile, in Britain there was nearly complete obliviousness to the epistemic issue exercising the Central Europeans--precisely because causality had not in the previous years been an issue for the British physicists.

Germany

In a study published some years ago I described the intellectual environment in Germany in the years immediately following her defeat in the First World War and connected the tone and content of that intellectual environment with the avowed desire among a considerable segment of the German-speaking Central-European theoretical physicists for a revolution of their science which would eliminate "causality" from its explanatory framework.[4] In order now to connect the reception of an acausal quantum mechanics with its origins, and provide a foil to the British, it is necessary for me to summarize parts of that study.

The Intellectual Environment

Prior to and during the First World War German physicists had viewed their science, if not themselves, as closely allied with, and essential to, Germany's technically advanced industry, and with the economic and military power which that industry ensured. Confident of their value in the eyes of the public, they were self-assured, even arrogant, vis-à-vis their colleagues in the humanities and social sciences. Then, however, Germany's completely unexpected military and industrial collapse at the end of 1918 brought an immediate and extreme public reaction against the industrial-scientific idols.

This wave of anti-technologic Lebensphilosophie was a celebration of "life," intuition, unmediated and unanalyzed experience; it was a rejection of reason and logical analysis because allegedly inseparable from positivism-mechanism-materialism and because, as fundamentally disintegrative, unsatisfying of the "hunger for wholeness." Naturally, spiritualism and astrology were among its vulgar expressions. The pervasive "life" rhetoric notwithstanding, the mood of this period was distinctly pessimistic. The most popular work of Lebensphilosophie was Oswald Spengler's Untergang des Abendlandes, in which, along with forecasts of the decline of Western civilization, many pages were spent proving that even mathematics and physics were completely culture-bound and must therefore share the general fate. In the vocabulary of Lebensphilosophie there were two characteristic words: one--Anschaulichkeit, intuitiveness--had strongly positive connotations; the other--Kausalität, causality--was emphatically pejorative. And the epitome of the abstract, unintuitive, and causal mode of apprehending reality was that of the theoretical physicist. Thus, overnight, the physicists and the mathematicians--but especially the theoretical physicists--found themselves in a thoroughly hostile intellectual environment.

Adaptations to the Environment

How did German physicists and mathematicians respond to this circumstance? At the ideological level--i.e., in their professed justifications of scientific activity, their epistemologic stance, and, generally, in their élan, their esprit, their confidence in the future of their science--there was an astonishingly fast, far-reaching, and unanimous adaptation to the Zeitgeist, to the lebensphilosophisch milieu, including an espousal of Spenglerian pessimism. The adaptive efforts did not stop at the ideological level. For as there is no clear separation between mood, motivation, and metaphysics on the one hand, and scientific activity and opinion on the other, very soon broad movements developed in mathematics and in theoretical physics to reconstruct the foundations of these disciplines. In mathematics this was "intuitionism," a doctrine first proposed by L. E. J. Brouwer a dozen years earlier, but which seized the Germans only now in their altered milieu. In theoretical physics this was, inevitably, the renunciation of "causality."

In the seven years between the end of the First World War and the appearance of Heisenberg's acausal matrix mechanics, an impressive group of German--including Austrian-German--physicists published proposals, arguments, or exhortations for the abandonment of causality in their science. The list, in temporal order, includes Franz Exner, Hermann Weyl, Richard von Mises, Walter Schottky, Walther Nernst, Erwin Schrödinger, Arnold Sommerfeld, Hans Albrecht Senftleben, and Hans Reichenbach. I have not listed the Dane Niels Bohr, who joined this movement and necessarily had a strong influence within it, nor have I listed other physicists--in particular, Max Born--who have left us indications in private papers that they too were strongly inclined to abandon causality. In order to establish certain essential connections, and the non-existence of others, I will say a few words about Weyl and about von Mises.

Hermann Weyl, in his early thirties when the war ended, was recognized as one of the broadest and most talented mathematicians of his generation. At Göttingen circa 1910, Weyl had been on the fringes of, and had drawn his wife from, the circle of enthusiastic disciples around Edmund Husserl. In the postwar period Husserl's doctrine of pure phenomenology degenerated into the existentialism of Heidegger and others--i.e., into one of the more esoteric of the various systems of Lebensphilosophie. The first explicit intrusion of Weyl's philosophic proclivities into his scientific work was an attempt, begun in 1917, to put the continuum of real numbers on an intuitionist foundation. After the war Weyl became the principal champion in Germany of Brouwerian intuitionism. But it was not merely the ontologic and methodologic basis of mathematics which Weyl felt to be in urgent need of reform. Drawn by the theory of relativity to theoretical physics, Weyl was among the first to come out against causality.

Weyl came to this revolutionary position within a year of the Armistice--in opposition not only to his own views and scientific efforts prior to 1918, but also as he supposed, to the unanimous view of his colleagues. He was moved to repudiate causality not as the result of any scientific problems, but because the character of the fundamental physical theories was incompatible with the subjective, intuitive, and "for our entire experience fundamental, unidirectionality of time"--a basic tenet of existentialism. The solution which Weyl found was to reerect the classical theories on an intuitionistically conceived continuum. Thus, "the rigid pressure of natural causality relaxes, and there remains, without prejudice to the validity of natural laws, room for autonomous decisions, causally absolutely independent of one another, whose locus I consider to be the elementary quanta of matter. These 'decisions' are what is actually real in the world."[5]

How is this indeterminism compatible with the validity of natural laws, and in particular with that thorough-going determinism which characterizes Einstein's theory of gravitation, and Weyl's own extension of it to include electricity? The answer lies in the analytic character which Weyl at first attributed to physics as a whole:

> Physics . . . does not deal at all with the materiality, the contentness, of reality; rather, what it recognizes is solely the formal constitution of reality. It has for reality the same significance as formal logic for the realm of truth.[6]

Soon, however, Weyl restricted this characterization to the laws of classical field theory, to gravity and electricity, thus allowing that now,

16   Paul Forman

with quanta and atomicity, physics was finally in touch with the substance of reality--an acausal reality, if Nature followed Weyl's way.

The quanta which Weyl spoke of in his first manifesto, published in the spring of 1920, are simply elementary particles of matter; only some months later did Weyl invoke the quanta of action (or energy) of the quantum theory as compelling him to say "clearly and distinctly that physics in its present state is simply no longer capable of supporting the belief in a closed causality of material nature resting upon rigorously exact laws."[7] Thus it is perfectly clear that the quantum theory was for Weyl a post factum rationalization for a position whose adoption represented an actualization of his own intellectual/emotional proclivities by contact with a Zeitgeist of the corresponding character.

Richard von Mises, a few years older than Weyl, was one of the leading applied mathematicians of Weimar Germany. Son of a high-ranking Jewish civil servant of the Austro-Hungarian Empire, his conservative nationalism was reinforced by the experience of being expelled by the French from Strassbourg early in 1919--along with all the other German professors at the University. His political leanings notwithstanding, von Mises, like Weyl, was attuned to contemporary philosophic, literary and artistic movements. Between the spring of 1920 and the autumn of 1921 he converted to an explicitly Spenglerian pessimism about the future of science, repudiating not merely causality, but "the entire scheme of 'physical explanation.'"[8] Although von Mises cited the quantum theory--i.e., the admittedly incomplete and often erroneous theory employed before the Heisenberg-Schrödinger quantum mechanics--as calling for such a thorough reconstruction, he made a point of stressing, in a paper read to his physicist colleagues, that the same conclusion must be drawn from his own field of classical mechanics, properly considered.[9]

These cases, the conversions of Weyl and von Mises, establish the influence of Lebensphilosophie in general, and of Spengler in particular. They show that considerations extrinsic to atomic physics had already predisposed some physicists to look for--indeed long for--a quantum mechanics cancelling causality.

Disciplining the Renegades

The social incentives for the production of anti-causal manifestos were so great that there would surely have been many more published by German theoretical physicists were there not disciplinary forces inhibiting such apostasy. "Apostasy" is indeed the appropriate word, for in the absence of any clear conception of how indeterminacy could be combined with lawfulness, a repudiation of causality was inevitably a repudiation of the scientific enterprise. And indeed, if a manifesto was to echo the prevailing pessimism, and give promise of a reform of physics in line with the prevailing anti-rationalism, the penitent physicist had necessarily to renounce the traditional epistemic ambitions of his science.

The multiplication of such manifestos was restrained by the disapprobation of some of the older, authoritative figures in the field. Undoubtedly often in private, and occasionally even in public, they reasserted the indispensibility of causality and chastised those colleagues who so rashly and wantonly abandoned it. The most censurious was Wilhelm Wien, professor of experimental physics at the University of Munich, who, although extremely flexible in adapting other tenets to the Zeitgeist, drew the line at causality. The most consistent and outspoken in his adherence to causality was Max Planck, professor of theoretical physics at the University of Berlin. By his side stood Albert Einstein, who, while avoiding counter-manifestos, seldom lost an opportunity to state publicly as well as privately his antipathy to any departure from strict causality. The third principal theoretical physicist in Berlin, Max von Laue was undoubtedly in accord with his two confreres. Elsewhere, too, the frowns of senior and influential colleagues-- and experimentalists were generally senior to theorists--must have restrained many an impulse to the expression of anti-causalist inclinations.

Seizing Upon the New Theory

This then was the situation in Germany into

which Werner Heisenberg launched in the summer of 1925 a new quantum mechanics. Heisenberg, age 24, a student of Sommerfeld, Born, and Bohr--all anti-causalists to a lesser or greater degree-- was himself a product of the German youth movement, which exuded Lebensphilosophie. The acausal character of Heisenberg's "Quantum-theoretical Reinterpretation of Kinematic and Mechanical Relationships" was implicit in his approach.[10] Renouncing the goal of picturing the inner workings of the atom, Heisenberg proposed as descriptive variables new mathematical entities--infinite matrices, as it soon appeared. Although in general the physical interpretation of these quantum mechanical variables was not immediately clear, the first which Heisenberg set up, and from which all the others were then constructed, was essentially and intentionally a table of probabilities, namely those for the atom to make a transition between any two of its discrete energy levels.

Heisenberg had been shuttling back and forth along the Göttingen-Copenhagen axis during the previous year. His work and views were closely related to those of Niels Bohr and Max Born, his mentors at these termini, and to those of Wolfgang Pauli whom he would visit in Hamburg en route. Quickly accepted and further developed at Göttingen, the matrix mechanics was, by the end of 1925, recognized as an important advance throughout the German-speaking Central-European physics community, and abroad where there was good rapport with this geographical-cultural center of theoretical atomic physics--notably, Leiden, Cambridge-Engl., Cambridge-Mass., Pasadena.

The Unanschaulichkeit, the unintuitiveness, and the implicit acausality of the matrix mechanics were repugnant to the conservative faction in the German physics community. As Erwin Schrödinger said early in 1926 in putting forward an alternative, a continuous, causal, wave mechanics: "we really cannot change the forms of thought, and what cannot be understood within them cannot be understood at all. There are such things--but I do not believe the structure of the atom is one of them."[11] Schrödinger, in his quantum mechanics of the hydrogen atom, identified the electron with the wave-like solution of a differential equation of familiar form. And, as we would expect, it was received enthusiastically by the causalists among the German physicist--Willy Wien, who published Schrödinger's papers, Max Planck, Albert Einstein, and others of lesser stature.[12] Outside of Germany it was adopted far more widely and enthusiastically than was the matrix mechanics.[13]

The question immediately arose: what is the relation of these two seemingly so different-- indeed antithetic--theories, each of which gave the same, correct, solution to a variety of problems. Schrödinger himself was able to show, in the spring of 1926, that the elements of the infinite arrays of matrix mechanics could be calculated directly from his ψ-functions, the solutions to his wave equation. Mathematically, the two theories were equivalent. This was an important result, but its effect was subversive of Schrödinger's interpretation of his own continuum theory. The acausalists soon came to see that the wave mechanics could give no more complete an account of atomic processes than their own matrix mechanics, which disavowed such pictures from the beginning.[14]

Max Born, as a result of his association with James Franck, had a standing interest in treating theoretically the collision of an electron with an atom. Unsuccessful in getting hold of the problem with the unwieldy apparatus of infinite dimensional matrices, Born turned, in the summer of 1926, to Schrödinger's wave mechanics. With it he succeeded in treating the problem as a diffraction phenomenon, and obtained a definite mathematical expression for the effect of the collision. How was one to interpret this expression and Schrödinger's ψ-functions generally? There was, as Born said, just one and obvious and natural interpretation, namely that ψ (soon corrected to $|\psi|^2$) was the probability of finding the electron and atom in any particular state after their collision.[15]

On obtaining these results Born's first reaction was to carry the good news not merely to his fellow atom-theorists but to the scientific public at large. He immediately wrote up his preliminary results in a summary, and generally comprehensible form, intending to publish them in Die Naturwissenschaften--the Science of Weimar

Germany.[16] In that short article Born went much farther in drawing general epistemic conclusions than these results alone would have required or could have justified:

> Thus the Schrödinger quantum mechanics gives and entirely definite answer to the question of the effect of a collision, but there is no causal relation involved. One gets no answer to the question 'what is the state after the collision,' but only to the question 'how probable is a prescribed effect of the collision'. Here the whole problematic of determinism intrudes itself. From the standpoint of our quantum mechanics there is no quantity which causally determines the effect of a collision in the individual case; moreover experiment thus far gives no evidence that there are inner properties of atoms which condition a definite outcome of the collision. Should we hope subsequently to discover such properties (perhaps phases of the motions in the atom). . .? Or should we believe, that the concordance of theory and experiment in the incapability of specifying conditions for a causal sequence of events is a preestablished harmony, which rests upon the non-existence of such conditions? I myself am inclined to abandon determinedness in the atomic world. But what is a philosophic question, for which physical arguments alone are not decisive.
> In practice, in any case, the indeterminism subsists for the experimental as well as for the theoretical physicist.

Clearly Born had a strong inner need to come before the public as an acausalist and free himself from the odium, and his science from the onus, of determinism. Although prudent in bowing to philosophy--indeterminism "is a philosophical question for which physical arguments alone are not decisive"--Born was reckless in presuming to speak not to, but for his fellow physicists. When, later that summer, Born published the details and refinements of his quantum mechanical calculation of the collision process he was far less categoric about the non-existence of causal variables; "it appeared to me at first improbable" that such existed, he said, apologetically, and explicitly left it up to the individual physicist to decide the question how he would.[17] Evidently Born ran into static from causalists, and, as he was a timorous man, ever anxious to avoid antagonizing his conservative colleagues, he backed away.

Werner Heisenberg was a much tougher person, and, unlike his Jewish colleagues, he had no special reason for keeping a low profile.[18] As his confidence in his brainchild grew through the development of a consistent probablistic interpretation of the matrix mechanics by P.A.M. Dirac and by Pascual Jordan, and through the success of the new theory with many previously intractable problems,[19] he became determined to draw from it a conclusive argument for the failure of causality. This he sought to do in his famous uncertainty principle paper, written in the spring of 1927.[20] Inconsistently, Heisenberg insisted that "physics is supposed only to describe formally the interconnection of perceptions," intending thus to nail those who would appeal to a causal reality behind the experimental appearances. But then, far from accepting the agnostic positivism which this position implied, Heisenberg sought to legislate what is _really_ the case on the basis of certain characteristics of his formal description. Naturally, his categoric conclusion, that "the invalidity of the law of causality is established definitively by the quantum mechanics," did not logically follow[21]--which circumstance testifies once again to the urgency of a theoretical physicist's need to justify his discipline and himself by pandering to the ideological demands of his cultural milieu.

But is there not some difficulty here with our interpretation? The quantum mechanics conceived by Heisenberg and developed into matrix mechanics in conjunction with Born and Jordan, arose under the sign of Unanschaulichkeit. It was a deliberately abstract theory, for its inventors had renounced the possibility of picturing in space and time the intra-atomic processes with which the theory had to deal. Were then the physicists whom I represent as so susceptible to the Zeitgeist in respect of causality, careless of its demands in respect of Anschaulichkeit, intuitiveness? In fact that was not the case, or rather only briefly so under the influence of the overwhelming internal difficulties in theoretical atomic physics.

In the course of the summer of 1926 Born recognized in his probablistic interpretation of Schrödinger's wave function a <u>via media</u> which, while renouncing "the causal definiteness of the individual events," nonetheless allowed "the retention of the familiar conceptions of space and time, in which the events take place in an entirely normal way."[22] Heisenberg was unhappy about Born's deviation from the quantum-mechanical creed.[23] At first he tried to dismiss the problem by arguing that "the hypothesis of the atomistic structure of matter is, from the outset, <u>unanschaulich</u>."[24] But soon he too apprehended the necessity of clearing their theory of the stigma of <u>Unanschaulichkeit</u>, and the problem gradually became his principal preoccupation. In an essay presenting the program and achievements of "Quantenmechanik" in <u>Die Naturwissenschaften</u> in the autumn of 1926, Heisenberg used <u>Anschauung</u>, <u>anschaulich</u>, or some other form from this root, once in every 250 words, sixteen times in all, concluding that "thus far there is still some essential feature missing in our picture."[25]

The result of Heisenberg's continued struggle with the problem through the winter of 1926/27 was his uncertainty principle paper, which he titled "On the Intuitive [<u>anschaulich</u>] Content of the Quantum-Theoretic Kinematics and Mechanics." His solution was to redefine the quality "intuitive," <u>anschaulich</u>, so as to make it predicable of his irremediably unpictorial quantum mechanics. Namely, Heisenberg equated "intuitive" to "satisfactory" in a strictly positivist sense: "From an intuitive theory in this sense one ought thus demand only that it be without contradiction with itself and that it permit the prediction without ambiguity of the results of all imaginable experiments in its domain."[26] Would it not have been much simpler, and perhaps also less obscurantist, to have made a positive virtue of the theory's <u>Unanschaulichkeit</u>? That Heisenberg was unable to do so, shows once again the overwhelming force of an epistemic predilection which, while not deriving exclusively from the <u>Zeitgeist</u>, carried its full authority.

Britain

I have contended that the reception of an acausal quantum mechanics by German physicists can be understood as the continuation of a long-standing controversy about the desired, or anticipated, character of the still-to-be-discovered theory of atomic processes. This controversy, extending from the early 1920s into the years immediately following the introduction of matrix and wave mechanics, owed its existence to a Zeitgeist at once anti-causal and anti-scientific. Although the point at issue was the concept of causality, the controversy was, basically, one between those physicists seeking to place their science in the service of, and those wishing to keep it insulated from, their wider intellectual milieu. Crucial in linking physicists and Zeitgeist, the source of the pressure to adapt, was, I have argued, the negative valuation of abstract physical science among the educated public at large. Yet, however well these theses may explain and integrate the facts of the German case, their validity as a general mechanism can only be tested by cross-cultural comparisons. I therefore offer in the following pages a description and analysis of the situation in Britain. My characterizations I think essentially correct, even though they result from only superficial investigations.

Science and Reason

In striking contrast with Germany, the intellectual mood in immediate post-war Britain contained relatively little irrationalism. High-brow attitudes certainly, and even middle-brow attitides on the whole, remained far from the anti-intellectualism of the continent. Not a trace of it was to be found in R. B. Haldane's extremely popular Reign of Relativity, first published in 1921.[27] Viscount Haldane, learned statesman in the service of higher learning, began his philosophical-historical treatise with a characterization of the contemporary intellectual temper. This he saw as critical and anti-authoritarian; he neither spoke irrationalism himself, nor did he hear or see irrationalism about him. The war experience had, manifestly, increased his own faith in science as a practical and ethical discipline, and his general commitment to "conclusions based on reasoned knowledge."[28] Although in Britain, as in Germany, fortune-telling was extremely popular among those facing the uncertainties of post-war life, it was only the late 1920s which saw in Britain a reaction of popular philosophers against "barren intellectualism," a turn toward eastern mysticism and the "intuitive approach."[29]

Again, in contrast with the Germans' reaction against physical science as a god that failed, there is every indication that the prestige of science was extremely high in Britain--higher than before the war, higher than was pleasing to some contemporary culture-critics. "Science had gradually become the faith of numerous cold-blooded people who had no use for revealed religion . . ." Robert Graves observed from a distance of two decades, thus confirming, but with a different valuation, Haldane's characterization.[30] The attention to, and support of, applied science, begun during the war, continued in the post-war period without any such ideological break as occurred in Germany.[31] In all anticipations of the future, whether optimistic or pessimistic, natural science and its technical applications were accorded a very large role.[32] "Epoch-making" discoveries of a thoroughly practical sort were standard newspaper fare,[33] and the demand for popular expositions of the results of fundamental physical research--and not merely of relativity-- was very high. In 1926 a reviewer for Nature observed that:

> Whatever properties may ultimately be assigned to the atom, there is one which cannot be omitted--its power to seize and captivate the human mind. In fact, if we judged by the output of the printing press in the last few years, we might not unfairly assume that no sooner does any one fall within the sphere of influence of this radiating personality than he is seized with an irresistible determination to go home and write a book about it. Nor is the proselytising zeal confined to the pure physicist, whose protegé the

the atom may be presumed to be. We
have books on the atom, some of them
quite well done, by chemists, by mathematicians, by technicians, and by journal-
ists, and addressed to all sorts and
conditions of readers. Thus we have
"Atoms for Amateurs," "Atoms for Adepts,"
"Atoms for Adolescents," "Atoms for
Archdeacons," "All about Atoms for Anybody"--these are not the exact titles,
but they indicate the scope of the
volumes well enough--in fact, there seems
to be a determination that no class of
reader shall be left without an exposition of the subject suited to his condition and attainments. As these volumes
continue to pour forth--there are two
fresh ones before us as we write--we must
assume that they find purchasers and
readers. If we add to these the enormous
output of serious scientific contributions
from the many laboratories engaged in
investigating the structure and properties of the atom, it is clear that this
infinitesimal particle exerts an attraction unique in the history of science
over the minds and imaginations of many
types of men.[34]

Such popular expositions were indeed bought. In 1929 the entire printing of James Jeans' Universe Around Us, 7500 copies, was sold out in one month. The following year the initial printing, 10,000 copies, of his Mysterious Universe was sold on the day of publication, and the book continued to sell as fast as it could be printed and bound: 1000 copies a day for a month.[35]

Far from continuing to decline in stature in the eyes of artists and literary intellectuals, physicists seem to have gained a measure of respect which they had long ceased to expect from that quarter. To be sure, in the first three decades of the century literary circles were less anti-intellectualist in Britain than on the continent. A Bertrand Russell, a Maynard Keynes, a Julian Huxley could be at home in Bloomsbury or in the group surrounding Lady Ottoline Morrell. It was rather D. H. Lawrence, with his deep anger at science, and his belief in the power of "the

dark loins of man," who was regarded in such
company as <u>outré</u>. And in this respect there seems
to have been no significant difference between the
wartime and the early postwar years.[36] Against
the limited following Lawrence gained, one must
place J. W. N. Sullivan's testimony (1927) that
"there can be no doubt that the prestige of science
has greatly increased in recent times." Before
the war the view of Nietzsche, Dostoievsky, and
Tolstoi,

> that the man of science was not a human
> being . . . became very popular with
> artists of all kinds. . . . It is evident
> that the position today is rather differ-
> ent. It has become different since the
> war. . . . The change was, I believe, due
> to Einstein . . . the respect of imagina-
> tive people for science in general has
> greatly increased . . . the result appears
> to have been disastrous. At a time when
> the physicists are abandoning materialism,
> the artists are accepting it.[37]

Given these circumstances, the morale of the
British physicists ought to have been high. And
so it appears to have been. In March 1924, welcoming their distinguished guests, including the
Duke of York and Prime Minister Ramsey MacDonald,
to the celebration of the fiftieth anniversary of
the Physical Society of London, F. E. Smith gave
it voice: "Gentlemen, we are not only proud of
the past, but confident of the future," etc., etc.[38]
It is impossible to imagine the corresponding
German association, the Deutsche Physikalische
Gesellschaft, entertaining guests of equivalent
rank in German society and government. Moreover,
should it, impossibly, have had the opportunity
to do so, it is impossible to imagine its spokesman so unqualifiedly confident. Thus, according
to the model I have advanced to link the physicists' ideology and conceptual predilections to
the <u>Zeitgeist</u>, the British, in contrast with the
Germans, should have felt little inclination to
alter the one or the other in any radical or revolutionary way.

<u>Mind Over Matter</u>

But <u>were</u> there any predilections of the

British mind of this period which were contrary to the established goals, methods, or worldpicture of the science of physics? If by "mind" we understand the highbrow synoptic-synthetic thinkers, the answer is, curiously, "yes." R. B. Haldane summed it up fairly well as a repudiation of the Victorian scientist's bifurcation of the world into an objective reality of matter and motion and a subjective world of qualities and values which are merely mental. On the contrary, for Haldane and most of the other British philosophers, the world

> exhibits mechanistic features, but it also has biological aspects not less important. It discloses the shaping influences of ends, and it possesses colour and beauty and value . . . there is a single whole within which fall matter and mind alike. . . . such is at least the view which is beginning to be insisted on in the twentieth century, even in scientific circles.[39]

A virtual litany of like convictions and anticipations fills the two volumes of personal credos published in 1924 and 1925 as <u>Contemporary British Philosophy</u>. In most cases they are expressions of a comprehensive panpsychism or hylozoism.[40]

The rise of analytic philosophy--and the concomitant resignation of the philosopher's grand pretensions--has so colored our perspective that it is difficult to recognize the situation as it was in the early 1920s: the established philosophers were still overwhelmingly metaphysical idealists, who, straight out, "define philosophy as the systematic study of the ultimate nature of reality," and who declare that ultimate reality to be mind.[41] Although hindsight enforces the qualification "<u>still</u> . . . metaphysical idealists," in fact these men seem not to have yet had an inkling of their fate. Nothing could have astonished them more than to learn how completely they, and their endeavours, have disappeared from the philosophers' history.[42] For as Haldane suggests, they were convinced that history was on their side. Above all, the recent development of physical science (Relativity!) seemed to many of them evidence of this sea change, for which in the early 1920s A. N. Whitehead came to be the most applauded, though perhaps least radical, theorist.⁴³

This characteristic anticipation of, or program for, the transformation of science may be seen as the British version of the revolution in Wissenschaft which Fritz Ringer showed to be so urgently desired by German humanists and social scientists in the decades after 1890.⁴⁴ Both originated as a reaction against scientific naturalism, but equally as a reaction against the relative decline in "relevance" and cultural influence of the humanistic scholar in the industrialized mass societies emerging in both countries in the late nineteenth century.

Yet while "Kausalität" became the principal target of the German ideologist, "causality" seems to have come only slowly, lately, and uncertainly within the sights of the British. Panpsychism, for example, may be elected to facilitate an alteration of the course of nature by mind, and thus provide an escape from physical determinism. It may also, following Spinoza, with equal logic provide the basis for a thorough-going determinism, psychical as well as physical. In fact, the British idealist philosophers of the period appear to have been divided roughly equally between acausalists and causalists.⁴⁵ More to the point, recent developments in physics were not seen as bearing upon this issue. Although on the whole British philosophers paid much more attention, and gave much greater weight, to the results of scientific research than did their opposite numbers in Germany, and although most were concerned to bridge the gap between the mental and physical worlds, and although many sought to exclude mechanical causation from the organic world, nonetheless in the early 1920s not one British philosopher, so far as I am aware, drew from contemporary physics ammunition for an assault upon causality.⁴⁶

This is the more surprising as there are indeed some indications of a gradual increase in Britain, between 1919 and 1925, of preoccupation with, and antipathy toward, causality. Haldane, for example, uses the word completely unselfconsciously and unpejoratively in The Reign of Relativity (1921), but not so in his contribution to Contemporary British Philosophy (1924). The most striking document in this connection is the collection of essays, Science, Religion and Reality (1925), edited by Joseph Needham. Here a series of scholars, some older and distinguished, some younger and soon to be distinguished, each drawing upon his own special field, unencumbered by scholarly apparatus, but respectful of their readers, gave their individual perspectives on the relation of Wissenschaft to religion and an ultimate reality.[47]

The collection was profiled by Arthur James Balfour, our second example of that uniquely British phenomenon, the statesman metaphysician. The essence of the conflict between science and religion lay, according to Lord Balfour, not in the factual but in the emotional sphere, in the character of the contemporary physical world picture: "the very lucidity of the new conceptions helps to bring home to us their essential insufficiency as a theory of the universe. . . . No man really supposes that he personally is nothing more than a changing group of electrical charges." His plea for free will, for recognition that human action "constitutes spiritual invasion of the physical world," was certainly a volte-face for the Chancellor of Cambridge University. Yet Balfour still vacillated on the issue of physical determinism, using the word "causality" now pejoratively, now unpejoratively, finding the divergence between prediction and observation of the flow of physical events now necessary and intrinsic, now merely "because our knowledge of natural processes is small, and our power of calculation feeble."[48]

Needham himself came forward here with that combination of metaphysical panpsychism and physical mechanism which he reiterated so often in the following years.[49] And Charles Singer, although giving indeterminists no encouragement, concluded his survey of the historical conflicts between science and religion by recognizing that "the tyranny of determinism" was an urgent problem for some scientists too.[50] One such was yet another contributor to this collection, who, however, declined to accept the traditional refuge, i.e. the Spinozist combination of panpsychism and determinism. But let us defer for a few pages the unique case of Arthur Stanley Eddington.

Summing up our view of the intellectual environment for the science of physics in post-war Britain: here, as in Germany, the deterministic world-picture of the physicists had some high-brow opposition reaching back a full generation, and increasing in intensity in the period considered. But in Britain, in contrast with Germany, the irrationalism which put steam into the issue of causality, was never quite acceptable in high-brow circles, and became popular at middle-brow only in the late 1920s. Further, whatever may have been the causes of the manifestations of anti-determinism among British synoptic-synthetic thinkers in the years following the First World War, the general lack of focus of this sentiment-- as indicated, <u>inter alia</u>, by the absence of any anti-shibboleth corresponding to "Kausalität"--and the continued strongly positive valuation of physical science, effectively neutralized the conditions under which, on our model, any widespread or far-reaching influence upon working physicists could be expected. Moreover, in the British case the anti-determinist sentiments began to appear so late that they could scarcely have had an impact in the period considered. But let us look at British physicists and see how our expectations are met.

Before Quantum Mechanics

Outside of Cambridge, where R. H. Fowler mediated contact with Central European work, the British physicists were not very active participants in the problems of theoretical spectroscopy and quantum statistics. They had, however, been continuously involved with problems of emission, absorption, and dispersion of radiation, and thus continuously puzzled by the contradiction between the wave theory of light and its particle-like properties.[51]

Charles G. Darwin, the great man's grandson and a leading figure among theoretical physicists in Britain, made several proposals, over a period of more than a decade, for approaching these problems by altering the classical equations of mechanics and abandoning the conservation of energy in atomic processes. But of a failure of causality, of unambiguous determination of atomic events, there is no serious thought in his mind.[52]

Confronting the logical contradiction between continuous extension and local concentration of radiant energy, Owen Richardson, one of Britain's most able physicists from the school and in the style of J. J. Thomson, suggested "it may be that it is impossible consistently to describe the spacial distribution of radiation in terms of three-dimensional geometry."[53] This was a more radical thought, and Bohr cited it as a precursor of his own doubts "whether the detailed interpretation of the interaction between matter and radiation can be given at all in terms of a causal description in space and time of the kind hitherto used for the interpretation of natural phenomena."[54] But Richardson's suggestion certainly need not, and in his own mind almost certainly did not, imply any failure of causality.

I have mentioned Darwin and Richardson because they were recognized leaders of theoretical physics in Britain in the early 1920s and because they have been cited as anticipators of an acausal quantum mechanics[55]--which they were not. There is, however, one British physicist who was.

Norman Robert Campbell, though no more than highly capable as physicist, was surely one of the most unfettered critical minds of his generation, and among the most acute methodologists in the sciences. Early in 1926 he published under the title "Time and Chance" thoughts which had evidently been with him for some time.[56] He proposed that "time, like temperature, is a purely statistical conception, having no meaning except as applied to statistical aggregates" of atoms, "that the ultimate magnitude, the statistical average of which is a temporal magnitude, is the probability . . . of a transition," and consequently that chance must be accepted as fundamental and irreducible in the course of nature. Apparently Campbell was prompted to publish his proposal, even though confessedly unable to make a working theory of it, by the appearance of the first papers on matrix mechanics. He found Heisenberg's positivism an attractive approach to atomic dynamics, but in the theory itself he seems to have recognized no affinity with his own. Indeed, although his proposal is for a fundamentally acausal theory, he himself never even uses the words "causality" or "determinism."

Evidently, all of these Britons--Darwin, Richardson, and also Campbell--present fundamentally different cases than the German theoretical physicists authoring manifestos against causality. Not only do their doubts and their proposals differ conceptually and semantically from the Germans', but also the form, rhetoric, occasion, audience, etc. of the documents advancing them show the British to be thinking only of their fellow physicists and without concern for their public image.

But let us now turn to the case of A. S. Eddington, and first of all to the manifesto--essentially of the same character as the more respectable of the German Bekenntnisse--which Eddington published in Science, Religion, and Reality. Although Lord Balfour, in his "Introduction," had found nothing in contemporary science to meliorate the conflict between human freedom and physical determinism, Eddington, speaking for "the Domain of Physical Science,' undertook to do just that.[57]

The core of Eddington's position, that the field theories of physics provide merely formal, logical, indeed tautological, connections between the "entities of the world," is strikingly similar to Hermann Weyl's, whose "conversion," said Eddington, "is very recent--as indeed is my own." But where existentialist Weyl had taken these entities, whose "intrinsic essence" lies outside the province of physics, to be the "autonomous decisions" localized in the elementary particles, the quanta of matter, Eddington, panpsychist and Friend, made these entities the elementary particles themselves, animated by "Mind, the Logos." The properties and behavior of the "entities of the world," the as yet undiscovered laws of atomic structure and of the quantum, "may be true laws of governance"--by which Eddington did not mean that these "transcendental" laws will be deterministic. On the contrary, because physics here, finally, makes contact with consciousness, "it may be that the normal laws are such that they can be set aside by human free will."[58]

Superadded to this already somewhat confused position are largely gratuitous elements of existentialism, in particular "actuality" as "another undoubted fact of experience which is left out of the scheme of theoretical physics," and of which account can be taken only by means of, and along with, consciousness.[59]

As with Weyl, Eddington's position in 1925 was the result of a gradual development extending back some half-dozen years. His earliest statements on the question appeared in 1920. These are so essentially connected with the character of Einstein's theory of gravity, and its generalization by Weyl, that it is difficult to imagine Eddington having come to such views much more than a year before.[60] In his initial position, admirably clear and unencumbered, there was no existentialism, no indeterminism, not even any panpsychism. It was, simply, that mind is no part of nature, views nature from outside, and constructs from the four-dimensional manifold of "point-events," which nature is postulated to be, a world whose order is mind's own. This being so it is possible, even likely, that the laws which are intrinsic to nature--those of atomicity and quanta--may prove to be "irrational."[61]

"Behavior whose laws are irrational," Eddington observed in retrospect, "was perhaps as near to the conception of undetermined behavior as the thought of the time could reach."[62] But Weyl, as we saw, in a German-speaking Central-European milieu arrived at indeterminism very quickly, and with little apparent difficulty. In the following years Weyl, whose work in relativity was the most important stimulus for Eddington's own, undoubtedly played an important role also in the development of Eddington's indeterminism and of his philosophic views generally. In particular, Weyl is the obvious source for Eddington's existentialist preoccupation with "actuality" and the unidirectionality of time, which is otherwise thoroughly uncharacteristic of British philosophy in this period.[63] Weyl is, in fact, the only person to whom Eddington refers in "The Domain of Physical Science," citing and quoting him repeatedly.

Eddington's case is similar to that of the

German acausalists in that his manifestos were not a response to scientific difficulties but an expression of religious-philosophical yearnings. Eddington differs, however, from most of his German counterparts in that he, like Weyl, deserves to be taken seriously. He struggled continually with these epistemic problems, and "audience reactions" were evidently not uppermost in his mind. His case is unique in Britain, but he was not entirely alone. Indeed, it is impossible to think of Eddington in this period, the mid-1920s, without thinking immediately of J. H. Jeans. Whether or not Jeans himself had strong views on free will and determinism--and I doubt that he did--it would have been out of character for him not to seek to outdo Eddington on this issue too. Ironically, the opportunity which he chose to tantalize an audience with the prospect that "the deadly inevitability of cause and effect has ended" was the award of the gold medal of the Royal Astronomical Society, early in 1926, to Albert Einstein.[64] And again, in President Jeans' address the only contemporary other than Einstein to be mentioned is Hermann Weyl--both for his work in relativity and for his suggestion that the quantum will change the universe into "a drama in which all the actors choose their actions as the play proceeds." Jeans is a puzzle; he seems not to have repeated this performance.

Although the anti-causal inclinations of an Eddington (or a Jeans) are most pertinent to the principal question here addressed, they were not characteristic for their milieu. Far more typical for British natural-philosophical thought in this period is that interpretation of the conceptual situation in physics advanced by Lancelot Law Whyte in 1927 in <u>Archimedes, or the Future of Physics</u>, namely that "in order to straighten out its atomic problems physics will have to take a hint from biology."[65] This notion, casually stated in the language of the work-a-day world, had come to Whyte two years before as a most powerful experience, a veritable revelation.

> 3 am August 21st 1925. An idea the implications of which are so tremendous that they appal me . . . A revolution of thought comparable to that of Einstein but wider, more fertile & synthetic in its

> effect. It indeed would be the fitting & perhaps the only fitting crown to all the dreams in my being. . . . How the stress came:--lying in the bed--thoughts of masturbation--an appeal to enter God's being, to dive into the Depths of Christ . . . the idea. That just as the Solution of Relativity demanded a fundamental reconsideration of the so-called limits of Science & their absorption into Science & reconstruction & a new understanding of them, So the solution of the Relativity-Quantum problem might involve the problem of life in such a way as to throw real light on the relation of Religion, Art & Science.[66]

Whyte soon recognized that his illumination was not unique. On the contrary, "the last few years," he wrote in 1927, "constitute another critical period," similar to that in which the idea of evolution by natural selection emerged simultaneously in different minds,

> since an idea, which when made precise will transform scientific thought, has already come independently to many thinkers. Since 1922 many scientists have felt that in studying the emission and absorption of light physics has come near to the problem of life.

Here Whyte cited Whitehead, first of all, then Eddington who "comes near to the same idea," and thirdly, Weyl. "It has also been expressed by others quite independently, though I do not know of other published references."[67]

The point particularly to be noted is that while Whyte anticipates a revolution in science, indeterminism receives no explicit attention, let alone a leading role; Whyte is simply unconcerned with that aspect of Weyl's and Eddington's views. And this seems characteristic. In July, 1924, the program of the joint meeting of the Aristotelian Society and the Mind Association included a symposium on the quantum theory, chaired by Whitehead, with contributions by J. W. Nicholson, D. Wrinch, F. A. Lindemann, and H. Wildon Carr.[68] The very

36   Paul Forman

title of the symposium--"The Quantum Theory: How far does it modify the mathematical, the physical and the psychological concepts of continuity?"-- shows how very far the British were from focusing on causality. Of the four speakers only Nicholson raised the issue; after stressing the strictly probablistic character of the current theory, he ascribed that characteristic to its incompleteness.[69]

After Quantum Mechanics

What then happened, how did the British physicists respond, when, in the latter half of 1925 and the first half of 1926 the papers on matrix and wave mechanics appeared in German journals? In particular how did they react to the intrinsically acausal character of the matrix mechanics and to Born's acausal interpretation of the wave mechanics? If the hypothesis which I advanced in discussing the reception of quantum mechanics by German physicists is general, then here in Britain too the reactions to quantum mechanics should be inferrable from the positions previously taken on the likely or desirable character of an adequate atomic theory, and particularly from prior alignments respecting causality. And if "causality" was previously not an issue--as, by and large, it was not in Britain? Then we should expect that in the near term, the first couple of years, the British physicists would remain oblivious to the problematic of physical interpretation and, especially, to any broader epistemic implications.

Indeed, that is just the way it was. Apart from translations of expositions by Max Born and Pascual Jordan (translations prepared, and their publication likely instigated, by J. Robert Oppenheimer) there was in the pages of Nature through 1927 nothing showing any awareness of the epistemologic issues[70]--nothing except the one-paragraph summary of Eddington's fifth Gifford lecture. But let's look briefly at the British physicists previously mentioned.

In February 1927, O. W. Richardson addressed the Physical Society of London, as its President, on "The Present State of Atomic Physics." Although he gave a couple of pages to the matrix mechanics and emphasized its mathematical equivalence to the wave mechanics, Richardson spoke as an avowed partisan of the latter theory: "the electron is regarded as a train of waves."[71] That there was anything problematic in this interpretation Richardson did not even hint. He said not a word about probabilities.

Again, in February 1927, C. G. Darwin published an extension of the formalism of wave mechanics to encompass the spin of the electron. He proposed "assimilating the electron to a transverse rather than a longitudinal wave." As in his previous proposals for modifying dynamics, Darwin was thinking about tinkering with the mathematical form of the equations--to be sure, the most fundamental equations--but of the interpretive issues he seems not to have had an inkling.[72]

And once again, in February 1927, R. H. Fowler, who of all British physicists had the closest rapport with atomic theory in Central Europe, provided Nature with a short exposition for the scientific public of "Matrix and Wave Mechanics." Fowler stressed the equivalence of these two systems of calculation, their completeness and finality: "We have at last a general dynamical method to apply to any atom, which is capable of yielding us by direct calculation any result for which we may ask."[73] That there were important classes of results for which one may not even ask--such as the position of an electron, or the time of an atomic transition--Fowler either failed to see, or thought unimportant to say.

Not even N. R. Campbell--I think it fair to say--had in the spring of 1927 yet recognized the indeterminism of the new quantum mechanics. Responding in the pages of Nature to Pascual Jordan's provocative question, "Will it ever happen that the time of a quantum jump is undetermined?," Campbell replied, "Certainly."[74] But his certainty derived entirely from his own view that time is a statistical concept which has meaning only for an aggregate of atoms. He saw no such implication in the new quantum mechanics, and to the end of the year seems to have remained cool, if not uncomprehending, toward Heisenberg's uncertainty principle.[75]

There was, however, as I indicated, one exception to this general obliviousness toward the

epistemic bearing of the new quantum mechanics. In
the first months of 1927 Eddington was delivering
at the University of Edinburgh the prestigious
Gifford Lectures, which obliged him to relate his
science to broader questions of philosophy and religion.[76] The fifth of these Eddington devoted to
atomic physics and its bearing upon free will, declaring that, as a result of the new quantum mechanics, "all the determinism is removed from the
laws of physics . . . whatever view we may take of
free will on philosophical grounds, we cannot
appeal to physics against it."[77] In recognizing
and seizing upon the acausal character of the new
theories Eddington is highly exceptional--probably
unique--among British physicists. Relative, however, to our model for receptivity, Eddington is
no exception, but rather confirms our expectations.[78]

In the middle term and long term the British
gradually came around. (Eddington found the process infuriatingly slow.[79]) The Solvay Congress
of October 1927 was an important educational
experience which opened the eyes of the participating British physicists to the inescapably
acausal character of the quantum mechanics.[80]
They strongly favored the wave mechanics because
the mathematics was familiar and the atomic processes were rendered in some sense pictorially.
The statistical character of the wave function
was unwelcome, but gradually accepted.[81]

In the examining the reception of an acausal
quantum mechanics I have dealt exclusively with
the near term, the first two years, and I have
described the attitudes and preoccupations, established in the preceding years, with which the
physicists met this innovation. I argued that
only in Germany was the indeterminism of the
theory immediately recognized and seized upon by
significant number of physicists, and that that
was an expression of their wish to achieve for
their science a more favorable regard by the
public. In Britain, by contrast, where the intellectual environment placed the physicists under
no pressure and causality had not previously been
a clear and important issue, the epistemic bearing
of the new theory was simply overlooked, and the
more congenial of its formalisms was adopted uncritically.

References

[1] Stephen G. Brush, The Kind of Motion We Call Heat: A History of the Kinetic Theory of Gases in the 19th Century, 2 vols. (Amsterdam, 1976), and "Irreversibility and Indeterminism: Fourier to Heisenberg," Jour. of the Hist. of Ideas, 1976, 37: 603-630. These publications give, respectively, general background and a qualification of my generalization.

[2] Friedrich Hund, Geschichte der Quantentheorie (Mannheim, 1967); Max Jammer, The Conceptual Development of Quantum Mechanics (New York, 1966).

[3] Quoting Max Planck, Kausalgesetz und Willensfreiheit. Öffentlicher Vortrag gehalten in der Preuss. Akad. d. Wiss. am. 17. Februar 1923 (Berlin, 1923), reprinted in Planck, Vorträge und Erinnerungen (Stuttgart, 1949), 139-168, esp. 140.

[4] P. Forman, "Weimar Culture, Causality, and Quantum Theory, 1918-1927: Adaptation by German Physicists and Mathematicians to a Hostile Intellectual Environment," Historical Studies in the Physical Sciences, 1971, 3:1-115.

[5] Weyl, "Das Verhältnis der kausalen zur statistischen Betrachtungsweise in der Physik," Schweizerische Naturforschende Gesellschaft, Verhl. (1919), Teil II, 152-153; Schweizerische Medizinische Wochenschrift, 1920, 50:737-741. Only this latter publication is reprinted in Weyl's Gesammelte Abhandlungen, 4 vols. (Berlin, 1968), 2: 113-122.

[6] Weyl, Raum, Zeit, Materie, 1st ed. (Berlin, 1918), 227.

[7] Ibid., 4th ed. (Berlin, 1921), 283-284; preface dated Nov. 1920.

[8] R. v. Mises, Naturwissenschaft und Technik der Gegenwart. Eine akademische Rede mit Zusätzen (Leipzig, 1922), 19.

[9] R. v. Mises, "Über die gegenwärtige Krise der Mechanik," Zeitschr. f. angewandte Math. u. Mech., 1921, 1:425-431, and Naturwiss, 1922, 10: 25-29.

[10] Heisenberg, "Über quantentheoretische Umdeutung kinematischer und mechanischer Beziehungen," Zeitschr. f. Phys., 1925, 33:879-893, translated in B. L. van der Waerden, ed., Sources of Quantum Mechanics (New York, 1969), which also translates the following papers of and with Max Born and Pascual Jordan developing Heisenberg's scheme into a consistent and comprehensive calculus of matrices.

[11] E. Schrödinger, "Quantisierung als Eigenwertproblem (Zweite Mitteilung)," Annalen der Physik, 1926, 79:489-527 (509). The entire series of papers was collected as Abhandlungen zur Wellenmechanik (Leipzig, 1928) and translated as Collected Papers on Wave Mechanics (London, 1928).

[12] Letters on Wave Mechanics: Schrödinger, Planck, Einstein, Lorentz, compiled by K. Przibram, trans. and introduced by Martin J. Klein (New York, 1967); Schrödinger's letters to W. Wien, 1926-27, of which copies are deposited in the Archive for History of Quantum Physics, Berkeley, Philadelphia, Copenhagen.

[13] In Britain, in particular, apart from Dirac's work and G. Birtwistle's The New Quantum Mechanics (Cambridge, 1928), the texts and applications of quantum mechanis were of the wave mechanics exclusively.

[14] Heisenberg to Jordan, München [1926] Jul. 28: "Vor ein paar Tagen hab ich hier zwei Vorträge von Schrödinger gehort und bin seitdem von der Unrichtigkeit der von Schrödinger vertretenen physikalischen Interpretation der Qu. M. felsenfest überzeugt." (Archive for History of Quantum Physics.)

[15] M. Born, "Zur Quantenmechanik der Stossvorgänge (vorläufige Mitteilung)," ZS. f. Phys., 1926, 37:863-67; "Quantenmechanik der Stossvorgänge,"

ibid., 1926, 38:803-827; "Das Adiabatenprinzip in der Quantenmechanik," ibid., 1926, 40:167-192; "Zur Wellenmechanik der Stossvorgänge," Ges. d. Wissensch., Göttingen, Nachr. (1927), 146-160. These papers are reprinted in Max Born, Zur statistischen Deutung der Quantentheorie, ed. Armin Hermann (Stuttgart, 1962).

[16] Born, ZS. f. Phys., 37:863, received 1926 June 25, included a note apologizing to his fellow physicists, the readership of the Zeitschrift für Physik, for the form of this initial publication. Born explained that the paper was intended for Die Naturwissenschaften but lack of space prevented its acceptance.

[17] Born, ZS. f. Phys., 38:826, received 1926 July 21.

[18] Compare, for example, the photographs of Born and Heisenberg in Armin Hermann, Werner Heisenberg, 1901-1976 (Bonn-Bad Godesberg: Inter Nationes, 1976), esp. those on pages 20 and 48.

[19] See Hund, Gesch. d. Qu. Th. (1967) and Jammer, Conceptual Dev. of Q. M. (1966).

[20] W. Heisenberg, "Über den anschaulichen Inhalt der quantentheoretischen Kinematik und Mechanik," ZS. f. Phys., 1927, 43:172-198, received 1927 Mar. 23. Quotations from p. 197.

[21] ". . . einfach ein logisches Versehen." Hugo Bergmann, Der Kampf um das Kausalgesetz in der jüngsten Physik (Braunschweig, 1929), 39.

[22] Born, loc. cit., note 17.

[23] Heisenberg, loc. cit., note 14.

[24] Heisenberg, "Quantentheoretische Mechanik," Deutsche Mathematiker-Vereinigung, Jahresber., 1927, 36:24*-25*. Abstract of lecture at the Naturforscherversammlung, Düsseldorf, 1926 Sep. 23.

[25] Heisenberg, "Quantenmechanik," Naturw., 1926, 14:989-994.

[26] M. Born and W. Heisenberg, "La mécanique des quanta," Electrons et photons. Rapports et discussions due cinquième conseil de physique [Solvay] . . . 1927 (Paris, 1928), 143-181, on p. 144. Having achieved this "quantentheoretische Umdeutung," so to speak, of Anschaulichkeit, Heisenberg was free to deprecate the "populäre Anschaulichkeit" of Schrödinger's wave mechanics: op. cit., note 20, 196.

[27] Richard Burdon, Viscount Haldane, The Reign of Relativity (1st ed., London, 1921 May; 3rd ed., 1921 August).

[28] Haldane, 3rd ed., p. 4.

[29] Robert Graves and Alan Hodge, The Long Week-End. A Social History of Great Britain 1918-1939 (London, 1940), 23, 202-203.

[30] Ibid., 91.

[31] Henry Frank Heath and A. L. Hetherington, Industrial Research and Development in the United Kingdom, a Survey (London, 1946).

[32] E.g., J. B. S. Haldane, Daedalus, or Science and the Future (London, 1923), and the reply by Bertrand Russell, Icarus, or The Future of Science (London, 1924).

[33] Graves and Hodge, The Long Week-End (1940), 92-93.

[34] "The Atom Again," Nature, 1926, 118:365.

[35] S. C. Roberts, "Memoir," in E. A. Milne, Sir James Jeans. A Biography (Cambridge, 1952), x-xi.

[36] Julian Huxley, Memories (London, 1970), 114, 160; Bertrand Russell, Autobiography, 2 vols.

(London, 1967), passim; R. F. Harrod, The Life of John Maynard Keynes (London, 1951), Chs. 5 and 6.

[37] J. W. N. Sullivan, Gallio, or The Tyranny of Science (London, 1927), 8-16. Of this negative valuation there was scarcely a hint in the collection of occasional pieces, Aspects of Science, which Sullivan published two years earlier.

[38] Physical Society, London, The Physical Society of London 1874-1924. Proceedings at the Jubilee Celebration Meetings (London, 1924), 2.

[39] Haldane, Reign of Relativity, 19.

[40] J. H. Muirhead, ed., Contemporary British Philosophy: Personal Statements (First Series) (London, 1924); . . . (Second Series) (London, 1925).

[41] J. M. E. McTaggart, ibid. (First Series), 251.

[42] E.g., the chapters on philosophy in C. B. Cox and A. E. Dyson, eds., The Twentieth-Century Mind. History, Ideas, and Literature in Britain, 3 vols. (London, 1972). G. J. Warnock, English Philosophy since 1900 (London, 1958), used his first brief chapter, "The Point of Departure," to make exactly this point, and then proceeded to give a history in which the metaphysics and idealism were completely omitted.

[43] A. N. Whitehead, The Concept of Nature (Cambridge, 1920) and numerous subsequent publications. Often assimilated to Whitehead's views were these of Samuel Alexander, whose Space, Time, and Deity, The Gifford Lectures . . . 1916-1918 (London, 1920; repr. 1927), was again highly regarded. Neither was an anti-causalist in the period or sense here treated. Haldane (Reign of Relativity, 117) quotes, and seemingly misconstrues, Whitehead's dictum that "causal nature is a metaphysical chimera" (Concept of Nature, 32). Whitehead was here speaking not against causality, but, like Husserl, against the hypostatization of

44   Paul Forman

a physical world devoid of qualities which is the cause of our perceptions of qualities.

[44] Fritz K. Ringer, Decline of the German Mandarins. The German Academic Community, 1890-1933 (Cambridge, Mass., 1969).

[45] Carveth Read and C. Lloyd Morgan, for example, were panpsychists espousing a thorough-going determinism: Muirhead, Contemp. Brit. Phil. (First Series), 278, 352-54.

[46] The British forewent a unique opportunity, which in Germany would unquestionably have led to a colorful nosegay of anti-causal manifestos: the lectureship founded in the Scottish universities by Lord Adam Gifford to treat Natural Theology in the widest sense of the word. As Rudolf Metz, A Hundred Years of British Philosopy (London, 1938), 779, observed, "this stimulating seed fell upon fertile soil. The greater part of the output of speculative thought in Great Britain since 1888, when the first Gifford Lectures were given, bears the name of this magnanimous foundation."

[47] Joseph Needham, ed., Science, Religion, and Reality (London and New York, 1925), with contributions by A. J. Balfour, Bronislaw Malinowski, Charles Singer, Antonio Aliotta, A. S. Eddington, J. Needham, J. W. Oman, Wm. Brown, Clement C. J. Webb, and Wm. R. Inge.

[48] A. J. Balfour, "Introduction" to idem, pp. 1-18; quotations from pp. 15, 17, 13, 17 respectively. Balfour had been moving ideologically very rapidly in the previous year or two. In his recent Gifford Lectures, 1922-23, he had shown himself still sympathetic to the scientific world view and still prepared to concede "that every belief is without exception causally determined, and, in the last resort, determined by antecedents which are not beliefs, nor indeed psychical events of any kind, but belong wholly to the non-rational world of matter and motion." Theism and Thought (London, 1923), 21.

[49] J. Needham, "Mechanist Biology and the

Religious Consciousness," in Science, Religion, and Reality (1925), 219-258; The Sceptical Biologist (New York, 1930).

[50] C. Singer, "Historical Relations of Religion and Science," in Science, Religion, and Reality (1925), 85-148; on p. 148.

[51] This latter circumstance is evident in Roger H. Stuewer, The Compton Effect (New York: Science History Publ., 1975); the former in Jammer; and P. Forman, "The Doublet Riddle and Atomic Physics circa 1924," Isis, 1968, 59.158-174.

[52] C. G. Darwin, "The Theory of Radiation," typescript, 54 p., dated 1912 Aug. (Archive for History of Quantum Physics, microfilm 36), esp. Ch. IV, "The Conditions for a Solution," 37-46; Darwin, "A Quantum Theory of Optical Dispersion," U. S. National Acad. of Sci., Proc. 1923, 9:25-30, communicated 1922 Dec. 1, and in still briefer form in Nature, 1922, 110:841-842; "The Wave Theory and the Quantum Theory," Nature, 1923, 111:771-773.

It is true that in an unpublished paper, "The Basis of Physics," dated July 1919--quoted by Jammer, 171--Darwin had considered that, "It may be that it will prove necessary to make fundamental changes in our ideas of space and time, or to abandon the conservation of matter and electricity, or even in the last resort to endow electrons with free will." That Darwin's appeal to acausality as a last resort should be regarded as largely rhetorical, is effectively emphasized by John Hendry in his typescript "Quantum Theory and Causality before 1926." (I am grateful to Dr. Hendry for communicating his work to me prior to its publication.) So far were physicists from any serious thought of acausality at this date that even Bohr, for whose eye Darwin had drafted these notes, understood Darwin's reference to free will as simply an expression of "the often seen sentence that the electrons cannot know the final state of transition and adapt its frequency to this beforehand. . . ." (Bohr to Darwin, draft 1919 July, incomplete and unsent. Archive for History of Quantum Physics, microfilm BSC 1.)

[53] O. W. Richardson, The Electron Theory of Matter, 2nd ed. (Cambridge, 1916), 507-508.

[54] N. Bohr, H. A. Kramers, J. C. Slater, "The Quantum Theory of Radiation," Philos. Magazine, 1924, 47:785-802, on p. 790.

[55] Ibid.; Jammer, 171.

[56] N. R. Campbell, "Time and Chance," Philos. Magazine, 1926, 1:1106-1117, dated 1926 Feb. 18. Cf., Campbell, "Atomic Structure," Nature, 1920, 106:408-409; 1921, 108:170.

[57] Eddington in J. Needham, ed., Science, Religion, and Reality (1925), 193-218. In choosing his title Eddington was likely echoing, and thus implicitly rebutting, Ernest Wm. Hobson's 1921/22 Gifford Lectures, The Domain of Physical Science (Cambridge, 1923). Hobson, passing up this ideal opportunity for repudiating causality, declared (p. 98) that "we are not acquainted with barriers which will prevent ever larger tracts of phenomena from being correlated with deterministic descriptive schemes."

[58] Ibid., 211, 217-18.

[59] Ibid., 211.

[60] Eddington, "The Meaning of Matter and the Laws of Nature according to the Theory of Relativity," Mind, 1920, 29:145-158; Space, Time, and Gravitation (Cambridge, 1920), Preface dated May 1; "The Philosophical Aspect of the Theory of Relativity," Mind, 1920, 29:415-422. Eight years later Eddington recalled that "the idealistic tinge in my conception of the physical world arose out of mathematical researches on the relativity theory. Insofar as I had any earlier philosophical views, they were of an entirely different complexion." Nature of the Physical World (Cambridge, 1928), preface. For the several questions addressed in this essay, Herbert Dingle's The Sources of Eddington's Philosophy (1954) is quite useless.

[61] It is likely that at this earliest stage Weyl was already an important influence upon Eddington. Weyl had expressed essentially this same view of the field laws in Raum, Zeit, Materie, which first appeared in 1918 and in its third edition late in 1919. Weyl is mentioned by Eddington in 1920 in both physical and logical juxtaposition to Eddington's exposition of these views.

The animus against causality seems to have taken hold of Eddington in the summer of 1920. In the first of his articles in Mind, published in April, Eddington argued that the laws of mechanics, gravity, and electricity are imposed by the mind upon the world. He did not suggest that there is anything oppressive in this circumstance or anything liberating to be anticipated from the discovery of "the actual order of nature," "the genuine laws of a possibly irrational world." The second of his articles in Mind, published in October, concluded on the contrary, that "this emancipation . . . is likely to be hailed with relief." And here again a direct influence by Weyl is certainly possible: his essay on 'the relation of the causal to the statistical viewpoint in physics' was published in August, 1920.

[62] Eddington, Relativity Theory of Protons and Electrons (1936) as quoted by L. Susan Stebbing, Philosophy and the Physicists (London, 1937), 190.

[63] This preoccupation had not yet possessed Eddington when, in May, 1922, he delivered the Romanes Lecture, The Theory of Relativity and its Influence on Scientific Thought (Oxford, 1922), 16-18. On the contrary he there placed himself fully behind the conventional Minkowskian view that Relativity had rendered the concept "now" completely arbitrary.

[64] J. H. Jeans, "Space, Time, and the Universe," Nature, 1926, 117:308-311. Einstein was almost certainly not present at this 1926 Feb. 12 meeting of the RAS.

[65] L. L. Whyte, Archimedes, or The Future of Physics (London, 1927), 9.

[66] L. L. Whyte, "Notes on scientific ideas, 1925-27" (Boston University, Mugar Library, Whyte Papers, box 38, folder 7e).

[67] Whyte, Archimedes (1927), 9, 95.

[68] Aristotelian Society, Proceedings, suppl. vol. 4 (1924), 19-49.

[69] Ibid., 22. Special mention should be made of Oliver Lodge as he held such a special place among British scientist of this period. "His bodily appearance," J. H. Muirhead opined, "is probably better known than that of any other man of our time outside the field of politics," and his outspoken belief in spiritualism made him notorious. Was he, on this latter account perhaps, an indeterminist. Far from it. Early in 1927 it was clear to him that "the present tendency admittedly is to feel . . . that the power of prediction is limited not only by our capacity, but by the nature of things, and that the uniformity of nature can be interfered with by the real agency of self-determination and free-will." (Modern Scientific Ideas, 10-11). But Lodge himself still held fast to the "faith that there is a reign of law and order." To quote once again John Henry Muirhead's Reflections by a Journeyman in Philosophy (London, 1942), 116-118, "there was more than a grain of truth in Samuel Alexander's remark to me as we once left Mariemont after a long talk with him, 'It would be an odd thing if spiritualists should turn out to be the last surviving materialists.'"

[70] M. Born, "Physical Aspects of Quantum Mechanics," Nature, 1927, 119:354-357; P. Jordan, "Philosophical Foundations of Quantum Theory," Nature, 1927, 119:566-569. Literal English translations of the titles of the original German publications in Die Naturwissenschaften would have been, respectively, "Quantum Mechanics and Statistics" and "Causality and Statistics in Modern Physics." The departures from the German titles are indicative of that very difference in focus between Germany and Britain which this essay is chiefly intended to emphasize. Born's paper was an adaptation of that which he presented at the

BA meeting in Oxford in August, 1926. Much as we would like to know what Born actually said on that occasion, the pertinent fact is that whatever he may have said seems to have remained without effect upon his auditors.

Among the more striking of the misapprehended opportunities to comment in Nature upon the epistemic issues were: the editorial attempting a general description of the "new points . . . raised" by Bohr's "Atomic Theory and Mechanics," appended to the 1925 Dec. 5 issue, points "which seem likely to be of such general importance" (116:809-810); the editorial on "The New Physics," 1926 Dec. 18, rejoicing that "the transition from the apparently unknowable to the knowable, and from the knowable to the known, is not only rapid, but is also undergoing a constant acceleration" (118:865-867); H. S. Allen's notice, 1927 Jan. 15, of recent progress in quantum theory (119:77-79).

[71] O. W. Richardson, "The Present State of Atomic Physics," Physical Society, London, Proceedings, 1927, 39:171-186.

[72] C. G. Darwin, "The Electron as a Vector Wave," Nature, 1927, 119:282-284. In response Jakov Frenkel wrote Darwin, from Leningrad, 1927 Mar. 16: "Your attempt to deal with the electron as with a vector wave is very interesting indeed. You will excuse me I hope for a bit of criticism: . . . What physical meaning is to be attached to your [vector wave] functions f and g?" (AHQP).

[73] R. H. Fowler, "Matrix and Wave Mechanics," Nature, 1927, 119:239-241.

[74] N. R. Campbell, "Philosophical Foundations of Quantum Theory," Nature, 1927, 119:779.

[75] Inferred from Campbell's correspondence with L. L. Whyte, 1927 Nov.-Dec. Whyte himself had nothing against causality or determinism, and was unimpressed by Heisenberg's uncertainty principle (Boston Univ., Mugar Library, Whyte Papers).

[76] Eddington's lectures were revised and printed as The Nature of the Physical World

(Cambridge, 1928).

[77] Brief report of Eddington's 5th Gifford lecture, 1927 Feb. 18, in Nature, 1927, 119:328.

[78] It is consistent with the frivolity of Jeans' one anti-causal escapade, that when acausality arrived in earnest he had difficulty in accepting it: The Mysterious Universe (Cambridge, 1930), 31-32, together with Jeans' silence on the issue between 1926 and this late date.

[79] Eddington, "The Decline of Determinism," Mathematical Gazette, 1932, 16:66-80, as reprinted in Smithsonian Institution, Annual Report (1932), 141-157.

[80] Dirac's conversion occurred on the spot: Electrons et photons, 261.

[81] E.g., Henry T. Flint, Wave Mechanics (London, 1929), who was recalcitrant: Geo. P. Thomson, The Wave Mechanics of Free Electrons (New York, 1930), and H. S. Allen, Electrons and Waves (London, 1932), who were reluctantly bending.

Part II

QUANTUM PHYSICS IN ITS HISTORICAL CONTEXTS

Paul Forman and the Environment and Practice of Quantum History

David C. Cassidy*

In the autumn of 1967 Paul Forman presented his doctoral dissertation, 'The Environment and Practice of Atomic Physics in Weimar Germany: A Study in the History of Science', to the University of California, Berkeley.[1] The subtitle insisted the genre was history, not sociology, in answer to a question raised at the time. During the decades that followed, the history of science and the historiography of quantum physics underwent profound transformations and especially in that period. As the history of physics rose rapidly from anecdote, recollection and reliance on published papers to professional status as an influential component of the history of science, so also did the sociology of science. In ways reminiscent of the formation of quantum mechanics during the 1920s, the environment and practice of quantum history were closely entwined. My remarks are intended only as an overview of the latter. They inevitably reflect my own experiences and observations as a graduate physics student who began the transition to history of physics in 1972.

Historians of physics, their histories and the materials of their research were few, scattered and widely diverse when, in 1961, an essential project was launched that played a major role in moving the field toward maturity. Sources for History

---

* Natural Science Program, Hofstra University, Hempstead, NY 11549; david.cassidy@hofstra.edu.

The following abbreviations are used: AAAS, American Association for the Advancement of Science; AHQP, Archives for the History of Quantum Physics; AIP, American Institute of Physics; *HSPS, Historical Studies in the Physical Sciences* (later, *Historical Studies in the Physical and Biological Sciences*); ICOS, International Catalogue of Sources; SHQP, Thomas S. Kuhn, J.L. Heilbron, Paul Forman, Lini Allen, *Sources for History of Quantum Physics: An Inventory and Report* (Philadelphia: American Philosophical Society, 1967).

[1] Paul Forman, 'The Environment and Practice of Atomic Physics in Weimar Germany: A Study in the History of Science' (PhD dissertation, University of California, Berkeley, 1967).

of Quantum Physics (SHQP) ran for three years, with follow-up projects and the addition of supplementary sources which continue to this day. Its purpose was 'to find and preserve primary source materials for the study of the history of quantum physics'.[2] For American sources, it worked in collaboration with a similar project sponsored by the American Institute of Physics.

With funding from the National Science Foundation, Thomas S. Kuhn, then at the University of California, Berkeley, and two Berkeley graduate students who had recently left physics for history — John L. Heilbron and Paul Forman — set out in search of unpublished correspondence, manuscripts and interviews of participants, all pertaining to the development of quantum physics in the period 1898–1933. By the conclusion of the project Kuhn and assistants, with a small staff, had recorded and transcribed over 175 interviews with 95 physicists. The project had by then located roughly 50,000 letters and manuscripts, a portion of which they had catalogued and photographed on microfilm. Copies were deposited in three locations: Berkeley, Philadelphia and Copenhagen. In 1967, the project published an *Inventory and Report* of the now Archive for the History of Quantum Physics (AHQP); 280 physicists were identified as having contributed to quantum physics and from whom interviews and/or letters had been obtained.[3] Institutional records, often preserved without encouragement, were beyond the scope of the project.

Similar to an earlier era when the discovery and accessibility of new manuscripts helped push the knowledge of nature to a new level, SHQP exerted an enormous impact on the historiography of quantum physics, especially regarding the development of quantum mechanics in the period 1919–27. The project not only located the essential primary sources for this history on two continents, but made those materials readily accessible in the United States and abroad to anyone able to travel to one of the archive locations. Even a graduate student just entering the profession (such as myself) could be among the first to revel in the unpublished letters of Heisenberg, Pauli, Bohr, Born, Schrödinger, Sommerfeld and many others as these physicists struggled to comprehend the atom, radiation and their quantum properties in the years leading up to and during the emergence of quantum mechanics.

While SHQP provided the unpublished sources for a mature quantum historiography, the means, aims and content of the discipline had yet to be defined in historians' terms. The introduction to the project's *Inventory and Report* informs us that 'no member of the staff had previously been concerned with the history of quantum physics' and the only sources from which to learn quantum history were

---

[2] 'Activities and Procedures', in *SHQP*, 1–10, on 1.
[3] Ibid., 3 and 'Author Catalog of Principal Sources', in *SHQP*, 11–12, on 11.

limited to a few surveys written by and for physicists.[4] Only as the project got underway did the pioneering and now classic histories by Martin J. Klein, Max Jammer and a few others begin to appear.[5]

In his preface to the *Inventory and Report*, John A. Wheeler, who had helped gain the support of the American Physical Society for the project, wrote that the 'revolution in theoretical physics' brought about by relativity theory and quantum mechanics is 'one of the greatest achievements of the human mind'. But, he lamented, very little is known of its history.[6] Now that several of the key figures in this achievement had recently died, he continued, a physicist must be concerned about this history, since 'the immortality of his heroes is at stake'. A project was desperately needed 'to capture the great dialogs and the great moments before they fade from memory'.[7]

'Interviews were the project's primary *raison d'être*', Kuhn and his staff wrote in their introduction to the *Inventory*, 'and our main efforts were devoted to their acquisition'.[8] But now, three years and 175 interviews later, they went on to inform the users of this inventory: 'Though interviews can produce unique and valuable information, contemporary manuscript and published records will continue to be the most significant resource for future historians'.[9] Forman's paper 'Alfred Landé and the Anomalous Zeeman Effect, 1919–1921', published in 1970, went even further.[10] Landé's innovations regarding the Zeeman effect did not follow the direct and logical path laid out by memory or even by the published papers. Rather, they entailed an intense and convoluted process revealed only through a study of his letters from that period. To drive home the point, Forman appended to his paper transcriptions of every scientific letter from and to Landé in February and March 1921.

The significance of letters is obvious to most science historians today, but it was apparently not so then. Nor has it been obvious to many physicists interested in

---

[4] 'Activities and Procedures' (ref. 2), 2.
[5] Max Jammer, *The Conceptual Development of Quantum Mechanics* (New York: McGraw-Hall Book Company, 1966). Martin J. Klein's early works include 'Max Planck and the Beginnings of the Quantum Theory', *Archive for History of Exact Sciences* 1 (1962): 459–79; 'Einstein's First Paper on Quanta', *The Natural Philosopher* 2 (1963): 59–86; and 'Einstein and the Wave-Particle Duality', *The Natural Philosopher* 3 (1964): 1–49. See also Armin Hermann, *Frühgeschichte der Quantentheorie (1899–1913)* (Mosbach in Baden, Germany: Physik-Verlag, 1969).
[6] John Archibald Wheeler, preface to *SHQP*, v.
[7] Ibid., vi.
[8] 'Activities and Procedures' (ref. 2), 6.
[9] Ibid. See also J.L. Heilbron, 'Quantum Historiography and the Archive for History of Quantum Physics', *History of Science* 7 (1968): 90–111.
[10] Paul Forman, 'Alfred Landé and the Anomalous Zeeman Effect, 1919–1921', *HSPS* 2 (1970): 153–261.

physics' history. Forman attempted to enlighten the latter in 1983 in a long historiographic section within his review of a multi-volume physicists' history of quantum physics. For the history of quantum mechanics, he wrote, the written letters are so extensive and insightful that recollections are practically superfluous. Even the scientific papers are inadequate because they fail to reveal the human interactions that led to the papers. But more important, these letters, when read across the temporal distance that separates historians from their subject, can provide one of the crucial elements for achieving historical 'truth': *independence*, 'in particular, independence from the judgments of the participating physicists, who so quickly upon the discovery of quantum mechanics laid down what were thenceforth regarded as the essentials of its historical development'.[11] Gaining access to, and making use of, contemporary manuscript sources was key but so too was sociology, the study of the human and social forces behind the physics. It was a goal that Forman had pursued from the start. Taking a sociological approach, he argued, meant intellectual independence from physicists' constructs and practices, and it meant relating the constructs and practices to the physicists' institutional and cultural environment.

These are recurring themes throughout Forman's historical and historiographic writings, and he was one of the first — and the most radical — to make use of them. It is difficult, as we know, to find causal connections between a cultural environment and the intellectual concepts and positions that arise within it. Yet an increasing emphasis on the necessity of independence for achieving mature science history and the need to bridge the divide separating environment and practice, internal and external, seemed for Forman and others at the time to have run parallel to events occurring within the turbulent environment of that era.

At least two environmental factors converged during the late 1960s and early 1970s. The first was the bursting of the job bubble for physicists toward the end of the sixties. David Kaiser has shown that the rapid escalation in the numbers of physics PhDs in the United States following Sputnik had by then produced such a glut of job seekers that annually, 'droves of new physics PhDs slid into the worst job shortage the nation has ever seen', far worse for physicists than during the Great Depression.[12] Federal budget cuts of defence research at universities, beginning in 1969, exacerbated the situation. In 1971, 1,053 physicists competed for just 53 jobs advertised at the annual meeting of the American Physical Society.[13]

---

[11] Paul Forman, review of *The Historical Development of Quantum Theory*, by Jagdish Mehra and Helmut Rechenberg, *Science* 220 (1983): 824–27, on 825.

[12] David Kaiser, 'Cold War Requisitions, Scientific Manpower, and the Production of American Physicists after World War II', *HSPS* 33, no. 1 (2002): 131–59, on 151.

[13] Ibid. *See also* Daniel J. Kevles, *The Physicists: The History of a Scientific Community in Modern America* (Cambridge, MA: Harvard University Press, 1971/1995), 414–16.

Seeing their career prospects dimmed after the years of training required for entering the profession, worried and restless graduate students and recent PhDs began looking to other fields for their future. At the same time, research in physics seemed uninspired, insular and overly controlled compared to the excitement, freedom and ready opportunities for truly original research offered by the history of physics, even if job opportunities there were no better. Similar developments were occurring in other sciences; the few programs and departments offering retraining and doctorates in the history of science were soon overwhelmed and new ones gradually emerged. Individuals educated in graduate physics found themselves immediately at home in the history of quantum physics. These circumstances helped contribute to the remarkable growth of historical studies in quantum history, especially quantum mechanics, and to the rapid professionalisation of the discipline in that period.

The second environmental factor, closely related to the first, began to arise shortly after the SHQP project completed its task in July 1964. By September, Kuhn and Heilbron had left Berkeley for positions on the East Coast. (Heilbron returned to Berkeley three years later.) Heilbron's masterful dissertation, 'A History of Atomic Structure from the Discovery of the Electron to the Beginning of Quantum Mechanics', completed in 1964, provided the foundation for subsequent historical studies of quantum physicists' efforts to improve and expand upon atomic models and to comprehend the failure of such models during the years just prior to the advent of quantum mechanics.[14] Forman remained at Berkeley to continue work on Weimar physics for his dissertation (he went to the University of Rochester in 1967). He was in Berkeley when the Free Speech Movement exploded on the campus during the autumn of 1964, contributing eventually to a rising cultural critique of science, especially physics, among disaffected science and physics students, recent doctorates and younger faculty. During the summer, many of the Berkeley student activists had been involved in Civil Rights activities in the South. Their attempts to promote these and other causes on campus were prohibited by a dean who insisted only the approved political parties were permitted to infringe on the academic sanctity of the campus. The students finally won, but by spring 1965 the protest had shifted to a bigger issue: the massive build-up of forces ordered that year for the Vietnam War.

Over the next several years, energised protesters at Berkeley and elsewhere began to raise questions about the role of universities in the support of federal policies such as the war. The growing arsenals of technological weapons, from

---

[14] J.L. Heilbron, 'A History of Atomic Structure from the Discovery of the Electron to the Beginning of Quantum Mechanics' (PhD dissertation, University of California, Berkeley, 1964). *See also* J.L. Heilbron, 'The Kossel-Sommerfeld Theory and the Ring Atom', *Isis* 58, no. 4 (1967): 450–85.

nuclear bombs to napalm, indicated to them that scientists and universities had lost control of their work and lost sight of the consequences of their research. Such sentiments found expression in several scientists' movements at the time. Among the most vocal was Science for the People. One of its most public events was the near disruption of the December 1970 meeting of the American Association for the Advancement of Science in Chicago, attended by Heilbron and a number of other prominent physics historians.[15] Forman, however, was absent for medical reasons. Many of the non-technical sessions during the meeting addressed issues of the day, but that did not stop the protesters, from anti-war and feminist groups to Science for the People, all of whom engaged to varying degrees in disruptive tactics. Behind the activists' tactics was the urgent call for scientists and their meetings to recognise the political and cultural environment around them and to consider critically the motives of their funding sources and the uses made of their work. In line with Forman's scholarly work, one protester at that meeting urged more loudly and categorically: 'It is time to stop saying that science stands outside of society. Science is a social activity just like being a policeman, a factory worker or a politician'.[16] But scientists, the public and the mainstream media generally agreed with the *Washington Post*, which editorialised in response to the chaotic meeting: 'It should not be beyond the power of scientists to restore reason to its normal throne at their conventions'.[17]

Dismantling this illusory throne of reason unsullied by the dirty world of politics, exposing the fight for funding and careers, and chronicling the adaptation of science to social agendas are, of course, prominent in Forman's work and the work of others. We can observe these and related themes gradually emerging during that period, in particular in Forman's early and influential contributions to the history of quantum mechanics. His first scholarly work, initially written in 1965 but not published until 1968, was a definitive study — for both the methodology and the content of quantum history — titled 'The Doublet Riddle and Atomic Physics *circa* 1924'.[18] External factors made no appearance. But the significance of the

---

[15] The meeting attendees in the history and philosophy of science are listed in Walter G. Berl's 'A Brief Guide to the 1970 AAAS Annual Meeting', *Science* 170 (1970): 873–99, on 891–92. Events at the meeting are reported in Phillip M. Boffey, 'AAAS Convention: Radicals Harass the Establishment', *Science* 171 (1971): 47–49. For a discussion of the events at the meeting, see Kelly Moore, *Disrupting Science: Social Movements, American Scientists, and the Politics of the Military, 1945–1975* (Princeton: Princeton University Press, 2008), 165–69.
[16] Quoted in Moore, *Disrupting Science* (ref. 15), 167.
[17] Editorial, 'The Scientific Approach to Controversy', *Washington Post*, 1 Jan 1971.
[18] Paul Forman, 'The Doublet Riddle and Atomic Physics *circa* 1924', *Isis* 59, no. 2 (1968): 156–74, originally presented at the annual meeting of the History of Science Society, 1965. In 1966, the revised version was awarded the Schumann Prize.

paper is that, by relying heavily upon the letters in AHQP, in addition to published papers, Forman uncovered a previously little-known yet substantial dimension to the struggles of physicists in the years immediately preceding the emergence of quantum mechanics. He revealed the central role of spectroscopy in sharpening the sense of failure of the old quantum theory. The riddles of atomic spectra formed an essential element of the internal history through the advent of quantum mechanics and electron spin in 1925. Equally important, he showed that once physicists had spin and the new mechanics in hand, the doublet riddle and related problems quickly faded from memory as a 'big mistake', to be recovered decades later only through independent study of the letters and the unravelling of the complicated complex line spectra. Even if external factors were at work, there was still much to be learned about the highly intensive internal work of physicists to find the new physics during the years just prior to quantum mechanics and spin.

In the preface to his dissertation in 1967, Forman explained that, along with the practice of atomic physics during the Weimar period, he was including elements of the political, economic and institutional environment of the physics profession, which, he wrote, 'usually make no appearance at all in histories of scientific advance, but which were nonetheless highly significant factors in the life and work of the practicing physicist'. In discussing political motives, he announced that he would make 'no distinction between internal and external factors'. However, for the physics, he explained that he still felt 'obliged to maintain some such distinction to avoid entering the sociology of knowledge'.[19] In the last third of the work he offered as a case study an account of Landé and his discovery of a semi-empirical rule for the anomalous Zeeman effect in which, Forman showed, the direction of his work, although not its content, 'was, in fact, a response to the exigencies of German academic careers'.[20]

In 1970, as anti-war unrest and the social critique of physics reached new heights and public attention at the AAAS meeting, Forman was prepared to offer more extensive and, by then, more developed remarks on the historiography of physics history in the preface to his previously mentioned paper on Landé and the Zeeman effect. There are two goals, he writes, to be achieved through the study of a scientific innovation. The first concerns internal practice: 'The primary aim of the historian of science is to reconstruct the science — i.e., the social activity and the material and ideational artifacts associated with it — of the past'.[21] It is not simply to reconstruct the 'orthodox doctrines' of the past but also to determine the

---

[19] Forman, 'The Environment and Practice' (ref. 1), iv.
[20] Ibid., 357.
[21] Forman, 'Alfred Landé' (ref. 10), 155.

extent to which those doctrines were held through a careful study of all contemporary written sources.

Forman's second task for the physics historian was environmental. It entailed 'the excision of a slice of the scientific life of that place and period'. Such an excision provides 'the potential for displaying the interlocking of conceptual and social factors', or, 'in the jargon of our discipline', it is well suited 'to demonstrate the untenability of the internalist-externalist distinction'. His words growing more heated, Forman declared such a distinction

> so foreign to contemporary historical scholarship that we must regard its persistence among historians of science as one of the more blatant ideological atavisms testifying to our phylogenetic (and frequently ontogenetic) connection with the sciences. For if one asks how this particular physicist came to this particular problem at this particular time and why he communicated these particular results in these particular forms, how indeed can one avoid regarding an innovation as the outcome of the motivated acts of an acute man, working in a particular social environment for his own advancement as well as for the progress of science? The fiction of an autonomous development in the world of scientific ideas has been maintained by systematically mistaking description for explanation, and by systematically refusing to look for or at contrary evidence.[22]

Forman's prescriptions encompassed an historiographic tension within physics history and among physics historians: innovations and discoveries were made by individuals and groups working on highly technical problems possessing their own internal aims and logic, but the individuals' work and motives and outlook, especially in quantum mechanics, were conditioned by their community affiliations and contingent upon their social-cultural-political environment. Capturing that tension at the moment of an emerging innovation within an excised slice of scientific life — i.e., a cultural cross section of a time and place — was an aim especially well suited for establishing a fully professional historiography of scientific biography, a genre still mired in the legendary hagiography of the past. It was an ideal that I, arriving after the publication of this paper, attempted to achieve in my own biographical work. In Forman's paper, Landé's professional life, combined with his work on spectroscopy, now served as 'a concrete picture of the reciprocal interaction of self-interest and the discipline's interest as they are channeled and constrained by the political environment, by economic circumstances, by the structure of academic life and careers, by the organization and

---

[22] Ibid., 157.

mores of the community of physicists, by accepted physical theories, and by experimental facts'.[23]

Forman's paper appeared in the second volume of the journal *Historical Studies in the Physical Sciences*, founded by Russell McCormmach in 1969. It was wholly in line with the agenda McCormmach had enunciated in his foreword to the first volume:

> I hope that this journal will encourage the application of the techniques of intellectual history to problems of the modern period ... Having said this, I want to stress that this journal will by no means be exclusively concerned with the physical scientists' intellectual heritage and equipment ... It is one of the purposes of this journal to stimulate the study of the social function of the physical sciences and the professional role of their practitioners ... The vision of the history of science that relegates the historiographic traditions of internal and external history to mutually exclusive roles is sterile, obstructing the synthesis of the intellectual and social history that must come.[24]

In volume three of *HSPS* Forman took the largest step yet toward realising the envisioned synthesis when he published an exhaustively researched study of how a physical concept, acausality, can be so interlocked with social factors that, he argued, a hostile cultural environment can induce an entire community of physicists and mathematicians to adapt to that environment for professional and personal self-interest through acceptance of such a concept. If Forman was right, the divide separating not just environment and practice, but the external culture and the internal concepts of physics, had been definitively bridged.[25] The argument has been celebrated, discussed and debated ever since as the 'Forman Thesis'.

By the early 1970s Forman had helped open and establish the mature discipline of quantum history — both in content and in methodology — through his insistence on tapping the riches of AHQP in order to look beyond the published papers to the physicists and their formative ideas. The results indicated much remained to be done regarding the internal history, the external environment, and

---

[23] Ibid., 157–58.
[24] Russell McCormmach, foreword to *HSPS* 1 (1969): vii–viii.
[25] Paul Forman, 'Weimar Culture, Causality, and Quantum Theory, 1918–1927: Adaptation by German Physicists and Mathematicians to a Hostile Intellectual Environment', *HSPS* 3 (1971): 1–115. The work and influence of Heilbron, Forman and McCormmach are further discussed in Lewis Pyenson's 'Three Graces', in *The Strength of History at the Doors of the New Millennium*, ed. Ignacio Olábarri and Francisco J. Caspistegui (Pamplona, Spain: Ediciones Universidad Navarra, 2005), 261–335.

the combination of the two. By then, the rapidly expanding community of quantum historians was increasingly ready to take up the task. Similar to the founders of quantum mechanics, their work became a cooperative venture that even involved the practice of exchanging long letters in the style of the correspondence they were studying. Crucial to both quantum mechanics and its history was a ready means of communicating results. For quantum history *HSPS* helped serve this purpose. Moreover, during the next decade it set the standard for history of physics as a whole through the high quality of its carefully edited papers resting upon the riches of AHQP and other archives, and through McCormmach's masterful historiographic forewords to each volume extending over the range of physics history.

Studies appearing in *HSPS* included, in addition to those mentioned, the famous 'three-men work' by Forman, Heilbron and Weart on the international physics community, circa 1900, as well as a number of fundamental quantum histories that helped uncover the route to quantum mechanics after World War I.[26] Many of the quantum papers there and elsewhere followed the newly opened spectroscopic trajectory into that history. Others focused on the failure of quantum atomic models, the origins of matrix mechanics, Schrödinger's route to wave mechanics, the wave-particle dualism and the rise of the Copenhagen interpretation.[27] Although many of these papers were decidedly internalist in nature, despite McCormmach's hopes, some (and I include myself among them) saw this as the necessary foundation for exploring the environmental aspects of that history.

As these publications and the riches of AHQP reached wider audiences, the quantum history community expanded internationally during the second half of the seventies, spurred on by the launch of projects to publish Pauli's complete scientific correspondence and to produce a history of CERN.[28] Quantum history

---

[26] Paul Forman, J.L. Heilbron and Spencer Weart, 'Physics circa 1900: Personnel, Funding, and Productivity of the Academic Establishments', *HSPS* 5 (1975): 1–185.

[27] Some of the quantum works appearing in editions of *HSPS* edited by McCormmach include, chronologically, J.L. Heilbron and Thomas S. Kuhn, 'The Genesis of Bohr's Atom', *HSPS* 1 (1969): 211–90; V.V. Raman and Paul Forman, 'Why Was It Schrödinger Who Developed de Broglie's Ideas?' *HSPS* 1 (1969): 291–314; Martin J. Klein, 'The First Phase of the Bohr-Einstein Dialogue', *HSPS* 2 (1970): 1–39; Edward MacKinnon, 'Heisenberg, Models, and the Rise of Matrix Mechanics', *HSPS* 8 (1977): 137–88; Daniel Serwer, '*Unmechanischer Zwang*: Pauli, Heisenberg, and the Rejection of the Mechanical Atom, 1923–1925', *HSPS* 8 (1977): 189–256; Paul Hanle, 'Indeterminacy before Heisenberg: The Case of Franz Exner and Erwin Schrödinger', *HSPS* 10 (1979): 225–69; D. Cassidy, 'Heisenberg's First Core Model of the Atom: The Development of a Professional Style', *HSPS* 10 (1979): 187–224.

[28] Wolfgang Pauli, *Wissenschaftlicher Briefwechsel mit Bohr, Einstein, Heisenberg u. A.*, ed. Armin Hermann, Karl von Meyenn and Victor F. Weisskopf (New York: Springer-Verlag, 1979–); Armin Hermann, John Krige, Ulrike Mersits and Dominique Pestre, eds., *History of CERN*, 3 vols. (Amsterdam: North Holland Publishers, 1987–1996).

conferences during the late seventies that featured Italian and American historians, including Heilbron and Forman, led to postdoctoral exchanges and the development of a strong contingent of Italian historians in the field that continues to this day. Similar developments occurred in Japan. In each of these countries the support of physicists and physicists' organisations was essential, ironically, to the launching of an independent quantum history.

By the early 1980s the flood of quantum histories reached a plateau, after which the production of new such work slowed to a trickle.[29] In addition to the attraction of other sciences with less researched histories, the reasons for this shift were, again, mainly disciplinary in nature. The maturation of the history of science by the early eighties meant that full graduate programs were required to enter the field. Such programs were now more frequently offered in history departments, where social and science history blended naturally and students were trained for future jobs in similar departments. With equilibrium re-established in the supply and demand for physics jobs, physics graduates were less likely to leave physics for history. Further, this meant that fewer historians of science were sufficiently trained in physics to handle quantum mechanics, let alone the next frontier to open in the 1980s: the development of quantum field theory and its applications to nuclear and high-energy physics (cosmic rays) during the 1930s.

After 1980, the discipline of quantum history as a whole seemed to enter a consolidation phase for the next decade and a half, during which, in hindsight, it laid the foundations for a new burst of research by a new generation then entering the field. While some moved on to the thirties or to institutional histories, for the twenties it was a period of monographs, compilations and compendia, such as Mehra's and Rechenberg's encyclopedic volumes.[30] In addition, with the AHQP's assembly of individuals' papers, together with the availability of institutional records (though scattered across two continents), a host of book-length biographies

---

[29] For instance, Mara Beller, 'Pascual Jordan's Influence on the Discovery of Heisenberg's Indeterminacy Principle', *Archive for History of Exact Sciences* 33, no. 4 (1985): 337–49; Cathryn Carson, 'The Peculiar Notion of Exchange Forces — I: Origins in Quantum Mechanics, 1926–1928', *Studies in History and Philosophy of Modern Physics* 27, no. 1 (1996): 23–45.

[30] Jagdish Mehra and Helmut Rechenberg, *The Historical Development of Quantum Theory*, 6 vols. (New York: Springer-Verlag, 1982–2001). A sampling of monographs: Bruce R. Wheaton, *The Tiger and the Shark: Empirical Roots of Wave-Particle Dualism* (New York: Cambridge University Press, 1983/1991); John Hendry, *The Creation of Quantum Mechanics and the Bohr-Pauli Dialogue* (Hingham, MA: D. Reidel, 1984); Olivier Darrigol, *From c-Numbers to q-Numbers: The Classical Analogy in the History of Quantum Theory* (Berkeley, CA: University of California Press, 1992); Mara Beller, *Quantum Dialogue: The Making of a Revolution* (Chicago: University of Chicago Press, 1999); Helge Kragh, *Quantum Generations: A History of Physics in the Twentieth Century* (Princeton: Princeton University Press, 1999).

appeared, such as those examining Schrödinger, Heisenberg, Kramers, Dirac and the four men who made quantum electrodynamics.[31] Some authors attempted to unite the environment and practice of physics within their biographical works; others summarised the results of earlier internal studies in authoritative textbook accounts in today's notation for those trained in today's physics. Above all, this was a period of collecting, editing and publishing the papers and correspondence of the great men of the quantum revolution in projects of widely varying size, financing and editorial scope.[32]

The advent of the Internet and the World Wide Web rendered a rich trove of search aids, bibliographic references and, in some cases, original sources readily accessible via any online computer. Many of the publishers of the physics journals of the quantum era, as well as those in which many of the secondary historical papers appeared, have made their earlier issues available to scholars on the Web. The International Catalogue of Sources (ICOS) for history of physics, maintained by the AIP's Center for History of Physics, offers a convenient entry into the riches of published and unpublished source materials. It has incorporated the SHQP *Inventory* and its supplementary collections, which have continued to grow over the decades, currently reaching 355 microfilm reels and 774 paper files.[33] The AHQP repositories have also multiplied. Scholars can now access many of the unpublished materials of quantum history in at least 19 locations around the world. The future posting of the entire content of this archive on the Web is one of the announced goals of the recently founded international project History and Foundations of Quantum Physics, sponsored by the Max Planck Institute for

---

[31] Max Dresden, *H.A. Kramers: Between Tradition and Revolution* (New York: Springer-Verlag, 1987); Walter Moore, *Schrödinger: Life and Thought* (New York: Cambridge University Press, 1989); Helge Kragh, *Dirac: A Scientific Biography* (New York: Cambridge University Press, 1990); David Cassidy, *Uncertainty: The Life and Science of Werner Heisenberg* (New York: W. H. Freeman, 1992); Sylvan S. Schweber, *QED and the Men Who Made It: Dyson, Feynman, Schwinger, and Tomonaga* (Princeton: Princeton University Press, 1994).

[32] Niels Bohr, *Collected Works*, 12 vols., ed. Léon Rosenfeld, Erik Rüdinger and Finn Aaserud (Amsterdam: Elsevier, 1972–2007); Albert Einstein, *The Collected Papers of Albert Einstein*, ed. John Stachel et al. (Princeton: Princeton University Press, 1987–); Werner Heisenberg, *Gesammelte Werke* [Collected Works], 9 vols., ed. Werner Blum, Helmut Rechenberg and H. P. Dürr (Berlin: Springer-Verlag, 1984–1993); H.A. Lorentz, *The Scientific Correspondence of H. A. Lorentz*, vol. 1, ed. A.J. Kox (New York: Springer-Verlag, 2008).

[33] For ICOS and related resources, see http://www.aip.org/history (accessed 20 Aug 2009). The Office for History of Science and Technology at Berkeley maintains AHQP as well as the Quantum Physics Database, a finding aid. The Deutsches Museum in Munich has an online listing and finding aid for the Sommerfeld Correspondence at http://www.lrz-muenchen.de/~Sommerfeld (accessed 20 Aug 2009).

History of Science, together with the Fritz Haber Institute of the Max Planck Society, both in Berlin.[34]

As a result of differing educational and employment practices among nations, recent decades have also brought forth an increasing international community of quantum historians, many of whom are trained in physics and are now associated with the Berlin quantum-history project. Compared with their predecessors of the early Forman era, historians working in today's environment appear more motivated to examine the conceptual content of their subject than to integrate the practice of physics with its historical milieu. Some utilise perspectives informed by contemporary intellectual history and philosophy of science, while others pursue an internalistic approach that relies, thus far, almost entirely upon the published papers. Carefully unravelling the internal arguments of the great conceptual results is an essential first step toward achieving an understanding of the subject. But if the past teaches any lesson, it is that the mature history will require taking the next step: an attempt to answer not just what occurred, but also how and why those arguments and results arose among the physicists who brought them forth. As both groups progress in their work, while drawing more fully upon the readily available sources and methods, their work and results will surely expand and hopefully converge.

Such a convergence will offer contemporary historians much to keep them busy. Some of the early pioneers of quantum history, including Kuhn and co-workers, profited greatly from the historiography of the Scientific Revolution. It has served as a rich source of historiographic innovation throughout the history of science, and it probably remains so today. Even a cursory reference to that historiography reveals a number of gaps in the history of quantum physics and many topics that require more work. Although SHQP had identified at least 280 significant contributors to quantum mechanics, all of the available histories have focused on roughly the same 20 or so big names. Very little is known about how the other 260 contributors fit into that history, individually and collectively, and little is known about the many other contributors who were not included among the select 280. Similar to the Scientific Revolution, the development of quantum mechanics was the product of an intense community effort put forth by a large number of remarkably gifted individuals. Yet the emphasis of quantum history so far has been largely biographical. We need to know more about how to encompass the individual and the community, the macro- and the microhistory.[35] If letters, rather than meetings, were the primary mode of discourse for quantum physicists, how did they function in creating and maintaining the

---

[34] For the quantum-history project, see http://quantum-history.mpiwg-berlin.mpg.de/main (accessed 20 Aug 2009).

[35] This is among the problems listed by Peter Galison in his 'Ten Problems in the History and Philosophy of Science', *Isis* 99, no. 1 (2008): 111–24.

community structure and the creativity of its participants? And, once again, we might ask in what ways contemporary sociology and the social history of science could help us now to comprehend the creation of new quantum knowledge.

Some work has been done on the role of scientific schools, such as those in Munich, Göttingen and Copenhagen, in the formation of quantum mechanics, but there is still room for more.[36] How exactly did these schools differ from each other, what characteristic perceptions and approaches did they cultivate, and what was their influence on the work of their members? Although our subject is the history of quantum physics, the focus so far has been overwhelmingly on quantum theory, to the neglect of quantum experiments. We still know very little about the 'material culture' of the experimental data and equipment and interactions with theorists that helped push those theories forward. Even some aspects of the theories themselves leave room for more work. Among them are the roles played by other 'big mistakes' in the formation of quantum mechanics and a more complete comprehension of the nature and extent of the 'crisis state' that preceded quantum mechanics. Little has been done on the history of the old quantum theory since the publication of Kuhn's study on the origins of the quantum discontinuity in 1979.[37]

John Heilbron has suggested the quantum history of the early *HSPS* era may now be as dead as the old quantum theory.[38] Nevertheless, the lessons of a half-century of quantum history and historiography, combined with the ready availability of a wealth of primary and secondary sources of all types in the hands of a new generation of historians in the emerging environment of a new century, may now have prepared us to answer more fully than ever before the question posed by Paul Forman in 1970: How did the innovations that led to the development of quantum mechanics come to occur at that time, in that place, by those people, and in this form?

Acknowledgments

I am very grateful to Paul Forman and John Heilbron for their helpful comments and suggestions. I am also grateful to the participants in the first conference of the Quantum History Project (HQ1), July 2007, at the Max Planck Institute for History of Science in Berlin, where I first presented some of my remarks.

---

[36] For instance, see Michael Eckert, *Die Atomphysiker: Eine Geschichte der theoretischen Physik am Beispiel der Sommerfeldschule* (Braunschweig: Vieweg, 1993); Suman Seth, *Crafting the Quantum: Arnold Sommerfeld and the Practice of Theory, 1890–1926* (Cambridge, MA: MIT Press, 2010).
[37] Thomas S. Kuhn, *Black-Body Theory and the Quantum Discontinuity, 1894–1912* (New York: Clarendon Press, 1979).
[38] J.L. Heilbron, e-mail message to author, 23 Mar 2009.

Culture and Mechanics in Germany, 1869–1918: A Sketch

Richard Staley*

Paul Forman's famous account of the sudden acceptance of acausal physics amidst the tumults of Weimar culture is marked by the historical specificity of its claims. After World War I, Forman thought that an unusual time had pressed unusual demands, and that physicists and mathematicians in German-speaking Europe had bowed to a hostile cultural environment by accepting an acausal quantum mechanics before such acceptance was warranted by firm experiment. This was very much a Zeitgeist argument, but writing in a period during which a distinction between internal and external factors too readily set science apart from the society within which it was developed, Forman's argument and guiding assumptions were strongly shaped by assumed differences between science and culture. Here, I want to help explain the phenomena that Forman localised so sharply in time and space by arguing that several features of the cultural environment he depicted were structured by longer-term dynamics that had their origins as much within the physics discipline as outside it. In particular, I will assert that some of the discourse discussed by Forman represented yet another phase in a longstanding dialogue concerning the relations between culture and mechanics.[1]

The main symptom of the extraordinary cultural moment Forman diagnosed was the 1918 publication and success of Oswald Spengler's infamous *Der Untergang des Abendlandes*, a book that railed against the deadening effect of

---

* Department of the History of Science, University of Wisconsin-Madison, 210 Bradley Memorial Building, 1225 Linden Drive, Madison, WI 53706–1528; rastaley@wisc.edu.

    The following abbreviations are used: *PZ, Physikalische Zeitschrift; HSPS, Historical Studies in the Physical Sciences.*

[1] Paul Forman, 'Weimar Culture, Causality, and Quantum Theory, 1918–1927: Adaptation by German Physicists and Mathematicians to a Hostile Intellectual Environment', *HSPS* 3 (1971): 1–116.

causal physics on the one hand, and saw the sciences as reflecting the character of the specific epoch in which they found expression on the other. Spengler had his quarrels with current physics, but also with current ways of understanding history. Indeed, he regarded them both as tarred with the same brush and thought of himself as introducing new methods to the latter.[2] Forman judged Spengler's book a flight from reason and regarded physicists who accepted aspects of his views about history, or who advocated indeterminism in physics, as currying favour by softening the implications of their discipline. When first asked to revisit Forman's work, I built upon John Heilbron's identification of a movement in physics he called 'descriptionism', to argue that both Spengler and the physicists who articulated similar approaches to the place of physics in culture were drawing on a relatively common perspective developed within the physics community circa 1900, not outside it.[3] Figures such as Ernst Mach, Karl Pearson and Henri Poincaré had earlier argued that science offered economical descriptions or the best possible image of phenomena rather than final causal understandings. In addition to this emphasis on description, they stressed links between physics and neighboring disciplines, and sometimes even urged an understanding of all laws as statistical rather than causal. Indeed, Spengler's 1904 doctoral dissertation had drawn on Ernst Mach and Wilhelm Ostwald in presenting what he described as an 'energetic' account of the Greek philosopher Heraclitus. Spengler scholars have argued many of the themes of his later, sensationally successful book had been prefigured in this little-noticed doctoral work.[4]

The present paper extends this argument by sketching a few principal features of the relations between mechanics and culture as these were articulated by a range of commentators in the German-speaking realm from the late nineteenth century to World War I. Historians of physics have often argued that a new attention to the foundations of mechanics, together with the formulation of alternative approaches based on electromagnetic theory, helped prepare the ground for the development of relativity theory but they have rarely explored the extent to which

---

[2] Oswald Spengler, *Der Untergang des Abendlandes: Umrisse einer Morphologie der Weltgeschichte*, vol. 1, *Gestalt und Wirklichkeit*, 4th unchanged ed. (Munich: Oskar Beck, 1918/19), 1–10.
[3] J.L. Heilbron, '*Fin-de-Siècle* Physics', in *Science, Technology and Society in the Time of Alfred Nobel*, ed. Carl Gustaf Bernhard, Elisabeth Crawford, and Per Sörbom (Oxford: Nobel Foundation, 1982), 51–73; Richard Staley, 'The *Fin de Siècle* Thesis', *Berichte zur Wissenschaftsgeschichte* 31 (2008): 311–30.
[4] Oswald Spengler, *Heraklit: Eine Studie über den energetischen Grundgedanken seiner Philosophie* (Halle, Germany: Hofdruckerei C.A. Kaemmerer & Co., 1904); John Farrenkopf, *Prophet of Decline: Spengler on World History and Politics*, Political Traditions in Foreign Policy Series (Baton Rouge: Louisiana State University Press, 2001), 14–16.

physicists writing on mechanics joined a debate with significantly broader implications crossing several disciplines and attracting the attention of diverse cultural commentators. Here, I will show that treatments of mechanics often engaged with new understandings of the place of the sciences in the developmental history of society; and in part because of their relevance to historical materialism, they were taken up by historians, philosophers, political economists and sociologists, as well as physicists. While broad-ranging discussions of mechanics began in the 1860s and took various forms over the next 40 years, there was an efflorescence of debate on culture and mechanics in Germany on the eve of World War I, with books by Ostwald on the energetic foundations of the cultural sciences, by Mach on *Kultur und Mechanik*, and Walter Rathenau's *Zur Kritik der Zeit* and *Zur Mechanik des Geistes*, all published between 1909 and 1915. I suggest that collectively these works may help to explain the terms in which Spengler developed his distinctive analysis of science and culture.

The Critical History of Mechanics and Relative Knowledge of History

Historians of science have typically focused on a relatively small number of textbooks and histories of mechanics that appeared in the late nineteenth century as examples of a new critique of the disciplinary foundations of physics. In particular, the works published by Gustav Kirchhoff, Ernst Mach and Heinrich Hertz between 1876 and 1894 are well known for the unease they expressed toward central features of mechanics as it had been developed since Newton. Kirchhoff's textbook on mechanics abjured the search for final causal explanations in favour of understanding mechanics as offering rather the most complete and simplest possible description of motion (a feature on which Heilbron focused in developing his characterisation of descriptionism). Kirchhoff believed mechanics provided the reductive key to all natural science, and traced mechanics itself back to the fundamental concepts of space, time and matter.[5] While Ernst Mach believed similarly that science offered economical descriptions, his 1883 account of the historical development of mechanics critiqued even those fundamental concepts by offering a pointed attack on the absolutes of Newtonian theory. All

---

[5] Gustav Kirchhoff, preface to *Vorlesungen über mathematische Physik: Mechanik* (Leipzig: B.G. Teubner, 1876). *See also* Christa Jungnickel and Russell McCormmach, *Intellectual Mastery of Nature: Theoretical Physics from Ohm to Einstein*, vol. 1, *The Torch of Mathematics, 1800–1870* (Chicago: University of Chicago Press, 1986), 303; Christa Jungnickel and Russell McCormmach, *Intellectual Mastery of Nature: Theoretical Physics from Ohm to Einstein*, vol. 2, *The Now Mighty Theoretical Physics, 1870–1925* (Chicago: University of Chicago Press, 1986), 126; J.L. Heilbron, '*Fin-de-Siècle* Physics' (ref. 3), 53.

apparently absolute phenomena — the absolute space and time that Newton described in establishing the grounds for his mechanics — were in fact relative measures, and even inertial resistance might be explained by all the other masses of the universe.[6] For Heinrich Hertz, in lectures on mechanics that were published posthumously in 1894, the anthropomorphic elements of the understanding of force were problematic. He offered a way of dispensing with the concept of force by invoking hidden masses.[7] These discussions were undoubtedly of great importance in cultivating what became a widespread debate on foundations within the physics community.[8] But the circumstances surrounding the publication of one of the earliest comprehensive histories of mechanics will show that mechanics had been brought under question earlier, and that it attracted attention well beyond the discipline.

In April 1869 the philosophical faculty of the University of Göttingen advertised a prize that indicates they regarded mechanics to be ripe for scrutiny, and thought that historical and critical methods were an appropriate way to develop new insight into the discipline. The inaugural prize from the Beneke Stiftung would reward a critical history of the general principles of mechanics, stipulating the work should begin with Galileo's contributions.[9] Given the novelty of requesting such a treatment of a natural science, the instructions were exhaustive and carefully distinguished the two sides of the task. The historical work would involve indicating when, by whom and in relation to which particular problems each of the different essential principles of mechanics had been first discovered and expressed, as well as tracing the circumstances that led to changes in the original expression of the principles, or justified drawing specific principles together into a general principle. The critical task involved examining the logical foundation, mathematical formulation and empirical support of each of these mechanical principles. It was left to the writer's taste and judgment to determine whether the historical and critical sides should be amalgamated or treated separately.

---

[6] Ernst Mach, *Die Mechanik in ihrer Entwickelung: Historisch-kritisch dargestellt* (Leipzig: F.A. Brockhaus, 1883), 202–28. Theodore M. Porter focuses on Mach (and Karl Pearson) in describing the importance of descriptionist physics to the social sciences in 'The Death of the Object: *Fin de siècle* Philosophy of Physics', in *Modernist Impulses in the Human Sciences, 1870–1930*, ed. Dorothy Ross (Baltimore: Johns Hopkins University Press, 1994), 128–51.

[7] Heinrich Hertz, *Die Prinzipien der Mechanik in neuem Zusammenhange dargestellt* (Leipzig: J.A. Barth, 1894).

[8] See Jungnickel and McCormmach, *Intellectual Mastery*, vol. 2 (ref. 5), chap. 24; Suman Seth, 'Crisis and the Construction of Modern Theoretical Physics', *British Journal for the History of Science* 40 (2007): 25–51.

[9] Eugen Karl Dühring, *Kritische Geschichte der allgemeinen Principien der Mechanik* (Berlin: T. Grieben, 1873), iii–viii.

The end-point stipulated by the faculty is revealing: the account should be carried as far as considering the new conception of natural forces, their mode of action and the transition between their forms, which had been stimulated by investigations of the mechanical equivalent of heat.[10] Mechanics had been applied to a new realm and transformed in the process.

Five entries were received and the faculty described the characteristics of the winning manuscript in glowing terms. Its author was already on his way to becoming well known within the German academic community and, after first publication in 1873, the book went through several editions. It will perhaps be surprising, however, that this was the author's first foray into physics. Eugen Karl Dühring had originally practised law, but in 1864 became a *Privatdozent* in philosophy and politics at the University of Berlin. There followed a wide-ranging series of publications that moved from very general philosophical concerns, expressed in his books *Natürliche Dialektik* and *Der Werth des Lebens*, to studies of political and economic theory and advocacy of socialism in works such as *Capital und Arbeit* and *Kritische Geschichte der Nationalökonomie und des Socialismus*.[11] In 1869, Dühring's *Kritische Geschichte der Philosophie von ihren Anfängen bis zur Gegenwart* concluded with reflections that presaged his concern with mechanics. Dühring wrote that the growing importance of scientific forms of thought hitherto had little influence on comprehensive philosophical formulations and underlined the care required to understand the proper limits of concepts such as the indestructibility of matter; the difference between concepts of force and mechanical force; and the need to recognise clearly the significance of the very different forms in which forces made their appearance. But he described the universal insight of the conservation of mechanical force (understood within these intellectual limits) as being of epoch-making importance.[12] The call for a history of mechanics, then, coincided with a principal element of Dühring's intellectual interests — and it was

---

[10] Ibid. The rubric is discussed at iii–iv, their evaluation at iv–vi and the circumstances of the Beneke prize at vi–viii. *See also* Alberto A. Martínez, *Kinematics: The Lost Origins of Einstein's Relativity* (Baltimore: Johns Hopkins University Press, 2009), 90–95.

[11] Eugen Karl Dühring, *Der Werth des Lebens: Eine philosophische Betrachtung* (Breslau: Eduard Trewendt, 1865); Eugen Karl Dühring, *Natürliche Dialektik: Neue logische Grundlegungen der Wissenschaft und Philosophie* (Berlin: E.S. Mittler und Sohn, 1865); Eugen Karl Dühring, *Capital und Arbeit* (Berlin: A. Eichhoff, 1865); Eugen Karl Dühring, *Kritische Geschichte der Nationalökonomie und des Socialismus von ihren Anfängen bis zur Gegenwart*, 4th newly revised and expanded ed. (Leipzig: C.G. Naumann, 1871/1900). *See also* Eugen Karl Dühring, *Sache, Leben und Feinde: Als Hauptwerk und Schlüssel zu seinen sämmtlichen Schriften*, 2nd completed and expanded ed. (Leipzig: C.G. Naumann, 1882/1903).

[12] Eugen Karl Dühring, *Kritische Geschichte der Philosophie von ihren Anfängen bis zur Gegenwart* (Berlin: L. Heimann, 1869), 501–03.

to be the submission of a seasoned practitioner of critical histories rather than that of a physicist by profession that won the approbation of the Göttingen faculty.

If critical history was a well-established genre in Germany, developed first in biblical criticism and studies of antiquity and now increasingly important in philosophy, what it might mean for science was an open question. Dühring regarded the history of mechanics as an indispensable basis for a general history of science — that was at present simply a bold idea, lacking both the requisite materials and analytical perspectives. The primary model for his own endeavour — to represent clearly concepts often hidden in the luxuriance of symbolic operations — was provided by the historical sections of Lagrange's analytical mechanics. Dühring moved chronologically from the work of Galileo, Huygens and Newton through the general formulations and analytical developments realised in Lagrange's work, to thermodynamics. His chapters examined principle after principle in comprehensive detail, discussing, for example, the principle of the conservation of living forces and of virtual velocities, while paying continual attention to the philosophical assumptions underlying mechanical principles. Philosophically, Dühring's body of work is important in Germany because it helped initiate a re-evaluation of Kant by integrating Kant into readings of Hume and Locke, in the framework of Dühring's own positivist approach.[13] He contributed, too, to the separation of kinematics as a fundamental basis for the discipline of mechanics.[14] However, while widely read, with second and third editions following in 1877 and 1887, his study of mechanics was to be overshadowed by two polemical debates that engaged Dühring equally in party and academic politics. Importantly, both disputes turned in part on Dühring's perspectives on mechanics (sometimes as these were expressed in his philosophical and political studies).

Like Dühring, Friedrich Engels had developed his socialism in conjunction with an avid study of the natural sciences. From January 1877, Engels combatted Dühring's increasing success in German socialist circles by attacking his revolutions, or over-turnings (*Umwälzungen*), in philosophy, political economy and socialism in a series of articles first published in the Social Democratic Workers' Party paper *Vorwärts* (as well as other forums), before being collected together as *Herrn Eugen Dühring's Umwälzung der Wissenschaft* in 1878.[15] While Engels had many bones to pick with particular features of Dühring's accounts of physics and

---

[13] Klaus Christian Köhnke, *The Rise of Neo-Kantianism: German Academic Philosophy between Idealism and Positivism*, Ideas in Context (Cambridge: Cambridge University Press, 1991), 22, 244–46.

[14] Martínez, *Kinematics* (ref. 10), 90–95.

[15] Helena Sheehan, *Marxism and the Philosophy of Science: A Critical History: The First Hundred Years* (Atlantic Highlands, NJ: Humanities Press, 1993), 28–30.

biology, the philosophical stakes involved are illustrated most clearly when he takes up the question of whether Dühring's faith in final and ultimate truths across science and morality is justified. Engels thought certain results in the exact sciences are eternally true, but he held this to be far from the case for the majority of their results. Mathematics had tasted the tree of knowledge with the introduction of variable magnitudes and the infinite; astronomy and mechanics were still worse off, while physics and chemistry were sciences swamped by hypotheses. In contrast, Engels charged, Dühring claimed ultimate truth for his philosophical system and regarded moral truths to have the same validity as mathematical insights (when their ultimate foundation was clearly understood). If precious little in the exact sciences could claim finality, Engels argued in the human sciences, knowledge is 'essentially relative, in as much as it is limited to the investigation of interconnections and consequences of certain social and state forms which exist only in a particular epoch and among particular peoples and are by their very nature transitory. Anyone therefore who sets out to hunt down final and ultimate truths — genuine, absolutely immutable truths — will bring home but little, apart from platitudes and commonplaces of the sorriest kind'.[16]

Engels scorned Dühring's understanding of morality in favour of recognising the class basis to moral law, but he had great sympathy with Dühring's struggles against academic authority. Dühring clearly had a jaundiced view of academic politics that was initially represented indirectly in his advocacy of J.R. Mayer's priority in establishing the mechanical theory of heat and the general concept of energy. Mayer, the amateur, had struggled against the indifference of the academy without receiving proper acknowledgment from the authorities — a situation perhaps not unlike Dühring's own position in Berlin. But in the second edition of his history of mechanics in 1877, Dühring made the point about Mayer's priority tell much more pointedly against the contributions of the Berlin physicist Hermann Helmholtz (whom he also denigrated for advocating the piquant nonsense of non-Euclidean geometries). Dühring railed too against what he described as the 'secondary figures' of Berlin mathematicians, and attacked the professorial caste in general. In conjunction with earlier attacks on Adolf Wagner, the Berlin professor of political economy, Dühring had bitten the hand that feeds far too savagely to be ignored, and in July 1877 the Prussian Ministry of Education and the University of Berlin remoted him.[17]

---

[16] Friedrich Engels, *Anti-Dühring: Herr Eugen Dühring's Revolution in Science*, 3rd ed. (Moscow: Foreign Languages Pub. House, 1878/1962), 122–25, on 125. *See also* Martínez, *Kinematics* (ref. 10), 94–95.

[17] David Cahan, 'Anti-Helmholtz, Anti-Zöllner, Anti-Dühring: The Freedom of Science in Germany during the 1870s', in *Universalgenie Helmholtz: Rückblick nach 100 Jahren*, ed. Lorenz Krüger (Berlin: Akademie Verlag, 1994), 330–44, on 336–40.

Thus the circumstances surrounding the publication of the first critical history of mechanics indicate that the proper understanding of disciplinary foundations could be equally significant for socialist materialism and academic politics, supporting Christopher Herbert's argument that philosophical relativism in radical thought formed a significant backdrop to developments within physics, one largely overlooked by current scholarship.[18] That a similar discursive breadth accompanied mechanics through to the early twentieth century is illustrated by the ethnographic framework within which Mach developed his now better-known critical history of mechanics, and by controversies over the fruitfulness of his philosophical approach that emerged simultaneously within the German physics community and Russian political circles shortly before World War I.

Culture and Mechanics

Already we have seen that Mach's 1883 account of the development of mechanics critiqued Newton's understanding of absolute time and space (in contrast to Dühring's interest in absolutes).[19] Distinct from the closely technical focus of Kirchhoff's lectures on mechanics, and going an illustrative step further than Dühring, Mach set his reflections in a general historical framework that began with images of ancient mechanical experience depicted on Egyptian and Assyrian tombs, and contrasted this instinctive, irreflective knowledge to the investigative apprehension of scientific knowledge, which had come much later (Fig. 1).[20] Mach thought a clear assessment of current science required recognising the context in which mechanics had originally been developed. One major lesson was that theoretical suppositions were always based on the intellectual refinement of prior practical experience, but Mach was also interested in demonstrating how the religious tenor of earlier periods had given a theological guise to the contents of

---

[18] Christopher Herbert focuses on Britain in *Victorian Relativity: Radical Thought and Scientific Discovery* (Chicago: University of Chicago Press, 2001).

[19] Mach's preface notes that Dühring's '*schätzbare*' book had not influenced his own thoughts on the subject, which had already been essentially completed and published before 1883, but allowed that at least in reference to the negative side of the criticism one would find points of agreement — Mach, *Mechanik in ihrer Entwickelung* (ref. 6), vii. Mach's approach also stood in contrast to Carl G. Neumann's discussions of an 'alpha body' that could define an absolute frame of reference. *See* Martínez, *Kinematics* (ref. 10), 95–98.

[20] Mach, *Mechanik in ihrer Entwickelung* (ref. 6), 1–7, esp. Fig. 1. Mach had earlier written that we should not let go of the guiding hand of history, which had made everything and could change everything. *See* Ernst Mach, *Die Geschichte und die Wurzel des Satzes von der Erhaltung der Arbeit*, 2nd reprint ed. (Leipzig: J.A. Barth, 1872/1909), 3.

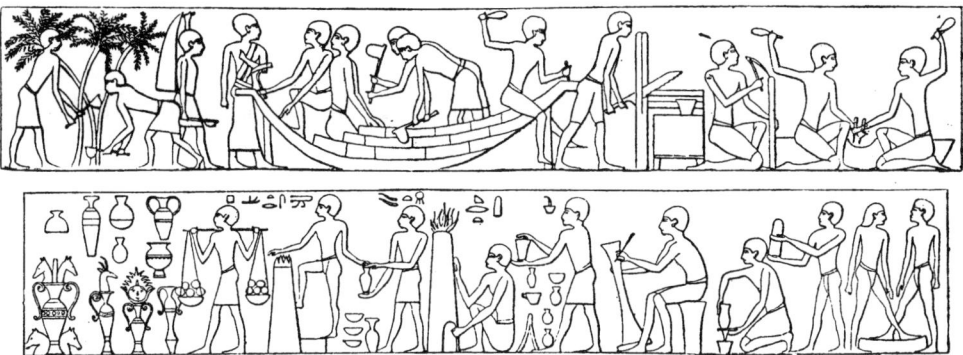

Fig. 1: The first image in Mach's *The Science of Mechanics: A Critical and Historical Account of its Development* illustrated ancient mechanical experience.

Source: Ernst Mach, *Die Mechanik in ihrer Entwickelung: Historisch-kritisch dargestellt*, 6th ed. (Leipzig: F.A. Brockhaus, 1908).

mechanics, which was only gradually sloughed off over time. The present concept of force was a fetishist reminder of the earlier period — Mach concurred with the anthropologist E.B. Tylor on this score.[21]

Thus Mach depicted mechanics as a product of development; at the same time he sought to prevent his modern audience from investing it with too much significance. As the simplest and earliest science, understanding mechanics was indispensable for a full comprehension of present science, but it should not thereby be regarded as more fundamental than any other science. Rather, physical phenomena were unified in themselves. Simultaneously thermal, magnetic, electrical and chemical as well as mechanical, they would always escape the classifications imposed by the abstractions of mechanics, and which approach would go deepest was presently unclear.[22] Mach insisted on the artificiality of mechanical conceptions, and it is revealing that his study concludes by considering the relations among mechanics, physics and physiology. The seventh edition of *Mechanik in ihrer Entwickelung* in 1912 took the additional step of returning the reader to a complex interplay between savage and contemporary experience, the labour relations of ancient Egypt with slaves erecting the pyramids and the way the twisted form of an old fire-drill worn by its cord may have led some pre-Archimedean Greek to think of constructing a screw. The final paragraph shows

---

[21] Mach, *Mechanik in ihrer Entwickelung* (ref. 6), 435. Mach was referring to Edward Burnett Tylor's *Primitive Culture: Researches into the Development of Mythology, Philosophy, Religion, Art, and Custom* (London: J. Murray, 1871).
[22] Mach, *Mechanik in ihrer Entwickelung* (ref. 6), 467.

that Mach was inspired by the reawakening of his son Ludwig's childhood facility with mechanical work:

> My son vividly describes how in an ethnographical museum the dynamical experiences of his youth again vividly came to life; how they were awakened again by the perceptible traces of the work on the objects exhibited. May these experiences be used for the finding of a universal genetic technology [*einer allgemeinen genetischen Technologie*], and perhaps, by the way, lead a little deeper into the understanding of the primitive history of mechanics.[23]

In 1915 Mach took up both this aim and byproduct in the book *Kultur und Mechanik*.[24] When he did so, he was joining a field of discourse that thematised the relations among technology, science and culture that was especially vibrant in this period. Engineers, scientists, sociologists and historical economists alike participated, sometimes crossing and sometimes defending disciplinary turf. The historian Eduard von Meyer's 1906 book *Technik und Kultur* prompted spirited attacks from engineers, and especially Friedrich Dessauer (who defended the integrity of technology by giving it a spiritual dimension); and four years later Werner Sombart took the same title, *Technik und Kultur*, for a key paper delivered to the first meeting of the German Society for Sociology.[25] For his part, Wilhelm Ostwald lectured on science and culture and outlined a tripartite division of disciplines into those in which order, energy and life were the principal concepts, regarding energy to play a subsidiary but significant role in physiology, psychology and the cultural sciences. Outlining a moral imperative toward the minimisation of entropy, Ostwald drew the ire of Max Weber in transferring the methods of the physical sciences into the social sciences.[26]

---

[23] Ernst Mach, *Die Mechanik in ihrer Entwicklung: historisch-kritisch dargestellt*, 7th improved and expanded ed. (Leipzig: Brockhaus, 1883/1912), 487.

[24] Ernst Mach, *Kultur und Mechanik* (Stuttgart: W. Spemann, 1915).

[25] See the discussion in Eric Schatzberg, '*Technik* Comes to America: Changing Meanings of *Technology* before 1930', *Technology and Culture* 47 (2006): 486–512, on 495; and Jeffrey Herf, *Reactionary Modernism: Technology, Culture, and Politics in Weimar and the Third Reich* (Cambridge: Cambridge University Press, 1984), 133–34 .

[26] See Wilhelm Ostwald, *Energetische Grundlagen der Kulturwissenschaft* (Leipzig: Alfred Kröner Verlag, 1909). The essay 'Kultur und Wissenschaft' was published in Wilhelm Ostwald, *Der Energetische Imperativ* (Leipzig: Akademische Verlagsgesellschaft, 1912), 30–54. See also R.J. Deltete, 'Wilhelm Ostwald's Energetics 1: Origins and Motivations', *Foundations of Chemistry* 9, no. 3 (2007): 3–56; R.J. Deltete, 'Wilhelm Ostwald's Energetics 2: Energetic Theory and Applications, Part I', *Foundations of Chemistry* 9, no. 3 (2007): 265–316; C. Hakfoort, 'Science Deified: Wilhelm Ostwald's Energeticist World-View and the History of Scientism', *Annals of Science* 49 (1992): 525–44. Weber's reaction is described in Robert Proctor's *Value-Free Science? Purity and Power in Modern Knowledge* (Cambridge, MA: Harvard University Press, 1991), 143–44.

The work of the German-Jewish industrialist, author and later politician Walther Rathenau illustrates the metaphorical purchase of the concept of mechanics in such discourse. In a series of books that discerned the spirit of his time (*Reflexionen, Zur Kritik der Zeit* and *Zur Mechanik des Geistes* before World War I),[27] Rathenau offered a portrait of mechanism that Ernst Mach described as providing an extremely brilliant and stimulating consideration of industry, technology and the economy.[28] Rathenau argued that while mechanisation had its origins in the production of goods and the need to feed ever-increasing populations, it had grown to reach into every sphere of life:

> To the economist it appears as mass production and distribution of goods; to the industrialist as division of labor, accumulation of labor, and manufacture; to the geographer as the development of means of transportation and communication, and colonization; to the technician as the control of natural forces; to the scientist as the application of the results of research; to the sociologist as the organization of labor; to the business man as enterprise and capitalism; to the politician as realistic economic and political statecraft.
>
> But common to all these forms of expression is one spirit, which separates them in a definite and peculiar fashion from the modes of life of earlier centuries: an impulse of specialization and abstraction, of standardized thinking devoid of surprise and humor, of complicated uniformity; a spirit which seems to justify the name "mechanization" even when applied to the sphere of emotion.[29]

For Rathenau the deadening effect of mechanisation had to be combated by the birth of the soul.

If Mach's account of culture and mechanics would have special valence in this environment, he also wrote on the heels of two different disputes which together illustrate that the attention paid to mechanics as a fulcrum for the discussion of philosophical attitudes toward science was as broadly based as it was longstanding in time. In 1909, just as Engels had 30 years earlier taken up the challenge of Dühring's growing popularity, Vladimir Lenin responded to the increasing importance of Machian thought among Russian socialists with a polemical attack on what he saw as an erosion of the philosophical foundations of materialist thought. Machians stressed the relativity of our knowledge and Lenin thought relativism condemned one to 'absolute skepticism, agnosticism and sophistry, or to subjectivism.' Accusing Mach of having slipped into idealism by way of relativism, Lenin offered instead the objective reality of Marx's and Engels' materialist world outlook, exemplifying his arguments with a discussion of the diverse treatments of space and time provided by Dühring, Engels and

---

[27] Walther Rathenau, *Reflexionen* (Leipzig: S. Hirzel, 1908); Walther Rathenau, *Zur Kritik der Zeit* (Berlin: S. Fischer, 1912); Walther Rathenau, *Zur Mechanik des Geistes* (Berlin: S. Fischer, 1913).
[28] Mach, *Kultur und Mechanik* (ref. 24), 6.
[29] Rathenau, *Zur Kritik der Zeit* (ref. 27), 51–52.

Mach.[30] Mach regarded his second opponent as far more important. His fellow physicist Max Planck critiqued Mach's view that science offered merely economical descriptions of sensation and presented an alternative historical perspective in which the development of science traced an increasing departure from specific sense perceptions toward deanthropomorphised measurements and general principles, reaching toward a physics unified in relation to its details and to physicists of 'all places, all times, all peoples, all cultures'.[31]

In addition to underlining the centrality of mechanical experience before intellectual knowledge, perhaps the most direct way that Mach responded to the perspectives of Lenin and Planck in 1915 was by thoroughly integrating mechanics with broader realms of historical experience. Mach drew on metallurgy and chemical technology as well as readings in archeology, ethnography, philology and music history. For Mach technology was genetic in being inherently developmental, requiring such a close practical relationship between thought and tool, that were all our machines to be swept away, even knowing as we do now the heights of mechanical civilisation, we would have to reconstruct the very first steps of building a wooden screw and move through countless intermediate stages:

> We would have to build machine upon machine, in long chains without any link missing, one machine completing the other; and each would have to work a while, and produce, in order to fulfill its place in the chain and bring us to the final goal.[32]

---

[30] Vladimir Ilyich Lenin, *Materialism and Empirio-Criticism: Critical Comments on a Reactionary Philosophy*, 2nd ed. (New York: International Publishers, 1927), 176–89; Herbert, *Victorian Relativity* (ref. 18), 16; John Eric Marot, 'Marxism, Science, Materialism: Toward a Deeper Appreciation of the 1908–1909 Philosophical Debate in Russian Social Democracy', *Studies in East European Thought* 45 (1993): 147–67. For Mach's response, see John T. Blackmore, *Ernst Mach: His Work, Life, and Influence* (Berkeley: University of California Press, 1972), 232–46. Graham argues that the issue later plagued Boris Hessen's advocacy of relativity in the Soviet Union and helped shape his famous address on Newton. See Loren R. Graham, 'The Socio-Political Roots of Boris Hessen: Soviet Marxism and the History of Science', *Social Studies of Science* 15 (1985): 705–22, on 710–11.

[31] Planck and Mach debated their epistemologies in Max Planck, 'Die Einheit des physikalischen Weltbildes', *PZ* 10 (1909): 62–75; Max Planck, *Acht Vorlesungen über theoretische Physik gehalten an der Columbia University in the City of New York im Frühjahr 1909* (Leipzig: Hirzel, 1910), quote on 6; Ernst Mach, 'Die Leitgedanken meiner naturwisseschaflichen Erkenntnislehre und ihre Aufnahme durch die Zeitgenossen,' *PZ* 11 (1910): 599–606; Max Planck, 'Zur Machschen Theorie der physikalischen Erkenntnis: Eine Erwiderung', *PZ* 11 (1910): 1186–90. See Blackmore, *Ernst Mach* (ref. 30), 217–27; John Blackmore, 'Ernst Mach Leaves "The Church of Physics"', *British Journal for the Philosophy of Science* 40 (1989): 519–40; J.L. Heilbron, *The Dilemmas of an Upright Man: Max Planck and the Fortunes of German Science*, with new afterword (Cambridge, MA: Harvard University Press, 2000), 44–60.

[32] Mach, *Kultur und Mechanik* (ref. 24), 85. For a sensitive perspective on Mach's book, in a comparative framework that relates Exner, Mach and Boltzmann and explores the work of Spengler, see Michael Stöltzner, 'Franz Serafin Exner's Indeterminist Theory of Culture', *Physics in Perspective* 4, no. 3 (2002): 267–319, on 294–97.

In the midst of war Mach's book thus emphasised that civilisation required the sustenance and cautionary lessons of its actual history, with all its losses. This was an implicit corrective to colleagues who might overestimate the ease of maintaining current circumstances and see essential differences between Western and primitive cultures, for in such a situation our own capabilities would be far outstripped by those living simpler forms of life beside us. In contrast, William Thomson imagined an interstellar traveller easily able to make glass and rule parallel lines to build a grating and recover the meter from the wavelength of light, though he started with nothing but his books.[33] Mach would insist that the coming generation needed its footing in the exact sciences, yes, but it also needed to experience the lessons of history and cultivate inner cultural values if it wished to win a new epoch as the Hellenic Greeks had prepared our own (with the screw among their possessions) (Fig. 2).[34]

Conclusion

Oswald Spengler's career and interests surely made him a keen observer of the field of discourse I have sketched here, and this discourse likely helped set the terms in which he accounted for his era, and established an audience for his idiosyncratic views after the war. As a student, Spengler took full advantage of the openness of the German university system, tasting lectures across a wide range of disciplines, abjuring specialisms and reading Darwin, Goethe and Nietzsche. His doctoral dissertation was written under the supervision of Alois Riehl, whose neo-Kantian philosophy and interest in science has been described as a continuation of Dühring's approach.[35] Spengler later commented that Dühring's *Wert des Lebens* was rarely mentioned but exercised the greatest influence on the succeeding generation.[36] Spengler was equally familiar with descriptionist approaches, and argued the best guide to understanding Heraclitus's conception of cosmic existence was the energetic theory of Mach and Ostwald.[37] Exploring differences between modern and Greek conceptions, Spengler had begun lines of thought that he would elaborate dramatically in the course of World War I. His enormously

---

[33] William (Lord Kelvin) Thomson, 'Electrical Units of Measurement', in *Popular Lectures and Addresses,* vol. 1, *Constitution of Matter*, Nature Series (London: Macmillan, 1889), 73–136, 107–19.
[34] Mach, *Kultur und Mechanik* (ref. 24), 86.
[35] Köhnke, *Rise of Neo-Kantianism* (ref. 13), 244–46.
[36] Spengler, *Untergang des Abendlandes* (ref. 2), 523.
[37] On Spengler, see Anton Mirko Koktanek, *Oswald Spengler in seiner Zeit* (Munich: Beck, 1968); H. Stuart Hughes, *Oswald Spengler: A Critical Estimate* (New York: Charles Scribner's Sons, 1952); Farrenkopf, *Prophet of Decline* (ref. 4). Jeffrey Herf discusses Spengler's attitude toward socialism and his treatment of the machine and technology in Herf, *Reactionary Modernism* (ref. 25), chap. 3.

Fig. 2: The cover of Ernst Mach's *Kultur und Mechanik*, illustrated by his son Felix Mach.
Source: Mach, *Kultur und Mechanik* (ref. 24).

successful but always controversial book would found cultural commentary on an expansive history of East and West; but it would also go beyond a merely causal approach to history. Spengler thought history had drawn its methods from the science that had most completely disciplined the methods of cognition: physics. His own work would study the 'world as history' rather than the world as nature and it would elevate analogy to establish what he described as a morphology of world history.[38]

In Spengler's view the sciences and culture of a particular period went through a common life cycle of growth and decay, having more in common within a given period than with their successors across periods and characterised by moments of extreme change. The way Spengler resolves a discipline into the epoch in which it is expressed is illustrated by a contrast that he draws between the different perspectives that would be taken on mechanics by a physicist, and by a skeptic who has tracked down the psychology of the scientific conviction in mechanical explanation: 'To the one, present-day mechanics is a logical system of clear, unambiguous concepts and of relations as simple as they are necessary; while to the other it is an illusion characteristic of the structure of the Western-European spirit'.[39] Spengler wrote that from the point of view of the researcher, present-day physics might have a determinate theme,

> but as an historical phenomenon, in its object, methods and results, physics is an expression and actualization of a particular soul-character, an element of a macrocosm, and every one of its results is a symbol. What it is that physics, which exists solely in the spirit of particular members of a culture, believes to find through these results, was already established in the form and manner of the search.[40]

Spengler regarded causal physics as coextensive with law, but associated causality with rigidity (and death). The principle of causality was a late and 'somewhat artificial' possession of the 'energetic intellect of higher cultures'.[41] Spengler's history enabled a prophetic vision of decline: he regarded Western European physics as having reached the limits of its possibilities and thought he saw 'a deep and utterly unconscious skepsis' in 'the rapidly increasing use of enumerative and statistical methods, which aim only at the probability of the results,

---

[38] Spengler, *Untergang des Abendlandes* (ref. 2), 5–10, esp. 9. Farrenkopf argues that Spengler later changed his philosophy of history, significantly in his 1931 book on prehistory, *Der Mensch und Technik*; Farrenkopf, *Prophet of Decline* (ref. 4), chap. 10.
[39] Spengler, *Untergang des Abendlandes* (ref. 2), 528.
[40] Ibid.
[41] Ibid., 165.

and forgo in advance the absolute exactitude of the laws of nature, as one understood it in hopeful earlier generations'.[42]

I hope this account has done enough to demonstrate that mechanics was worth fighting over in the late nineteenth and early twentieth centuries, and that its role as a disciplinary foundation for physics also helped make its status a vital concern for philosophers and materialists alike, in arguments and debates that moved between academic and party politics and made questions of relative and absolute knowledge central to doctrines of space and time, political philosophy and philosophical history. The question of the relations among culture and science was also of particular interest in Germany at the dawn of World War I, and this discursive field should help us appreciate both the register in which Spengler wrote, and the great interest his work sustained on its publication. The debate he joined was structured by longstanding concerns as much as it was rendered especially critical by postwar conditions.

Acknowledgments

I would like to thank those involved in the Forman Revisited conference for stimulating me to explore the relations between John Heilbron's and Paul Forman's important studies in the paper 'The Fin de Siècle Thesis'. As a development of that argument, the present discussion of culture and mechanics first took shape in a paper on physics as a historical science, delivered to the 2009 Chicago-MPI workshop on the historical sciences. I would like to thank the organisers and participants for their critical interest, and Lynn Nyhart and Robert Brain for many valuable suggestions. Finally, I am especially grateful to Cathryn Carson for her help in refining this essay out of those rather different beginnings, and to Suman Seth and Mike Shank for their helpful readings.

---

[42] Ibid., 596–97.

The Establishment of a Network of Reactionary Physicists in the Weimar Republic

Stefan L. Wolff *

In his well-known article, 'Weimar Culture, Causality, and Quantum Theory, 1918–1927', Paul Forman posited a thesis that has become rather controversial. He maintained the advancement of the quantum theory in Germany was facilitated by physicists' adaptation to the cultural milieu.[1] One aspect of this thesis deals with the connection between a physicist's attitude toward causality and his 'general intellectual-political orientation'.[2] As it was, the supporters of causality were, for the most part, politically conservative and reactionary — Einstein being one of the few exceptions. In later studies, Forman also attempted to prove a close connection between the directions of physical research and the political orientation of the researchers.[3]

---

* Forschungsinstitut des Deutschen Museums, Museumsinsel 1, D-80538 Munich, Germany; s.wolff@deutsches-museum.de.

  The following abbreviations are used: ASP, Arnold Sommerfeld Papers, NL 089, DMM; BAB, Bundesarchiv Berlin; DMM, Deutsches Museum, Munich; DNVP, Deutschnationale Volkspartei; DPG, Deutsche Physikalische Gesellschaft; GSPK, Geheimes Staatsarchiv Preussischer Kulturbesitz Berlin; HS, Handschriften; HSP, Hans Schimank Papers, Institut für Geschichte der Naturwissenschaften, Universität Hamburg; JSP, Johannes Stark Papers, SBPK; JZP, Jonathan Zenneck Papers, NL 053, DMM; NL, Nachlass; SPKB, Staatsbibliothek Preussischer Kulturbesitz, Berlin; WWPB, Wilhelm Wien Papers, SBPK; WWPM, Wilhelm Wien Papers, NL 056, DMM.

[1] Paul Forman, 'Weimar Culture, Causality, and Quantum Theory, 1918–1927: Adaptation by German Physicists and Mathematicians to a Hostile Intellectual Environment', *Historical Studies in the Physical Sciences* 3 (1971): 1–115. A critical evaluation is found, for example, in Helge Kragh, *Quantum Generations* (Princeton: Princeton University Press, 1999), 153–54.

[2] Forman, 'Weimar Culture' (ref. 1), 113.

[3] Paul Forman, 'The Helmholtz-Gesellschaft: Support of Academic Physical Research by German Industry after the First World War', unpublished manuscript; Paul Forman, 'Scientific Internationalism and the Weimar Physicists: The Ideology and Its Manipulation in Germany after World War I', *Isis* 64 (1973): 151–80; Paul Forman, 'The Financial Support and Political Alignment of Physicists in Weimar Germany', *Minerva* 12 (1974): 39–66, on 60.

In this investigation it will first be discussed how a group of German physicists active in the Deutsche Physikalische Gesellschaft (DPG) gradually banded together beginning in 1914, partly out of a desire to counter the dominance of members located in Berlin. Furthermore, most of these physicists were avid supporters of nationalistic and reactionary politics. A stable core developed within this group through a nexus of personal relationships, which I refer to hereafter as the 'network'. Network is used here in the social-science sense to describe a grouping without hierarchy, in which members felt bound to one another through friendship or at least trusted one another enough to act in concert on issues of science policy.[4] This network stood against a more loosely organised, moderately liberal-conservative group. In the period under study here, physicists in Germany acted on decidedly political and ideological grounds, and topics of heated debate between these different camps concerned not only the scientific journal publishing enterprise but also value-laden issues such as the manner of citation in scientific articles. In this way, scientific aspects were linked to political ones. The confrontations reached such an intensity that an organisational split among German physicists was several times under serious consideration.[5]

The network did not limit political involvement to the academic realm, but extended its activities to party politics. During the Weimar period, the party in which these activities played out was mostly the Deutschnationale Volkspartei (DNVP), which was overwhelmingly hostile to the new democracy and entered into government coalitions under the leadership of the Zentrumspartei only during a short period in the mid-1920s. Ideological quarrels over the theory of relativity and quantum theory were likewise sometimes fought along the battle lines of such politically defined fronts. The dominating ideology in the network was characterised by an aggressive chauvinism, *völkisch* ideas featuring a more or less pronounced anti-Semitism and after 1918, an anti-democratic attitude.

---

[4] Jens Aderhold, *Form und Funktion sozialer Netzwerke in Wirtschaft und Gesellschaft* (Wiesbaden: Verlag für Sozialwissenschaften, 2004).

[5] W. Wien to Arnold Sommerfeld, 4 May 1915, WWPM; also printed in Michael Eckert and Karl Märker (eds.), *Arnold Sommerfeld Wissenschaftlicher Briefwechsel*, vol. 1: 1892–1918 (Berlin: GNT-Verlag, 2000), 494–95. Franz Himstedt to Walther Nernst, 15 July 1925; and Himstedt to 'Vorstandsmitglieder der Deutschen Physikalischen Gesellschaft und an die Vorsitzenden der Gauvereine', 8 August 1925, both in WWPM. Max von Laue to Sommerfeld, 20 July 1925, HS 1977-28/A, 197, DMM.

The individual members of the network appear to have been considerably less open to modern scientific developments, in line with Forman's thesis. Thus, the paper will discuss in conclusion whether and in what way the theory of relativity and quantum theory played a role in the individual research of the 'network' physicists. This question is more specific than the Forman thesis, which connects political attitudes to general preferences in science but does not necessarily restrict its discussion to the physicists' own research.

The Bylaws of the German Physical Society and the Outbreak of War

The Deutsche Physikalische Gesellschaft (DPG) was established as a national organisation in 1899, so extending the existing Berliner Physical Society to the entire German Empire.[6] Until 1914, the numerical dominance of the 'Berliners', who made up about half of the total membership, was further strengthened by the organisation's bylaws (i.e., the governing rules) traditionally composed in their favour.[7] This situation provided the breeding grounds for future conflicts, with physicists living outside of Berlin feeling increasingly disadvantaged by the bylaws. Their weaker position within the DPG was even reflected in the membership lists: the list of Berlin members was given first, followed by a second list labelled *Auswärtige* (or 'out-of-towners'), in which the non-Berliners were grouped together with foreign members. What began as a problem of regionalism within the organisation became increasingly acrimonious as it assumed the character of a general political and ideological conflict.

In light of the biennial cycle in which the conventions of natural scientists (*Naturforscherversammlungen*) took place, at the beginning of 1914 plans were initiated to hold a special congress for physicists in the off-years. This led to open controversy within the DPG. Wilhelm Wien, who held the Würzburg chair, and Wilhelm Hallwachs from the Technische Hochschule Dresden wanted, on the one

---

[6] For the history of the German Physical Society, see Wolfgang Schreier and Martin Franke, 'Geschichte der Physikalischen Gesellschaft zu Berlin 1845–1900', F9–F59; and Armin Hermann, 'Die Deutsche Physikalische Gesellschaft 1899–1945', F61–F105, both in *Physikalische Blätter* 51 (1995), 150 Jahre Deutsche Physikalische Gesellschaft, ed. Theo Mayer-Kuckuk.

[7] The membership lists from the end of 1913 show a total of 670 members. Of these, 309 are individual members with Berlin addresses and 347 have out-of-town addresses. Among the out-of-towners, 65 are located outside Germany and Austria-Hungary (including 16 from Switzerland, 11 from the United States and 8 from the Netherlands). Members List in *Verhandlungen der Deutschen Physikalischen Gesellschaft* 15 (1913).

hand, to resist any attempt to convene future physicist congresses exclusively in Berlin. On the other hand, the two men saw this as an opportunity to push through a general organisational reform within the DPG. In view of the 'closed ranks of Berlin physics',[8] Wien wrote a circular to several colleagues as early as January 1914 in an effort to coordinate further efforts with a broader base of support.[9] He went so far as to consider establishing a new organisation, which he would then allow the weakened Berliners to join under changed conditions. The name 'Helmholtz-Gesellschaft' had already been selected for this new organisation (later to be used in a different context).[10] However, Wien was forced to accept that the new organisation could not be realised just yet, one reason being that the idea was not received as warmly among physicists as he had hoped.

In May it began to be obvious that the Berliners, led by the current chairman, Fritz Haber, were ready to make concessions.[11] Wien was convinced their concession to decentralise the organisation would create more favorable organisational prerequisites to bring about secession at some later point, if necessary.[12] Finally, in July 1914, the group associated with Wien and Hallwachs accepted a draft version of the bylaws, although they signalled their remaining reservations by declining to sign the document directly. However, on the basis of these new bylaws, they expressly declared their willingness to abandon the establishment of a new association.[13] But when war broke out, the planned passage of the new bylaws at the coming *Naturforscherversammlung*, which had been cancelled, had to be postponed. There was a consensus therefore that the reorganisation of the DPG could only take place after a peace agreement, which was generally expected soon.[14] (Table 1 lists the individuals involved in this correspondence.)

Although the conflict seems at first glance to have been about organisational issues, such as the location of board meetings and membership dues, the political and ideological components that had existed subliminally soon surfaced. In connection with the controversies among intellectuals during World War I known as the *Krieg der Geister*, it was once again Wien who wanted to mobilise his

---

[8] As stated in Wilhelm Hallwachs to W. Wien, 6 February 1914, WWPM.

[9] Circular from W. Wien to Egon von Schweidler, Woldemar Voigt, Friedrich Paschen, Ferdinand Braun, Philipp Lenard, Otto Wiener, Walter König and Ernst Lecher, 21 January 1914, WWPM; Franz Himstedt was added to the group in July: Himstedt to W. Wien, 14 July 1914, both in WWPM.

[10] W. Wien to Hallwachs, 7 February 1914, WWPM.

[11] Haber to Hallwachs, 27 May 1914; and W. Wien to Haber, 26 June 1914, both in WWPM.

[12] W. Wien to Hallwachs, 11 July 1914, WWPM.

[13] Ibid.; W. Wien to Fritz Haber, 13 July 1914; and an undated circular from W. Wien, with reference to the draft version of the bylaws and their refusal to sign, all in WWPM.

[14] Haber to W. Wien, 17 July 1914; and W. Wien to Hallwachs, 21 April 1915, both in WWPM.

Table 1: Correspondents during the efforts to decentralise the DPG in 1914.

Name	Biographical dates	Location in 1914
Ferdinand Braun	1850–1918	Strassburg (Strasbourg)
Wilhelm Hallwachs	1859–1922	Dresden
Franz Himstedt	1852–1933	Freiburg
Walter König	1859–1936	Gießen
Ernst Lecher	1856–1926	Vienna
Philipp Lenard	1862–1947	Heidelberg
Friedrich Paschen	1865–1947	Tübingen
Egon v. Schweidler	1873–1948	Innsbruck
Woldemar Voigt	1850–1919	Göttingen
Wilhelm Wien	1864–1928	Würzburg
Otto Wiener	1862–1927	Leipzig

colleagues at the end of 1914 around an initiative by Johannes Stark.[15] He relied in part on those colleagues with whom he had already communicated on the bylaws, but also attempted to involve several Berliners. Although Wien had failed to grab the reins of organisational leadership before the war, he now apparently strove to become a sort of leader in political and ideological matters among German physicists. He urged his colleagues in Germany and German-speaking Austria to join him in issuing a statement against the 'unjustified English influence that has infiltrated German physics'. Specifically, he suggested 'that the English should no longer be given greater consideration in the citation of literature than our own fellowmen, as has happened many times'.[16] Wien thus subjected the selection of cited literature to a restriction that was not justified by scholarly reasons, but solely by political ones. This rejection of the so-called *Engländerei* or Anglo-fixation, which revealed a tendency toward intellectual autarky while lamenting the supposedly inadequate recognition of German achievements, corresponded with a widespread mood in Germany following the outbreak of war. Thus, Wien's 'Proclamation', which had been signed by 16 physicists, received a positive response in the press, after 700 copies were printed and sent in March 1915 to all universities in Germany and Austria. The public perceived this as evidence that Germany was also under attack on the intellectual front and Wien's action seemed justified because inaction

---

[15] S.L. Wolff, 'Physicists in the "Krieg der Geister": Wilhelm Wien's " Proclamation"', *HSPS* 33 (2003): 337–68; a German version of this paper appears in *Acta Historica Leopoldina* 48 (2007): 41–62.

[16] Circular from W. Wien with the 'Proclamation', 22 December 1914, ASP, 59, also printed in Eckert and Märker, *Arnold Sommerfeld*, vol. 1 (ref. 5), 487–89.

would be interpreted as weakness. However, Wien was largely turned down by the Berliners — he had not been able to convince any of those whom he addressed to sign. He interpreted this as evidence 'that a split does actually exist within German physics'.[17] (See Table 2 for a list of the signatories.)

Table 2: Signatories of the proclamation by Wilhelm Wien in 1915 (in cooperation with Lenard).

Name	Biographical dates	Location in 1915
Ernst Dorn	1848–1916	Halle
Franz Exner	1849–1926	Vienna
Wilhelm Hallwachs	1859–1922	Dresden
Franz Himstedt	1852–1933	Freiburg
Walter König	1859–1936	Gießen
Ernst Lecher	1856–1926	Vienna
Philipp Lenard	1862–1947	Heidelberg
Otto Lummer	1860–1925	Breslau
Gustav Mie	1868–1957	Greifswald
Franz Richarz	1860–1920	Marburg
Eduard Riecke	1845–1915	Göttingen
Egon v. Schweidler	1873–1948	Innsbruck
Arnold Sommerfeld	1868–1951	Munich
Johannes Stark	1874–1957	Aachen
Max Wien	1866–1938	Jena
Wilhelm Wien	1864–1928	Würzburg
Otto Wiener	1862–1927	Leipzig

This gap could no longer be seen as a division over organisational matters alone; the discord also included political and ideological issues. Such a situation, in which each line of conflict reinforced the other, had already occurred much earlier in the case of the Deutsche Chemische Gesellschaft.[18] From this perspective, the Berlin physicists appeared to be more liberal and cosmopolitan, and the high percentage of Jews in the academic world of the new Imperial capital added another dimension to this antagonism. In the heat of the larger struggle, anti-Semitism remained latent at first and was seldom openly articulated. However, in the private correspondence between the cousins Max and Wilhelm Wien, for example, such attitudes were expressed unfiltered. So in 1906, Max Wien believed that the 'great physics of Berlin' would 'become Jewified [*verjudeln*]' if Heinrich Rubens, a baptised Jew,

---

[17] W. Wien to Sommerfeld, 4 May 1915, WWPM.
[18] Alan J. Rocke, *The Quiet Revolution: Hermann Kolbe and the Science of Organic Chemistry* (Berkeley: University of California Press, 1993), 353–57.

was appointed to the chair of the university;[19] in December 1918, the two Wiens decided to undertake a winter excursion 'despite the lack of snow and the profusion of Berlin Jews to be expected' at the location they had selected.[20] Even if such anti-Semitism was only expressed privately, it indeed extended to professional matters such as filling university chairs. When contemplating the impending vacancy of a professorship in mathematics in 1911, Max Wien considered the 'Jewish question' to be a menacing one, stating that 'so far, our faculty is still undefiled [*noch rein*]'.[21]

An open attack against the Berlin physicists, which indeed invoked a vocabulary of this sort, was launched in 1918 by Stark, who belonged to the group slowly beginning to form in the wake of the events described here. Stark accused the journal *Die Naturwissenschaften* and its editor-in-chief, Arnold Berliner, who — as his name presumably indicated — came from a Jewish family, of showing an inappropriate deference toward foreign countries. Already at the end of 1914, a controversy had arisen between the two men because Berliner had refused to review Lenard's anti-English pamphlet 'England und Deutschland zur Zeit des großen Krieges' as Stark had demanded.[22] Pretending that an unnamed colleague had allegedly blamed him for submitting contributions to a journal produced in an 'international-Jewish spirit', Stark insisted in August 1918 on a major article about the hostile attitude toward Germany among scientific circles in England and France. As to be expected, Berliner rejected this proposal and Stark interpreted this in *völkisch* terminology as evidence that *Die Naturwissenschaften* placed 'internationalism and business' above the 'interests of the German people [*die Interessen des Deutschtums*].' After that, he refused to give *Die Naturwissenschaften* any further support, even indirectly.[23] Stark was surely targeting not just the journal and its editor-in-chief, but the entire, closely associated scientific community in Berlin.

Settling the Bylaws

During the course of the war, there appeared to be some movement on the organisational front, which is why Arnold Sommerfeld wrote to Wilhelm Wien in

---

[19] M. Wien to W. Wien, 2 August 1906, WWPM. Rubens was appointed to the chair at the Berlin University.

[20] M. Wien to W. Wien, 19 December 1911, WWPM.

[21] M. Wien to W. Wien, 13 October 1911; and M. Wien to W. Wien, 26 April 1914, both in WWPM.

[22] Arnold Berliner to Johannes Stark, with handwritten notations by Stark, 8 December 1914 and 16 December 1914, JSP.

[23] Incomplete copy of a letter from Stark to Berliner, 13 August 1918, as well as a copy of the letter from 17 August 1918, and both in WWPM. Berliner to Stark, 15 August 1918, 29 August 1918, JSP. Stark to W. Wien, 8 September 1918, WWPM.

August 1918 that he felt there were no grounds for the evidently ongoing threat of a 'schism' within the DPG.[24] Hence it came as a surprise when the Berliners founded *Zeitschrift für Physik* late in 1919. They justified this step as an economic necessity to preserve the DPG journal *Verhandlungen*. Its lengthier scientific articles would be transferred to *Zeitschrift für Physik*, and therefore the whole matter was presented more as a split of the old *Verhandlungen* than the creation of a new journal.[25] However, a belief emerged that competition would develop with the traditional *Annalen der Physik* edited by Planck and Wien, as well as with *Physikalische Zeitschrift*. Wien viewed this as an affront and, for a while, considered resigning from the DPG.[26] But in other matters, he noticed that his critique was already yielding consequences. Early in 1920, even before the first *Naturforscherversammlung* since the end of the war, initial steps toward decentralisation were taken by establishing regional associations (*Gaue*). However, the key decisions on fundamental bylaw issues were to be reserved for the meeting that autumn in Bad Nauheim.

In February 1920, long before the meeting, Philipp Lenard of Heidelberg cooperated with Wien in sending out a circular to a select group of colleagues in the DPG. In it, the two men argued for fresh guidelines in the area of reviewing new publications in physics.[27] Direct ideological links can be drawn from this circular and the earlier 'Proclamation' against *Engländerei*. According to these 'Heidelberg-Würzburg' proposals, the business of reviewing was to be transformed 'from a Berlin enterprise into a German one' and had to 'guarantee above all that German literature be given greater consideration than foreign'.[28] In order to ensure this goal, Lenard recommended the appointment of intermediate editors as supervisors who would, in certain cases, even supplement citations accordingly. Only in this way, he maintained, could it be guaranteed that the DPG 'would be regarded, and would act, as a truly German enterprise'.[29] In the case of the *Verhandlungen*, Lenard argued for terminating the publication because it was 'non-German according to the spirit of the content'.[30] By this time, he had clearly positioned himself as anti-Semitic, at least vis-à-vis Wien, by using the expression

---

[24] Sommerfeld to Wien, August 1918, WWPM, reprinted in Eckert and Märker, *Arnold Sommerfeld*, vol. 1 (ref. 5), 602–04.

[25] Karl Scheel to Sommerfeld, 7 December 1919, ASP, 18, Folder 3, 7.

[26] W. Wien to Sommerfeld, 12 December 1919, ASP, 18, Folder 3, 7.

[27] W. Wien to Lenard, 12 February 1920; and Lenard to W. Wien, 17 February 1920, both in WWPM.

[28] Karl Mey *et al.* to Sommerfeld, 28 April 1920, and 'Bericht über eine gemeinsame Sitzung von Vertretern der Deutschen Physikalischen Gesellschaft und der Deutschen Gesellschaft für technische Physik am 28.4.1920', ASP, 18, Folder 3, 8.

[29] W. Wien to the DPG, undated fragment of circular (with Gustav Mie's signature), 18 April 1920, WWPM.

[30] Lenard to W. Wien, 24 April 1920, WWPM.

'Aryan' in connection with another organisational issue, expressing his wish to 'reassert the purely German (Aryan) spirit in German physics'.[31] By September, 54 members had signed Lenard's petition,[32] which also addressed the issue of extending the decentralisation of the DPG so that the local Berlin group would no longer be granted any special privileges. Demands were spelt out by Wien in a four-point list.[33] Briefly, the list dictated that the members of Berlin constitute only one *Gauverein* among the others, *Verhandlungen* in their present form will be suspended, and that only the managing director, the editor of one of the journals and both recording clerks (*Schriftführer*) had to be Berliners. Furthermore, the list included new rules in regard to how the board should be composed and also specified the connection between conferences, member conventions and elections.

In light of the chauvinistic choice of words and the aggressive language used by Lenard, it might seem surprising that he received so positive a response to his proposals. Even the DPG *Vorstand* seriously attempted to discuss them. At the time, Lenard was by no means an isolated maverick. He clearly articulated the general insecurity felt by many in the bourgeoisie, who saw their entire value system threatened after the collapse of Imperial Germany. The publisher Georg Hirzel, who was concerned about his journal *Physikalische Zeitschrift*, articulated such anxieties and compared the state of affairs in the DPG to that of the general political situation: 'Should science now become just as fragmented as the political parties and throw aside all things appreciated, steadfast, and old?'[34] Lenard infused his views in addition with an anti-Semitism that attributed to the Jews nothing less than a catalytic effect on the impending destruction he perceived. He lamented to Wilhelm Wien in August 1920, 'it is indeed as if nearly *everything* that stands in high esteem, has stood the test of time and has the dignity of age for us is to be sent to the devil. A kind of mass suicide of the German spirit whereby the Jews act now as a medium'.[35]

To push through the demands of Lenard and Wien, a small, select group of the petition's signatories met in Bad Nauheim the day before the

---

[31] Lenard to W. Wien, 17 April 1920 (referring to Stark's foundation of a 'Fachgemeinschaft akademischer Physiker'), WWPM. On the use of the term 'Aryan' in the anti-Semitic sense, see for example Houston Stewart Chamberlain, introduction to *Arische Weltanschauung*, 3rd ed. (Munich: F. Bruckmann, 1915).

[32] According to Lenard, as found in Lenard to Sommerfeld, 2 September 1920, ASP, 18, Folder 3, 8.

[33] 'Zum Programm für die Neuordnung der Deutschen Physikalischen Gesellschaft', included in W. Wien to Mie, undated, Folder 5853 (copies of correspondence between W. Wien and Mie), copy no. 58a-b; W. Wien, draft letter to the DPG in the expectation that the organisational issues would be clarified in Bad Nauheim, 4 June 1920; and W. Wien to Lenard, 18 May 1920, all in WWPM.

[34] Georg Hirzel to Sommerfeld, 14 April 1920, Debye Papers, Archiv der Max-Planck-Gesellschaft, Berlin.

[35] Lenard to W. Wien, 14 August 1920, WWPM.

Naturforscherversammlung to agree on a coordinated procedure or, as Lenard put it, on how 'to present the reform proposals in a generally satisfactory form to the entire assembly'.[36] In addition to Max Wien, Hallwachs, Franz Himstedt and Gustav Mie, who had already participated in the correspondence concerning the 'Proclamation', the others invited to this exclusive gathering were chair holders Clemens Schaefer from Marburg; Jonathan Zenneck from the Technische Hochschule in Munich; Friedrich Krüger from the Technische Hochschule in Danzig; Ernst Gehrcke from the Physikalisch-Technische Reichsanstalt (PTR), who also taught as an associate professor at the University of Berlin and Stark's collaborator, Ludwig Glaser. All of these men were securely in the right wing of the political spectrum, with mostly pronounced *völkisch* opinions.[37] (See Table 3

Table 3: Participants at the preparatory meeting in Bad Nauheim in 1920 (listed according to age).

Name	Biographical dates	Location in 1920
Franz Himstedt	1852–1933	Freiburg
Wilhelm Hallwachs	1859–1922	Dresden
Philipp Lenard	1862–1947	Heidelberg
Wilhelm Wien	1864–1928	Würzburg/Munich
Max Wien	1866–1938	Jena
Gustav Mie	1868–1957	Halle
Jonathan Zenneck	1871–1959	Munich
Johannes Stark	1874–1957	Greifswald
Friedrich Krüger	1877–1940	Danzig
Ernst Gehrcke	1878–1960	Berlin (PTR)
Clemens Schaefer	1878–1968	Marburg/Breslau
Ludwig Glaser	1889–?	Würzburg

---

[36] Lenard to Sommerfeld, 2 September 1920, ASP, 18, Folder 3, 8. The invitation appears in Lenard to W. Wien, 14 September 1920, WWPM.

[37] Lenard to W. Wien, 14 August 1920, WWPM, includes the list of those invited. The names of Gehrcke, Glaser, Max Wien and Krüger are underlined with an accompanying note: '... especially count on those underlined, as far as I can tell!'; W. Wien to Lenard, 18 August 1920, WWPM; in addition to Gehrcke and Weyland, Glaser was one of three authors to publish in the series 'Schriften aus dem Verlage der deutschen Naturforscher zur Erhaltung reiner Wissenschaft', which only appeared in 1920; for more on this, see Klaus Hentschel, *Interpretationen und Fehlinterpretationen der speziellen und der allgemeinen Relativitätstheorie durch Zeitgenossen Albert Einsteins* (Basel: Birkhäuser, 1990), 140. On Glaser, *see also* Alan D. Beyerchen, *Wissenschaftler unter Hitler* (Frankfurt: Ullstein, 1982), 243. On Gehrcke, see Milena Wazeck, 'Einsteins Gegner' (PhD dissertation, Humboldt Universität, 2007), 259–70. Dieter Hoffmann, 'Ein Experimentalphysiker als antitheoretischer Sammler', in *Cut and Paste um 1900: Der Zeitungsausschnitt in den Wissenschaften*, ed. Anke te Heesen (Berlin: Vice Versa, 2002), 70–80.

for a list of the participants.) Stark and Wilhelm Wien helped Lenard co-organize the meeting.[38] Originally, Lenard wanted to extend invitations to Paul Weyland and Sommerfeld as well. Just prior to this meeting, Weyland had gained public attention as the organiser of a rally held in the Berlin Philharmonic Hall against the theory of relativity, at which he himself appeared as one of the speakers, along with Gehrcke. The fact that Lenard even considered inviting Sommerfeld, despite their well-known differences, probably indicated his desire to alleviate the shortage of theoretical physicists in the circle.[39]

The petitions by Wien and Lenard had already been discussed in late April at a joint meeting of representatives from the DPG and the Deutsche Gesellschaft für technische Physik, at which an attempt was made to blunt their point by suggesting concrete ways to improve the quality of the reviewing.[40] Lenard did not find these at all sufficient and turned to the full assembly for a decision. The organisational demands that Wien had submitted passed by a slim majority of 48 to 40, with the board members abstaining. Wien was also elected as the new chairman.[41] Lenard was less successful. His more far-reaching, yet vague demands for 'stronger protection of the temporal priority of German researchers' met with resistance, and even Wien did not support them any longer. Lenard was very disappointed, seeing that his proposals 'met with resistance + rejection in Nauheim by the Jews with their [Karl] Scheel ... Wien did not improve anything.'[42] For his part, Wien considered the reforms a success. He used the existing polarisation among German physicists to achieve the decentralising outcome he sought, with Jena hosting in 1921 the first physicists' congress held independently of the biennial *Naturforscherversammlung*. As their later correspondence shows, however, the disagreement between Lenard and Wien over Nauheim did not disturb their relationship in the long term.

A small episode illustrates that the tension within the DPG continued to exist and that the editorial staff of the journals, with the influence they had over their

---

[38] Arne Schirrmacher (ed.), Philipp Lenard: *Erinnerungen eines Naturforschers* (Heidelberg *et al.*: Springer, 2010), 233–234.

[39] Lenard to W. Wien, 28 August 1920, WWPM. On Paul Weyland, see Andreas Kleinert, 'Paul Weyland, der Berliner Einstein-Töter', in *Naturwissenschaft und Technik in der Geschichte*, ed. Helmuth Albrecht (Stuttgart: Verlag für Geschichte der Naturwissenschaften, 1993), 199–232.

[40] Mey *et al.* to Sommerfeld, 28 April 1920, ASP, 18, Folder 3, 8.

[41] Wien Familienchronik 1914–28, unpublished, private property of Waltraud Wien, Munich, entry from 23 September 1920, p. 183; W. Wien to Lenard, 25 November 1920, WWPM; 'Geschäftssitzung der Deutschen Physikalischen Gesellschaft in Bad Nauheim am 21 September 1920', Verhandlungen der Deutschen Physikalischen Gesellschaft 3, no. 1 (1920): 84.

[42] Handwritten notations by Lenard on a letter from W. Wien to Lenard, 20 December 1920, WWPM. *See also* 'Geschäftssitzung' (ref. 41), 86.

content, came to take on nearly strategic importance. In 1923, Rudolf Seeliger, an associate professor from Greifswald close to the circle of participants in the preparatory meeting in Bad Nauheim, expressed his concern that, just like the newspapers, one of the scholarly journals could soon land in 'Jewish hands'.[43] With the help of the publisher Hirzel, Seeliger wanted to plot to replace his Jewish colleague Max Born on the editorial board of *Physikalische Zeitschrift*. Born immediately suspected 'a systematic attack by the group Lenard, Stark, and Seeliger'.[44] In light of the eventual failure of the plot, Sommerfeld mockingly noted, 'the Seeliger Putsch is a good counterpart to the Hitler Putsch'.[45] However, Seeliger in fact achieved his goal the following year when a dispute between Born and Hirzel enabled him to join the editorial board.[46]

For a while it appeared the two camps had reached a truce. Then the conflict welled up again, as is revealed by the vehemence with which a controversy in 1925 played out. It was triggered by a paper published — in English — in *Zeitschrift für Physik*. By this time the critics who had from the beginning viewed the new *Zeitschrift* as a serious competitor to the traditional and soon almost second-rank *Annalen* had been proved correct; the *Zeitschrift für Physik* had risen to become the most important scientific journal of the discipline in Germany and many foreign authors had already published there.[47] However, the presence of the English language was unprecedented. According to the editor and managing director Karl Scheel, this was 'an oversight caused by work overload'.[48] In this case, everyone agreed foreign languages were not to be permitted. Yet the episode was occasion enough to revive the old conflicts. In light of the attacks by the 'Lenardians', as Max von Laue called what he saw as a cohesive faction,[49] the Physical Society of Berlin sent out a circular in which it protested strongly against those who wanted to turn this episode into a 'scandalous affair'.[50] On the other

---

[43] R. Seeliger to W. Wien, 9 October 1923, WWPM.

[44] Born to Sommerfeld, 24 October 1923, ASP, 6.

[45] Sommerfeld to Born, 10 November 1923, ASP, 1. Also, Sommerfeld to Laue, 27 November 1923, HSP.

[46] Sommerfeld to Laue, 27 November 1923, HSP; and Debye to Sommerfeld and Himstedt, 5 January 1924, ASP, 18, Folder 3, 7.

[47] Reviewing the developments in retrospect, Sommerfeld to Grüneisen, 12 June 1934, ASP, 18, Folder 3, 10.

[48] Laue to Sommerfeld, 20 July 1925, HS 1977-28/A,197, DMM. The article in question was by R.N. Ghosh, 'On the Electrical Conductivity of Heated Gas', *Zeitschrift für Physik* 32 (1925): 113–18. On this, *see also* Hermann, 'Deutsche Physikalische Gesellschaft' (ref. 6), F89–F90.

[49] Laue to Sommerfeld, 20 July 1925, HS 1977-28/A,197, DMM.

[50] As found in Meitner to Sommerfeld, 31 July 1925, Meitner Papers, Churchill Archives Centre, Cambridge University.

hand, Max Wien — in his capacity as chairman of the DPG and a successor of his cousin Wilhelm Wien in this position — expressed his dissatisfaction to the members of its journal committee, writing in particular of the 'lack of national dignity' that he felt this episode had demonstrated.[51] The ensuing tension between him and the managing office in Berlin prompted him finally to resign the chairmanship in July 1925.[52] For their part, Walther Nernst and Laue were adamant that the matter had not been politicised at the initiative of Berlin.[53] At first, Max Wien's predecessor and vice chair Himstedt tried to mediate, warning of the danger of a split within the DPG. After his efforts failed, he also resigned his post.[54] Nernst was of the opinion that the entire matter was not particularly important for the community of German physicists in the face of greater concerns, but Himstedt disagreed: 'Much more important than physics for many, hopefully for most colleagues, is the protection of German dignity.[55] Lenard had already terminated his membership in the DPG because he considered himself to have become 'dirtied [*verunreinigt*]' by the course of events.[56] Laue correctly assessed the situation when he held the 'prevailing tensions and contrasts' in German physics responsible for these disputes.[57]

The Network and its Positioning Within Politics at Large

Over the course of these events from 1914 to 1925, a relatively stable core developed within this reactionary group of German physicists, a core that sought to exert influence through coordinated actions in the manner described above. In the following section, we will more closely examine this core, which was made up of nine full professors, all of whom had attended the preparatory meeting in Bad

---

[51] M. Wien to Mitglieder der Zeitschriftenkommission der Deutschen Physikalischen Gesellschaft, 12 June 1925, WWPM; *see also* M. Wien to Sommerfeld, 11 May 1925, ASP, 18, Folder 3, 7; M. Wien to W. Wien, 30 May 1925, WWPM; W. Wien to Planck, July 1925, WWPM and Planck to W. Wien, 25 July 1925, JSP.

[52] M. Wien to Mitglieder des Vorstands and Vorsitzende der Gauvereine, 10 July 1925, WWPM.

[53] Nernst and Laue to Himstedt and M. Wien, 25 July 1925, WWPM.

[54] Himstedt to Nernst, 15 July 1925 and 29 July 1925, WWPM; M. Wien to Himstedt, 5 August 1925, WWPM; M. Wien to W. Wien, 1 August 1925, WWPM, reporting on Himstedt's attempt to mediate; Himstedt to Mitglieder des Vorstands der Deutschen Physikalischen Gesellschaft und Vorsitzende der Gauvereine, 8 August 1925, WWPM.

[55] Himstedt to Nernst, 15 July 1925, WWPM.

[56] M. Wien to Lenard, 11 May 1925, WWPM; Lenard to M. Wien, 20 May 1925, WWPM. Max Wien's appeal to remain in the society was in vain. M. Wien to Lenard, 29 May 1925, WWPM. *See also* Beyerchen, *Wissenschaftler* (ref. 37), 139.

[57] Laue to Sommerfeld, 20 July 1925, HS 1977-28/A, 197, DMM.

Nauheim. After Stark voluntarily withdrew from his professorship in 1922, he attempted to return to university with the help of those colleagues with whom he was still in contact, so he is counted among this core group. (See the list in Table 4.) Although also in attendance in Bad Nauheim, Glaser and Gehrcke are not included here because they had very little contact with the others. They remained on the sidelines; even though they were at universities, their positions were of minor rank as *Privatdozent* and associate professor, respectively. Hallwachs too is not considered here since he died in 1922.

Table 4: The 'Network' (listed according to age).

Name	Biographical dates	Location starting in 1920	Party affiliation
Franz Himstedt	1852–1933	Freiburg	DNVP
Philipp Lenard	1862–1947	Heidelberg	NSDAP
Wilhelm Wien	1864–1928	Munich	No party affiliation
Max Wien	1866–1938	Jena	DNVP
Gustav Mie	1868–1957	Halle/1924 Freiburg	Member of SS sponsor circle
Jonathan Zenneck	1871–1959	Munich	DNVP; SA
Johannes Stark	1874–1957	Greifswald/external	NSDAP
Friedrich Krüger	1877–1940	Danzig/1921 Greifswald	NSDAP supporter
Clemens Schaefer	1878–1968	1920 Marburg/ 1926 Breslau	Fatherland party

The members of this core group were closely associated with one another largely through an array of personal contacts and in some cases friendships, forming what we have termed a network. A close connection existed between the cousins Wilhelm and Max Wien since their youth. Wilhelm Wien was quite a close friend of Mie,[58] with whom he had sometimes gone on vacation even before the war.[59] Since 1920 he also considered Himstedt a friend.[60] Despite occasional discord, Wilhelm Wien continued to be the most important contact for Lenard. His son Karl even studied under Lenard for a while.[61] Stark also maintained a steady correspondence with Wilhelm Wien, although professional and personal

---

[58] According to Mie's autobiography, *Aus meinem Leben*, quoted in Gunter Kohl, *Relativität in der Schwebe: Die Rolle von Gustav Mie*, preprint no. 209, Max-Planck-Institut für Wissenschaftsgeschichte, Berlin, p. 19.
[59] Mie to W. Wien, 4 February 1913, WWPM.
[60] W. Wien to his wife, Luise Wien, 23 September 1920, WWPM.
[61] W. Wien to his son, Karl Wien, 23 May 1925, in Wien Familienchronik (ref. 41), 278–79.

reservations remained.⁶² With his appointment to Munich, Wien was able to keep in direct contact with Zenneck. In other words, Wilhelm Wien was in intense personal and written contact with six of the other eight men. Following a conference in Danzig in 1925, Krüger, Zenneck and Max Wien undertook an excursion together to Rügen, which illustrates that the communication among the three men went beyond strictly professional contact.⁶³ Furthermore, close friendly relations had developed between Max Wien and Zenneck, as is evident in Wien's expressed gratitude to Zenneck for the article the latter wrote in 1927 on the occasion of his sixtieth birthday, an article that — as Max Wien emphasised — 'only a true friend could have written in this way'.⁶⁴ Schaefer alone was not close to any of the others.

One essential feature of every network is the mutual support its members provide, which in the case of this network became evident with regard to university appointments, among other things. For example, the network tried to help keep Himstedt in his professorship when his retirement was impending. Lenard considered Himstedt and Wilhelm Wien his most important allies in countering the 'harmful influence [*schädlichen Einfluß*]' of Planck.⁶⁵ Wilhelm Wien wanted to send a petition to the ministry regarding the matter, but Himstedt felt it was a thoroughly hopeless undertaking. He thought of Zenneck, Lenard, Gehrcke, Max Wien and Mie as possible successors — all members of the network except for Gehrcke, who was closely linked, as mentioned above.⁶⁶ Zenneck actually received the appointment, which he turned down to remain in Munich.⁶⁷ In the end, it was Mie who went to Freiburg.⁶⁸

Despite Stark's difficult personality, the network attempted to bring him back into the academic world that he had voluntarily left in 1922. A vacancy in Marburg in 1926 seemed to offer a good opportunity. Stark was recommended by Schaefer, who held the chair but was leaving to take up an appointment in Breslau. Lenard,

---

⁶² About 50 letters of correspondence between Stark and Wilhelm Wien, including copies, can be found in WWPM.
⁶³ M. Wien to W. Wien, 23 September 1925, WWPM.
⁶⁴ M. Wien to Jonathan Zenneck, 24 January 1927, JZP, Box 47.
⁶⁵ Lenard to W. Wien, 25 August 1922, WWPM.
⁶⁶ Himstedt also mentioned that Lenard placed Gehrcke first on his proposed list of candidates. *See* Himstedt to W. Wien, 26 January 1923 and M. Wien to W. Wien, 29 November 1923, both in WWPM.
⁶⁷ Zenneck to Bayerisches Ministerium für Unterricht und Kultus, 14 June 1923, Bayerisches Hauptstaatsarchiv, Munich, Folder MK43350. Himstedt had already inquired of W. Wien whether Zenneck would accept such an appointment: Himstedt to W. Wien, 26 January 1923, WWPM.
⁶⁸ Sommerfeld to Mie, 22 April 1924, reprinted in Michael Eckert and Karl Märker, eds., *Arnold Sommerfeld Wissenschaftlicher Briefwechsel*, vol. 2: 1919–51 (Berlin: GNT-Verlag, 2004), 159–60.

Wilhelm Wien and Zenneck also tried to intervene on Stark's behalf with the Prussian Ministry of Culture.[69] But their efforts were in vain. Max Wien thought that ultimately Stark's own actions were to blame and that his irrational behaviour made it nearly impossible for anyone to help him. Wien put it simply: 'He is completely crazy [*völlig verrückt*]'.[70]

When Lenard, for his part, submitted his request to retire on 3 January 1927, in Heidelberg, he initially wanted Stark to be appointed his successor; Wilhelm Wien and independent of Wien — Zenneck supported this idea in their written appraisals. Lenard's idea met with insurmountable resistance, however, because the faculty had quite different ideas and favoured the Nobel-prize winners James Franck and Gustav Hertz. They also named Hans Geiger as a possible candidate. Lenard himself then suggested a number of physicists who would, in his mind, continue research in a manner similar to his own: besides Stark these were Schaefer, Gehrcke and Lenard's former students, Carl Ramsauer and August Becker. The appraisals written by Wilhelm Wien and Zenneck, as well as a letter by Himstedt, who also got involved, had no influence on the decision of the faculty. This prompted Lenard finally to withdraw his retirement request on 16 April 1927. The reason he gave for the withdrawal was the 'urgent request by a number of colleagues' in his field, naming, in particular, Wilhelm Wien, Himstedt and Zenneck.[71] Once again, a good many of the network members had been involved.

Such mutual preferences were also revealed when the DPG received an official invitation to the fiftieth anniversary celebration of the London Physical Society in February 1924. Chairman Himstedt named Wilhelm Wien and Zenneck as delegates. Laue's intervention on behalf of Nernst, who would have represented the more liberal Berlin physicists, was unsuccessful.[72]

Apart from such involvement, nearly all of the network members demonstrated not only an above-average interest in politics in general but were active in party politics. Wilhelm Wien, who described himself as 'national and

---

[69] Lenard to Preussisches Kultusministerium, 18 September 1926; Zenneck to Lenard, 6 October 1926; and Lenard to Zenneck, 7 October 1926, all in JZP, Box 40.

[70] M. Wien to W. Wien, 31 December 1926, WWPM.

[71] Lenard to W. Wien, 29 March 1927, 9 April 1927, WWPM; Lenard to Zenneck, 29 March 1927, 2 April 1927, JZP, Box 40; Zenneck to Lenard, 4 April 1927, JZP, Box 40; Lenard to Ministerium für Kultus und Unterricht, 'Sondergutachten', 1 April 1927, 16 April 1927, JZP, Box 40. M. Wien thought Gehrcke insufficiently qualified: M. Wien to W. Wien, 7 April 1927, WWPM. *See also* Charlotte Schönbeck, 'Physik', in *Die Universität Heidelberg im Nationalsozialismus*, ed. Wolfgang U. Eckart, Volker Sellin and Eike Wolgast (Heidelberg: Springer, 2006), 1087–149, esp. 1093–94.

[72] Himstedt to W. Wien, 23 February 1924, 2 March 1924 and 4 March 1924, all in WWPM.

conservative', was the only one who kept his distance from any political party.[73] During the war, Schaefer had been a member of the Vaterlandspartei (Fatherland Party), a right-wing movement founded by Admiral Tirpitz in reaction to the peace resolution passed by the Reichstag in 1917. According to later statements, he felt close to the DNVP, but did not join the party.[74] Max Wien and Zenneck also found their way into this mostly anti-republican party, while Himstedt even advanced to the rank of a board member.[75] In November 1932, Himstedt and Max Wien signed a leaflet by the Deutsche Ausschuss (German Committee), which was closely associated with the DNVP. The leaflet was entitled 'Mit Hindenburg für Volk und Reich' (With Hindenburg for the People and the Empire) and referred to German and Christian values in calling for voters to support the re-election of Hindenburg as President.[76] Lenard publicly declared his support for Hitler in 1924 but did not join the National Socialist party until 1937, while Stark became a member as early as 1930.[77] During the World War I, Mie supported a decidedly annexationist policy. He called for the control of Belgium and was among the 352 university teachers who signed the petition authored by the theologian Reinhard Seeberg and addressed to the Imperial Chancellor, in which such demands were made.[78] During the early years of the Weimar Republic, Mie initially sympathised with the national-liberal Deutsche Volkspartei but expressed his disappointment in the party after 1921. What he lamented at the time was the absence of an outstanding personality in German politics.[79] Otherwise, there is no evidence of his involvement in any party activities. In 1933, he became a supporting member (*Fördermitglied*) of the SS.[80] Krüger joined national and *völkisch*-oriented associations such as the Deutscher Reichsbund Kyffhäuser and the Reichsverband der

---

[73] So Wien put it in a letter to the editor of the *Tägliche Rundschau*, a newspaper to which he and his parents had subscribed since its founding in 1881 and which advocated this political line. Wien to the editor, 14 February 1916, WWPM.

[74] From the card file of all university instructors, 'Meldungen der Hochschulen zur Anlage der Kartei gemäss Runderlass vom 13.12.1934': Schaefer record, BAB, R 4901/13275, microfiche 171.

[75] M. Wien record, BAB, R 4901/13280, microfiche 220; Zenneck record, BAB, R 4901/13281, microfiche 227. On Himstedt's board activity, see Himstedt to Wien, 28 August 1924, WWPM.

[76] Deutscher Ausschuss, 'Mit Hindenburg für Volk und Reich', copy in GSPK, I HA Rep. 76, Kultusministerium Va no. 10046, pp. 75–6.

[77] Beyerchen, *Wissenschaftler* (ref. 37), 138 (on Lenard), 160 (on Stark).

[78] W. Wien to Mie, 6 November 1915 and 8 February 1916, WWPM; see Bernhard vom Brocke, 'Wissenschaft und Militarismus', in *Wilamowitz nach 50 Jahren*, ed. William Calder, Hellmut Flashar and Theodore Lindken (Darmstadt: Wissenschaftliche Buchgesellschaft, 1985), 649–719, esp. 689, 711.

[79] Mie to W. Wien, 26 October 1921, WWPM.

[80] Mie record, BAB, R 4901/13271, microfiche 139.

Kriegsteilnehmer-Akademiker after the war.[81] When writing recommendations for appointments to university positions, he raised the issue of whether a potential candidate might possibly be Jewish.[82] Sometimes such a question was not easy to answer, but the markedly anti-Semitic volumes titled 'Statistik des Judentums an den deutschen Universitäten' (An Enumeration of Jews at German Universities) promised answers. Krüger had helped support them financially out of his own pocket starting around 1930.[83] About the same time, he began to make contributions to the NSDAP.[84] Against this backdrop, the speech he gave on the occasion of Hitler's fiftieth birthday on 20 April 1939 as part of the Greifswald University lecture series does not appear to be a concession to the *Zeitgeist* only, but with high probability expresses his own conviction that the 'Führer' had earned praise for having 'dedicated all his strengths to reawakening and reinvigorating our Fatherland, to restoring the *Reich* to its earlier size and grandeur.'[85] Zenneck also gave public speeches at that time, which showed such an agreement with official politics.[86]

Anti-Semitism belonged to the milieu in which the network was embedded. Often Lenard and Stark openly expressed their anti-Semitism, while in the case of the two Wiens, as was pointed out earlier, such sentiments were reserved for more

---

[81] Krüger record, BAB, R 4901/13269, microfiche 117.

[82] For example, see Krüger to Hans von Mangoldt, 26 January 1915, ASP, 10. He (inaccurately) describes Herbert Freundlich as 'most certainly Jewish.' Freundlich was a Protestant with one Jewish paternal grandparent.

[83] Krüger record, BAB, R 4901/13269, microfiche 117. The editor of the volumes was Achim Gercke, who served as an expert on race research at the Reich Interior Ministry during the National Socialist regime. Apparently, Krüger belonged to a larger circle of supporters who organised themselves in the 'Kreis der Freunde und Förderer der deutschen Auskunftei'. Between 1928 and 1932, Gercke published eight volumes, each of which had the title *Der jüdische Einfluss an den deutschen Hohen Schulen: Ein familienkundlicher Nachweis über die jüdischen und verjudeten Universitäts- und Hochschulprofessoren*. In addition, he collected data on Jews in public life for many years, under the rubric 'Archiv für berufsständische Rassenstatistik'. See Rudolf Kummer, 'Achim Gerckes Judenkartei', Kummer (Reichserziehungsministerium) to Reichsministerium des Inneren, 2 July 1937, BAB, BDC-REM PA, Kummer, pp. 42–46; available online at: http://homepages.uni-tuebingen.de/gerd.simon/Judenkartei.pdf (accessed 11 December 2009). *See also* Diana Schulle, *Das Reichssippenamt* (Berlin: Logos-Verlag, 2001).

[84] Krüger record, BAB, R 4901/13269, microfiche 117.

[85] Friedrich Krüger, *Wissenschaft und Leben*, Greifswalder Universitätsreden 52 (Greifswald: L. Bamberg, 1939), 3.

[86] Zenneck gave a speech at Bayerische Akademie der Wissenschaften on 16 June 1937, which was printed in Jonathan Zenneck, *Wissenschaft und Volk* (Munich: Verlag der Bayerischen Akademie der Wissenschaften, 1938); he also delivered the inaugural speech at the annual meeting of DPG on 1 September 1940, which was printed in *Verhandlungen der Deutschen Physikalischen Gesellschaft* 21 (1940): 31–34.

confidential environments. Yet the expression of anti-Semitic opinions was not limited to occasional remarks but permeated their entire correspondence, most certainly when the matter at hand was university appointments. Max Wien, for example, complained that Sommerfeld had offered them the (supposed) 'Jew' Landé for Jena, when they already had Auerbach, whom they considered enough in this respect.[87] Zenneck often perceived unwanted 'Jewish forces' at work in the appointment processes; for example, when Gustav Hertz received a professorship in Halle in 1926.[88] In 1933, he defended the regulations stipulated by the so-called 'Law for the Restoration of the Professional Civil Service' (*Gesetz zur Wiederherstellung des Berufsbeamtentums*) vis-à-vis English colleagues who accused the Germans of actions against 'non-Aryans'. Citing specific examples, Zenneck sought to defend the necessity of the 'non-Aryan paragraph' of the law, because in his view it was improving situations that he considered untenable. There were institutes, he maintained, 'in which all assistants, ... faculties in which the majority of the professors, and ... companies in which the larger number of the directors and engineers were Jewish'.[89] The aforementioned support that Krüger gave to anti-Semitic publications can only be interpreted as an expression of his convictions. Schaefer stated in 1933 that he had not admitted any non-Aryan students since 1926.[90] No documentation of any statements of this sort by Himstedt and Mie has been found.

Thus we have seen how, between 1914 and the initial phase of the Weimar Republic, the antagonism between the capital Berlin and the rest of the Empire led to the formation of the network and the emergence of a 'right wing' within the DPG, which was active in organisational affairs. Anti-Semitic resentments were used strategically here against targets in Berlin, including Albert Einstein and Arnold Berliner, editor of *Die Naturwissenschaften*. This appears to correspond with a general societal development in Germany, as parts of the established bourgeoisie felt their property and value system threatened after the war by the new Weimar Republic.

---

[87] M. Wien to W. Wien, 8 October 1919, WWPM. Felix Auerbach was Jewish. Landé's grandparents were Jewish, but he did not consider himself Jewish. Landé, interview by C. Weiner, 1973, American Institute of Physics, Washington, DC, and Elke Brychta, Anna-Maria Reinhold, Arno Mersmann, ed., *Mutig, streitbar, reformerisch: Die Landés — Sechs Biografien, 1859–1977* (Essen: Klartext-Verlag, 2004).

[88] Zenneck to Himstedt, 4 April 1927, JZP, Box 38.

[89] Zenneck to Karl Kiesel, 16 August 1933, JZP, Box 15. On this point, *see also* S.L. Wolff, 'Die Ausgrenzung und Vertreibung von Physikern im Nationalsozialismus — welche Rolle spielte die DPG?' in *Physiker zwischen Autonomie und Anpassung*, ed. Dieter Hoffmann and Mark Walker (Weinheim: Wiley-VCH Verlag, 2007), 91–138, esp. 94.

[90] Schaefer to the Kurator der Universität und der Technischen Hochschule Breslau, 22 June 1933, GSPK, I HA Rep 76 Kultusminsterium Va Sekt 4 Tit IV no. 51.

Network Members' Fields of Physical Research

Let us now turn to the question of connections between political and ideological positions, on the one hand, and orientations within physical research, on the other. Within the latter sphere we might include issues such as the exercise of influence over appointments and journal publications, as well as involvement in organisations dedicated to the financial support of physical research and thus guided it in specific directions. For example, Wilhelm Wien attained a key position in the 'Helmholtz-Gesellschaft', which had been founded with the help of conservative-reactionary Rhineland heavy industry and primarily supported experimental projects including technology.[91] Beyond matters of this sort, we may ask whether any direct impact on the scientists' own research is evident. Such an impact would contradict the self-conception of natural science, which prides itself on a methodology in which subjectivity has no place, at least in principle. However, exactly in regard to the interpretations of theoretical physics, some leeway existed and still exists in which ideological positions can play a role. In the period under study, the theory of relativity and the quantum theory became matters onto which controversial ideological debates were projected. This circumstance enables us to examine whether a connection between political thought and scientific practice for the fields of research affected by these theories really existed and, in cases where this connection can be confirmed, to investigate the character of the connection.

Wilhelm Wien was one of the few remaining universalistic physicists with competence in experimental as well as theoretical physics. His scientific activities demonstrate clearly how evaluations and directions in physics were influenced, perhaps even changed, by political-ideological attitudes. In 1922, he believed that physics was in the midst of a crisis due to the questions posed by the general theory of relativity and by the internally self-contradictory statements of quantum theory. Planck, in response, pointed out to him that a completely new situation had arisen and, as a result, the earlier criteria of analysis could no longer claim to be unconditionally valid:

> It is true, physics used to be simpler, more harmonious, and therefore also more satisfying. One had beautiful theories and could trust them. Today, things are different. New ideas have appeared, not as a superfluous luxury but as inexorable consequences of new facts, and the old views can no longer be upheld without any revision.[92]

---

[91] Paul Forman, 'The Helmholtz-Gesellschaft: Support of Academic Physical Research by German Industry after the First World War', unpublished manuscript; Paul Forman, 'The Financial Support and Political Alignment of Physicists in Weimar Germany', *Minerva* 12 (1974): 39–66, esp. 60–64.
[92] Planck to W. Wien, 13 June 1922, WWPB.

Wien balked at such views. Yet his attitude toward the general theory of relativity appears to have been ambiguous. In 1913, 1918 and 1919, he had proposed Lorentz and Einstein together for the Nobel Prize, praising in particular the explanation of the perihelion precession of Mercury as a major success of the generalisation to accelerated systems.[93] After the war Wien changed his mind. On the recommendation of Lenard, he even received the anti-relativist Weyland in August 1920. However, he subsequently warned Lenard that Weyland, whom he had not known previously, was a 'hothead' who could 'harm [their matter] by exaggerated anti-Semitism'.[94] Quite different from Lenard, Wien did not reject the general theory of relativity in principle. However, he was no longer willing to consider the existing observations of light's deflection as sufficient verification.[95] By this time, the topic had left the confines of academia and become a political issue. Initially, Wien limited his open opposition to supposed transfers of the idea of relativity to areas outside of physics. Controversial public discussion seemed increasingly pointless to him. Therefore, he recommended, 'not to make oneself into a supporter or an opponent of the theory, but to regard it in the only manner which is appropriate to science, namely, as a way to recognize the characteristics of the laws of nature, [a way] which can be true or false'.[96] But evidently Wien did not remain as neutral as he appears here. His scepticism must have grown, perhaps even into outright dislike. Otherwise, the ether drift experiments of the American physicist Dayton Clarence Miller, which in the summer of 1925 seemed to confirm the existence of the ether, would hardly have been sufficient for Wien to reject both theories of relativity so readily: 'should the accuracy of the Miller's experiments be verified, of which there is little doubt, then the theory of relativity, the special as well as the general, would be done with'.[97]

Wien had a difficult time with the development of the quantum theory, as did so many of his generation. In the face of the 'overrunning' [*Überwucherung*] of atomic physics by statistics, he felt that one of the theoretical foundations on which he relied was deeply shaken. Wien feared that physicists were in danger of losing sight of the 'ultimate aim of theoretical physics, the understanding of finite causal connections'.[98] Because experiment indicates, after all, where causality exists, it appeared to him impossible 'that processes in nature that are accessible

---

[93] W. Wien to the Nobelkomitee für Physik, 16 January 1918, WWPM.
[94] W. Wien to Lenard, 6 August 1920, WWPM.
[95] W. Wien, *Die Relativitätstheorie vom Standpunkt der Physik und Erkenntnislehre* (Leipzig: Ambrosius Barth, 1921), 16–17.
[96] Ibid., introduction.
[97] W. Wien to Schrödinger, 16 September 1925, WWPM.
[98] Wien, *Die Relativitätstheorie* (ref. 95), 9.

to physical observation and measurement should permanently elude the representation of their causal connection in the form of natural laws'.[99] Although Wien esteemed Niels Bohr, he was sceptical of Bohr's readiness to abandon the existing foundations of physics.[100] He deeply mistrusted the approach of Werner Heisenberg because that exclusively theoretical-oriented physicist had, in Wien's opinion, 'not the slightest idea about experiments'.[101]

To Wien, therefore, and all those who shared his view of the 'disagreeable state of theoretical physics', the theory of Schrödinger opened up new prospects.[102] In August 1926, Max Wien wrote his cousin Wilhelm that Schrödinger's ideas had brought about 'a kind of redemption'.[103] This topic held a firm grip on Wilhelm Wien. To the very end, he continued to maintain 'that the elementary processes ... can be determined by causal laws'.[104]

While Wien remained critical of general relativity and quantum theory, his project of a comprehensive handbook of experimental physics, which he as editor almost single-handedly carried forward starting in 1924, was essentially focused on returning to experimental facts. Although the *Handbuch der Experimentalphysik* was not directly conceived to counter the modern theories, it certainly competed with the *Handbuch der Physik* published by Springer.[105] Nearly all the authors in Wien's handbook originated from German-speaking areas. In a letter of Erwin Schrödinger we learn that this handbook received a nickname:

> Your letter reassured me to a degree because in local Jewish circles — I move in such almost exclusively(!) — news had spread that the "Aryan" handbook (as I heard it jokingly called by another party) was about to go to pieces. All else aside, I would have been deeply depressed if we had already reached the point that a larger enterprise within our science could no longer be undertaken without a pronounced predominance of the Jewish element.[106]

Wilhelm Wien's reactionary political convictions thus stand in direct connection to his reservations about modern scientific theories. This is illustrated particularly by

---

[99] W. Wien, *Vergangenheit, Gegenwart und Zukunft der Physik: Rede gehalten beim Stiftungsfest der Universität München am 19. Juni 1926* (Munich: Hueber, 1926), 11.
[100] W. Wien to Schrödinger, 23 October 1926, WWPM.
[101] Ibid.
[102] W. Wien to Schrödinger, 23 May 1925, WWPM.
[103] M. Wien to W. Wien, 22 August 1926, WWPM.
[104] W. Wien and C. Müller, 'Wärmestrahlung', in *Handbuch der Experimentalphysik*, vol. 9.1, ed. W. Wien (Leipzig: Akademische Verlagsgesellschaft, 1929), 345–484, esp. 391.
[105] The demarcation between classical and modern physics is by no means clear. See Richard Staley, 'On the Co-Creation of Classical and Modern Physics', *Isis* 96 (2005): 530–58.
[106] Schrödinger to W. Wien, 17 September 1925, WWPM.

the fact that he changed his evaluation of the theories of relativity after the end of the war without the emergence of new evidence that would have provided him with a scientific basis for doing so. At the start of his career, Wien's work had been chiefly theoretical and he still harboured the aspiration to understand the entire field of physics. In his function as an organiser of the scientific enterprise (editor of *Annalen der Physik* and a handbook), as well as in lectures and review articles, he occupied an extraordinarily broad field of topics in which he actively grappled to the last with fundamental problems of physics. In Wilhelm Wien's case, ideological and political attitudes were clearly manifested within scientific research. Yet this broad spectrum of scientific work makes Wien singular in the circle of physicists studied here.

In the case of Friedrich Krüger, an obituary written by his colleagues allows us to surmise that he deliberately avoided topics associated with quantum theory: '[He] could not take any keen interest in the most recent development of our science and in any case did not decisively intervene here'.[107] As a student of Walther Nernst, Krüger was particularly active in fields straddling the border between physics and physical chemistry. With this background, it would have seemed logical to pursue the connection to quantum theory established by his teacher. This justifies the impression that, on the contrary, Krüger instead tried to avoid this connection in his own research.

The biographies of Lenard and Stark have already been treated extensively elsewhere. Here it is necessary only to point out again that both men not only adhered to the extreme political positions described above but also adamantly rejected the modern physical theories.[108]

Three further members of the network cannot be analysed in detail, due on one side to a lack of sources, and on the other to the fact that their research scarcely overlapped with modern theories. Himstedt, the eldest member of the group, dealt primarily with electrical phenomena, X-rays and radioactivity. During the Weimar Republic, he was no longer very active scientifically apart from some lectures on astronomical topics. Max Wien and Jonathan Zenneck worked mainly in the area of applied and technical physics. Other than some historical and biographical subjects, Zenneck's almost 200 publications treated only such topics.[109] These men were not, or in Himstedt's case, were no longer, required to take a stand

---

[107] Otto Reinkober and Rudolf Seeliger, 'Friedrich Krüger', *Physikalische Zeitschrift* 41 (1940): 480–85.

[108] A classic in this regard is Beyerchen, *Wissenschaftler* (ref. 37), in which chapters are dedicated to Stark and Lenard.

[109] On Zenneck, see Walter Dieminger, *Jonathan Zenneck*, Deutsches Museum, Abhandlungen und Berichte 29 (Munich: Oldenbourg, 1961), 1–44; S.L. Wolff, 'Jonathan Zenneck als Vorstand im Deutschen Museum', in *Das Deutsche Museum in der Zeit des Nationalsozialismus*, ed. Elisabeth Vaupel and Stefan L. Wolff (Göttingen: Wallstein-Verlag, 2010), 78–126.

on the new theoretical foundation of physics for their own research. Their scientific preferences had been implicitly articulated in the above-mentioned controversies, but are not immediately evident in their own research.

For the remaining two men, Mie and Schaefer, their political and physical positions cannot be clearly pinpointed across the entire period from World War I to 1933. Their scientific interests were broader than those of Himstedt, Zenneck and Max Wien and included theoretical physics, even though their main professional appointments were in experimental physics. Mie's PhD dissertation had taken up a mathematical topic; his chief area at first had been in theoretical physics. The Greifswald chair, which he occupied starting in 1902, conferred upon him the responsibility for experimental physics as well. Usually his students worked experimentally, while Mie attempted in his research to establish links between theoretical and experimental physics.[110] His work on a theory on matter led him to a theory of relativity for gravity, which resulted in personal contacts with Einstein during 1913–20. He seemed open to new developments in quantum theory. Mie had a very religious background and increasingly avowed a worldview influenced by Protestantism.[111] The interaction between Mie's political-ideological attitude and his scientific activity escapes simple categorisation. Nor does this appear possible in the case of Schaefer. As an experimentalist, he worked chiefly in the area of optical phenomena. Through his teaching of theoretical physics, which resulted in a multi-volume textbook, he analysed modern theories without any recognisable ideological reservations.[112]

Conclusion

The network of reactionary physicists formed in response to the dominance of the 'Berliners' in the DPG and to what was seen as insufficient consideration of German achievements in the scholarly literature. These issues indeed possessed the capacity to mobilise a majority of opinion among physicists for a limited period, which enabled the network to carry off organisational reforms within the DPG. In their coordinated activities in the internal politics of the discipline, including professional appointments, the members of the network also articulated their similar preferences in research.

---

[110] According to Mie's autobiography, *Aus meinem Leben*, quoted in Kohl, *Relativität* (ref. 58), 18.
[111] Gustav Mie, *Naturwissenschaft und Theologie* (Leipzig: Akademische Verlagsgesellschaft, 1932). Gustav Mie, *Die Denkweise in der Physik und ihr Einfluß auf die geistige Einstellung des heutigen Menschen* (Stuttgart: Enke, 1937).
[112] See Clemens Schaefer, *Einführung in die theoretische Physik*, vol. 3.2, *Quantentheorie* (Leipzig: Walter de Gruyter, 1937).

In this essay, the question was posed whether these physicists' political-ideological attitudes, expressed in part via their assessments of the theory of relativity and quantum theory, had a direct impact on the scientific work — a more narrowly defined question when compared to Forman's overall thesis. The findings leave us with a rather differentiated answer. In the case of four physicists from the nine-member network (Wilhelm Wien, Lenard, Stark and Krüger), the answer is affirmative. For the oldest physicist in the network (Himstedt), the question was no longer pertinent to the topics with which he dealt. Likewise, attitudes toward fundamental theoretical problems, as far as they existed, had no impact on the wide field of applied and technical physics (Max Wien and Zenneck). Furthermore, it is not always possible to condense complex ideological views and changing political orientations into unequivocal statements (Mie and Schaefer). This shows not that Forman's thesis is false, but that this way of posing the question does not always apply.

At issue is not a simple generational divide in which the older physicists fell out of touch with modern developments. Several of the important physicists from Berlin, such as Planck and Nernst or the presidents of the Physikalisch-Technische Reichsanstalt Emil Warburg and Friedrich Paschen, belonged to the same age groups as those in the network. They were quite receptive to the new developments in physics. But politically, they represented the more liberal element among German physicists.[113]

While the network's successes contributed significantly to the politicisation of German physicists, many times it failed as well. Often, the network was on the defensive with regard to appointments. This prompted Max Wien to lament in 1926 that the influence exerted by Berlin and Göttingen in this regard reached even to southern Germany.[114] Just as unsuccessful was the attempt to return Stark to a university post, and another setback was the failed effort to put forward a successor for Lenard who could have hoped to gain the approval of the faculty. Only by refusing to retire for a few more years was Lenard able to maintain the status quo. This reveals an important deficit of the network. Since its members

---

[113] Although Planck had some reservations about quantum theory, these were not of the same calibre as those expressed by the opposition in the network. On Nernst, see Diana K. Barkan, *Walther Nernst and the Transition to Modern Physical Science* (Cambridge: Cambridge University Press, 1999). On Warburg, see S.L. Wolff, 'Emil Warburg – mehr als ein halbes Jahrhundert Physik', *Physikalische Blätter* 48, no. 4 (1992): 275–79. On Paschen, see Edgar Swinne, *Friedrich Paschen als Hochschullehrer* (Berlin: D.A.V.I.D. Verlagsgesellschaft, 1989). Gehrcke pointed out in a derogatory way that Paschen had signed the election petition of Willy Hellpach, the presidential candidate of the liberal Deutsche Demokratische Partei: Gehrcke to Stark, 8 October 1925, JSP.

[114] M. Wien to W. Wien, 17 January 1926, WWPM.

were comparatively distant from the developments in modern physics, in part because of the lack of theorists among them, the network was not in a position to sufficiently extend its organisational strength to professional research issues. The network of reactionary physicists lacked a scientific perspective on the future, which explains its dwindling importance by the end of the Weimar Republic.

The bogeymen perceived by the network disappeared after 1933 with the National Socialist regime. The republic had ceased to exist; the antagonism between Berlin and the rest of the country had died out and after the expulsion of so-called non-Aryans, academic anti-Semitism lost its target. The network of reactionary physicists, which had to cope with the loss of its leading light when Wilhelm Wien died in 1928, would have needed a new basis to justify its existence. But the question was no longer relevant anyway because a split had occurred in the small group. Lenard and Stark decided to dedicate themselves to a '*Deutsche Physik*' (literally, 'German physics') and met with the adamant resistance of their former allies, Max Wien and Zenneck.[115]

Paul Forman has pointed out many times the close correlation between scientific and political positions in the Weimar Republic. Even though it cannot be shown that each member of the network rejected modern physical theories, they all participated in the political-professional activities that favoured and fostered other directions in research. To this extent, the existence of this network confirms Forman's thesis.

Acknowledgments

I thank Michael Eckert, Ulf Hashagen and Dieter Hoffmann for helpful discussions and Paul Forman especially for his help and advice.

---

[115] On 'German physics', see, for example, Michael Eckert, 'Die Deutsche Physikalische Gesellschaft und die 'Deutsche Physik', in Hoffmann and Walker, *Autonomie und Anpassung* (ref. 89), 139–72. Despite his opposition to 'German physics', Zenneck remained true to his earlier views when he wanted to keep Sommerfeld off the *Zeitschrift für Physik* editorial staff in 1936 because of Sommerfeld's earlier 'pro-Semitic activity'. See Hoffman and Walker, *Autonomie und Anpassung* (ref. 89), 152.

Philosophical Rhetoric in Early Quantum Mechanics 1925–27: High Principles, Cultural Values and Professional Anxieties

*Alexei Kojevnikov**

'I look on most general reasoning in science as [an] opportunistic (success- or unsuccessful) relationship between conceptions more or less defined by other conception[s] and helping us to overlook [danicism for "survey"] things.'

Niels Bohr (1919)[1]

This paper considers the role played by philosophical conceptions in the process of the development of quantum mechanics, 1925–1927, and analyses stances taken by key participants on four main issues of the controversy (*Anschaulichkeit*, quantum discontinuity, the wave-particle dilemma and causality). Social and cultural values and anxieties at the time of general crisis, as identified by Paul Forman, strongly affected the language of the debate. At the same time, individual philosophical positions presented as strongly-held principles were in fact flexible and sometimes reversible to almost their opposites. One can understand the dynamics of rhetorical shifts and changing strategies, if one considers interpretational debates as a way

---

* Department of History, University of British Columbia, 1873 East Mall, Vancouver, British Columbia, Canada V6T 1Z1; anikov@interchange.ubc.ca.

The following abbreviations are used: AHQP, Archive for History of Quantum Physics, NBA, Copenhagen; *AP, Annalen der Physik*; *HSPS, Historical Studies in the Physical Sciences*; NBA, Niels Bohr Archive, Niels Bohr Institute, Copenhagen; *NW, Die Naturwissenschaften*; PWB, Wolfgang Pauli, *Wissenschaftlicher Briefwechsel mit Bohr, Einstein, Heisenberg a.o.*, Band I: 1919–1929, ed. A. Hermann, K.V. Meyenn, and V.F. Weisskopf (New York: Springer, 1979); RKZ, Ralph Kronig Nachlass, Eidgenössische Technische Hochschule, Zurich; *SHPS, Studies in the History and Philosophy of Science*; WWM, Wilhelm Wien Nachlass, Deutsches Museum, Munich; *ZP, Zeitschrift für Physik*.

[1] Niels Bohr to C.G. Darwin, draft of a presumably unsent letter, around July 1919, NBA.

of justifying property claims over the emerging new revolutionary theory, at a time when its major concepts were *in statu nascendi*, while personal relations and institutional hierarchies between its major contributors were still developing and negotiable.

Ehrenfest, Darwin, and Pauli on the Forman Thesis

In a June 1919 letter to his friend and colleague Niels Bohr, Paul Ehrenfest, professor of theoretical physics at the University of Leiden, commented insightfully on changes in the European intellectual climate after World War I:

> [I]t is remarkable that precisely here, in the circles of men having much to do with technology, production, industry, patents etc., opinions develop so uniformly about perspectives of culture. Overall there is building up an uncannily intensive reaction *against rationalism* ... If I am not entirely mistaken, in the next 5–10 years we will see the following happening at the institutes of higher learning (including technical!). Professors raised as relatively *rational* and disciplined individuals will despairingly and uncomprehendingly face the complaints and demands of a relatively "*mystical*" student body. At the same time, scientifically less clear but personally warmer teachers will gain the main influence over students.[2]

Ehrenfest's observation remarkably and almost literally supports the core claim of Paul Forman's article of 1971 that, in the immediate wake of World War I, a strong wave of reaction against rationalism and favouring more mystical lines of thought swept through not just the intellectual public in general, but also such professionals as engineers and exact scientists previously expected to strongly resist such trends.[3] It is worth noting that the letter was from a physicist in the Netherlands to his colleague in Denmark, signifying the mood did not remain confined to Germany and Austria, the countries who had lost the war, but also affected at least the neighbouring neutral countries. Although not abandoning his personal rationalistic convictions, Ehrenfest appeared to defer to the opinions of the younger, as academics often do with the newest intellectual fashions.

Ehrenfest's letter also contains a strikingly self-conscious recognition and expectation that professors would adapt knowingly, rather than unreflectively, to

---

[2] Paul Ehrenfest to Niels Bohr, 4 June 1919, NBA (emphasis in the original).

[3] Paul Forman, 'Weimar Culture, Causality, and Quantum Theory, 1918–1927: Adaptation by German Physicists and Mathematicians to a Hostile Intellectual Environment', *HSPS* 3 (1971): 1–115.

the direction of the prevailing intellectual wind. He did not state it explicitly in this letter, but his other correspondence of the time reveals quite clearly that, in his own field of theoretical physics, he admired and regarded Bohr as precisely the kind of professor who would resonate with and inspire the younger generation of students.[4] The quoted description, indeed, sits well with characteristic hagiographic references in existing recollections regarding the way Bohr's charisma influenced younger students. Bohr appears there as a kind of philosophical guru, whose thoughts were too profound to be understood or even expressed clearly, but this only helped them to be tremendously inspiring. Whether or not Ehrenfest's letter thus contained implicit advice to Bohr, and whether or not Bohr accepted the hint or arrived at similar ideas on his own, around the same time he was already inclined 'to take the most radical *or rather mystical* views imaginable' with regard to the daunting problem of the quantum interaction of matter and radiation. That much Bohr admitted himself while composing a reply to his British friend and colleague Charles Galton Darwin.[5]

Darwin's letter and his enclosed manuscript suggested several possible ideas of this kind, including acausality. Darwin 'felt that the fundamental basis of physics is in a desperate state. The great positive successes of the quantum theory have accentuated ... also the essential contradiction on which it rests.' As a remedy, he proposed 'to knock away the props of classical physics one by one and find, after a particular one has been removed, that our difficulties have become reconciled. It may be that it will prove necessary to make fundamental changes in our ideas of time and space or to abandon the conservation of matter and electricity or even as a last forlorn hope to endow electrons with free will.' Personally, Darwin preferred to think that 'contradictions in physics all rest on the exact conservation of energy' and his favourite solution thus involved 'denying that conservation is anything more than statistical.'[6] Four years later, this acausal hypothesis would become Bohr's preferred choice, too, but in 1919 he was not yet ready to decide which, if any, of the mentioned 'mystical' hypotheses could be endorsed.

Ehrenfest's 1919 formulation of the Forman thesis *avant la lettre* makes an important addition to it by specifying the effective milieu whose demands made

---

[4] For example, Ehrenfest to A.F. Joffe, 6 September 1920, in *Erenfest–Ioffe: Nauchnaia Perepiska, 1907–1933* (Leningrad: Nauka, 1973), 146–150.

[5] Bohr to Darwin, July 1919, NBA (ref. 1): 'or rather mystical' is inserted into the sentence above the line. *Cf. also* Bohr to Ehrenfest, 22 October 1919, NBA.

[6] Darwin to Bohr, 20 July 1919, with an enclosed manuscript 'A Critique of the Foundations of Physics', NBA.

professors adapt — students in auditoriums. Indeed, students may well have been the immediate audience many professors cared about the most. In this respect, however, Bohr's situation was rather peculiar and differed from Ehrenfest's. From early on in his professorship at the University of Copenhagen, Bohr delegated the task of giving general lectures to his assistants and did not teach undergraduate students. With only a couple of exceptions, he did not train doctoral students either. Almost all of the young physicists who worked with him received their degrees elsewhere and usually came to his institute as postdoctoral fellows. Arguably the most outspoken among them, Wolfgang Pauli, has also left a well-documented manuscript record that reflects the development of his views on quantum acausality.

Pauli's probably earliest extant comment on the unpredictability of quantum behaviour can be found in his letter written in Copenhagen in June 1923 to Arnold Sommerfeld in Munich. The remark, however brief, is quite provocative, since it compared Pauli's personal feeling of insecurity and professional anxiety with the uncertainty of a microscopic object, thus suggesting a possible additional motivation for the development of acausal thought. The letter primarily dealt with Sommerfeld's proposal that Pauli obtain his Habilitation and become a *Privatdozent* at the University of Munich. Pauli declined the tempting, if belated, offer from his former teacher and doctoral adviser, hoping instead that Bohr, upon returning from an American trip, would invite him again for another stay in Copenhagen. He did understand, however, that tiny Denmark could offer him only temporary support but no long-term professional perspective:

> I will certainly not be able to remain here [in Copenhagen] forever and must sooner or later habilitate at one of the German universities ... Bohr's [possible] offer, however, makes me leave the question of my Habilitation open, for now ... The only thing certain is that I will still spend the coming semester in Hamburg ... *What happens later I know as little as an electron knows in advance where it will jump in 10^{-8} seconds* (I have only described the forces deflecting me from Munich, but ... of course, very strongly attractive forces come from Munich as well).[7]

At the moment of writing, Pauli's professional future looked very uncertain indeed. A year earlier he had accepted a temporary appointment in Copenhagen to help Bohr write papers in German, which nearly coincided with an abrupt collapse of the German mark and the start of hyperinflation. Pauli was unable even to

---

[7] Wolfgang Pauli to Arnold Sommerfeld, 6 June 1923, in PWB, p. 94 (emphasis added).

purchase a railway ticket to travel to Copenhagen from Hamburg and had to ask Bohr to send him an advance in Danish crowns.[8] Erwin Schrödinger, who had just recently obtained a professorship in financially secure Switzerland, congratulated Pauli on his escape and suggested that he not return to Germany in the foreseeable future.[9] They understood each other as fellow Austrians: both knew how hyperinflation had decimated science in their home country immediately after World War I and both had then left Austria to seek better career opportunities in the still relatively stable economy of Germany. Now that the financial collapse had caught up with Germany, too, both anticipated the worst on the basis of their Austrian experiences and sought professional appointments elsewhere.

Returning to Hamburg after his Copenhagen year, Pauli was relieved to discover that his worst fears had not materialised. Indeed, inflation did not damage science in Germany as severely as it had in Austria. Research and publishing continued at a fast pace, even if a prohibitive exchange rate had destroyed many possibilities for foreign travel and subscriptions to foreign publications. Scientific infrastructure and laboratories built during the imperial period were still far better than anywhere else in Europe. Salaried professors maintained tolerable incomes. The negative effects could be described as structural rather than outright destructive. Arguably in the most difficult situation were younger academics, like Pauli, in a career stage between doctorate and first professorial appointment. Formerly they had typically taught at universities as independent *Privatdozenten*, but inflation made this professional class almost extinct, or at least at a practical level indistinguishable from lower assistants, as it was no longer possible to sustain one's livelihood on 'soft money' such as students' fees.[10]

Pauli's metaphorical comparison of the uncertainty of his career trajectory with that of a quantum particle is very appropriate. Formerly, the professional paths of younger academics in Germany resembled trajectories of classical particles determined by the power fields of their professors. Acting in this fashion, Max Born asked for and received Sommerfeld's permission earlier in 1923 to hire the latter's student Heisenberg for a year. In a similar deal, Pauli also worked for Born in Göttingen during the previous year, and as of January 1923 both professors

---

[8] Pauli to Bohr, 5 September 1922, and Bohr to Pauli, 8 September 1922, both in PWB, pp. 63–4.
[9] Erwin Schrödinger to Pauli, 8 November 1922, in PWB, p. 69.
[10] Pauli to Bohr, 16 July 1923, in PWB, p. 102. For more on the effects of inflation on academic life and the fate of *Privandozenten*, see Paul Forman, *The Environment and Practice of Atomic Physics in Weimar Germany: A Study in the History of Science* (PhD dissertation, University of California, Berkeley, 1967), 206–37. For observers' remarks on the disappearance of *Privatdozenten*, see Alexi Assmus, 'The Creation of Postdoctoral Fellowships and the Siting of American Scientific Research', *Minerva* 31 (1993): 151–83, on 178.

assumed Pauli would eventually return to Munich as Sommerfeld's *Privatdozent*.[11] Had Sommerfeld actually made such an offer to him earlier, his invitation would have been hard to decline, but in the meantime inflation had set in and undermined the resources of German professors, sending Sommerfeld, and later also Born, to visiting professorships in the United States as a way to improve personal finances. Younger academics had a much harder time surviving the intermediate state of negative finances, which was what *Privatdozent* became. Some effectively worked for their professors as lower-level assistants. Others, like Pauli, were able to become quasi-independent, abandon the notion of determined classical paths and experiment with less predictable professional jumps via metastable states such as temporary positions or postdoctoral fellowships abroad.

Metaphors are dangerous, however, as they encourage reasoning in both directions. Pauli's anxiety over his personal career at the height of the German crisis could quite plausibly make him more inclined to think of electrons in similar terms and by 1925 arrive at an important conclusion that the notion of their classical trajectories inside atoms should be completely abandoned, which in turn inspired Heisenberg onto the path leading to the new quantum mechanics.[12] No matter how tempting, however, one cannot assume a direct transition from Pauli's indeterministic remark of 1923 to his advocacy of probabilistic quantum mechanics three years later. The problem is that in the meantime he also made contradictory pronouncements in no uncertain terms: 'I definitely believe that *the probability concept should not be allowed in the fundamental laws of a satisfying physical theory.* I am prepared to pay any price for the fulfilment of this desire, but unfortunately I still do not know the price for which it is to be had'. The above declaration of faith sprang not from the pen of Einstein in his 'God does not throw dice' mood, but from Pauli writing to Bohr in November 1925.[13] It contradicts much of what Pauli is otherwise known for, but at the time was made as seriously and sincerely as, later on, he would express his probabilistic convictions.

We can understand this quote in the context of its precise timing and reference. Pauli was describing to Bohr the promise of Heisenberg's new matrix mechanics but obviously taking critical aim at, without the need to mention explicitly, Bohr's earlier failed attempt to introduce the acausal principle into the foundations of quantum theory — the Bohr-Kramers-Slater theory of 1923–24. The subversive idea that energy in atomic interactions may be conserved only

---

[11] Max Born to Sommerfeld, 5 January 1923, quoted in Jagdish Mehra and Helmut Rechenberg, *The Historical Development of Quantum Theory*, vol. 2 (New York: Springer, 1982), 73.

[12] For a detailed account of the development of Pauli's ideas, see John Hendry, *The Creation of Quantum Mechanics and the Bohr-Pauli Dialogue* (Dordrecht: Reidel, 1984).

[13] Wolfgang Pauli to Niels Bohr, 17 November 1925, in PWB, p. 260 (emphasis added).

statistically, rather than strictly, had been discussed earlier by Darwin and a few other authors in letters and unpublished manuscripts. By 1923, Bohr's opposition to the rising popularity of light quanta developed to the point of desperation when he became ready to accept and endorse the idea in a formal publication, and endure the associated risks. Once Pauli left Copenhagen for Hamburg and was no longer directly employed by Bohr, he could not hold back anymore with his opposition to the 'reactionary Copenhagen putsch'. To his relief, by 1925 experimentalists decisively refuted the Bohr-Kramers-Slater proposal.[14] Feeling vindicated in his devotion to the strict validity of the law of energy conservation, Pauli was also ready, as the above quote shows, to reject the fundamentality of the probabilistic approach altogether.

His views would change once again in less than a year. In the summer of 1926, Pauli introduced the probabilistic understanding of Schrödinger's wave function and remained ever after a proponent of the statistical interpretation of quantum mechanics. Pauli's flip-flops on the issue of causality, however, convey an important lesson, namely, that philosophical pronouncements of quantum physicists, no matter how strongly expressed, should not be taken as general and long-term commitments, but as context-dependent and flexible. As a matter of fact, such drastic shifts on fundamental issues and principles were not characteristic of Pauli alone, and not only with regard to the question of causality. Rather, they can be regarded as a distinctive feature of the early quantum philosophy in general.

The Problem with Quantum Philosophy

Not long before his death in 1962, Bohr confessed to Thomas Kuhn that he had hardly any hope of achieving an understanding between quantum physicists and philosophers. He expressed the complaint in, for Bohr, unusually strong and categorical terms: 'I think it would be reasonable to say that no man who is called a philosopher really understands what one means by the complementarity description'.[15] As if they were aware of this charge, philosophers retaliated some 30 years later in a volume devoted to the assessment of Bohr's contribution to philosophy. In equally strong words, Don Howard expressed doubts 'whether or not Bohr's philosophy of physics can be given a coherent interpretation'. As Howard summarised the problem, 'There was a time, not so very long ago, when Niels Bohr's influence and stature as a philosopher of physics rivalled his standing as a physicist. But now

---

[14] Pauli to Hendrik Anthony Kramers, 27 July 1925, in PWB, p. 234.
[15] Bohr, interview by Kuhn, 17 November 1962, AHQP.

there are signs of a growing despair — much in evidence during the 1985 Bohr centennial — about our ever being able to make good sense out of his philosophical views'.[16]

The *noblesse oblige* of the professional philosopher, however, did not permit Howard to give up:

> I think that the despair is premature .... What is needed at the present juncture is really quite simple. We need to return to Bohr's own words, filtered through no preconceived philosophical dogmas. We need to apply the critical tool of the historian in order to establish what those words were and how they changed over time. We need to assume, at least provisionally, that Bohr's words make sense. And we need to apply the synthetic tools of the philosopher in order to reconstruct from Bohr's words a coherent philosophy of physics.[17]

In the main part of this paper I will take up the historical part of Howard's advice and follow the twists and turns of quantum philosophy during the years 1925 to 1927, from Bothe and Geiger's refutation of the Bohr-Kramers-Slater theory and Heisenberg's first paper on matrix mechanics to the Solvay congress of 1927 and the first open disputes between Einstein and Bohr. Simultaneous with the invention of quantum mechanics itself — to which nearly 200 authors contributed over this period — a half-dozen physicists were developing competing philosophical interpretations of the not yet completed theory.[18] Ordered by age, this group included Einstein, Bohr, Born, Schrödinger, Pauli, Heisenberg and Jordan. Altogether, they expressed quite a variety of conflicting philosophical views, which can be grouped around four main issues: *Anschaulichkeit-Unanschaulichkeit* (roughly translated as visualisability-unvisualisability), continuity-discontinuity, the wave-particle dilemma and causality-acausality.

For an historian analysing these views, the main difficulty lies not in the lack or paucity of sources, but on the contrary, in their intimidating overabundance and contradictory nature. That most of the above-mentioned participants and also others

---

[16] Don Howard, 'What Makes a Classical Concept Classical? Toward a Reconstruction of Niels Bohr's Philosophy of Physics', in *Niels Bohr and Contemporary Philosophy*, ed. J. Faye and H.J. Folse (Dordrecht: Kluwer, 1994), 201–29, on 201.

[17] Ibid. *See also* Henry J. Folse, 'Niels Bohr and the Construction of a New Philosophy: Essay Review', *Studies in the History and Philosophy of Modern Physics* 26 (1995): 107–16, on 108; Don Howard, review of *Niels Bohr: A Centenary Volume*, ed. A.P. French and P.J. Kennedy, *Annals of Science* 44 (1987): 196–98.

[18] For a scientometric overview of the emerging field of quantum mechanics, see A. Kozhevnikov and O. Novik, 'Analysis of Informational Ties in Early Quantum Mechanics (1925–1927)', *Acta historiae rerum naturalium necnon technicarum* 20 (1989): 115–59.

on their behalf continued the dispute in some form for many years after 1927 further complicates the situation. They kept commenting, explaining and restating their positions, usually without acknowledging that their views continued to shift as the times and situation changed. Not only did the authoritative spokesmen of quantum mechanics disagree with each other, sometimes openly and sometimes subtly, but, as the Pauli example above has demonstrated, even the extant record of an individual prolific writer contains mutually contradictory philosophical declarations which could only be understood within the statements' short-term context.

In order to reduce unavoidable confusion and make sense of changing allegiances, the following analysis imposes two strong chronological restrictions on the use of sources. First, it generally avoids using post-1927 texts in which physicists explained and reinterpreted their earlier views, such as the famous recollections by Bohr, Heisenberg and Born. These later accounts were developed in the context of continued post-1927 disagreements over the foundations of quantum mechanics and they tend to add more contradictions than clarifications if used as sources of historical information about the earlier periods. Second, even within the period of 1925–27, I will impose a finer time scale. The theory developed so quickly that its basic principles underwent fundamental changes approximately every six months. Statements concerning its interpretation also changed at a corresponding rate and it makes little sense to use, say, Heisenberg's pronouncements of spring 1927 for the purpose of understanding what he thought and meant in the fall of 1925, or, conversely, in relying on his initial programmatic statements of 1925 as valid for the resulting mature quantum mechanics. One can, however, describe the state of quantum philosophy at a given stage of characteristic six-month lengths by using only those historical sources which come from that same time period and find, on the one hand, a sufficient number of such sources, and on the other, a significant reduction in contradictions among them.

The following reconstruction depends very heavily on the wealth of the historiography of quantum physics and the insightful observations and interpretative ideas from the existing literature on the topic. Let me mention at the outset only those studies to which I owe the most. Paul Forman in several papers, including the classic 'Weimar Culture, Causality, and Quantum Theory', demonstrated how the ideologically laden concepts of *Anschaulichkeit,* acausality and *Individualität* entered physicists' discourse even prior to 1925 and were subsequently ascribed to quantum mechanics.[19] In another paper, John L. Heilbron described the post-1927

---

[19] Forman, 'Weimar Culture' (ref. 3); *see also* Paul Forman, 'Kausalität, Anschaulichkeit, and Individualität, or, How Cultural Values Prescribed the Character and the Lessons Ascribed to Quantum Mechanics', in *Society and Knowledge: Contemporary Perspectives in the Sociology of Knowledge and Science*, ed. Nico Stehr and Volker Meja (New Brunswick, NJ: Transaction Books, 1984), 333–48.

spread of the Copenhagen philosophy with its characteristic 'combination of imperialism and resignation'.[20] I shall concentrate on the intermediate period in the hope of establishing a bridge between these two works. John Hendry in his book on the Bohr-Pauli dialogue presented a history of quantum mechanics in the making, along with its philosophy, from a more or less Copenhagen perspective. In contrast, Mara Beller, in a series of papers and a resulting book, developed a critique of the historical myth and of the Copenhagen orthodoxy.[21] For my reconstruction, I use many of their important findings, but also disagree on some points. The reasons for my disagreements are generally twofold — restrictions on the use of sources explained above and my neutral stance on philosophical issues. While admiring physicists' earlier and later interpretations of quantum mechanics as exciting intellectual achievements, I do not feel committed, at least for the purpose of this study, to any of their interpretations in particular.

One of the main conclusions of the analysis below may ultimately disappoint philosophers, namely that having fulfilled the first, historical part of Don Howard's advice and made some sense of the physicists' philosophical statements, it would become clear that the last, philosophical part of his proposal — i.e., to synthesise from them a 'coherent philosophy of physics' — is unrealisable. Physicists' shifting views on philosophical issues can be explained historically, in their own local times and contexts, but taken together as a set they constitute a self-contradictory body of propositions that allows for a variety of irreconcilable interpretations. Overall, the philosophical discourse of quantum physicists appears opportunistic in the sense of Niels Bohr's quote in the epigraph to this paper. Physicists made philosophical statements as if announcing strongly-held principles, but they also kept changing them rather easily, sometimes to almost the opposite in the course of a single year. Furthermore, they also used those statements as rhetorical resources in their intradisciplinary rivalry, in some cases overstating the existing differences, or downplaying and hiding them away, due to tactical reasons and personal relationships.

Mara Beller came to a similar conclusion in her argument 'against the possibility of a consistent version of the Copenhagen interpretation', namely that

---

[20] J.L. Heilbron, 'The Earliest Missionaries of the Copenhagen Spirit', *Revue d'histoire des sciences* 38 (1985): 195–230.

[21] Hendry, *Creation* (ref. 12); Mara Beller, *Quantum Dialogue: The Making of a Revolution* (Chicago: University of Chicago Press, 1999), and her earlier papers: 'Matrix Theory before Schrödinger: Problems, Philosophy, Consequences', *Isis* 74 (1983): 469–91; 'Born's Probabilistic Interpretation: A Case Study of 'Concepts in Flux', *SHPS* 21 (1990): 563–88; 'The Birth of Bohr's Complementarity: The Context and the Dialogues', *SHPS* 23 (1992): 147–80; 'Schrödinger's Dialogue with Göttingen-Copenhagen Physicists', in *Erwin Schrödinger: Philosophy and the Birth of Quantum Mechanics*, ed. Michel Bitbol and Olivier Darrigol (Gif-sur-Yvette, France: Editions Frontières, 1992), 277–306.

'philosophical pronouncements by quantum physicists are most adequately understood as local, shifting, and opportunistic'. In her view, 'numerous inconsistencies in the Copenhagen interpretation of quantum physics' cannot be explained 'on the basis of the conceptual evolution alone' but 'are of a psychosocial origin'.[22] Rather than placing the chief blame on a particular interpretation, I see in such inconsistencies a general pattern of behaviour of the entire group, regardless of the sides they took in the controversy. The point is therefore not that a particular version of quantum philosophy is unsatisfactory, but that the entire interpretational debate was something else dressed up in philosophical garb. The professional philosophers' feeling of despair comes not from the deficiency of their 'synthetic tools', but from the *a priori* assumption that some consistent and coherent doctrine was hiding behind the pronouncements by physicists. For a philosopher, dropping this assumption would amount to admitting that the discourse was not philosophical in the strict sense. In the conclusion, therefore, I will have to switch the mode of analysis from history of ideas to cultural history in order to understand what kind of activity it was, if not philosophical.

Matrix Mechanics (Fall 1925)

Familiar concepts and images of classical physics were not faring well in the atomic domain. In quieter and more positive times scientists could have remained more tolerant of the developing contradictions, but those who shared the existential experiences of life in Europe during the second decade of the twentieth century were accustomed to seeing crises and revolutions in every venue of life, including science. Often they were more willing than reluctant to read existing problems as signs of foundational crises.[23] The quantum theory of the atom developed since 1913 by Bohr and Sommerfeld with co-workers indicated a radical solution for one such crisis at the price of revising some basic and proven postulates of classical mechanics and electrodynamics. After spectacular successes in understanding and calculating atomic spectra of hydrogen, the theory also encountered problems, in particular, in attempts to generalise it to the case of multi-electron atoms. Again, in some other epoch, ours for example, physicists would have been more inclined to see the glass as half-full rather than half-empty, or at least allow the adolescent theory a little more time to prove itself. In the radical 1920s, however, revolutionary proposals themselves, and not just traditional

---

[22] Mara Beller, 'The Rhetoric of Antirealism and the Copenhagen Spirit', *Philosophy of Science* 63 (1996): 183–204, on 183, 185; and Mara Beller, 'The Conceptual and Anecdotal History of Quantum Mechanics', *Foundations of Physics* 26 (1996): 545–57, on 545, 550.

[23] On 'crisis' in science, see Forman, 'Weimar Culture' (ref. 3), 26–29.

beliefs, were subject to heightened degrees of criticism. By 1924, circles of physicists around Bohr in Copenhagen and Max Born in Göttingen came to the conclusion that the quantum theory of the atom, too, no matter how young and radical, had entered a state of serious crisis.

To find another radical solution, they were prepared for further sacrifices in the most basic principles of physics. 'Most basic' to them meant philosophical, and being 'philosophically minded' constituted a praise within this circle. It was not quite obvious, however, what exactly had to be sacrificed. The list of possible and tried victims included, but was not limited to: 1) ideas of space and time, 2) energy conservation, 3) causal description, 4) the concept of electromagnetic field and 5) continuity of kinematics.[24] After a number of unsuccessful attempts, they found much promise in a July 1925 paper by Heisenberg[25] and collaborated on the theory, which became known as matrix mechanics. It existed in its original form until the beginning of 1926 with its own characteristic set of philosophical preferences.

Unanschaulichkeit. The first and most distinctive on the list, as demonstrated by Beller's analysis,[26] suggested abandoning the usual ideas about space and time. Our common visual intuitions, one could argue, relied on human experiences in the macroscopic world with objects roughly the size of our own, but did not have to remain valid within the microscopic domain of the atom. Trying to make sense of atomic phenomena with the help of such inadequate intuitive visual (*anschauliche*) representations could be the chief source of contradictions encountered within the quantum theory of the atom. Different formulations of this idea were provided by Bohr (complete space-time representation of atomic processes is impossible), Born (geometry fails within the atom), Heisenberg (positions and trajectories of the electron in the atom do not exist) and Pauli (abandonment of the mechanical, spatial-temporal representation of the stationary state of the hydrogen atom). To build a new theory from the ground up, it had 'first to throw away visual representations of the atom', the *Anschaulichkeit*.[27] Not necessarily rejoicing about this feature, Heisenberg, Pauli, Born, Jordan and Dirac accepted *Unanschaulichkeit* as the basic and necessary premise of the new theory.

---

[24] See Hendry, *Creation* (ref. 12), 20, 29, 31, 33, 36, 37, 55 and 64, for relevant quotations from various authors; *see also* PWB, and Darwin to Bohr, 20 July 1919, NBA (ref. 6).

[25] Werner Heisenberg, 'Über quantentheoretische Umdeutung kinematischer und mechanischer Beziehungen', *ZP* 33 (1925): 879–93.

[26] Beller, 'Matrix Theory' (ref. 21).

[27] Pauli to Bohr, 12 December 1924 and 17 November 1925, both in PWB, pp. 188, 260; Werner Heisenberg, "Über quantentheoretische Kinematik und Mechanik", *Mathematische Annalen* 95 (1926): 683–705, on 684.

Discontinuity. In matrix mechanics, the atomic world was *unanschaulich* in large part because of its fundamental discontinuity. 'In processes at microscopic dimensions of space and time, a discontinuous element plays the dominant role', which could not be adequately expressed and represented with the usual, continuous space-time conceptions.[28] Matrix mechanics inherited not only the discontinuous energy states of Bohr's early atomic theory, but also Born's 1924 program of a 'truly discontinuous theory' which proposed to consistently replace all continuous physical concepts with discrete sets. In matrix mechanics, the transition from classical to quantum theory was achieved accordingly by substituting continuous variables with discrete matrices.

The following two philosophical issues did not play such a major role at the matrix mechanics stage as they had and would in some earlier and later versions of quantum theory. Their very absence is significant, nevertheless.

No waves, no particles. Since matrix mechanics and wave mechanics competed with each other, some commentators tended to assume matrix mechanics favoured corpuscular ontology over waves. Beller rightly criticised this view, but her assertion that matrix mechanics 'was thoroughly permeated by wave-theoretical concepts'[29] is equally untenable (she supported this claim mostly with quotes from the earlier period of the Bohr-Kramers-Slater theory). Both waves and particles were visual representations and thus unsuited for an *unanschaulich* theory. Only outside of the atom did radiation consist of waves while electrons were corpuscles, but in the inside, the electron and its radiation together were represented by a discontinuous and unvisualisable set of matrix elements. Neither pictures of waves nor of particles were useful for its description. The only exception to this attitude came in Born's American lectures of winter 1925, where he tried to combine, somewhat artificially, the *Unanschaulichkeit* of matrix mechanics with Einstein's wave theory of matter, and suggested (even before wave mechanics) the existence of some undulatory process within the atom.[30]

No time, no acausality, no statistics. The idea of acausality together with the statistical conservation of energy had been tried earlier in the Bohr-Kramers-Slater theory of 1923–24. Bohr turned to that risky hypothesis in a last desperate attempt to save the wave theory of electromagnetic radiation from the abhorrent (to him) notion of light quanta. At that juncture, Schrödinger welcomed the acausal idea, while Pauli and Einstein criticised it (the latter not yet doing so as a matter of philosophical principle, but because he had already tried it earlier without much

---

[28] Werner Heisenberg, 'Quantenmechanik', *NW* 14 (1926): 989–94, on 989.
[29] Beller, 'Matrix Theory' (ref. 21).
[30] Max Born, *Problems of Atomic Dynamics* (Cambridge, MA: MIT Press, 1926), 69–70.

success). Born and James Franck did not feel happy about it, either, but did not want to contradict Bohr and were trying to say something polite, if vague. Heisenberg, Bohr's formal employee during that year, appeared to accept the idea on the surface, but likely not in his heart: in his papers, he used the language and approach of the Bohr-Kramers-Slater theory, but carefully avoided its most dangerous assumption.

Refuted by Bothe and Geiger's experiment in 1925, the idea seemed to be totally discredited and did not appear in matrix mechanics at all. Pauli's comment in November 1925 to this effect, which strongly rejected the very use of probability in fundamental physical theory, has been quoted above. Heisenberg distanced his new approach from the discredited attempt by purging the very word 'probability' from his matrix mechanics papers. Instead of 'probability of [atomic] transitions', he consistently used 'intensities of emitted radiation'. The two phrases can be used interchangeably in our times, but in the context of 1925 physicists were quite sensitive to this choice of words.

Handling disagreements. The authors of matrix mechanics did not agree on some other interpretational issues. The most serious of these concerned the definition of the basic quantities of the new theory. Born defined them mathematically simply as matrix elements, thus deviating from Heisenberg's original (and not entirely satisfactory) physical definition of them as amplitudes of emitted radiation. What Heisenberg took for the most important physical postulate of matrix mechanics does not even appear in the core of the theory in Born and Jordan's presentation. They only introduce it as an auxiliary assumption ('Heisenberg's *Annahme*') for the purpose of calculating intensities of spectral lines at the very end of their paper.[31]

This discrepancy helps to explain why Heisenberg disliked Born's matrices and was unhappy about the very name 'matrix mechanics'. He contrasted his 'physical' approach to the 'mathematical' one of the Göttingen physicists and had Pauli and probably Bohr on his side. Heisenberg struggled (largely unsuccessfully) to insist on his interpretation of the theory while collaborating with Born and Jordan on the famous *Dreimännerarbeit* of November 1925. From Göttingen, he wrote to Pauli:

> I tried as hard as I could to make the theory more physical, but am only half-satisfied with the result. I am still quite unhappy about the whole thing and was so glad to hear that, with regard to mathematics and physics, you are completely on my side. Here I am in a milieu that thinks and feels exactly the opposite and I

---

[31] A. Kozhevnikov, 'Electrodynamics in Matrix Mechanics: Discord in Interpretation of the Theory' (in Russian) (Moscow: Institute for History of Science and Technology, 1987).

worry whether I am just too stupid to understand mathematics ... I always feel irritated when the theory is called matrix mechanics and for a time seriously wanted to cross the word "matrix" completely out of the paper and replace it with, for example, "quantum-theoretical variable". (After all, "matrix" is one of the dumbest mathematical words.)[32]

Despite these private complaints, conflicts did not go public. The authors of matrix mechanics chose to collaborate on the new theory. They advanced their diverging interpretations in separate publications, but did not explicitly set them against each other and avoided discussing their disagreements in public.

Wave Mechanics (Spring 1926)

Schrödinger's first paper on wave mechanics of January 1926 cautiously emphasised formalism rather than interpretation. As another precaution, he made a friendly gesture toward matrix mechanics in mentioning that both theories had one basic feature in common: the abandonment of the notion of electron trajectories.[33] The statement was hardly sincere, because the reasons for this abandonment were very different in the two theories. In wave mechanics, the electron did not have a definite position not because of *Unanschaulichkeit*, but because it was represented by a continuous wave and spread out in three-dimensional space. Once Schrödinger had become more confident of the success and power of his theory, he did not need the protective rhetoric any longer and fully engaged in the interpretation business. In March, he established a mathematical connection between the basic formulae of the two theories and proclaimed them 'mathematically equivalent'.[34] This was an understatement — wave mechanics was certainly much more powerful in calculations than matrix mechanics — but the implication was that the criterion for choosing between the two should be interpretation rather than formalism. At that stage, Schrödinger was confident his interpretation had to be preferred.

Complete restoration of Anschaulichkeit. Wave mechanics' main philosophical advantage appeared in the rehabilitation of *Anschaulichkeit*. Not only did the usual three-dimensional geometry remain completely valid on the microscopic

---

[32] Heisenberg to Pauli, 16 November 1925, in PWB, p. 255. Dirac subsequently designed a special term 'q-numbers' for quantum variables.

[33] Erwin Schrödinger, 'Quantisierung als Eigenwertproblem (Erste Mitteilung)', *AP* 79 (1926): 361–76.

[34] Erwin Schrödinger, 'Über das Verhältnis der Heisenberg-Born-Jordanschen Quantenmechanik zu der meinen', *AP* 79 (1926): 734–56.

scale but even the motion of the electron within the atom could be represented pictorially (the difference from classical theory being the visual image was of a vibrating string instead of a moving corpuscle). The space-time visualisation of microscopic processes once again became possible.

Continuity. In wave mechanics, discrete energy levels are obtained as solutions of a continuous wave equation. One could still, in principle, choose which particular aspect to emphasise as fundamental — continuity or discontinuity — and the question turned into a heated debate in 1926. Bohr wanted to welcome wave mechanics but insisted it should be understood precisely as a description of discontinuous atomic states. Schrödinger, on the other hand, emphasised the continuity aspect alone, taking 'a departure from fundamental discontinuity' as his main philosophical slogan and programmatic goal.[35] For him, not only were discrete energy states artefacts of continuous undulatory processes, but quantum transitions themselves had to be explained as continuous changes from one vibrational mode to another, point particles had to be understood as wave packets and the very relationship between classical and quantum descriptions was to be conceived as 'the *continuous* transition from micro- to macro-mechanics'.[36]

Wave ontology. Although de Broglie's dualistic papers on waves and particles provided initial inspiration to Schrödinger, duality did not figure prominently in wave mechanics during its heyday in the spring of 1926. Schrödinger openly and obviously preferred waves to corpuscles as ontological reality. Radiation appeared in his theory in the form of classical electromagnetic waves. Electrons were seen as corpuscles only on the scale of lower resolution, whereas at the truly microscopic quantum scale they were wave packets of a finite size. (The difference is similar to that between the geometrical and the more fundamental wave optics.) Schrödinger hoped at the time to develop a theory in which all particle-related concepts would be replaced consistently by undulatory ones (for instance, energy would have to be replaced by frequency and the concept of quantum transitions by resonance). Such an ultimate field-like view had no need for wave-particle duality.

No statistics, no acausality. This is the only main philosophical feature that wave mechanics and matrix mechanics had in common. Their principal stakes were elsewhere, but both shared a definite dislike for statistical considerations and deliberately eluded the language of probabilities. Although Schrödinger initially supported the statistical Bohr-Kramers-Slater theory, its defeat must have affected him, too, for, just like Heisenberg, he consistently used the term

---

[35] Schrödinger to Wien, 18 June 1926, WWM.
[36] Erwin Schrödinger, 'Der stetige Übergang von der Mikro- zur Makromechanik', *NW* 14 (1926): 664–66 (emphasis added).

'intensities' instead of 'transition probabilities'. Moreover, he hoped to explain quantum transitions through a causal and continuous process: in a linear combination of vibrational modes, some coefficients would grow, while others would decrease in time, thus accounting for the gradual transition from one vibrational mode to another.

Reactions to wave mechanics. The rivalry between the two approaches has sometimes led commentators to assume the authors of matrix mechanics accepted Schrödinger's theory only reluctantly, after it found a very enthusiastic general reception among physicists. A distinction between happiness and quickness can provide a more accurate perspective. The captains of matrix mechanics were among the first to abandon the sinking vessel and to start using the new methods of wave mechanics, although in ways that often transcended the boundaries of Schrödinger's original intent.[37] Pauli was the quickest: he learned of the new achievement from Sommerfeld, and in April 1926, simultaneously with Schrödinger, developed the proof of the 'mathematical equivalence' of the two theories. Born was happiest: he easily and enthusiastically converted to wave ontology in his papers of summer 1926. Heisenberg was the unhappiest, but even he used wave functions in his June 1926 paper. Only Dirac was slow, first turning to Schrödinger's methods in August 1926.

Their reaction to the philosophy of wave mechanics was certainly much more critical, but even here some of Schrödinger's accomplishments could not be resisted. *Anschaulichkeit* had to be rehabilitated, at least partially. Much of matrix mechanics' former radical opposition to visualisation of atomic processes quietly disappeared from its authors' subsequent publications in the course of 1926. Besides wave mechanics, another visual concept also contributed to this change of heart: the proposal of the spinning electron gained quick acceptance, despite the initially sceptical reception by Pauli, Heisenberg and Bohr. At the end of the day, Euclidean geometry did not fail within the atom and visual pictures of microscopic processes proved, once again, their usefulness. *Unanschaulichkeit* retained some territory: quantum transitions, or mysterious jumps, avoided visualisation despite Schrödinger's initial hopes. But it became increasingly hard to insist on it as a grand philosophical principle, although Heisenberg (with some assistance from Bohr) continued his desperate struggle against visualisation until the spring of 1927 and his own paper on the indeterminacy principle. A better strategy was to hide the philosophical defeat by shifting the public debate to other issues of controversy.

The wave ontology appealed to at least some of the matrix people. Born, who had liked Einstein's idea of matter waves even earlier, subscribed to it

---

[37] Kozhevnikov and Novik, 'Analysis' (ref. 18).

enthusiastically. Bohr was also quite sympathetic and Pauli did not particularly object. Heisenberg was as unhappy about waves as just about all other physical ideas of wave mechanics. He wanted to deprive the wave function of its physical meaning as a wave and reduce it to a mere mathematical tool. Dirac also preferred particles to waves and the treatment of the wave function ψ as an abstract mathematical symbol.

The entire group united in opposition to Schrödinger's continuity claim. Born's only disagreement with Schrödinger was to insist that wave mechanics 'permits description not only of the stationary states, but also of quantum jumps'.[38] Pauli wrote to Schrödinger in May 1926: 'I have generally the strongest doubt in the feasibility of a consistent wholly continuous field theory of the de Broglie waves. One must probably still introduce into the description of quantum phenomena essentially discontinuous elements as well'.[39] The stated goal of Heisenberg's two papers of summer 1926 was to prove the essential discontinuity of atomic phenomena, even when described by the Schrödinger function. And famously, the dispute between Bohr and Schrödinger during the latter's visit to Copenhagen in September 1926 centred on their main disagreement on discontinuous quantum jumps.

Quantum Mechanics (Fall 1926)

While appropriating Schrödinger's wave mechanics, Born, Pauli, Heisenberg, Dirac and Jordan did not feel bound by his original interpretations but applied the theory quite liberally to new kinds of problems, thereby changing the meaning of its basic concepts. By generalising the approach to treat the multi-electron problem, Heisenberg and Dirac transformed ψ into a wave function in multi-dimensional space, which eroded its initial visual interpretation as a wave in ordinary space. By applying the method to the problem of scattering, Born, Pauli and Dirac changed ψ into a guiding field for particles and into a probability distribution, once again depriving it of its original physical meaning. By the end of 1926, Dirac and Jordan unified all these new accomplishments into a general scheme under the name of transformation theory and declared the (non-relativistic) quantum formalism completed. Their decisive synthesis brought about further shifts in philosophical positions.[40]

---

[38] Quoted in Beller, 'Born's Probabilistic Interpretation' (ref. 21), on 567. *Cf. also* Beller, 'Schrödinger's Dialogue' (ref. 21).
[39] Pauli to Schrödinger, 24 May 1926, in PWB, p. 326. (English translation from Hendry, *Creation* (ref. 12), on 86).

Limited Anschaulichkeit. The common perception that Schrödinger lost his philosophical struggle overlooks the major fact that he had basically won the battle for *Anschaulichkeit*. Objections to ordinary geometry, the usual ideas of space and time, and to visual pictures with either waves or particles disappeared. Born used all these notions in his papers on scattering in wave mechanics. Pauli made a further concession and a reversal of his earlier cherished beliefs when he rehabilitated the notion of the 'position of the electron within the atom', the probability of which was now determined by the wave function. Without an open admission of failure, the initial programmatic claim of matrix mechanics was dropped and disappeared from the discourse. However, the restoration of *Anschaulichkeit* did not become absolute: probabilistic arguments imposed restrictions on it. The theory permitted calculation only of the probabilities of electrons' positions and of the still-*unanschauliche* quantum jumps.

Symmetry between continuous and discontinuous representations. In Copenhagen in September 1926, while Bohr and Schrödinger conducted their very intense, principled and stubborn disputes about continuity and discontinuity in atoms, Dirac quietly worked on a paper that would render this entire polemic obsolete. Unlike the rest of the group, Dirac did not label his ideas 'philosophical', but his reformulation of the basic principles of quantum mechanics affected others' philosophical reasoning. Dirac developed a mathematical formalism in which both continuous and discontinuous quantum variables could be used in a relatively symmetrical fashion. His theory allowed transformations from one set of variables to another, thus putting them on an equal footing.[41] The continuity-discontinuity dilemma turned into a choice determined by simple mathematical convenience regarding which particular variables could work better for calculating one or another problem in atomic physics. It no longer made much sense to treat it as a matter of philosophical gravity.

Duality. Following Born's reinterpretation of the wave function as a guiding field for particles, both wave and particle visualisations of microscopic events began to be used, frequently and often interchangeably, in quantum mechanics. Some physicists preferred one over the other but the discipline as a whole demonstrated a rather promiscuous use of both corpuscular and wave pictures (partly justified by the transformation theory, although Dirac personally always gravitated toward particles). A physicist could use one or both of these visualisations as intuitively helpful pictures, but pushing the matter too far by asking for disciplined

---

[40] P.A.M. Dirac, 'The Physical Interpretation of the Quantum Dynamics', *Proceedings of the Royal Society A* 113 (1927): 621–41; Pascual Jordan, 'Über eine neue Begründung der Quantemmechanik', *ZP* 40 (1927), 809–38.

[41] Dirac, 'Physical Interpretation' (ref. 40).

usage or clear choice between them looked increasingly pedantic and old-fashioned. We may call such widespread carelessness and libertarian use of either wave or particle language with inconsistent switches from one to the other 'duality' to distinguish it from rarer occurrences of 'dualism', or serious statements about the ontological reality of wave-particle chimeras.[42]

Causality and statistics. With the erosion of earlier philosophical principles, a new, statistical idea was on the rise in the fall of 1926 through the contributions of Born, Pauli, Dirac and Jordan. In the corpuscular representation, the wave function determined probabilities of electrons' states and transitions. In August 1926, on the eve of his Copenhagen visit, Schrödinger explained in a letter to Wilhelm Wien his standing on the interpretational issues. Schrödinger rejected *a limine* 'Bohr's standpoint, that a space-time description is impossible', but showed somewhat more understanding for Born's emerging statistical picture:

> Today I no longer like to assume with Born that an individual process of this kind is "absolutely random", i.e., completely undetermined. I no longer believe today that this conception (which I championed so enthusiastically four years ago) accomplishes much. From an offprint of Born's last work in the *Zeitschr.f.Phys.* I know more or less how he thinks of things: the *waves* must be strictly causally determined through field laws; the *wave functions*, on the other hand, have only the meaning of probabilities for the actual motions of light or material particles. I believe that Born overlooks that — provided one could have this view worked out completely — it would depend on the taste of the observer which he now wishes to regard as real, the particle or the guiding field. There is certainly no criterion for reality if one does not want to say: the *real* is only the complex of sense impressions, all the rest are only pictures.[43]

Schrödinger was thus prepared for a compromise, on positivistic terms, between the wave and the corpuscular, the causal and the statistical, interpretations of the theory. If one were inclined to accept waves as the ultimate reality, the fundamental laws of the theory would be causal. If fundamentality of particles were assumed, their laws of motion would be probabilistic. Schrödinger preferred the former option, but was willing to put up with those who gravitated toward the

---

[42] Alexei Kojevnikov, 'Einstein's Fluctuation Formula and the Wave-Particle Duality', in *Einstein Studies*, vol. 10, *Einstein Studies in Russia*, ed. Yuri Balashov and Vladimir Vizgin (Boston: Birkhäuser, 2002), 181–228.

[43] Schrödinger to Wien, 25 August 1926, WWM. I am thankful to Cathryn Carson for a copy of the original text. The English translation is partially borrowed from Walter J. Moore, *Schrödinger: Life and Thought* (Cambridge: Cambridge University Press, 1989), 225–26.

latter. Born's position at the time, as expressed in his July 1926 paper on probabilistic scattering, seemed compatible. He personally liked the corpuscular and acausal picture rather than the one with waves and causality, but regarded this still as a matter of philosophical preference, not a scientific conclusion: 'I myself am inclined to renounce determinism in the world of atoms. But that is a philosophical question for which physical arguments alone are not decisive'.[44]

Philosophies of Compromise (1927)

In the fall of 1926, three centres could compete for leadership in the new quantum mechanics. In Copenhagen, Bohr was still rather silent in public (and had not published much at all since the failure of the Bohr-Kramers-Slater theory). But he hired Heisenberg as a lecturer, who kept on publishing important papers, attracting new visitors to the institute, and on the philosophical front still defended the remains of the matrix mechanics agenda (*Unanschaulichkeit* and discontinuity). Schrödinger promoted wave mechanics and the ideas of wave ontology and continuity in Zurich. In Göttingen, where the whole thing started, Born was determined to maintain momentum despite the damaging loss of Heisenberg to Copenhagen and, together with Jordan, was developing the probabilistic version of quantum mechanics. The following year, new philosophies appeared which drew upon the earlier ideas in more complex and mixed ways.

Born's move toward acausality. Approximately once a year Bohr invited a distinguished visitor to his Copenhagen institute. Extending such an invitation to Schrödinger indicated Bohr's interest in an agreement, cooperation and a possible deal, rather than a quarrel.[45] Indeed, during their week-long non-stop Copenhagen discussions, Bohr did not push hard on *Unanschaulichkeit* and was sympathetic to the wave mechanics in general and the wave ontology in particular. In return, he wanted Schrödinger to retreat on the continuity claim and accept the fundamental discreteness of atomic phenomena. A compromise along these lines would have included a fusion of the wave mechanics with discontinuous quantum states and jumps of Bohr's earlier atomic theory:

> A few weeks ago we had a visit by Schrödinger, which gave rise to much discussion regarding the physical reality of the postulates of atomic theory. I suppose you know that the wonderful results Schrödinger has arrived at has led to the

---

[44] Max Born, 'Zur Quantenmechanik der Stossvorgänge', *ZP* 37 (1926): 863–67. For a more detailed analysis, see Beller, 'Born's Probabilistic Interpretation', (ref. 21), and Nancy Greenspan, *The End of the Certain World: The Life and Science of Max Born* (New York: Basic Books, 2005), 139.
[45] Bohr to Schrödinger, 11 September 1926; Schrödinger to Bohr, 21 September 1926, NBA.

suggestion, taken up with great enthusiasm from various sides, that the ideas of discontinuity which underlie the interpretation hitherto given of the phenomena might be unnecessary. This appears, however, to be a misunderstanding, as it would seem that Schrödinger's results so far can only be given a physical application when interpreted in the sense of the usual postulates. Indeed they offer a most welcome supplement to the matrix mechanics in allowing to characterize the stationary states separately.[46]

Schrödinger, however, refused to accept discontinuity as stubbornly as Bohr insisted upon it. As we saw above, he preferred a compromise with Born rather than Bohr.

Born, for his part, resolutely declined Schrödinger's advances. As professorial wrangling often goes, he encouraged his *Privatdozent* Jordan to launch an open attack in print on the philosophy of wave mechanics. Schrödinger tried to smooth out the relationship and complained in a private letter to Born, which the latter ridiculed in his private circle. Early in 1927 Born and Jordan publicly proclaimed acausality as the most important philosophical lesson of quantum mechanics. Relying on the new formalism of the transformation theory, they explicitly criticised Schrödinger's wave ontology. On the other hand, their philosophy had room for Copenhagen's favourite discontinuity, thus making possible a compromise with Bohr.[47]

Schrödinger's move toward wave-particle dualism. In the fall of 1926, Schrödinger was named the second choice (after Sommerfeld, but before Born) in the search to fill the most prestigious chair of theoretical physics at the University of Berlin. After Sommerfeld declined as anticipated, Schrödinger accepted the offer and moved to Berlin (his former position in Zurich would subsequently become Pauli's). A win on prestige, however, eventually turned into an institutional disadvantage for Schrödinger. In subsequent years he worked in relative isolation, usually with only a couple of associates in Berlin, while much larger and more active research communities of younger students and postdoctoral visitors grew around Göttingen and Copenhagen (and also later around Heisenberg in Leipzig). In philosophical terms, Schrödinger moved toward an open critique of the statistical interpretation after Born had rejected a compromise: 'Personally I no longer regard this [statistical] interpretation as a finally satisfactory one, even if it proves useful in practice. To me it seems to mean a renunciation, much too

---

[46] Bohr to Kronig, 28 October 1926, RKZ; *Cf. also* Beller, Schrödinger's Dialogue' (ref. 21).
[47] Pascual Jordan, 'Kausalität und Statistik in der modernen Physik', *NW* 15 (1927): 105–10; Max Born, 'Quantenmechanik und Statistik', *NW* 15 (1927): 238–42; *See also* Beller, 'Born's Probabilistic Interpretation' (ref. 21), 572–73.

fundamental in principle, of all attempt to understand the individual process'. Eventually, he would also retreat from a strong wave ontology and, together with his *Privatdozent* Fritz London, embrace wave-particle dualism. London's lectures on wave mechanics in Berlin opened with a programmatic statement on the dual (wave and particle) nature of quantum objects.[48]

Heisenberg's move to indeterminacy. Born's acausality met with mixed reactions in Copenhagen. Heisenberg welcomed statistics as an argument against Schrödinger's philosophy, but both he and Bohr preferred to view it as a part of 'formalism' rather than the philosophy of quantum mechanics, which both intended to develop on their own. The resulting Göttingen-Copenhagen alliance, if it can be called that, formed out of convenience. On the basis of the shared 'formalism' of quantum mechanics, its major spokesmen advanced de facto diverging interpretational claims, but did not criticise each other's views in public, maintaining at least a posture of good cooperation.

Heisenberg, still Bohr's subordinate, refused to wait patiently. In his famous paper of March 1927, he argued that the statistical formalism led to a fundamental philosophical consequence: the unavoidable uncertainty in the simultaneous measurement of a particle's position and velocity. Although not quite so radical as the *Unanschaulichkeit* claim of the earlier matrix mechanics, it imposed a fundamental restriction on the visualisability of classical theories. In a letter to Ralph Kronig, Heisenberg summarised the combination of philosophical themes of his work as follows: 'I have recently done a paper about the visualizable content [*anschaulichen Inhalt*] of the (certainly discontinuous) quantum mechanics, based on the now completed scheme, which also presents my (or all of us here's) answer to the question: light quanta or waves. You will see it in the *Zeitschrift!*'[49]

Bohr's move toward complementarity. Bohr considered Heisenberg's uncertainty paper premature and they argued intensely over the manuscript. It appears that Heisenberg sent it out for publication without Bohr's approval, in a breach of existing institutional norms. An unwritten but strict rule in German universities at the time, and also in Copenhagen, required students, employees and visiting fellows to submit research papers to journals only with the permission of their professor and director of the institute where the work had been done. No matter

---

[48] Erwin Schrödinger, 'Der Energieimpulssatz der Materiewellen', *AP* 82 (1927): 265–72, on 272; cited following the translation in Erwin Schrödinger, *Collected Papers on Wave Mechanics* (Providence, RI: American Mathematical Society, 2003), 135–36; Fritz London, 'Quantenmechanik, insbesondere Anwendungen auf die Mehrkörperproblem u.d. Chemie', unpublished lectures at Berlin University, 1928–29, Fritz London Papers, Duke University Archive, Box 4.

[49] Werner Heisenberg, 'Über den anschaulichen Inhalt der quantentheoretischen Kinematik und Mechanik', *ZP* 43 (1927): 172–98; Heisenberg to Kronig, 8 April 1927, RKZ.

how justifiably famous, Heisenberg was not yet a professor, even if he was about to become one. Once his anticipated professorship at a German university materialised, he left Copenhagen for Leipzig. It took a couple of additional years to heal, but not completely, his somewhat strained relationship with Bohr.[50]

In terms of scientific papers, Bohr wrote and published almost nothing during the two years when quantum mechanics was being created. Perhaps still recovering from the fiasco of the Bohr-Kramers-Slater proposal, he occupied himself primarily with administrative matters: the expansion of his institute with the help of a Rockefeller grant and arrangements for an increasing number of foreign visitors. Now that both the construction of a new institute building and the formal edifice of the new theory was complete, he felt the urge to develop his interpretation of quantum mechanics. Bohr's writing proceeded, as usual, slowly and required a helper with whom he could collaborate on discussing the manuscript and dictation. Always struggling to arrive at definitive formulations and almost never fully satisfied, Bohr often went through multiple revisions and proofs while dictating his papers. With the help of Oskar Klein, he completed the manuscript by the end of 1927.

Bohr's interpretation is complex and difficult to understand, in part because it draws on everybody else's, as if trying to ensure all important contributors would find something in it they personally cherished: Bohr's own favourite discontinuity, Schrödinger's wave packets, Heisenberg's *(Un)anschaulichkeit* and indeterminacy, and Born's acausality. According to Bohr, there is a fundamental discontinuous, somewhat mystical, individuality (*Individualität*) at work in all microscopic processes and our imperfectly human means of comprehending it. When trying to make sense of atoms, one cannot help but alternate between visual space-time and causal-logical descriptions of events. Both derive from classical physics and our macroscopic experiences and are therefore not entirely suitable for describing the strange microscopic world. But experimental settings, insofar as they involve macroscopic instruments, make it essential that we use such classical languages. An unavoidable and uncontrollable disturbance of the microscopic object in the process of observation imposes limits on their applicability, however. An experimental set-up designed to investigate and determine the space-time picture of microscopic phenomena makes impossible their causal representation and vice versa. In attempting to combine them too literally in quantum physics, as classical physics was able to, one becomes mired in inevitable

---

[50] 'I was so unhappy last winter, how everything became estranged and how ungrateful I seemed towards you ... I hope you can forgive everything that I have done wrong'. Heisenberg to Bohr, 21 August 1927, NBA. *See also* Werner Heisenberg, *Liebe Eltern! Briefe aus kritischer Zeit 1918 bis 1945* (Munich: Langen Müller, 2003), 121–22.

contradictions: it is thus necessary to renounce the possibility of their simultaneous unlimited application within the quantum domain. Each representation separately is also insufficient for understanding the full range of possible experience with atoms, but every imaginable experiment can be accounted for in terms of one or other description. Though based on conflicting sets of notions, these representations should be taken not as mutually exclusive, but complementary — only their combined, alternating use can produce the fullest possible account of the microscopic world.[51]

Discussion

The preceding analysis reveals the intricate ways in which Weimar culture and its ideological values influenced the discourse of physicists during the creation of quantum mechanics. As Paul Forman has described in his analysis, the experience of general social crisis after the war affected scientists' mentality and inspired their talk about 'crisis in science'.[52] The latter notion often implied not merely the economic difficulties of the profession, but also crises in the conceptual foundations of existing knowledge. Scientists became much more willing, in comparison with relatively normal and stable times, to revise or entirely abandon fundamental principles and commitments of their respective disciplines. In the case discussed here, such culturally amplified criticisms were directed not only at basic concepts of classical physics, but even at some key assumptions of the quantum theory of the atom, which had only been around for a decade but was about to become labelled as, characteristically, the 'old' quantum theory.

Cultural concepts also guided the direction of scientists' criticisms and their search for new principles. Had, for example, a larger share of the debate about quantum phenomena taken place in Great Britain, the question of whether electrons have free will would have acquired a major prominence in the new theory. In Central Europe, the discussion centred instead around issues identified by Forman as carrying highly controversial, value-laden meanings within Weimar culture — *(Un)Anschaulichkeit* and *(a)causality* — both of which played leading roles in physicists' thinking and in their attempts to define the new principles of quantum theory. *Individualität* figured less prominently, but did make an appearance in Bohr's complementarity interpretation, essentially standing in for the indivisibility of quanta. The other two foci of the interpretational

---

[51] Niels Bohr, 'The Quantum Postulate and the Recent Development of Atomic Theory', *Atti del Congresso Internazionale dei Fisici*, vol. 2 (Bologna: Nicola Zanichelli, 1928), 565–88; For a critical analysis of its assumptions, see Heilbron, 'Earliest Missionaries' (ref. 20), esp. 199–200.
[52] Forman, 'Weimar Culture' (ref. 3).

controversy — continuity/discontinuity and wave/particle ontology — belonged to the general tradition of philosophising about physics. Within such a culturally framed debate, as the above analysis shows, quantum physicists found various, often competing and incompatible, ways to apply and translate these general concepts into the language and problems of their specific field.

The way they did this, and their philosophical discourse, appears to be simultaneously high-principled — they were utterly serious in making strongly worded philosophical statements; relatively undisciplined — their conceptions only 'more or less defined by other conceptions'; and opportunistic — the proclaimed principles kept changing too often. Having followed the twists and turns of their discussions, one can hardly avoid the impression that physicists acted as if compelled to hurry up in declaring general philosophical conclusions, which often happened to be premature, because the theory itself was still *in statu nascendi*. It appears proposing a philosophical interpretation was an invaluable act in itself, apart from the choice of a particular philosophy or the probable time for it to hold. Such a behavioural pattern, too, calls for a cultural interpretation.

Some background aspects of the phenomenon, at its most basic and obvious level, are not uncommon, but recognisable as typical and natural for a specific cultural group, the German or German-speaking academics, or *Gelehrte*. The culture of Germanic academe upheld the strong ideal of a scientific genius required to partially double as a philosopher. Culturally, a truly great scientist was expected not only to make discoveries in a special field of research, but to go into it in such depth as to contribute to a general philosophical outlook, and to such conclusions that would be meaningful to all members of educated culture, transcending narrow professionalism and disciplinary boundaries. To this widely shared belief we owe the abundance of printed talks and Habilitation speeches addressed to general academic audiences, in which German scholars discussed broader cultural meanings of their special field of study.[53] This genre of writing has provided some of the most valuable sources for Forman's work on Weimar culture and quantum acausality. For the purposes of current discussion, we can take it for granted as a well-established and entrenched ritual, which in the case of quantum mechanics, however, produced an atypical outcome.

---

[53] Hermann von Helmholtz often served as the role model and a typical example of such a combination; see Emil Warburg *et al.*, eds., *Helmholtz als Physiker, Physiologe und Philosoph* (Karlsruhe: Müllersche Hofbuchhandlung, 1922); Lorenz Krüger, ed., *Universalgenie Helmholtz. Rückblick nach 100 Jahren* (Berlin: Akademie-Verlag, 1994). For a discussion of the 'physicist as philosopher' phenomenon, see Cathryn Carson, *Heisenberg in the Atomic Age: Science and the Public Sphere* (Cambridge: Cambridge University Press, 2010).

The very scope of the debate was already unusual — it would be hard to point out another scientific development in which the existing genre of philosophising produced an intellectual fight of such intensity and inconsistency of positions among such a number of prominent participants. The sheer volume of polemical writings and philosophical commentary accompanying the creation of quantum mechanics can be compared, perhaps, with only a case from another culture — the controversy provoked by the first publication of Darwin's *Origins*.[54] Though wider than usual, the circle of those who participated in the interpretational polemics around quantum mechanics was still restricted. It included several recognised leaders, as well as a few unavoidable marginal authors and outsiders to the field, but characteristically not the mainstream contributors to its technical development, the almost 200 post-docs, assistants and PhD students who authored the majority of publications during the first two years of quantum mechanics. More than a decade ago I had an opportunity to meet in Göttingen with one of the last living members of that cohort, Friedrich Hund, and in the course of conversation inquired in passing about Hund's own position in the interpretational controversy. He surprised me at the time by replying straightforwardly that it was not his business, but then added, somewhat more expectedly, 'but, of course, Bohr was right'.[55] As a young assistant in Göttingen and subsequently a postdoctoral fellow in Copenhagen in 1925–27, Hund occupied himself with calculations of molecular spectra using quantum mechanics, but was not entitled to contribute to the public debate about its interpretation.

Besides its strong interest in philosophising, the academic culture that produced quantum mechanics was also extremely sensitive to questions of hierarchy, with both these concerns closely linked. After all, contributing to the generally important philosophical outlook was considered the attribute of a truly great scholar, not necessarily of an aspiring or rank and file researcher. In this respect it is somewhat unusual to find among the entitled participants not only ordinary professors and *Geheimräte*, but also Pauli, Heisenberg and Jordan — all extremely important, but still junior contributors to quantum mechanics. Taking a closer look, however, one can see the precariousness of their participation. Pauli was involved mainly in the informal then unpublished exchange of philosophical ideas, via private correspondence. Jordan essentially entered the public debate on behalf of his professor, Max Born. And even the recognised pioneer, Heisenberg, before he became a professor himself, had a hard time insisting on his right to publish an interpretation that would become known as the uncertainty principle. He did this despite Bohr's reservations only by violating the existing strict, if unwritten, subordination rules governing publication procedures.

---

[54] I am thankful to Simon Schaffer for this observation.
[55] I am grateful to Klaus Hentschel for the invitation to take part in his interview with Hund.

Participation in the philosophical discourse was thus a mark of prestige, privilege and status — recognition of not merely the social but also the intellectual hierarchy and a person's crucial contribution to the field. 'Perhaps it was also a battle over who did the whole thing first', admitted Heisenberg many years later.[56] In my view, the genre of philosophising did indeed provide physicists with a vehicle for making claims over the entire theory, but the claims were about property rather than priority. Competing philosophical interpretations did not reorder the chronology of individual contributions to the emerging field, but they reassigned the relative importance of those contributions for 'the whole thing'. Nobody questioned Heisenberg's credit as the author of quantum mechanics' first proposal, but he was deeply concerned about the decrease in its perceived value during the months when Schrödinger's interpretation rose in popularity. Similarly, nobody tried or could deprive Schrödinger of his authorship of the theory's central equation, but, depending on the interpretation, his contribution could be presented primarily as 'mathematical' (= technical) rather than 'philosophical' (= fundamental). And Bohr, by offering the last, if not final word on the developing interpretation secured his public reputation as the leader of the new theory, despite the fact that he did not publish on it during its development in 1925–27. In contrast, Schrödinger's failure to establish the prevailing philosophical interpretation signified his loss of control over the field.

The emphasis on what each participant considered his personal major contribution to quantum ideas may explain many of the consistencies and inconsistencies in their philosophical pronouncements. After having invented wave mechanics, Schrödinger abandoned his earlier flirtation with acausality in favour of the (causal) philosophy of continuity and *Anschaulichkeit*. Having reinterpreted the wave function probabilistically, Born and Pauli reversed their pronouncements about acausality and statistics to the affirmative position. Bohr persistently emphasised the fundamentality of discontinuity in quantum phenomena, obviously linked to the postulate of discrete states in his original model of the hydrogen atom of 1913. Einstein had expressed scepticism about quantum mechanics early on, even before it turned acausal, largely because it did not offer an answer to the crucial question — for him, in view of his earlier contributions to quantum physics — on the wave- or particle-like structure of light.[57]

One can imagine a different situation: a major scientific accomplishment belonging, more or less unquestionably, to one distinguished scientist. The ritual of philosophising would be performed in this case, too, as the privilege and duty

---

[56] Heisenberg, interview by T.S. Kuhn, 1963, AHQP, quoted in Beller, 'Conceptual and Anecdotal History' (ref. 22), on 556.

[57] Kojevnikov, 'Einstein's Fluctuation Formula' (ref. 42).

of a great scholar, but the leader's right to furnish his theory with a general interpretation would also likely have remained unchallenged. The creation of quantum mechanics, in contrast, was a real group effort, although not a team effort. No other great scientific innovation of the period, including relativity theory, had so many crucial and temporarily closely linked contributions from different authors, each with his own agenda and aspirations, and thus so many potential leaders at once. The existing genre of philosophising required quantum physicists to translate the meaning of their scientific accomplishment into the language of cultural and ideological values of the time. At the same time, it also offered a culturally approved and respectable form of public discourse within which they could implicitly, and therefore without losing face, debate their rival claims for the entire theory, which inspired them to develop several competing and incompatible translations. The intensity of scientists' philosophical disagreements corresponded to the unusually high level of intra-disciplinary competition; the latter started long before the theory was in any sense completed, as did the former. New and crucial contributions continued coming; even the most basic assumptions of quantum mechanics were still in flux, as well as the relationships between individual authors. Thus also rhetorical strategies kept changing, resulting in opportunistic shifts in announced philosophical principles between 1925 and 1927.

Contemporaries overwhelmingly perceived Bohr as the ultimate winner in the interpretational debate over the opposition from Einstein and Schrödinger. Philosophers who analyse the dispute today often find it hard to explain from a logical point of view why the Copenhagen philosophy had to be preferred to the arguments of its critics. From the criterion of better adaptation to the cultural values of the time, Schrödinger's *Anschaulichkeit* argument too does not seem to be much weaker than his opponents' acausality claim. How does one then account for the apparent victory of the Copenhagen interpretation in the 1920s? It appears to me that the debate constituted only a tip of the iceberg, albeit the most visible one, of the ongoing intra-disciplinary rivalry. Printed philosophical words by themselves could provide public justifications and rationalisations for the outcome, but not necessarily decide the competition over the new theory.

The latter depended more, I suggest, on mainstream contributors — younger students mentioned by Ehrenfest in his letter quoted in the beginning of this paper, postdoctoral visitors and assistants. These fellows, such as Friedrich Hund, as a rule did not participate directly in the philosophical polemics, but published the majority of papers and calculations, cited others' works and together constituted the decisive reference group. These mainstream contributors to the new theory were not aloof from philosophical arguments, but also influenced by professional opportunities, available problems to solve, financial considerations and the institutional authority of their professors. Their movements between places and the collective body of work submitted by them for publication from the Copenhagen and Göttingen institutes defined the perception of where the leaders of the new field were. On occasions, in particular during the mid-1920s, postdoctoral culture and its burgeoning activity could effectively dominate entire research institutions, acquiring an intellectual momentum of its own, rather than following professors' and directors' agendas. This 'postdoctoral revolution' in science and its impact on the development and character of quantum mechanics ultimately requires a separate analysis of its own.[58]

Acknowledgments

Support for research leading to this paper initially came from the Max-Planck-Institut für Wissenschaftsgeschichte, Berlin, and subsequently from NSF grant SES-9911008. My special gratitude goes to Finn Aaserud and the staff of the Niels Bohr Archive in Copenhagen for hospitality and invaluable advice. For further comments and discussions, I am thankful to Diana Barkan, Richard Beyler, Cathryn Carson, Paul Forman, Edward Jurkowitz, Jürgen Renn, Skuli Sigurdsson, Jessica Wang, Norton Wise and participants at the colloquia at Göttingen University, Cambridge University, Northwestern University, the Niels Bohr Institute and the Max-Planck-Institut für Wissenschaftsgeschichte.

---

[58] Alexei Kojevnikov, '"Knabenphysik": The Birth of Quantum Mechanics from a Postdoctoral Perspective', presented at the American Physical Society April Meeting, 11–15 April 2008, St. Louis, MO.

'The Shackles of Causality':
Physics and Philosophy in the Netherlands in the Interwar Period[†]

Kai Eigner and Frans van Lunteren*

The Dutch have a reputation of being an unphilosophic people. In the global history of philosophy the important contributions by the Dutch are few and far between .... In the domain of the sciences a philosophical approach to science and reality has never had a high status. The quest for an all-embracing view of reality and for a philosophical reflection on the results of the sciences is not a central concern in the Dutch intellectual tradition. An active scientist who is also a philosopher and who reflects on the philosophical implications of his work is even more an almost un-Dutch phenomenon.[1]

If there is truth in van Berkel's characterisation of Dutch science throughout the centuries, the period between the two world wars seems to depart from this

---

* Kai Eigner, Department of Philosophy, VU University of Amsterdam, De Boelelaan 1105, 1081 HV Amsterdam; k.eigner@ph.vu.nl. Frans van Lunteren, Department of Exact Science, VU University of Amsterdam; fh.van.lunteren@few.vu.nl.
  The following abbreviations are used: APE, Archive Paul Ehrenfest, Museum Boerhaave, Leiden, The Netherlands; ESC, Ehrenfest Scientific Correspondence, APE. *HSPS*, *Historical Studies in the Physical Sciences*.

[†] This paper was published earlier in Dutch as: 'Fokkers "Greep in de verte": Nederlandse fysica en filosofie in het interbellum', *Gewina* 26 (2003): 1–21. More recently, David Baneke has published a far more extensive study of Dutch academic intellectuals in the early twentieth century, which covers much of the same ground. See David Baneke, *Synthetisch denken: Natuurwetenschappers over hun rol in de moderne maatschappij, 1900–1940* (Hilversum: Verloren, 2008).

[1] Klaas van Berkel, *Citaten uit het boek der natuur* (Amsterdam: Bert Bakker, 1998), 241.

pattern. Van Berkel refers to the physicist Jacob Clay as 'the proverbial exception that proves the rule'. This follower of the Dutch philosopher Bolland was respected in circles of both philosophers and physicists for the way in which he linked science and philosophy during the interwar period. Closer study of Dutch physics during this time shows, however, that Clay was far from exceptional. Many physicists of his generation were interested in philosophical problems, often in connection with their own discipline. The concept of causality, for example, was analysed extensively. Some physicists were expressly opposed to the traditional conceptions of causality and appealed for this stance to both physical and extraphysical arguments.

The attitude of these physicists is remarkably similar to that of contemporary German scientists as described by Paul Forman.[2] Forman explains this attitude as a defensive reaction to the hostile cultural climate that emerged after the German defeat at the end of World War I. Physicists allegedly tried not only to improve the reputation of their discipline, but also to adapt the content of their science to the values that were held by the cultural elites of the Weimar Republic. This accommodation resulted in the physicists' projecting the prevalent sense of political and cultural crisis on to their own discipline, and it created an expectation of, and a call for, radical changes in the foundations of physics, in particular the abandonment of classical determinism.

Forman's by now classical thesis has attracted both approval and criticism. In particular the suggestion that the (indeterministic) quantum theory can be viewed as a result of a mental reversal among German physicists was a bridge too far for some critics. Others have pointed to the technical, purely physical reasons for the increasing scepticism about determinism within atomic physics.[3] Forman argued such technical factors were insufficient as an explanation, especially because the physicists were far from unanimous on this point. Insofar as they linked their doubts about causality with problems in atomic theory, often they referred to different problems.[4] This defence has not convinced all the critics and the disagreement about Forman's thesis continues to the present day.

It is therefore interesting to examine *whether* and, if so, *how* Dutch physicists motivated their views about causality. What technical arguments did they adduce for their views? Were there any *lebensphilosophische* leitmotifs in the intellectual

---

[2] Paul Forman, 'Weimar Culture, Causality, and Quantum Theory, 1918–1927: Adaptation by German Physicists and Mathematicians to a Hostile Intellectual Environment', *HSPS* 3 (1971): 1–115.

[3] See, for example, John Hendry, 'Weimar Culture and Quantum Causality', *History of Science* 18 (1980): 155–80.

[4] Forman, 'Weimar Culture' (ref. 2), 62.

background of the Dutch physicists that played a role and, if so, to what extent were they a reflection of cultural and social developments in the Netherlands? The topic is too complex to be treated exhaustively here. This exploratory contribution focuses on the physico-philosophical views of the Dutch physicist Adriaan Daniël Fokker. The choice is motivated by the fact that Fokker, more than anyone else in his time, integrated his philosophical and physical views into an original and coherent physical worldview. This worldview is discussed explicitly in Klomp's study on the reception of relativity theory in the Netherlands. Klomp shows that Fokker's ideas about the problem of causality were closely linked to his worldview.[5] At first sight it seems as if Fokker's view was primarily a rather idiosyncratic interpretation of the theory of relativity, which at the time was gaining popularity, but as we shall see, Fokker's views had far-reaching consequences for the content of physical theories.

If Fokker had, in these respects, a special position within the Dutch physics community, this was certainly not the case for his desire to reconcile the science of his time with philosophical and ideological views and values. Although Fokker's worldview was primarily founded on exact physical theories, certain distinguishing characteristics set it apart from nineteenth-century mechanism: an espousal of holism and concreteness and a rejection of materialism and determinism. So as to better understand Fokker's views and to make it plausible that his approach was not just an idiosyncratic preoccupation, some attention will be paid to a few of his contemporaries. Finally, we shall try to place the philosophical interests of Dutch scientists in the interbellum period in a wider context, which will take us back to the Forman thesis.

Relativity and Holism

After a short sojourn at the Delft Polytechnic, Adriaan Daniël Fokker studied physics in Leiden. In October 1913, he obtained his doctorate under Lorentz with a dissertation entitled 'On Brownian Motion in a Radiation Field'. He spent the next year with Einstein in Switzerland and with Rutherford and Bragg in England. After several years of army service during the war — in spite of Dutch neutrality, troops were kept on standby for the full duration — he worked as an assistant to Lorentz and Ehrenfest. In 1923, his growing reputation resulted in a professorship in theoretical and applied physics at the Delft Polytechnic. Four years later, he exchanged this position for that of custodian of the physics cabinet of Teyler's

---

[5] Henk A. Klomp, *De Relativiteitstheorie in Nederland: Breekijzer voor democratisering in het interbellum* (Utrecht: Epsilon, 1997). *See also* Frans van Lunteren, 'Natuurkunde en democratie', *Gewina* 21 (1998): 100–03.

Foundation in Haarlem. This meant that he was again working under Lorentz. After Lorentz's death in 1928, he succeeded him as curator of Teyler's physics cabinet and also as an extraordinary professor of physics in Leiden.[6]

During his time in Zürich Fokker assisted Einstein in his work on the general theory of relativity. As we shall see, it was this theory in particular that had a deciding influence on his physical worldview in subsequent years but it would be incorrect to ascribe his later views exclusively to Einstein's influence. His physical views were in several respects very different from Einstein's and he had already demonstrated in his dissertation an original and philosophical approach to physics. It is significant that in the foreword to his dissertation he not only expressed his thanks to Lorentz and Ehrenfest, both professors of theoretical physics, but also to the Leiden philosopher Gerard Bolland, the leading Dutch Hegelian philosopher and one of the most vociferous critics of modernity. Here, we glimpse one of the roots of Fokker's philosophical interest. The dissertation itself begins with a chapter of 'General Considerations', in which he describes the process of knowledge acquisition. He emphasises in particular the role of memory and the status of theoretical concepts such as that of force, which he characterises as a useful 'fiction'. He describes Einstein's theory of relativity as a Copernican revolution and embraces the relativity of simultaneity as a 'liberating' idea.[7]

It is remarkable that Fokker explicitly preferred Einstein's theory to that of his mentor, Lorentz. This was probably due to Ehrenfest's influence. Bolland, on the other hand, was to condemn the novel developments in physics a year later.[8] Most striking in the dissertation are two theses to be found at the end: 'The difficulty one might experience if one tries to imagine, without recourse to the ether hypothesis, that light emitted by moving bodies travels with the same velocity as that from a source at rest, is the result of ascribing real existence to the "emitted light"', and 'the meaning attached in physics to the word causality is in the process of changing'. These theses indicate the course that Fokker would take in his philosophical approach to physics.

Fokker became one of the most prominent advocates of Einstein's special theory of relativity. This theory heralded a break with a number of traditional notions from classical physics, such as the ether and absolute space and time. While

---

[6] Harry A.M. Snelders, 'Fokker, Adriaan Daniël (1887–1972)', in *Biografisch Woordenboek van Nederland*, vol. 3 (The Hague: Instituut voor Nederlandse Geschiedenis, 1989); Marijn van Hoorn, 'The Physical Laboratory of the Teyler Foundation (Haarlem) under Professor H.A. Lorentz, 1909–1928', *Bulletin of the Scientific Instrument Society* 59 (1998): 14–21.

[7] Adriaan D. Fokker, *Over Brown'sche bewegingen in het stralingsveld en waarschijnlijkheidsbeschouwingen in de stralingstheorie* (Haarlem: J. Enschedé en Zonen, 1913), 1–19.

[8] Klomp, *Relativiteitstheorie* (ref. 5), 41, 64.

Lorentz and many other Dutch physicists held on to these traditional concepts, Fokker was quickly converted to Einstein's ideas. In later attempts to generalise his theory of 1905 Einstein used an inference that the mathematician Minkowski had derived from Einstein's early work. Minkowski argued it was incorrect to view time and space as independent elements; instead they were inextricably linked in a four-dimensional universe. According to Einstein's general theory of relativity the presence of matter caused a curvature in this four-dimensional space-time, which is apparent in the motions of other material bodies.

Fokker was impressed by these ideas, but interpreted them in his own way. In a lecture he gave in Leiden in 1914 he suggested matter itself is nothing but a geometric entity. Then, in his inaugural lecture in Delft in 1923, he again denied any reality claims for matter. The atom was basically 'not an existing object, a thing, but an event, an ... occurrence'.[9] In an earlier lecture given to the society *Diligentia* in The Hague he characterised the 'outside world' as 'a history, full of events, nothing but events. Wherever there is a piece of reality, it is a reality of occurrences. One occurrence influencing another, succeeding each other, generating each other, connected with each other by links of cause and effect'.[10]

In Fokker's view the world is a single interconnected four-dimensional whole. The schemes of space and time in which observed events are located are human constructs, which do not correspond to reality. Just as we compose a three-dimensional image from the two-dimensional images produced by each of our eyes, so we experience what we see at two successive moments as a four-dimensional reality.[11] Reality consists only of events separated by four-dimensional intervals. The partition of space-time into space and time constructed by us is no more than a culture-dependent convention:

> Reality is like the bread that the baker supplies us: although we consume and enjoy bread only cut into slices, that is not a property of the bread. Cutting bread into slices is a local custom. There are nations that never cut bread, but only break it![12]

In relativity theory the length of the four-dimensional interval s that separates two events is usually defined as $s^2 = \Delta x^2 + \Delta y^2 + \Delta z^2 - (c\Delta t)^2$. If the two events are the emission and the reception of a light signal travelling in vacuum, then the

---

[9] Adriaan D. Fokker, *Moderne natuurkunde en techniek* (Eindhoven: Physica, 1923), 12.
[10] Adriaan D. Fokker, 'Grepen uit de relativiteitstheorie', *Natuurkundige Voordrachten: Voordrachten gehouden in de Maatschappij Diligentia te 's-Gravenhage* 51 (1923): 92–93.
[11] Adriaan D. Fokker, *Relativiteitstheorie* (Groningen: Noordhoff, 1929), 54.
[12] Adriaan D. Fokker, *Natuurkundige concepties van buitennatuurkundig belang*, inaugural lecture, Leiden (1928), 2.

interval has a length of zero. In Fokker's own words: 'The emission and reception of ... a light ray constitutes a zero interval. A further hypothesis is that all interactions of bodies in vacuum are transmitted via zero intervals'.[13] The four-dimensional intervals are, according to Fokker, the real intervals between events. If this interval has magnitude zero, the events are not separated but are in direct contact. Fokker took this literally:

> Is being separated by a zero interval still a separation? Or can we say with a somewhat contrived phrase that being separated by a zero interval is a non-separation? We locate a star in the galaxy that we see, at a distance of 900 parsec, three thousand years ago. But the interval is zero. The star was not just over there three thousand years ago, but it still exists now, and here.[14]

As a result of this so-called 'signal contact', fields are not needed in Fokker's worldview to explain action at a distance. Signal contact amounts to the same thing as direct contact. From the four-dimensional perspective, action at a distance requires no explanation so that it is possible 'to develop a point dynamics ... that does not avail itself of considerations concerning a field. In such a point dynamics no energy or momentum is emitted or derived from the field'.[15] In this way Fokker questioned the existence of gravitational fields and denied the existence of the ether and electromagnetic fields.[16] Light is not an electromagnetic field phenomenon, but a direct contact between two events. From this perspective it becomes clear what he meant by the thesis in his dissertation that real existence should not be ascribed to light.

Emitted light is a direct contact with an event in the future, just as received light is a signal contact with an event from the past. Because of the symmetry between past and future, the zero interval possesses a general reciprocity, a very important concept in Fokker's worldview. Where there is action there is, conversely, reaction. In Fokker's view this means also that, just as the past affects the future, an opposite effect on the past is caused by future events. When there is a causal connection there is no need to speak of an earlier state, the cause, which causes the later state, the effect. Suffice to say that two states

---

[13] Ibid., 6.
[14] Ibid., 7.
[15] Adriaan D. Fokker, 'Wederkeerigheid in de werking van geladen deeltjes', *Physica* 9 (1929): 33–42, on 36, 42.
[16] This topic was first explored in Adriaan D. Fokker, 'De localiseering der electromagnetische energie', *Werken van het genootschap ter bevordering van natuur- genees- en heelkunde te Amsterdam*, second series, 9 (1918): 327–34.

are connected. Causal connection is therefore nothing more than 'simply connection'.[17]

Causality and Determinism

This view of causality was in tune with the views of Fokker's contemporaries. Around the beginning of the interwar period there was indeed a shift in the meaning of the concept of causality that Fokker mentioned in his dissertation. The traditional notion of cause and effect was replaced by a functional relation.[18] Fokker, however, derived unexpected and far-reaching consequences from this innovation. In his inaugural lecture in Leiden in 1929, which had the significant title 'Physical Conceptions of Extraphysical Import', his opposition to a determinist interpretation of this functional connection was formulated in prose that had almost existentialist overtones:

> Because of the general reciprocity in physics, mutuality is to be expected of the zero interval ... In the past lies the present, in the present what shall be. We can also state the obverse: in the present lay the past, in what shall be, the present ... This refines the concept of causality. What is in the process of happening is not only produced by what has happened before, but also by the course that history will take subsequently. By this double-sided closure of causality, determinism is cancelled. The supposed shackles and straitjacket of causality are no more than formulas that attempt to express that there is a connection between events, and what this connection is ... Causality is double-edged. It shreds one-sided determinism no less than a one-sided faith in providence. There is only one history, the universe happens only once, and in no other way than it actually happens. There is only one fate, one lot. There is no lottery with lots. We are part of a single unexchangeable, unrepeatable lot. Are we to think of this lot as either fate or as destiny? That would leave out half of the equation.[19]

Fokker clearly did not advocate a customary form of indeterminism. His point is not that the state of a system at a certain moment leaves open several options for the state of that system at a later moment, but that it is impossible to separate these states and to view one as the effect of the other. His holistic vision does not

---

[17] Adriaan D. Fokker, *Filosofie in de natuurkunde: Voordrachten in Teylers Stichting* (The Hague: Nijhof, 1949), 122.
[18] G.J.L. Scheurwater, *Oorzaak en gevolg: Causaliteitsdiscussies in Nederland in de tweede helft van de 19de eeuw* (Delft: Eburon, 1999), 13.
[19] Fokker, *Natuurkundige concepties* (ref. 12), 12.

allow such a separation. It is the interconnected whole that determines the parts, not the other way round. A year after his Leiden lecture he again emphasised this point in an article about 'faith and science':

> When one is at the beginning of an event, it is impossible to know what kind of event is ahead. The development of the event has to be awaited. Yet, when in this context modern writers sometimes expressly refer to indeterminism, one can agree that the beginning is incapable of determining the end, but one should not think that there is no coherence or connection within the event. Rather there is precisely that connection as a result of which the event is an elementary whole, and in which the past can be viewed as determined by the future as much as the future by the past, left by right as much as right by left. If such a connection is to be denoted as causality, it is to be a causality that is refined, from a concept that is one-sided in a certain direction into an all-sided concept. It may be added that, by this refinement of causality, determinism, the one-sided determination by the past, has been sublimated into an all-sided determination by the whole itself.[20]

In popular essays he later formulated the same singular views of causality and ascribed an almost metaphysical significance to the zero intervals, 'which signify the presence of what is absent, the actuality of the past and the future', and in which we 'can easily see an image of the omnipresent, always existing, unbounded eternity of God'.[21] How seriously he always took the indissoluble connection of past and future is pointedly manifest in a letter that he wrote to Einstein in 1955. In it he asked:

> Why do we only remember past events, and not also future events? Is this determined biologically? What then is the purpose in life of this? Was the visionary Tiresias a person in whom this biologically determined inhibition worked insufficiently, so that he also experienced the zero intervals pointing in the other direction?[22]

These remarkable questions sprang from his well-known views, as is apparent from another passage, where Fokker again describes the zero interval as a signal contact:

> Zero interval means separation zero, which means contact! And the theory actually comprises action contacts with retarded and with advanced potentials, or by

---

[20] Adriaan D. Fokker, 'Geloof en natuurwetenschap', *De Smitse* 4 (1929): 295.
[21] Fokker, *Filosofie* (ref. 17), 101–02.
[22] Adriaan D. Fokker to Albert Einstein, 14 February 1955, Museum Boerhaave Leiden, Archive Adriaan D. Fokker, inventory number 267 e.

means of emitted and absorbed quanta! Just as Newtonian gravitation was an action at a distance, in this view we meet a contact at a distance, a telecontact. Interaction involving transfer of momentum and energy takes place even without weakening due to distance. In the traditional view fields are conceived with a critical velocity, because action at a distance can only be explained as transmission of a local action.[23]

Einstein reacted after five days with a letter in which he rejected Fokker's proposal. He wrote that he hoped relativity theory would be absorbed in a unified field theory. 'Then there will be no "things" and "interaction" between them, but only "fields"'.[24] Fokker didn't allow himself to be discouraged. In later work he called the zero interval a 'now-here-ray', the absorption of light a 'far-touching' or *telethigma*, and the emitted light a 'distant grasp', a *telehapsis*.[25]

Advanced and Retarded Potentials

That Fokker was serious about his eccentric views is also apparent from the fact that he tried to adapt existing theories to bring them into line with his views. Given a system of electric charges, Maxwell's equations enable the physicist to determine the resulting potentials for any point in space and at any moment in time. Mathematically there is for each point in space and moment in time a double solution. The so-called *retarded* potential at a certain point is determined by the states of charges present elsewhere at *earlier* times, so that all these states are separated from the position and time of the potential to be determined by, what Fokker called, zero intervals. However, there is also a solution in which the potential is determined by the states of the same charges at a *later* time, the so-called *advanced* potential. Because of traditional causality arguments, physicists habitually ignore the latter solutions as 'unphysical'. Fokker took a different view.

As early as 1921, Fokker proposed the inclusion of the advanced potentials in the physical description of electrodynamic systems, in an article entitled 'Stationary Electron Motions without Radiation Resistance'.[26] He suggested that

---

[23] Ibid.
[24] Klomp, *Relativiteitstheorie* (ref. 5), 60.
[25] Adriaan D. Fokker, *Tijd en ruimte, traagheid en zwaarte. Chronogeometrische inleiding tot Einsteins theorie* (Zeist: De Haan, 1960).
[26] Adriaan D. Fokker, 'Stationaire elektronenbewegingen zonder stralingsweerstand', *Physica* 1 (1921): 109.

his proposal offered a solution for a problem in Bohr's quantum theory. According to Bohr's theory of the atom of 1913, electrons move in circular orbits around a heavy, positively charged nucleus. But according to classical electrodynamics the electrons would continually radiate energy, decelerate and quickly be captured by the nucleus. To prevent this undesirable effect, Bohr simply postulated that, contrary to the classical theory, electrons in so-called stationary orbits do not radiate energy. Electromagnetic radiation is only exchanged during the saltational transition of an electron from one stationary state to another.

Fokker claimed he could explain why an electron in a stationary state doesn't lose energy. In his description of the atomic system he combined retarded and advanced potentials. By choosing as the real potential the average of the two, the terms corresponding to the radiation loss disappeared. However, he didn't explain how this seemingly elegant solution related to Bohr's quantum theory of radiation, which was the basis of the successful explanation of the spectral frequencies of the hydrogen atom.

After his appointment in 1928 as Lorentz's successor in Leiden, Fokker worked out his ideas about general reciprocity in physics. His inaugural lecture, 'Physical Conceptions of Extraphysical Import', can be viewed as a programmatic exposition.[27] The retarded and advanced potentials were, as zero intervals, integrated into the four-dimensional system of the special theory of relativity. The scientific papers in Dutch, German and French published by him after his appointment advocated a physical worldview in which reality consisted only of particles (or rather events) separated by relativistic intervals, characterised by general reciprocity.[28]

Fokker adduced two advantages in his approach. First, he pointed to the Lorentz invariance of the interaction between the particles realised in this way. As a result, Lorentz's electron theory was brought 'into line with the demands of relativity theory'. Secondly the introduction of advanced potentials made it possible to abolish action at a distance without having to introduce 'considerations concerning a field'.[29] Together with the electromagnetic field, light — which was after all no more than a periodic disturbance of that field — lost its raison d'être.

Although Fokker's theoretical proposal was not taken up, he was not the only one to attempt to get rid of fields in physics by using advanced potentials. It seems that the recluse and enigmatic Dutch physicist Tetrode independently hit upon the

---

[27] Fokker, *Natuurkundige concepties* (ref. 12).
[28] Adriaan D. Fokker, 'Wederkeerigheid'(ref. 15), 33–42; Adriaan D. Fokker, 'Ein invarianter Variationssatz für die Bewegung mehrerer elektrischer Massenteilchen', *Zeitschrift für Physik* 58 (1929): 386–93; Adriaan D. Fokker, 'Théorie relativiste de l'interaction de deux particules chargées', *Archives du Musée Teyler, Haarlem* 3, no. 7 (1931): 176–82.
[29] Fokker, 'Wederkeerigheid' (ref. 15), 42.

same idea, also criticising the notion of unidirectional causality.[30] Much later the American physicists Feynman and Wheeler worked out a similar scheme in the hope of solving other problems involving causality in quantum mechanics.[31] In Fokker's case, however, this approach was closely linked to his holistic and time-symmetrical view of reality in which reality could not be ascribed to matter or fields and neither time nor space.

Lebensphilosophie

It is perhaps somewhat daring to label Fokker's worldview as a form of *Lebensphilosophie*, or philosophy of life. There are no direct indications that Fokker felt attracted to, or was influenced by, German or French philosophers who are usually considered representatives of this movement. And yet there are many elements in Fokker's worldview that seem to legitimise the use of the term. *Lebensphilosophie* is not a well-defined concept, but it is usually characterised by the following attributes: an aversion to the scientific naturalism associated with the modern sciences and an appreciation of unanalysed, immediate human experience or intuition. This was often associated with a negative attitude toward abstract analysis, reductionism, determinism and materialism, and an emphasis on holism, freedom and time — or rather the *experience* of time — as elements constitutive of human existence. Viewed in this way, Fokker's philosophical views certainly qualify as *lebensphilosophisch*.

Initially the characteristics mentioned above seem to imply a downright negation of physics and certainly of the theoretical physics of which Fokker was a representative. However, in the specific accents that Fokker placed in his physical worldview we can easily recognise these characteristics. As we have seen, Fokker deprived the physical world of its material character by emphasising four-dimensional *events*, which together constituted reality. Events, which are so much closer to experience than spatial material objects, are the true substance of Fokker's relativistic world. In quantum physics he also found indications for the 'dematerialization of matter'.[32] He compared the quantum-mechanical picture of the atom with a 'dance', a 'way of moving'[33] without

---

[30] Hugo Martin Tetrode, 'Über den Wirkungszusammenhang der Welt: Eine Erweiterung der Klassischen Dynamik', *Zeitschrift für Physik* 10 (1922): 317–28.

[31] John A. Wheeler and Richard P. Feynman, 'Interaction with the Absorber as the Mechanism of Radiation', *Reviews of Modern Physics* 17, no. 2–3 (1945): 157–81, on 157, with references to Fokker and Tetrode.

[32] Fokker, *Natuurkundige concepties* (ref. 12), 15.

[33] Adriaan D. Fokker, 'Over te betwijfelen evidenties', *Het kouter* 1 (1936): 12–24, on 20.

material substrate: 'radiation together with the quantum has made us overcome materiality'.[34]

Fokker's vision can also be considered a rehabilitation of time, or rather of the experience of time. Whereas static three-dimensional space had been the traditional foundation of physics, Fokker emphasised the experience of time as an essential element of reality: 'The elementary fact [is] an occurrence, a process ... we live in the fundamental mystery of the stream of time'.[35] He characterises man as 'rooted and growing in the fluent reality of our life's events flowing past'.[36] The dynamic, process-like character of reality — of *experienced* reality — is a central element in Bergson's philosophy. According to Fokker, abandoning the supposed reality of the three-dimensional world was 'the first step towards fulfilling the task of understanding the phenomena dynamically, as a whole, instead of as a piecemeal compilation of successive static states'.[37]

This brings us immediately to the holistic character of Fokker's worldview. Reality is a concrete, one-off history, but this history is not, as in classical physics, determined by one momentary world state in the past or in the future. Such a state could not even be determined unambiguously because of the relativity of the concept of simultaneity. Only the entire history determines the whole. It is not by accident that Fokker expressed himself in similar terms about man: 'We happen as a history. Nobody knows us completely who doesn't know our history to the end. Our person is only complete when we die'.[38]

Fokker saw this all-embracing vision as a negation of the classical determinism and as a liberation from the 'supposed shackles and straightjacket' of causality. Interestingly, his own view of causality can be regarded as a synthesis of the traditional view and the teleological view, which explains a process from its final destination. The German historian Spengler, one of the most prominent and influential representatives of the *Lebensphilosophie*, saw fate as the real antithesis to the Faustian concept of causality characterising Western culture. A return to a teleological approach in historiography already raised many eyebrows and a reintroduction of teleological principles in physics would be inconceivable. Yet Fokker's claim that the present was in part determined by the future came close to such a move. As we have seen, he worked this idea out in a particular physical form by the introduction of advanced potentials.

---

[34] Fokker, *Filosofie* (ref. 17), 130.
[35] Adriaan D. Fokker, 'Relativistische studie: proeve van antwoord aan prof. dr. G. Heymans', *De Gids* 86, no. 4 (1922): 244–71, on 249; quoted in Klomp, *Relativiteitstheorie* (ref. 5), 56.
[36] Fokker, *Filosofie* (ref. 17), 101–02.
[37] Fokker, 'Relativistische studie' (ref. 35), 267; quoted in Klomp, *Relativiteitstheorie* (ref. 5), 57.
[38] Fokker, 'Evidenties' (ref. 33), 14.

Fokker's questioning of traditional causality is not the only conceivable common ground with Spengler's ideas. His remark quoted earlier to the effect that the classical concepts of space and time are culturally determined also has a Spenglerian flavour. In his *Untergang des Abendlandes* Spengler treats scientific views as culture-dependent expressions of particular civilisations. He regards civilisations as organisms that live through cyclical patterns of ascent, flourishing and decline. As Western civilisation was in decline, the same was true for Western science. Seemingly timeless certainties (views of time, space and causality) would give way to novel insights and in this way clearly advertise their culture-dependency.

Finally, it is remarkable how Fokker links an apparently abstract and esoteric theory such as relativity theory with visualisability. His emphasis on events had to do with the significance that he attached to concepts closely related to immediate experience. Fokker's talks for wide audiences, in which he tried to demonstrate empirically how the general theory of relativity could be abstracted directly from experience, are another example of this attitude.[39] As a teacher and as president of a national education committee he was strongly opposed to a formal, axiomatic presentation of physical theories as closed systems. Teaching had to be clear, ostensive, and as far as possible linked to direct experience.

Like many in his time Fokker ascribed a liberating effect to relativity theory. It was this theory that had liberated not only physics, but also philosophy from the straightjacket of the classical, supposedly self-evident notions of space and time, Newton's laws of motion and Newton's equally indubitable law of gravitation. Kantian certainties ascribed to our cognitive faculties had succumbed under the corrective pressure of experience. Experience and empiricism implied contingency and therefore freedom. In Fokker's words:

> Self-evident truths must be distrusted from time to time. Sometimes the belief in what seems self-evident obstructs the way to deeper insight. It is then important to view that self-evident truth in its restrictedness, narrowness, almost backwardness, however useful and inevitable it may be in its restricted validity, and to attempt to free our thinking from it.[40]

---

[39] Adriaan D. Fokker, 'Zwaarte en traagheid', *Voordracht Bataafsch Genootschap der Proefondervindelijke Wijsbegeerte te Rotterdam, Verslag der voordrachten* 2 (1924): 207–11; Adriaan D. Fokker, 'Schoolproeven als inleiding tot Einsteins gravitatietheorie', *Physica* 4 (1924): 149–55; Adriaan D. Fokker, 'Proeven over zwaarte en traagheid', *Handelingen van het Nederlandsche Natuur- en Geneeskundig Congres gehouden te 's-Gravenhage* 20 (1925): 162–64.
[40] Fokker, 'Evidenties'(ref. 33), 23.

Contemporary Dutch Scientists

Fokker was probably exceptional in the Netherlands in developing an all-embracing physico-philosophical vision, a view that in addition influenced his physical theorising. He was certainly not the only physicist with an interest in philosophy, though. Hendrik Antoon Kramers, appointed professor of theoretical physics in Leiden in 1934, noticed the same thing; witness the introduction of his inaugural lecture:

> More strongly than has been the case for many years, in modern professional literature and in modern textbooks writers "philosophize". We have to go back nearly a hundred years to encounter a similar phenomenon. That was the time of Romanticism, when German scholars in particular interlaced their technical writings with metaphysical discussions ... And our time teems with discussions of causality, determinism, about the subjective-objective opposition and what not.[41]

Kramers, who regarded Baudelaire's *Les Fleurs du Mal* as similar in spirit to theoretical physics, fits this portrait of an era very well himself.[42] In the early twenties, when he worked on Bohr's theory of the atom in Copenhagen, Kramers was only too willing to abandon the determinism of classical physics.[43]

It seems natural to link this philosophical interest with the revolutionary developments in the foundations of physics, in particular those concerning quantum mechanics. And yet this is not a sufficient explanation. The Dutch physicists who entered the philosophical debate during the interwar period had already shown an interest in this area before the war, in some cases even before Einstein's first work on the theory of relativity. In the last decade of the nineteenth century the Amsterdam physicists Kohnstamm and Van der Waals Jr. attended the lectures of the neo-Kantian philosopher Bellaar Spruyt. Together they edited and completed his unfinished *History of Philosophy* after his death.[44]

Several young Leiden physicists were also captivated by philosophy. Clay, who was mentioned above, had early on fallen under the spell of the philosopher Bolland. He also moved in the philosophical circles around the writer and psychiatrist Frederik van Eeden. During the war we also encounter another student of

---

[41] Ibid., 23.
[42] Hendrik B. Casimir, *Het toeval van de werkelijkheid* (Amsterdam: Meulenhoff, 1983), 178.
[43] Hans Radder, 'Kramers and the Forman Theses', *History of Science* 21 (1983): 165–82, esp. 172–73.
[44] Philip Kohnstamm, Johannes Diederik van der Waals, Jr., and G.H. Leignes Bakhoven, eds., *De geschiedenis der wijsbegeerte naar de dictaten van wijlen prof. C.B. Spruyt* (Haarlem: de Erven F. Bohn, 1904).

Lorentz's, the Utrecht physicist Ornstein, in those circles, introduced by the mathematician Brouwer.[45] Young Leiden physics students such as Coster, Burgers and Tinbergen instead took refuge in socialism. The wide cultural interests that Kramers entertained as a student actually made his teacher Ehrenfest doubt whether he was suited to a scientific career.[46]

In the interwar period we find all these inspired, enthusiastic minds in the, as yet scarce, physics chairs in the Netherlands.[47] In the same period Clay, Van der Waals Jr. and Kohnstamm were considered serious candidates for philosophy chairs.[48] Immediately after the war Kohnstamm argued for the creation of a major course in philosophy within the Amsterdam faculty of mathematics and sciences.[49] Kramers would later make a case for establishing a chair of methodology and philosophical analysis of the exact sciences.[50]

During and after the war Van der Waals Jr., Kohnstamm and Clay turned against the Kantian view of causality as a necessary connection between cause and effect.[51] In the case of the first two men similar motives played a role as with Fokker. Van der Waals Jr. felt that the old view of causality was incompatible with the idea of moral responsibility and could therefore not be combined with an ethical philosophy. In 1902 he called upon philosophers to safeguard the freedom of the human will by allowing for an element of spontaneity. For similar reasons the physicist Philipp Kohnstamm grasped every opportunity to undermine the thesis of the strictly deterministic character of physics. Before the war he had already expressed his doubts about the traditional dogma of causality and believed that he could adduce physical arguments for this view. For Kohnstamm his philosophical leanings, together with political commitment and his conversion to Christianity in 1914, formed the materials for the philosophical-pedagogical doctrine that he later developed, which he called Biblical personalism.[52]

---

[45] Dirk van Dalen, *L. E. J. Brouwer: Een Biografie* (Amsterdam: Bert Bakker, 2001), 378.
[46] Max Dresden, *H. A. Kramers: Between Tradition and Revolution* (New York: Springer, 1987), 92–94.
[47] Kohnstamm, Van der Waals, Jr. and Clay in Amsterdam; Ornstein and Kramers in Utrecht (in 1934 Kramers left Utrecht for Leiden); Burgers in Delft and Coster in Groningen.
[48] Van Dalen, *Brouwer* (ref. 45), 378.
[49] Ibid., 254
[50] Dresden, *Kramers* (ref. 46), 496. Eventually such a chair was established in Utrecht. The first person to hold the chair was J.B. Ubbink, a student of Kramers.
[51] G.J.L. Scheurwater, *Oorzaak en gevolg: Causaliteitsdiscussies in Nederland in de tweede helft van de negentiende eeuw* (Delft: Eburon, 1999), 279–81.
[52] For an extensive discussion, see Philip Kohnstamm, *Schepper en schepping: Een stelsel van personalistische wijsbegeerte op bijbelschen grondslag*, vol. 1, *Het waarheidsprobleem: Grondleggende kritiek voor het christelijk waarheidsbewustzijn* (Haarlem: H.D. Tjeenk Willink & Zoon, 1926).

It is not inconceivable that in his later years Kohnstamm influenced Fokker's philosophical views. The two physicists maintained regular contact. Not only did they share a deep interest in matters of education, they were also linked by family ties; Kohnstamm was Fokker's brother-in-law. It is therefore helpful to pay more attention to Kohnstamm's persistent opposition to determinism. Kohnstamm had publicly renounced determinism in physics as early as 1908. The occasion was his inaugural lecture at the University of Amsterdam. And while he borrowed his ammunition partly from Boltzmann's probabilistic interpretation of the second law of thermodynamics, he did not hide his strong aversion to naturalism. Moreover, he added that even if Boltzmann's views were mistaken and the laws of nature absolute, this would not affect human freedom.[53]

In 1914 Kohnstamm founded a new philosophical journal entitled *Synthesis*. Its aim was to reflect upon modern life in such a way as to overcome the old chasm between intellectualism and science on the one hand and intuition and faith on the other. In his article 'The Rise and Fall of the Concept of Law of Nature', published in 1916, he resumed his attack on determinism. Again he left no doubt about his deeper motives. If determinism were implemented consistently, man would be made into a machine, a vision that Kohnstamm branded a 'naturalistic nightmare'.[54] It was determinism, as he emphasised again later, that 'destroys the personality', because the personality 'exists only as long as it can develop new, unexpected aspects'.[55]

In his view, physics provided no support for determinism. Its laws were valid only approximately (Boyle's law and, as had become clear now, Newton's laws of motion), or merely limited freedom (the law of conservation of energy), and in both cases they were non-deterministic. His final conclusion was that:

> We are forced to give a place in our worldview, alongside abstraction to concreteness, alongside the general to the individual, alongside the law of nature to the constellation, alongside scientific thinking to historical thinking, alongside causality to purposiveness, alongside necessity to personality.[56]

A few years later he formulated this as follows: 'In freedom is found what is unique, historically individual, always new and unexpected, in short, life itself'.[57]

---

[53] Philip Kohnstamm, *Determinisme en Natuurwetenschap*, inaugural lecture, University of Amsterdam (1908).
[54] Philip Kohnstamm, 'Ontwikkeling en onttroning van het begrip natuurwet', *Synthese* 3 (1916): 53–132, on 80.
[55] Philip Kohnstamm, 'Over natuurwetten, wetmatigheid en determinisme', *Onze eeuw* 22 (1922): 292–336, on 312, 334.
[56] Kohnstamm, 'Ontwikkeling' (ref. 54), 131.
[57] Kohnstamm, 'Over natuurwetten' (ref. 55), 336.

Later he also found arguments against determinism in the theory of relativity. He consulted Fokker and from him he derived the view of the primacy of 'the knowledge of the four-dimensional whole of spacetime', which cannot be derived from a momentary 'three-dimensional given'. The theory of relativity had, Kohnstamm claimed, led to the rehabilitation of 'Time', which he considered part of 'the Essence of Man'.[58] While Fokker's writings regularly touch on the prevailing *Lebensphilosophie*, Kohnstamm's writings are steeped in them. Because of the overlap of the views of the two men, Kohnstamm's more open confessions provide additional support for the interpretation of Fokker's work presented above.

Kohnstamm also managed to convince several other Dutch physicists of his view that modern physics did not require or imply determinism. Among them was the young student Gerard Sizoo, who in 1921 backed up Kohnstamm's views in a talk for his fellow students and continued doing so when he became the first physics professor in the newly founded science department of the VU University of Amsterdam.[59] Another possible convert was Lorentz's Leiden successor Paul Ehrenfest. In 1928 he defended Kohnstamm's indeterminist campaign in a letter to the Groningen astronomer Kapteyn in the following words:

> Kohnstamm ..., driven by the needs of his heart, had already used the physical insights of those days for an acute criticism of simple-minded determinism. Just because I always saw clearly that Kohnstamm knows what physics is and what it is not, he, as few of my other friends, was able to help me understand views, that I initially opposed since I was full of misunderstanding and distrust.[60]

Philosophical leanings and even a penchant for *Lebensphilosophie* were not only the reserve of physicists. Forman himself refers to the intuitionism of the mathematician Brouwer, which gained great support in Germany as a welcome alternative to formalism.[61] Brouwer, with his strong mystical proclivities, viewed mathematics as more akin to psychology, philosophy and even theology than to the sciences, which were focused on the material world.[62] In his strongly anti-intellectualist pamphlet, *Life, Art and Mysticism* (1906), he specifically rejected

---

[58] Kohnstamm, *Schepper en schepping* (ref. 52), 293–94, 318.
[59] Ab Flipse, *'Hier leert de natuur ons zelf den weg': Een geschiedenis van Natuurkunde en Sterrenkunde aan de VU* (Zoetermeer: Meinema, 2005), 55, 118.
[60] Paul Ehrenfest to Jacobus C. Kapteyn, 12 February 1928, APE, ESC 6, Section 5.
[61] Forman, 'Weimar Culture' (ref. 2), 60–61.
[62] Van Dalen, *Brouwer* (ref. 45), 465.

rigid causality: 'You will recognize your *free will* ... through the walls of causality, "miracles" continue to glide and flow, visible only to the free and enlightened ... over and above physical causality you can see a clear direction in your own life's course, determined by the self ..."[63]

Like Brouwer, his later Amsterdam colleague Mannoury also played a central role in the philosophical circles mentioned above. In the Dutch life sciences there are likewise indications of leanings toward *Lebensphilosophie*. Bert Theunissen has pointed out that holistic views were prevalent among morphologists and zoologists. The Utrecht professor of histology and embryology Jan Boeke, for example, expressed his opposition to the 'analytical and materialistic' tradition of the previous century, which was blind to the 'harmony' in nature.[64]

A Philosophical Movement

If these examples are indicative of a widespread attitude among Dutch scientists, their views were in line with those among many self-styled Dutch philosophers. Shortly after the turn of the century several commentators signalled the emergence of a broad 'philosophical movement' in the Netherlands. Although this movement lacked homogeneity and a central organisation it showed several characteristics which, despite its eclecticism, suggest at least some coherence and therefore warrant the use of the term *movement* in the singular. Its most general characteristics were an espousal of subjective idealism, and a rejection of scientific naturalism and materialism and a focus on existential problems rather than abstract questions. Representatives were mostly philosophical dilettantes. With the exception of the self-made Hegelian philosopher Bolland, the academic philosophers, who were predominantly neo-Kantian, did not participate in the movement.

The central organ of the movement was the *Tijdschrift voor Wijsbegeerte* (Journal of Philosophy), founded in 1907, and its central spokesman the journal's chief editor and co-founder, Johan Bierens de Haan. His prime aim was to fuse Spinozism and idealism into a life-oriented system. This was to be his philosophical response to the problems of modern 'dieszeitig' society, in which specialisation, industrialisation, consumerism and big-city life had isolated 'the

---

[63] Luitzen E.J. Brouwer, 'Life, Art and Mysticism', *Notre Dame Journal of Formal Logic* 27 (1996): 389–429, 394 (emphasis in the original).

[64] Bert Theunissen, 'Jan Boeke en de harmonie van het organisme: Een case-study van de totaliteitsidee in de 20ste-eeuwse Nederlandse biologie', *Tijdschrift voor de geschiedenis der geneeskunde, natuurwetenschappen, wiskunde en techniek* 11 (1988): 58–74.

Eternal as an unreachable Jenseits'.[65] Among the other founders of the journal were the physicist Clay and the physicist and former Christian-anarchist Lodewijk Grondijs, another follower of Bolland. When war broke out, Grondijs immediately resigned from his job as physics teacher and became a war correspondent in Belgium, France and finally Russia, where he sided with the White army against the Bolsheviks. After the war he moved to Paris, where he studied Byzantinology, starting a new academic career in the latter field after his return to the Netherlands in 1928.[66]

The autodidact Bolland had managed to secure the Leiden chair of philosophy in 1896 by becoming an expert in German idealist philosophy. Around 1900 he single-handedly effected a revival of Hegelianism in the Netherlands by arranging a new edition of Hegel's works and by publishing his own neo-Hegelian *Pure Reason: A Book for the Friends of Wisdom*.[67] In the latter work he castigated modern society and revealed the illusory nature of the notion of progress, which he blamed for the demise of genius and the crippling of culture. Modern science filled him with disdain. Even Lorentz could not escape his scorn. His electron theory was little more than 'a fabrication, already known to the ancient Greeks'. Einstein's theory of relativity was dismissed as a 'a caper of mathematics, showing off as physics'. The eccentric, but charismatic Bolland attracted large crowds wherever he chose to utter his inimitable dialectical expressions.[68]

Another central figure in the movement was the writer-physician Van Eeden, mentioned above. As we have seen, several scientists moved in his circle, at least for some time, as did a number of writers and artists. What they shared was not a well-defined social or philosophical view and emerges perhaps most clearly in Van Eeden's plea for the foundation of an international academy for philosophy. This school should be:

> ... truly free, that is, not under the influence of any sect or party, but totally universal, seeking for the unity in all religions and complementing and broadening science with all those functions of the human mind that so far are not methodically practised in Western universities. This includes also mysticism and

---

[65] Siebe Thissen, 'De nalatenschap van Erasmus: Een wijsgerige beweging en haar cultuurtaak', *Wijsgerig Perspectief* 34, no. 5 (1994/1995): 167–71; and Siebe Thissen, 'Een wijsgerige beweging in Nederland en haar publieke rol (1850–1922)', *Krisis: Tijdschrift voor Filosofie* 60 (1995): 22–39.
[66] M.C. Jansen, 'Grondijs, Lodewijk Hermen (1878–1961)', Biografisch Woordenboek van Nederland, vol. 4 (The Hague: Instituut voor Nederlandse Geschiedenis, 1994).
[67] Gerardus J.P. Bolland, *Zuivere Rede: Een boek voor Vrienden van de Waarheid* (Leiden: Adriani, 1904).
[68] Willem Otterspeer, *Bolland: Een Biografie* (Amsterdam: Bert Bakker, 1995), 457, 489.

occultism, philosophy of religions and of the so-called fine arts. And all this directly linked with practical life.[69]

Indeed, neither the traditional churches nor the new scientific doctrines were able to fulfill the intellectual and spiritual needs of large parts of the higher echelons of Dutch society.[70] The years before the war were a golden age for synthetic movements that provided sense and meaning, like Spinozism, Hegelianism, utopianism, socialism, Christian anarchism, spiritism, Buddhism, theosophy and vegetarianism.[71] When the International School of Philosophy was founded in Amersfoort in 1916, it was expected to usher in a new age, in which the ills of modern society would be healed.

The leading representatives of the movement, men like Bolland, Bierens de Haan and Van Eeden, travelled through the country to give a series of lectures in various cities to mixed audiences. When Van Eeden gave a number of lectures in Hilversum and Amsterdam in the winter of 1901–02, he was confronted with 'total freethinkers or materialists, theosophists, strictly orthodox Catholics, orthodox Calvinists, modern Protestants, liberal Catholics, people without a specific religious persuasion, social democrats, free socialists, revolutionary and Christian anarchists, free-thinking democrats and liberal capitalists'. Local branches were established in various towns. As the example of Grondijs suggests, physicists were in the forefront of the movement. Both Clay, who acted as right-hand man to Bolland, and Kohnstamm presided over such local circles, in Leiden and Amsterdam respectively.[72]

The prominent place of physicists in the philosophical movement may seem surprising in the light of its seemingly anti-modern and anti-scientistic nature. In late nineteenth-century Europe the crisis of liberalism, manifesting itself in the so-called 'social question', and growing concerns about modern industrial society in general, often resulted in a fierce criticism of science. Science had, on the one hand, propelled the growth of industry and modern technology and, on the other, promoted a materialist worldview in which there was no place for culture, values and spirituality. Many European critics

---

[69] Frederik van Eeden, quoted in H.W. Schmitz, 'Hogeschool, Academie of School? De significi en de oprichting van de Internationale School voor Wijsbegeerte', in *Filosofie in Nederland: De Internationale School voor Wijsbegeerte als ontmoetingsplaats, 1916–1986*, ed. A.F. Heijerman and M.J. van der Hoven (Meppel: Boom, 1986), 14.

[70] Editorial note, 'Ter Inleiding', *Geschiedenis van de wijsbegeerte in Nederland* 9 (1998): 3–7.

[71] Jan Romein, *Op het breukvlak van twee eeuwen* (Leiden: Brill, 1967): 631–51. Romein points to the close link at the time between spirituality and socialism.

[72] Klaas van Berkel, *Citaten uit het boek der natuur* (Amsterdam: Bert Bakker, 1998), 243.

declared science to be bankrupt.⁷³ In the Netherlands, however, feelings did not become as heated as in other countries, such as France. Science was at times accused of one-sidedness and blindness to the mystery of things,⁷⁴ yet there was hardly any blanket rejection of science. Instead, spiritual and occult movements presented themselves eagerly as 'scientific'. As Van Eeden's words suggest, the idea was rather that science needed to widen its outlook — for example, by being more receptive to the immaterial, to purposiveness, and to truths and insights acquired by intuition.⁷⁵

As the fault was not with science in general, but with a one-sided view of science, those scientists who shared an aversion to crude materialism could easily join in with other cultural critics. The more so as recent developments in physics and biology confirmed their belief that science was gradually moving away from this outdated view of nature and that it provided support for idealism, holism and contingency. Having studied Lorentz's (!) principle of relativity, Van Eeden could announce triumphantly that materialism had finally been overcome. In fact, it was probably a similar perception that motivated scientists to adopt leading roles in the philosophical movement.

Actually, in the eyes of the physicists the abuse of modern science, often based on serious misunderstanding, was most prevalent in the new sciences of man, such as criminal anthropology and experimental psychology. An example in case was the doctrine of the Groningen philosopher-psychologist Gerard Heymans. The widespread desire for a synthesis of the material and the spiritual realm found satisfaction not only in a revived Spinozism, but also in numerous other monistic systems. Haeckel's 'World Soul' had its Dutch counterpart in Heymans' *psychic monism*. According to Heymans' teachings all knowing subjects are part of one embracing world consciousness and all processes they observe in the external world are ultimately psychical in nature. The laws of nature are only a reflection of the laws of consciousness, and it is the task of the psychologist to chart these laws. According to Heymans these laws were strictly logical in character, they provided a foundation for various Kantian self-evident

---

[73] Roy Macleod, 'The "Bankruptcy of Science" Debate: The Creed of Science and Its Critics, 1885–1900', *Science, Technology, and Human Values* 7 (1982): 2–15; see also J.L. Heilbron, 'Fin-de-Siècle Physics', in *Science, Technology, and Society in the Time of Alfred Nobel*, ed. C.G. Bernhard et al. (Oxford: Pergamon, 1982), 51–73.

[74] Mary Kemperink, *Het verloren paradijs* (Amsterdam: Amsterdam University Press, 2001), 207–20.

[75] Frank Huisman, 'Wie geneest: De strijd om de culturele autoriteit in de Nederlandse gezondheidszorg', in *De opmars van deskundigen: Souffleurs van de samenleving*, ed. Frans van Lunteren, Bert Theunissen and Rienk Vermij (Amsterdam: Amsterdam University Press, 2002), 109–14.

truths and inexorably determined the states of consciousness. They left no room for a genuinely free will.[76]

While the physicist Lorentz was captivated by Heymans' explanation of the remarkable parallelism of external phenomena and the theories produced by the human mind,[77] the next generation of physicists considered here was fiercely opposed to Heymans' views. To them, his deterministic idealism seemed as barren and suffocating as the materialism they detested. It ignored the open character of empirical science and was basically anti-religious and used to support a conservative view of society based on timeless and indubitable certainties. It isn't surprising, then, that Clay, Van der Waals Jr., Kohnstamm and Fokker turned first and foremost against Heymans' view of causality. On several occasions this resulted in polemical exchanges with the Groningen philosopher.[78]

But Heymans was not alone in his determinist view of man. In the late nineteenth-century the naturalist conception of man as a plaything of fortune, or rather of heredity and social environment, was not restricted to Dutch novelists. Modern theologians, pedagogues, physicians, anthropologists and jurists emphatically denied the very notion of free will. If early twentieth-century Dutch scientists were particularly eager to question causality and determinism in their own discipline, part of the reason may well be found in the radical determinism prevalent among Dutch representatives of the humanities in the late nineteenth century.

Fokker and the Forman Thesis

The war gave a new and powerful impulse to the aforementioned concerns about modernity. In this respect there are obvious parallels between the situation in Germany described by Forman and that in the Netherlands. Contrary to Germany, the Netherlands had not descended into an economic and political crisis immediately after the war. But the sense of despondency that got a grip on, in particular, Central European intellectuals and artists was felt there too. The war had made it clear how destructive modern technology could be, and the distrust in the exact sciences was reinforced. Dutch scientists and mathematicians such as Kohnstamm and Dijksterhuis drew attention to an anti-intellectual spirit in the country in the early twenties, just as did a somewhat dispirited Heymans.[79] The historian Huizinga wondered if the university,

---

[76] Klomp, *Relativiteitstheorie* (ref. 5), 71–80.
[77] Bert Theunissen and Henk Klomp, 'H.A. Lorentz' visie op wetenschap', *Gewina* 21 (1998): 1–14.
[78] Klomp, *Relativiteitstheorie* (ref. 5), 71–89.
[79] Ibid., 9–10, 87.

the last stronghold of Western civilisation, should not screen itself off from society.[80]

The cultural pessimism that emerged after the war was not short-lived. It would reach its climax in the 1930s after the worldwide depression and the takeover in Germany by the Nazis. The most prominent exponent of this sentiment in the Netherlands was the very same Huizinga with his deeply pessimistic *In the Shadow of Tomorrow* of 1935. However, many others joined him in this despondency, often showing signs of having been influenced by Spengler.[81]

The Dutch physicists had another reason to worry about their image. On the one hand, the growth of university staff and the rising costs of equipment required more and more financial support; on the other hand, the even faster increasing numbers of students demanded new types of careers for scientists and expansion of the existing ones, which were mainly available in the teaching profession. Moreover, the academic scientists increasingly felt the competition of the professional schools, such as the Delft Polytechnic and the Wageningen agricultural school, which were lobbying for academic status. All this implied that physics had to emphasise its practical significance and at the same time attempt to shed its materialistic connotations. As had been the case before the war, an aversion to materialism manifested itself even in the highest political circles.[82] In 1921, the Minister of Education, former clergyman J.T. de Visser, managed to seriously misapprehend Kohnstamm's plans for a philosophy chair within the science faculty, as is shown by the phrasing of his support:

> It has become clear to me that recently a scientific study has arisen, particularly in physics, which more than thus far wishes to emphasize, if I may use the expression, the psyche of plants and animals, and which thus, to use the terminology of the present day, over against a one-sided materialistic approach wants to do justice to a more idealistic approach even in the domain of physics.[83]

During the interwar period Dutch physicists were acutely aware of the need to create public support for their discipline. The Utrecht professor of experimental physics, Ornstein, emphasised the significance of science for technological and

---

[80] Ibid., 11.
[81] Frans Ruiter and Wilbert Smulders, *Literatuur en moderniteit in Nederland, 1840–1990* (Amsterdam: Arbeiderspers, 1996), 213–14.
[82] Frank Huisman, 'Wie geneest De strijd om de culturele autoriteit in de Nederlandse gezondheidszorg', in *De opmars van deskundigen: Souffleurs van de samenleving*, ed. Frans van Lunteren, Bert Theunissen and Rienk Vermij (Amsterdam: Amsterdam University Press, 2002), 110.
[83] J.T. de Visser, quoted in Van Dalen, *Brouwer* (ref. 45), 256.

industrial development and maintained intensive contacts with these sectors.[84] Leiden, which had close ties with the Philips Physical Laboratory, created an extraordinary chair for Holst, the director of this laboratory. In this way students could familiarise themselves with career options in industry.[85] Notwithstanding this element of self-interest, the new attitude was also inspired by a genuine social commitment that directed the scientific outlook beyond the confines of the academic world.

The theoretician Fokker also promoted the interests of Dutch physics. In the early 1920s, he initiated the creation of a Dutch physics journal, *Physica*. In addition, he was very active in the Dutch Physical Society, founded around the same time, of which he was president for many years. As president of a national committee for physics education he fought for the modernisation of physics teaching. He also made efforts to create links to society, exerting himself for the popularisation of physics, and often said that he thought it was 'of vital importance ... for our science that educated lay people would keep abreast of the results and the method of physics'.[86] Popularisation was therefore, in his eyes, 'an imperative duty in order to maintain the vital conditions of science, in so far as it is not focused on the purely technical'.[87] Fokker's numerous popular lectures and writings usually concerned relativity theory and his own worldview that was connected with it. He could latch onto the enormous public interest in Einstein's work in the 1920s.

Fokker's responsiveness to a critical attitude toward the sciences is apparent in the introduction of a series of popular lectures meant to illustrate the significance of science for society. These meetings were held late in 1939 in Teyler's Foundation, shortly after the beginning of World War II. Referring to Van Marum's belief in progress based on the sciences, he confronted his audience with the following questions:

> Have we more than a smile for the naivete of Van Marum's simplistic enthusiasm? Don't we see the bankruptcy of science? ... Has science only induced man to be reckless ... Is science at fault? Does it abstract too much? Is it blind to what

---

[84] Han G. Heijmans, *Wetenschap tussen universiteit en industrie: De experimentele natuurkunde in Utrecht onder W. H. Julius en L. S. Ornstein 1896–1940* (Rotterdam: Erasmus Publishing, 1994), 117–37.

[85] This was originally Ehrenfest's initiative and received Fokker's support. See Paul Ehrenfest to Adriaan D. Fokker, 3 October 1928, APE, ESC 4, Section 4; Ehrenfest, 'Verklaring der Hoogleraaren der Natuurkunde', 12 December 1928, APE, ESC 8, Section 4.

[86] Adriaan D. Fokker, review of *Over den wereldaether*, by J.D. van der Waals, Jr., *Physica* 3 (1923): 62–63, on 62.

[87] Klomp, *Relativiteitstheorie* (ref. 5), 54–55.

it can't understand? Does it ratiocinate too naively? Does it ignore human passions?[88]

The impression of doubt and uncertainty conveyed by these questions can hardly have been dispelled by the positive final verdict about science. In light of his sensitivity to the public recognition of his discipline it is not surprising that he emphasised those elements of his worldview in keeping with the values of the intellectual elite.

What conclusion can be drawn from all this? Forman undoubtedly deserves great credit for having uncovered a meaningful link between an anti-causal attitude among German physicists and anti-intellectual currents in academic circles in the Weimar Republic. In the Netherlands there seems to have been a similar resonance within the scientific community. The objection of some critics that the anti-causal stance of the physicists was primarily inspired by internal problems within atomic physics becomes less convincing in light of the Dutch situation. In the Netherlands the atomic theory played no role whatsoever in the early debates on causality. It required considerable mental acrobatics to derive anti-causal conclusions from a deterministic theory like relativity theory.

On the other hand, it emerges more clearly in the Netherlands that the roots of these sentiments are to be found in the period preceding the war and that they arose from a widespread sense of uneasiness about the naturalist perception of man and the world prevalent in elite social circles. Even Dutch scientists did not hide their aversion to the materialist and reductionist views, which they attributed to their nineteenth-century predecessors. Dutch scientists did not so much yield to a hostile climate, but rather took on leading roles in the anti-naturalist philosophical movement that swept through Dutch intellectual circles from 1900 onward.

It is as yet unclear to what extent the Dutch situation allows for generalisation beyond the Dutch cultural situation. Religion appears to have played a strong role in the Dutch qualms with regard to modernity, even among scientists. Yet, it is important to realise that religion itself was swept along by the new sensitivities that emerged around the turn of the century. The most fervent determinists in the nineteenth century could be found among Dutch modern theologians, who even argued that morality presupposed determinism. Around 1900, many Dutch liberal protestants turned away from modern theology, no longer able to stomach its intellectualism and determinism. A similar shift occurred in Dutch Spinozism, which

---

[88] Adriaan D. Fokker, 'Narede', in *De betekenis en de rol van de wetenschap in de maatschappij: Zatermiddagvoordrachten in Teylers Stichting te Haarlem op 18 en 25 November, 2 December 1939, met voor- en narede van Adriaan D. Fokker*, ed. Hugo R. Kruyt (The Hague: Nijhoff, 1940), 111–12.

in the hands of such philosophers as Bierens de Haan became less intellectual and more subjective, even mystical.

In Germany the contrast between Wilhelmine and Weimar culture was probably sharper than that between the prewar and postwar situation in the Netherlands. The fierce anti-intellectualism of the Weimar period was far more rare in the Netherlands. Nevertheless, studying German views of causality and mechanism or, more generally, modernity in the late Wilhelmine period might well be worthwhile. The remarkable responsiveness among German scientists to intellectual concerns in the Weimar Republic at least suggests the presence of an anti-modern prewar substrate, even in scientific circles. However this may be, it seems fair to conclude that a final judgment on the Forman thesis must await further empirical research.

Crisis, Measurement Problems and Controversy in Early Quantum Electrodynamics: The Failed Appropriation of Epistemology in the Second Quantum Generation

*Anja Skaar Jacobsen**

By following his tracks, both physically and professionally, as the young Belgian theoretical physicist, Léon Rosenfeld, moved among the European centres of modern physics, this paper probes the different scientific cultures in Paris, Göttingen, Zurich and Copenhagen during the period 1927–33, which has been dubbed the 'crisis in quantum theory'.[1] Within that framework the paper explores two controversies and their inextricable entanglement during this time of confusion and despair among the physicists involved with early quantum electrodynamics. There was the controversy that arose between competing research programs, one represented by Dirac, the other by Jordan-Pauli-Heisenberg, and the

---

* Niels Bohr Archive, Blegdamsvej 17, DK-2100 Copenhagen, Denmark; skaar@nbi.dk.

The following abbreviations are used: AHQP, Archives for the History of Quantum Physics, NBA, Copenhagen; *BCW, Niels Bohr Collected Works*, 13 vols. ed. Léon Rosenfeld, Erik Rüdinger and Finn Aaserud (Amsterdam: Elsevier, 1972–2008); BSC, Bohr Scientific Correspondence, NBA; *DCW, The Collected Works of P. A. M. Dirac, 1924–1948*, ed. R.H. Dalitz (Cambridge: Cambridge University Press, 1995); *HSPS, Historical Studies in the Physical Sciences* (later, *Historical Studies in the Physical and Biological Sciences*); NBA, Niels Bohr Archive, Niels Bohr Institute, Copenhagen; *PSC*, Wolfgang Pauli, *Scientific Correspondence with Bohr, Einstein, Heisenberg, a.o.*, 4 vols., ed. Armin Hermann, Karl von Meyenn and Victor F. Weisskopf (Berlin: Springer, 1979–2005); RP, Rosenfeld Papers, NBA; *SP*, Robert S. Cohen and John J. Stachel, eds., *Selected Papers of Léon Rosenfeld*, Boston Studies in the Philosophy of Science, vol. 21 (Dordrecht: Reidel, 1979).

[1] Helge Kragh, *Quantum Generations: A History of Physics in the Twentieth Century* (Princeton: Princeton University Press, 1999), 196; Abraham Pais, *Niels Bohr's Times, in Physics, Philosophy, and Polity* (Oxford: Clarendon Press, 1991), 364.

controversy over the interpretation of measurement, pitting Bohr-Rosenfeld against Landau-Peierls and Pauli. Both controversies involved disagreements over the meaning of a 'quantised field'. My study reveals that the interpretation of quantum theory was far from uniform in 1927 among the protagonists of what was later called the Copenhagen Interpretation. Rather, the views of those physicists continued to evolve within the context of relativistic quantum mechanics and quantum electrodynamics. Against this background, the usual narrative of non-relativistic quantum mechanics, climaxing with the Solvay Conference in 1927, is challenged. Furthermore, it is suggested that not only did the Göttingen-Copenhagen quantum physicists encounter opposition to their radical new epistemology of quantum theory by the realists de Broglie, Einstein and Schrödinger; they also faced the challenge of conveying the epistemological implications of the theory to an upcoming generation of physicists, who had not been trained in this kind of thought, and most of whom very soon decided they had better refrain from taking part in any such philosophical subtleties and instead stick to pragmatic instrumentalism.

Göttingen 1927–29: The 'Indeterminist School'

After graduating from the University of Liège in 1926, the Belgian physicist Léon Rosenfeld moved to Paris, where he continued his studies at the prestigious École Normale under the supervision of Louis de Broglie, Paul Langevin and Léon Brillouin. Rosenfeld's main interest was the combination of quantum theory and relativity, and for a while he worked with de Broglie. However, encouraged by Langevin, who thought Rosenfeld would become isolated in Paris, Rosenfeld soon arranged to go to Göttingen, where much of the new formalism and mathematical techniques of quantum mechanics were cast. He landed in this stronghold of mathematics and theoretical physics to become Max Born's assistant at the crucial time immediately following the Fifth Solvay Conference in October 1927, in Brussels. Rosenfeld was almost certainly completely overwhelmed by the radical new ideas in quantum theory, such as indeterminism and acausality (complementarity seems not to have been an issue), which were probably discussed in the corridors, at seminars and at the guesthouse where the young visiting physicists stayed. Rosenfeld attended Pascual Jordan's lectures on quantum statistics and it was also Jordan who, in a long letter, explained the new ideas to him. Jordan attempted to comfort the bewildered Rosenfeld when he wrote: 'Strictly speaking, these things are all so difficult, that one can never *completely* grasp them; one has always to be satisfied if one has reached clarity *approximately* at least'.[2] Rosenfeld continued for a while to discuss the new ideas by mail with de Broglie, who, however, also had a hard time reconciling himself with the ideas fostered in this 'indeterminist school', as he called it.[3]

However, apart from his exchange of letters with de Broglie and Jordan, there seems to be no indication that Rosenfeld occupied himself seriously with the epistemology of quantum physics until he started working with Niels Bohr in Copenhagen in 1930. In 1929, Rosenfeld went along with his friend Walter Heitler to the first of what were later called the Copenhagen Conferences. These conferences were organised by Niels Bohr and usually took place each Easter during the 1930s. They constituted a much more informal venue than the Solvay meetings. New theories were presented and discussed by physicists who were, or had previously worked in Copenhagen, or who were Bohr's friends.[4] As such, the meetings took place in an atmosphere favourable to the latest developments in quantum theory. At the meeting in 1929, Rosenfeld was introduced to Bohr's ideas by Bohr himself but that seems not to have helped him to get a firmer grasp of them. Later he recalled that he felt completely mystified during the session.[5] It was probably more important for Rosenfeld's career that in 1927–29 he had been introduced to the Göttingen culture of mathematics and physics. During this period he worked hard to catch up with the recent theoretical developments and in the process matured as a highly sophisticated mathematical physicist.

Zurich 1929–30: The Pauli-Heisenberg Quantum Field Theory

Rosenfeld's maturation as a mathematical physicist came in handy while he stayed with Wolfgang Pauli's group in Zurich during the academic year 1929–30. At that time Pauli and Werner Heisenberg had just published their first joint paper, 'Zur Quantendynamik der Wellenfelder', which introduced a comprehensive quantum

---

[2] Pascual Jordan to Léon Rosenfeld, 'Liège (1922–40): Göttingen (1928–29)', 31 December 1927, RP. (In German, the quote reads: 'Es sind ja im Grunde genommen diese Dinge alle so schwierig, dass man sie sich niemals *ganz* klar machen kann; man muss immer zufrieden sein, wenn man wenigstens einen *Komparativ* der Klarheit erreicht hat!') (Emphasis in the original.)
[3] Louis de Broglie to Rosenfeld, 'Correspondance particulière: de Broglie', 14 December 1927, RP.
[4] Peter Robertson, *The Early Years: The Niels Bohr Institute, 1921–1930* (Copenhagen: Akademisk Forlag, 1979), 136–38.
[5] Léon Rosenfeld, 'My Initiation', *SP*, xxxi–xxxiv, originally published in the *Journal of Jocular Physics* 2 (Copenhagen: Institute of Theoretical Physics, 1945); Léon Rosenfeld, 'Quantum Theory in 1929: Recollections from the First Copenhagen Conference', *SP*, 302–12; originally published as a monograph under the same title (Copenhagen: Rhodos, 1971).

field theory based on Jordan's idea of the quantisation of waves.[6] Rosenfeld was enrolled in this ambitious project, which also included the American physicist Robert Oppenheimer and the Swedish physicist Ivar Waller. Pauli and Heisenberg had proved their theory was Lorentz invariant, but the proof was so complicated that, according to Gregor Wentzel, Pauli is quoted as having said, 'I forewarn the curious'.[7] Rosenfeld's first contribution to this research program was to correct an error in that proof.[8] In accordance with Jordan's second quantisation scheme, both radiation and the electron were treated as *fields*. Matter was described by Paul Dirac's relativistic wave equation from 1928. Electromagnetic radiation was described by the quantised Maxwell field.[9] When Heisenberg and Pauli applied their pioneering quantum field theory to the case of electrodynamics they encountered several problems of both a formal and technical kind as well as with respect to apparent physical predictions. Their procedure was modelled on classical Lagrangian dynamics where the quantisation was effected by letting the conjugate field variables obey commutation relations. However, the quantisation of the charge-free electromagnetic field failed — one of the conjugate momenta was zero, and therefore the quantisation rules could not be applied. In their attempts to get around this problem, either gauge invariance or Lorentz invariance had to be sacrificed. On Pauli's suggestion, Rosenfeld made an attempt, which proved successful, at a more general mathematical framework of the theory taking symmetry

---

[6] Werner Heisenberg and Wolfgang Pauli, 'Zur Quantendynamik der Wellenfelder', *Zeitschrift für Physik* 56 (1929): 1–61. Later that same year they submitted their second paper, 'Zur Quantentheorie der Wellenfelder: II', *Zeitschrift für Physik* 59 (1930): 168–90. It is important to mention that in between Pauli and Heisenberg's two publications, the Italian physicist Enrico Fermi published an alternative approach to field quantisation. See his 'Sopra l'electrodinamica quantistica', *Rendiconti della R. Accademia dei Lincei* 9 (1929): 881–87. Through his review article, 'Quantum Theory of Radiation', *Reviews of Modern Physics* 4 (1932): 87–132, Fermi's theory later became quite influential among the new generation of physicists who considered Fermi's theory more easily apprehensible than Heisenberg and Pauli's. See Silvan S. Schweber, 'Fermi and Quantum Electrodynamics (QED)', *Proceedings of the International Conference 'Enrico Fermi and the Universe of Physics', Rome, September 29 October 2, 2001*, ed. C. Bernardini, L. Bonolis, G. Ghisu, D. Savelli and L. Falera (Rome: ENEA, 2003), 167–97, on 194–95; and Silvan S. Schweber, 'Enrico Fermi and Quantum Electrodynamics: 1929–1932', *Physics Today* (June 2002): 31–36. *See also* Olivier Darrigol, 'The Origin of Quantized Matter Waves', *HSPS* 16, no. 2 (1986): 197–253, on 239–40.

[7] Gregor Wentzel, 'Quantum Theory of Fields (until 1947)', in *The Physicists' Conception of Nature*, ed. Jagdish Mehra (Dordrecht: Reidel, 1973), 380–403, on 382; reprinted from Markus Fierz and Victor F. Weisskopf, eds., *Theoretical Physics in the Twentieth Century* (New York: Interscience Publishers, 1960), 48–77.

[8] Léon Rosenfeld, 'Bemerkung über die Invarianz der kanonischen Vertauschungsrelationen', *Zeitschrift für Physik* 63 (1930): 574–75.

[9] Heisenberg and Pauli, 'Quantendynamik'(ref. 6), 26–27.

properties into account. However, this highly sophisticated and comprehensive work seems merely to have served as a justification for Pauli and Heisenberg's more easily applied ad hoc procedures.[10]

From a physical point of view, the quantum field theory suffered a more serious difficulty, viz. the prediction of an infinite value for the self-energy of the electron. Rosenfeld went on to show that the self-energy of the light quantum arising from its gravitation field was also divergent.[11] These problems with quantum electrodynamics seemed to further block development. Pauli and Heisenberg's (and Jordan's) approach was criticised for being incredibly tedious and mathematically complicated, as well as unphysical — the latter because it included quantities that did not represent observables.[12] Dirac, in particular, disliked Jordan's second quantisation because the operators representing matter fields did not have a classical analogue. In Dirac's view only the corpuscular nature of *radiation* should be derived as a second quantisation, such as he had done in his pioneering quantum treatment of radiation in 1927.[13] As Pauli and

---

[10] Léon Rosenfeld, 'Zur Quantelung der Wellenfelder', *Annalen der Physik* 5, no. 1 (1930): 113–52 (an annotated English translation of this paper by Donald Salisbury appears as preprint no. 381 for Max Planck Institut für Wissenschaftgeschichte (MPIWG), and will be published in *Annalen der Physik*); Léon Rosenfeld, 'La théorie quantique des champs', *Annales de l'Institut Henri Poincaré* 2, no. 1 (1932): 25–91. According to Donald Salisbury, Rosenfeld's results in this paper prefigured what is nowadays referred to as constrained Hamiltonian dynamics, developed independently by Paul Dirac and Peter Bergman in the early 1950s. See Donald Salisbury, 'Léon Rosenfeld and the Challenge of the Vanishing Momentum in Quantum Electrodynamics', forthcoming in *Proceedings of the HQ2 Conference on the History of Quantum Physics*, 14–17 July 2008 (Utrecht, The Netherlands: Studies in History and Philosophy of Modern Physics, 2009). Wolfgang Pauli suggested Rosenfeld's work served to justify the more easily applied *ad hoc* procedures by Heisenberg and him in his, *Die allgemeinen Prinzipien der Wellenmechanik*, in *Handbuch der Physik*, vol. 24, part 1, ed. H. Geiger and K. Scheel (Berlin: Springer, 1933), 83–272, on 264. *See also* Pauli's mention of Rosenfeld's work to Oskar Klein in: Pauli to Oskar Klein, 25 January 1955, *PSC*, vol. 4, 63–64; and Klein to Pauli, 11 February 1955, *PSC*, vol. 4, 98.

[11] Léon Rosenfeld, 'Über die Gravitationswirkungen des Lichtes', *Zeitschrift für Physik* 65 (1930): 589–99; Léon Rosenfeld, interview by Thomas S. Kuhn and J.L. Heilbron, Copenhagen, 19 July 1963, AHQP, 8–9.

[12] Rudolf Peierls to Rosenfeld, 'Correspondence particulière: Peierls', 15 January 1931, RP; Lev Landau and Rudolf Peierls, 'Extension of the Uncertainty Principle to Relativistic Quantum Theory', in *Quantum Theory and Measurement*, ed. John Archibald Wheeler and Wojciech Hubert Zurek (Princeton: Princeton University Press, 1983), 465–76, originally published in German in *Zeitschrift für Physik* 69 (1931): 56; Max Delbrück to Rosenfeld, 3 May 1932, BSC, microfilm 18; Darrigol, 'Origin' (ref. 6), on 244, 246.

[13] Paul Dirac, 'The Quantum Theory of the Emission and Absorption of Radiation', *Proceedings of the Royal Society A* 114 (1927): 243–65; *DCW*, 231–55; Helge Kragh, *Dirac: A Scientific Biography* (Cambridge: Cambridge University Press, 2005), 131; Darrigol, 'Origin' (ref. 6), 239.

Heisenberg's theory also seemed to have exhausted the possibilities of Jordan's second quantisation, several of the physicists involved in quantum electrodynamics thought the problems of the theory could not be solved without the introduction of some revolutionary new ideas. As a result, Jordan stopped working on quantum electrodynamics, and even Heisenberg and Pauli decided to leave their theory behind and start over by taking correspondence arguments into account anew.[14]

Copenhagen 1930–33: The 'Small War'

In the fall of 1930 the young Russian physicist Lev Davidovich Landau visited Bohr's institute in Copenhagen as part of *his* tour around the centres of modern physics in Europe. Upon Bohr's return from the Solvay meeting in October, where the problems of quantum electrodynamics had been the subject of discussion between sessions, Bohr and Landau continued to discuss field measurements.[15] Bohr had developed a speciality in analysing idealised measurements with the purpose of exploring the meaning of concepts of a strictly quantum character, such as *stationary state*, *uncertainty relations*, *matter waves*, *photons* and *spin* that would vanish from the theory in the classical limit. He wanted to reach an understanding of these objects, unfamiliar in classical physics, by investigating how they could be logically justified in an imaginary measurement situation. To Bohr, it was irrelevant whether his idealised experiments could be realised in practice.[16] The purpose of his analyses was to pair the *definitions* derived from the quantum formalism with the possibility of *measuring* these properties in an idealised experimental set-up to arrive at an interpretation of them. In this way, his sole purpose when

---

[14] Marcello Cini, 'Cultural Traditions and Environmental Factors in the Development of Quantum Electrodynamics (1925–1933)', *Fundamenta Scientiae* 3, no. 3–4 (1982): 229–53, on 237–39; Darrigol, 'Origin' (ref. 6), 200. Enrico Fermi also lost interest in quantum electrodynamics; see Schweber, 'Fermi and Quantum Electrodynamics' (ref. 6). For the development of early quantum electrodynamics, *see also* Silvan S. Schweber, *QED and the Men Who Made It: Dyson, Feynman, Schwinger, and Tomonaga* (Princeton: Princeton University Press, 1994); Arthur I. Miller, *Early Quantum Electrodynamics: A Source Book* (Cambridge: Cambridge University Press, 1994).

[15] Niels Bohr to Werner Heisenberg, 19 November 1930, BSC, microfilm 20. This letter seems not to have been sent.

[16] Bohr, 'Solvay Conference', 28 October 1933, typewritten manuscript, NBA, on 10. *See also* Niels Bohr, 'Discussion with Einstein on Epistemological Problems in Atomic Physics', in *BCW*, vol. 7, 339–81, on 362; originally published in P.A. Schilp, ed., *Albert Einstein: Philosopher — Scientist*, vol. 7, *The Library of Living Philosophers* (Evanston, IL: Open Court, 1949), 201–41. Rosenfeld, 'On quantum electrodynamics', *SP*, 413–41, on 424; originally published in Wolfgang Pauli, Léon Rosenfeld and V. Weisskopf, ed., *Niels Bohr and the Development of Physics* (London: Pergamon Press, 1955), 70–95.

examining measurement problems in quantum theory was to provide meaning and limitations to the formalism.[17] The realists de Broglie, Albert Einstein and Erwin Schrödinger questioned Bohr's concept of *individual* quantum systems in these idealised measurements, not to mention Bohr's interpretation of their acausal, indeterministic behaviour. In de Broglie's opinion, the statements of quantum theory could only be understood statistically. Moreover, he was convinced 'under this statistical appearance a much more profound reality hides itself', viz., determinacy.[18] Other physicists had difficulty in understanding the use or purpose of Bohr's thought experiments. This did not seem to change no matter how many times he explained himself, whether in writing or in the lecture room.

One thought experiment that particularly appealed to Bohr was Heisenberg's often described gamma ray microscope experiment from the spring of 1927. However, he found that Heisenberg's analysis had not fully exhausted the experiment's potential for elucidating how the uncertainty relation between position and momentum of an electron appeared as a result of the limitation built into the description of both 'the agency of measurement and of the object'.[19] The disagreements between Bohr and Heisenberg over the interpretation of this thought experiment developed into a heated conflict that threatened to end their close

---

[17] This, however, was never a trivial point in Bohr's papers. See, for example, Niels Bohr, 'The Quantum Postulate and the Recent Development of Atomic Theory', *Nature* 121, suppl. (1928): 580–90, on 582–84; Niels Bohr and Léon Rosenfeld, 'On the question of measurability of electromagnetic field quantities', in *SP*, 357–400; originally published in German in *Det Kongelige Danske Videnskabernes Selskabs Mathematisk-fysiske Meddelelser* 12, no. 8 (1933): 3–65, trans. Aage Petersen; *BCW*, Vol. 7, 123–166; and Wheeler and Zurek, *Quantum Theory* (ref. 12), 479–522. Rosenfeld later emphasised that Bohr's aim was never to provide a 'theory of measurement.' See Rosenfeld, unpublished report on Louis de Broglie's, *La théorie de la mesure en mécanique ondulatoire*, requested by Pergamon Press Ltd., enclosed in correspondence between Miss S. Stratford-Lawrence and Rosenfeld, 26 January and 3 February 1959, 'Epistemology 1959–1964', RP. *See also* Peter Bokulich and Alisa Bokulich, 'Niels Bohr's Generalization of Classical Mechanics', *Foundations of Physics* 35, no. 3 (2005): 347–71, in which these issues are presented in a very clear way.

[18] de Broglie to Rosenfeld, 6 January 1928, RP; de Broglie to Rosenfeld, 25 March 1928, 'Correspondances particulière: L. de Broglie', RP. *See also* Paul Forman, 'Kausalität, Anschaulichkeit, and Individualität, or How Cultural Values Prescribed the Character and the Lessons Ascribed to Quantum Mechanics', in *Society and Knowledge: Contemporary Perspectives in the Sociology of Knowledge and Science*, ed. Nico Stehr and Volker Meja (New Brunswick, NJ: Transaction Books, 1984), 333–47, on 340–43.

[19] Bohr, 'Quantum Postulate' (ref. 17), on 582–84; Werner Heisenberg, 'Über den anschaulichen Inhalt der quantentheoretischen Kinematik und Mechanik', *Zeitschrift für Physik* 43 (1927): 172–98, on 197–98. *See also* Léon Rosenfeld, 'Men and Ideas in the History of Atomic Theory', *SP*, 266–96, on 291–92, originally published in *Archive for the History of Exact Sciences* 7 (1971): 69–90.

friendship. They attempted to bring Pauli to Copenhagen as a mediator but he was unable to come. Only at the time of the Solvay Conference in October 1927 had Heisenberg and Bohr seemingly reached an agreement about the interpretation.[20] In connection with Bohr's controversy with Landau and Rudolf Peierls a few years later, the roles were reversed, as we shall see below; it was now Pauli's turn, as a supporter of Landau and Peierls' views, to disagree with Bohr about interpretation, with Heisenberg attempting to mediate between the two.

With respect to quantum electrodynamics, Bohr was concerned with the use of the classical concept of a *field* in a quantum frame. He hoped that an investigation of the measurability of the field components would enable physicists to get a better grasp of the idea of a quantised field and how it differed from the classical conception.[21] The basis for this undertaking was Heisenberg's Chicago lectures in 1929, in which Heisenberg was the first to derive uncertainty relations for the electromagnetic field components (later to be corrected by Bohr and Rosenfeld); he traced the origin of these uncertainties in an imaginary measurement, as well as pondered over the meaning of field components as functions of time and space.[22] In a series of letters (of which only a few were actually sent), Bohr briefed Heisenberg about how his ideas and discussions with Landau developed. Bohr suggested using test bodies, meant to probe the electromagnetic field in a measurement situation, much bigger than the electron, in order to avoid the infinities introduced by the treatment of point charges.[23]

In late November 1930 Landau returned to Zurich, where he got Peierls, Pauli's assistant, involved in the considerations about measurability in a relativistic quantum frame.[24] Encouraged by Pauli, the two young men decided to publish

---

[20] David C. Cassidy, *Uncertainty: The Life and Science of Werner Heisenberg* (New York: W.H. Freeman, 1992), 240–46.

[21] Niels Bohr, 'Field Measurements', 1930–1931, handwritten manuscript, NBA; Niels Bohr, 'Maxwell and Modern Theoretical Physics', *Nature* 128, no. 3234 (1931): 691–92; *BCW*, vol. 6, 359–60; Niels Bohr to Werner Heisenberg, 25 December 1930, BSC, microfilm 20; Léon Rosenfeld, 'Quantum Theory' (ref. 5). *See also* Olivier Darrigol, 'Cohérence et complétude de la mécanique quantique: l'exemple de 'Bohr-Rosenfeld', *Revue d'Histoire des Sciences* 44, no. 2 (1991): 137–79, on 158–59.

[22] Werner Heisenberg, *The Physical Principles of the Quantum Theory*, trans. Carl Eckart and Frank C. Hoyt (New York: Dover, 1949), 47–54, originally published in German in 1930.

[23] Bohr to Heisenberg, 8 December 1930, BSC; Bohr to Heisenberg, 25 December 1930 (ref. 21).

[24] Landau to Peierls, 9 August 1930, Peierls Papers, ref. no. CSAC 52.6.77/c.179, Special Collections and Western Manuscripts, Bodleian Library, Oxford University. The Russian physicist Vladimir A. Fock was also originally involved. See Sabine Lee, *Sir Rudolf Peierls: Selected Private and Scientific Correspondence*, vol. 1 (Singapore: World Scientific, 2007), 137, 167–68, 192, 206–09, 224. Fock published his own results with Jordan, 'Neue Unbestimmtheitseigenschaften des electromagnetischen Feldes', *Zeitschrift für Physik* 69 (1931): 206–09.

their ideas and were soon ready to circulate a paper to Bohr and Heisenberg dealing with these issues. They introduced their paper, later published with the title 'Erweiterung des Unbestimmtheitsprinzips für die relativistische Quantentheorie', ('Extension of the Uncertainty Principle to Relativistic Quantum Theory') with some general reflections on what they regarded as reasonable premises of measurements in terms of repeatability and predictability. More importantly, they drew attention to the important fact that the measurement of a field by a test charge would add a fluctuation to the field emanating from the acceleration of the test charge caused by the field to be measured.[25] Hence, not only were non-commuting field quantities exposed to uncertainties; each *single* field component was exposed to an uncertainty because of the measurement. On this basis, they concluded electromagnetic field quantities could not be measured in the quantum domain. Paired with the problem of singularities and the prediction of negative energy states in Dirac's electron theory — Landau and Peierls firmly believed all these problems were closely interrelated and consequences of an inconsistent mathematical formalism — they were led to the dramatic conclusion that 'it would be surprising if the formalism bore any resemblance to reality'.[26] Hence, they opted for overthrowing the very foundations of the theory.

When receiving their manuscript, Heisenberg was at first quite sympathetic to Landau and Peierls' critique of the many 'unobservables' in relativistic quantum mechanics.[27] However, such positive signals were definitely not transmitted from Copenhagen. Contrary to Landau and Peierls, and despite its paradoxical predictions, Bohr had great faith in Pauli and Heisenberg's quantum field theory. He was not inclined to question its consistency as long as the formalism obeyed the necessary relativistic invariance and rested on correspondence arguments with classical electromagnetic theory. The fact that he occupied himself with interpreting the very quantum field theory is clear evidence of his faith in the theory. The last thing Bohr needed was this paper by Landau and Peierls, which, in his view, only confused the picture and appeared to pull the rug out from under the theory. Heisenberg was soon to agree with Bohr that Landau and Peierls' conclusion about the inconsistency of the theory was reached on the basis of a sloppy and naïve frame of argumentation and built on misunderstandings.[28] Bohr was quite upset

---

[25] Landau and Peierls, 'Extension' (ref. 12); Bohr to Heisenberg, 19 November 1930, BSC; Bohr to Heisenberg, 8 December 1930 (ref. 23); Bohr to Heisenberg, 25 December 1930 (ref. 21).

[26] Landau and Peierls, 'Extension' (ref. 12), 475. *See also* Peierls to Rosenfeld, 15 January 1931, RP.

[27] Heisenberg to Landau and Peierls, 26 January 1931, cited in Lee, *Sir Rudolf Peierls* (ref. 24), 220–21, on 220.

[28] Heisenberg to Bohr, 23 January 1931, BSC; Bohr to Heisenberg, 30 January 1931, BSC, microfilm 20.

because he found their 'scepticism' about quantum electrodynamics 'completely unfounded', as he wrote to Pauli later in the spring of 1931.[29] The enormous stir which Peierls and Landau's paper aroused in Copenhagen was jokingly referred to as 'our small war' by Bohr's close collaborator, Oskar Klein.[30] However, Bohr was eager to get an opportunity to discuss the issues with Landau and Peierls before they published their paper and the annual Easter conference was therefore rescheduled to the end of February so that both could be present.[31] In spite of Bohr's famously strenuous efforts (see Fig. 1) no consensus was reached on the meeting. Bohr continued developing his thought experiment.[32]

Fig. 1: George Gamow's later reconstruction of a drawing he made at the time illustrating the discussion between Bohr and Landau. Rosenfeld commented on this drawing in 'Quantum Electrodynamics' (ref. 16), 413.
Source: The Niels Bohr Archive.

---

[29] Bohr to Pauli, 21 March 1931, *PSC*, vol. 2, 68–69, on 68. *See also* Bohr to Pauli, 25 January 1933, *PSC*, vol. 2, 152–53, on 152.
[30] Klein to Landau and Peierls, 16 February 1931, BSC, microfilm 23.
[31] Peierls to Genia Peierls, 3 February 1931, cited in Lee, *Sir Rudolf Peierls* (ref. 24), 227–30, on 228.
[32] Bohr to Heisenberg, 13 March 1931, BSC, microfilm 20.

Pauli was not present at the meeting but visited Copenhagen later in the spring. He disapproved strongly of Bohr's criticism of his younger protégés' paper. Intimately connected with the problems his and Heisenberg's theory had encountered, he was of the firm conviction that Landau and Peierls' inferences for quantum electrodynamics were completely justified. Incidentally, the Landau-Peierls paper was not the only matter that Bohr and Pauli quarrelled about then. There was at the time a problem with respect to the theoretical explanation of radioactive beta-decay processes, since the continuous beta-spectrum seemed to disobey the laws of conservation. Bohr was once again prepared to give up the principle of conservation of energy (the first time had been in the 1924 Bohr-Kramer-Slater paper). This shows how far Bohr was willing to go in sacrificing the virtues of classical physics for the new quantum theory. Pauli considered Bohr's solution to be 'on the *completely wrong* track' and suggested instead the existence of a neutral particle (later named the neutrino) not yet detected which would ensure energy conservation in these processes.[33] In January 1931, Pauli wrote to Klein, '[i]t is of course clear to me that the [neutrino] hypothesis does not suit Bohr and the Bohrians. It gives me special pleasure just for that reason to discuss it'.[34] Hence, Pauli seems to have had unusually strong objections to Bohr's physics at the time.

Heisenberg made an attempt at mediating between Pauli and Bohr. He wrote to Pauli:

> Your critique of the Copenhagen physics is justified in as much as of course no-one knows anything for sure. Still, Bohr's critique of the Landau-Peierls' work is quite interesting. Bohr is in agreement with L-P's uncertainty relations. He considers the derivation of those messy in a few places, but that is not essential. The main critique is much more directed at the inferences which L-P draw from the uncertainty relations. According to Bohr, the uncertainty relations mean in no way that the relativistic wave mechanics is too narrow and must give way to a more general formalism. Rather Bohr says: also in the nonrelativistic wave mechanics only a small number of all operators are measurable ... So all in all Bohr is of the opinion that current wave mechanics (Dirac + quantum electrodynamics) constitute a *satisfactory* schema ... In details Bohr makes some quite nice remarks.[35]

---

[33] Pauli to Klein, 18 February 1929, *PSC*, vol. 1, 490 (emphasis in the original); Charles P. Enz, *No Time to Be Brief: A Scientific Biography of Wolfgang Pauli* (New York: Oxford University Press, 2002), 211–40, on 213–14; Pais, *Niels Bohr's Times* (ref. 1), 366–70.
[34] Pauli to Klein, 8 January 1931, *PSC*, vol. 2, 51; quoted in Pais, *Niels Bohr's Times* (ref. 1), 368–69.
[35] Heisenberg to Pauli, 12 March 1931, *PSC*, vol. 2, 66–67. (In German, the quote reads: Deine Kritik der Kopenhagener Physik ist insofern berechtigt, als natürlich niemand etwas Positives weiss.

During the year 1931, Pauli suffered severe personal difficulties related to his recent divorce and his mother's suicide some years earlier.[36] While this did not prevent him from working, it may have amplified his choleric mood and made him even touchier than usual. Bohr tried to appeal to his rational sense: 'We do understand each other, and you know it is not the intention to underestimate the difficulties'.[37] Bohr may have hoped that he could persuade Landau and Peierls not to publish, or that they would at least correct some passages in their paper where references to their discussions with Bohr revealed to Bohr that they had misunderstood him. The scenario is reminiscent of Bohr's heated dispute with Heisenberg about the publication of his 1927 paper on the uncertainty relation, which included the interpretation of the gamma-ray microscope experiment.[38] Moreover, Bohr may have hoped that he could at least count on Pauli's understanding. However, Pauli gave Landau and Peierls the green light to publish their paper, provided they corrected an error in their derivation of the uncertainty relations for the field strengths.[39] The paper was submitted from Zurich on 3 March 1931. Bohr could not understand what he later called Pauli's 'fanaticism about it all', and as a result Bohr stopped responding to Pauli's letters for nearly two years, until in January 1933 he felt entirely confident about the conclusion of his own examinations of the measurability of the electromagnetic field made in collaboration with Rosenfeld.[40]

It was soon after the Copenhagen meeting in 1931 that Bohr let Rosenfeld in on his considerations of field measurements. Bohr needed an expert on quantum field theory and asked Rosenfeld to be in charge of the calculations supporting the physical arguments. Rosenfeld started collaborating with Bohr in 1930, at the same time as he obtained a teaching position at the University of Liège. For the

---

Immerhin ist die Bohrsche Kritik der Landau-Peierls Arbeit ganz interessant. Bohr ist mit den Ungenauigkeitsrelationen von L-P einverstanden. Die Ableitung derselben hält er an einigen Stellen für schlampig, aber dieser Punkt ist nicht wesentlich. Die Hauptkritik richtet sich vielmehr gegen die Schlüsse, die L-P aus den U[ngenauigkeits]-R[elationen] ziehen. Bohr meint, dass die U-R keineswegs bedeuten, dass die relativ[istische] Wellenmechanik zu eng sei und einem allgemeineren Formalismus weichen müsse. Vielmehr sagt Bohr: auch in der nichtrelativ[istischen] Wellenmech[anik] ist nur ein kleiner Teil aller Operatoren messbar ... Bohr meint also, alles in allem, dass man in der jetzigen relativistischen Wellen Mechanik (Dirac + Quanten Elektro Dynamik) ein *befriedigendes* Schema habe .... Im Einzelnen machte Bohr einige ganz hübsche Bemerkungen.) (Emphasis in the original.)

[36] Enz, *No Time* (ref. 33), 211–40.
[37] Bohr to Pauli, 27 April 1931, *PSC*, vol. 2, 76. (In Danish, the quote reads: Vi forstaar jo hinanden, og Du ved, at det ikke er Hensigten at undervurdere Vanskelighederne.)
[38] Cassidy, *Uncertainty* (ref. 20), on 241–43.
[39] Pauli to Peierls, 3 July 1931, *PSC*, vol. 2, 91.
[40] Bohr to Pauli, 25 January 1933, *PSC* (ref. 29).

next ten years he commuted back and forth between the two European cities. During 1931 Rosenfeld gradually became more familiar with Bohr's way of thinking while assisting him with his Faraday lecture and report for the Rome nuclear physics conference in addition to working on their joint paper on the measurability of electromagnetic fields. In the fall of 1931 Rosenfeld accompanied Bohr to the conference in Italy, and during their return trip Bohr stayed in Belgium for a few days to continue their joint work. Rosenfeld later recalled this to be the time when he waved farewell to 'Laplacia' forever, a metaphor for his full conversion to indeterminism and acausality and Bohr's 'great truth that we are not only spectators, but actors in the drama of existence'.[41]

Very briefly, the basic conclusions and implications that Bohr and Rosenfeld eventually drew from their examination of measurability of the electromagnetic field were that fields should not, as in classical electrodynamics, be thought of as probed by point charges such as electrons, but rather by extended test bodies; field quantities should no longer be represented by point functions but functions of averaged space-time regions and finally, the zero-field fluctuations made it clear that the causal connection in the classical theory between field and its sources was lost in quantum theory.[42] On the basis of their examinations of an imaginary experiment, Bohr and Rosenfeld furthermore repudiated Landau and Peierls' claim that the Pauli-Heisenberg theory was inconsistent with respect to the indeterminacy of field components. As mentioned above, Bohr did not share Landau and Peierls' scepticism about the consistency of the theory; nor did he consider measurability examinations a possible means to disprove the formalism nor did he think it was possible to find a formalism free of singularities.[43] At the Solvay meeting in 1933 he repeated his view that 'we can hardly, from considerations of measurements, find arguments against the theory of the type of Dirac,

---

[41] Léon Rosenfeld, 'A Voyage to Laplacia', *Journal of Jocular Physics* 3 (Copenhagen: Institute of Theoretical Physics, 1955), 3–8; *SP*, 704–08, quote on 708.

[42] Bohr and Rosenfeld, 'Question' (ref. 17); Niels Bohr, 'Field and Charge Measurements in Quantum Theory', unpublished manuscript (1937), printed in *BCW*, vol. 7, 195–209, on 198; Rosenfeld, 'Quantum Electrodynamics' (ref. 16), 422.

[43] As for Bohr's scepticism about Landau and Peierls' conclusions and the 'abuse' of logical arguments, see Heisenberg to Pauli, 12 March 1931, *PSC* (ref. 35); Bohr to Pauli, 25 January 1933, *PSC* (ref. 29); Bohr to Pauli, 14 February 1934, *PSC*, vol. 2, 285–86; Bohr, 'Solvay Conference', NBA (ref. 16). According to Rosenfeld, Bohr 'had by [1932] come to the conviction that a "closed" formulation of quantum electrodynamics free from singularities was impossible, and that the singularities of the current formulation could only be removed by taking proper account of the non-electromagnetic interactions'. Léon Rosenfeld, 'Ehrenfest as I saw him (An autobiographical chapter)', May 1971, in RP: 'Supplement: History of Quantum Theory', 12–13. *See also* Rosenfeld, 'Quantum Electrodynamics' (ref. 16), 430.

and of the field theory. The importance of the discussion on measurement is more that we get to know a number of things about the character of the restrictions in the use of classical pictures'.[44] These imaginary measuring processes could *only* be used for providing a logical justification of the theory's statements, not for shooting down or arguing for modifications, of the formalism, for example in terms of what were later denoted 'hidden parameters'.[45] As Bohr wrote in hindsight to Pauli in January 1933, 'it is impossible to disprove a relativistically invariant formalism by simple relativity arguments'.[46] That said, had Bohr and Rosenfeld's examination of the measurability of field quantities *not* resulted in logical consistency with the predictions of quantum electrodynamics, one would suspect it would have caused some problems for the theory. Be that as it may, let us return to the spring 1932.

Copenhagen 1932–33: Dirac, 'The Ugly Duckling'

Like Landau and Peierls, Dirac was clearly of the opinion that a new approach to relativistic quantum mechanics was needed in order to overcome the paradoxes in quantum electrodynamics. He presented his bid at Bohr's Easter Conference in the spring of 1932.[47] However, he did not get the applause he may have hoped for. Up to that time, Dirac had not entered into elaborations on measurability of the observables in his quantum algebra, neither in his papers nor in his textbook, and he was criticised for that by Pauli.[48] His new paper published in March can be said to contain attempts at such considerations, most likely prompted by his discussions with Bohr during these years, but his considerations still did not satisfy Pauli (see quote below).[49] More important, his paper contained a rather sharp critique of the Pauli-Heisenberg quantum field theory. Dirac stressed, contrary to

---

[44] Bohr, 'Solvay Conference', NBA (ref. 16), 6.

[45] *See also* Bokulich and Bokulich, 'Niels Bohr's Generalization' (ref. 17).

[46] Bohr to Pauli, 25 January 1933, *PSC* (ref. 29). (In Danish the quote reads: det er umuligt at modbevise en relativistisk invariant Formalisme ved simple Relativitetsargumenter.) *See also* Bohr to Pauli, 14 February 1934, *PSC*, vol. 2, 285–86; Pais, *Niels Bohr's Times* (ref. 1), 361.

[47] Paul Dirac, 'Relativistic Quantum Mechanics', *Proceedings of the Royal Society A* 136 (1932): 453–64; *DCW*, 621–34.

[48] Enz, *No Time* (ref. 33), 252; Wolfgang Pauli, review of *The Principles of Quantum Mechanics*, by Paul Dirac, *Naturwissenschaften* 19 (1931): 188, reprinted in Ralph Kronig and Victor F. Weisskopf, eds., *Collected Scientific Papers by Wolfgang Pauli*, vol. 2 (New York: Wiley Interscience, 1964), 1397–98.

[49] Bohr to Heisenberg, 4 June 1930, BSC; Bohr to Dirac, 29 August 1930, BSC; Dirac to Bohr, 30 November 1930, BSC; Bohr to Heisenberg, 8 December 1930, BSC; Bohr to Heisenberg, 7 May 1931, BSC; Pauli to Dirac, 11 September 1932, *PSC*, vol. 2, 115–16.

what was the case in the Jordan-Pauli-Heisenberg approach, particles and fields should have different ontological status. He treated the electromagnetic field as a plane wave and the electron as a particle; he introduced what was later called the interaction picture and derived a wave equation that described in one dimension two electrons and their respective interaction with an electromagnetic field, including, implicitly, an interaction between the particles. However, he did not address the problem of the infinite self-energy of the electron, nor did his theory solve this problem.[50]

Rosenfeld may have been concerned that Dirac's new theory might interfere with the results of his and Bohr's ongoing investigations of the measurability of the electromagnetic field. However, after Dirac's presentation of his theory, Rosenfeld quickly succeeded in proving Dirac's new theory was mathematically equivalent to the Pauli-Heisenberg theory. Rosenfeld circulated his proof among the physicists present at the meeting to hear their opinion.[51] The ensuing correspondence among them gives the impression of a rather hostile attitude toward Dirac's work and the style of his work, as well as a somewhat teasing attitude toward his introverted personality, particularly among Pauli, Heisenberg and Rosenfeld.[52] Allegedly, Paul Ehrenfest was of the opinion that Dirac had been treated as the ugly duckling on this occasion in Copenhagen.[53] Pauli acted particularly hurt by Dirac's criticism of his and Heisenberg's theory. The following September he wrote him a blunt letter:

> Your recently published remarks in the *Proceedings of the Royal Society* concerning quantum electrodynamics was, to put it mildly, certainly no masterpiece. After a muddled introduction, which consists of sentences which are only half understandable because they are only half understood, you come at last, in an oversimplified one-dimensional example, to results which are identical to those obtained by applying Heisenberg's and my formalism to this example (this identity is immediately recognizable and has since been calculated in much too complicated a fashion by Rosenfeld). This end of your paper conflicts with your

---

[50] Bohr and Rosenfeld, 'Question' (ref. 17), 360; Pauli to Rosenfeld, 2 May 1932, *PSC*, vol. 3, 748–49.
[51] Léon Rosenfeld, 'Über eine mögliche Fassung des Diracschen Programms zur Quantenelektrodynamik und deren formalen Zusammenhang mit der Heisenberg-Paulischen Theorie', *Zeitschrift für Physik* 76 (1932): 729–34.
[52] Pauli to Rosenfeld, 2 May 1932, BSC (ref. 50); Pauli to Lise Meitner, 29 May 1932, *PSC*, vol. 2, 113–14; Pauli to Dirac, 11 September 1932, BSC (ref. 49); Heisenberg to Rosenfeld, 9 May 1932, RP; Rosenfeld to Dirac, 30 April 1932, Paul A. M. Dirac Collection, Florida State University Library, Tallahassee, Florida; Paul Dirac to Léon Rosenfeld, 6 May 1932, 'Correspondance particulière: Dirac', RP; reprinted in Kragh, *Dirac* (ref. 13), 136.
[53] Rosenfeld, 'Ehrenfest', RP (ref. 43), 12.

assertion, stated more or less clearly in the introduction, that you could somehow or other construct a better quantum electrodynamics than Heisenberg and I.[54]

Later that fall Dirac published a second paper with the Russian physicists Boris Podolsky and Vladimir A. Fock, in which he maintained his new quantum electrodynamics based on an interacting particle picture was an improved departure from that of Heisenberg-Pauli on *physical* grounds. He considered Rosenfeld's equivalence proof 'obscure' and gave a new simplified proof of the equivalence between the two formalisms.[55] Pauli eventually considered the Dirac-Fock-Podolsky paper 'a great improvement' of Dirac's theory because of the mathematical elegance with which a manifestly covariant formulation of quantum electrodynamics was achieved.[56] Indeed, if Dirac had been treated like the ugly duckling in 1932, his new refined quantum electrodynamics was soon recognised as a beautiful swan by previous opponents. In 1935, Rosenfeld summarised the development of quantum electrodynamics to date in the following way:

> As Heisenberg and Pauli first showed, quantum theory of radiation can be directly extended to an invariant and logically closed 'quantum electrodynamics', which summarises the quantised electromagnetic field in the most general sense and the matter field associated with the point model as well as their interaction. The proof of the invariance of the field equations and the quantum conditions in their original form demands rather complicated considerations, but Dirac later found a very beautiful presentation of the theory, in which its invariance appears immediately.[57]

Dirac's electrodynamics, together with Fermi's, became the starting point for the further development of QED after the war, whereas Jordan's, Pauli and

---

[54] Pauli to Dirac, 11 September 1932, *PSC* (ref. 49), 115. *See also* Pauli to Rosenfeld, 2 May 1932, *PSC* (ref. 50); Pauli to Meitner, 29 May 1932, *PSC* (ref. 52), 114.
[55] Paul Dirac, V.A. Fock and B. Podolsky, 'On Electrodynamics', *Physikalische Zeitschrift der Sowjetunion* 2, no. 6 (1932): 468–79, on 468; *DCW*, 635–48, on 637.
[56] Pauli to Dirac, 13 May 1933, *PSC*, vol. 2, 160–61; Kragh, *Dirac* (ref. 13), 137–39.
[57] Léon Rosenfeld, 'Kvanteteori og Feltfysik', *Fysisk Tidsskrift* 33 (1935): 109–21, on 114. (In Danish the quote reads: Som Heisenberg og Pauli først paaviste, kan Kvanteteorien for Straalingen, ved at lade den indskrænkede Betingelse falde, umiddelbart udbygges til en invariant og logisk afsluttet 'Kvanteelektrodynamik', der sammenfatter baade det kvantiserede elektromagnetiske Felt i dets almindeligste Forstand og det til Punktmodellen svarende Materiefelt samt deres Vekselvirkning. Beviset for Feltligningernes og Kvantebetingelsernes Invarians i disses oprindelige Form kræver temmelig komplicerede Betragtninger, men senere fandt Dirac en meget smuk Fremstillingsmaade af Teorien, hvori dens Invarians umiddelbart træder frem.)

Heisenberg's, and Rosenfeld's papers were forgotten.[58] Pauli may be held partly responsible for that development. In his comprehensive and influential *Handbuch der Physik* article, *Die allgemeinen Prinzipien der Wellenmechanik*, published in 1933, Pauli relegated his own and Heisenberg's field theory to a footnote. In addition, the 'peculiar' second quantisation technique was merely mentioned in passing and it was suggested the concept of matter waves ought to be used with caution; they differed essentially from light waves 'because the functions ψ and ψ^* are symbolic quantities, not themselves directly observable, and they contain the quantum of action'.[59] As for Pauli's enthusiasm for Dirac's new theory, it came too late to be included in the article.[60]

It was during the period from the fall of 1932 through 1933 that Bohr and Rosenfeld worked most intensely on their joint paper. Bohr did not pay much attention to Dirac's new theory.[61] In the Bohr-Rosenfeld paper, Dirac's new approach was briefly mentioned but only to say that it had not solved the problem with the infinite self-energy of the elementary particles. In addition, Bohr and Rosenfeld did not need a *quantum* interaction theory of the electron and the electromagnetic field since in their investigations it sufficed to use charged test bodies that could be described classically. Thus, they restricted themselves to the pure field theory.[62]

Epilogue

Because Bohr did not respond to Pauli's letters from the spring of 1931 until January 1933, Pauli was excluded from closely following the development of the Bohr-Rosenfeld paper.[63] In his *Handbuch der Physik* article he introduced Landau and Peierls' ideas about repeatability and predictability of measurements, which led him to his ideas of measurements of the first and second kind. Pauli also referred to John

---

[58] Schweber, *QED* (ref. 14), 12, 56, 73; Schweber, 'Fermi and Quantum Electrodynamics' (ref. 6); Cini, 'Cultural Traditions' (ref. 14), 250. In his *Selected Papers on Quantum Electrodynamics* (New York: Dover, 1958), Julian Schwinger took his starting point in Dirac's and Fermi's pioneering papers, while Pauli and Heisenberg's papers were omitted because of their considerable length.

[59] Pauli, *Die allgemeinen Prinzipien* (ref. 10), 198–200; Darrigol, 'Origin'(ref. 6), 247–48; Kragh, *Dirac* (ref. 13), 137–38.

[60] Wolfgang Pauli to Paul Dirac, 10 April 1933, *PSC*, vol. 2, 157.

[61] Bohr to Dirac, 14 November 1932, BSC, microfilm 18; Bohr to Pauli, 25 January 1933, *PSC* (ref. 29); Bohr to Ralph H. Fowler, 28 October 1932, BSC, microfilm 19. At the Solvay meeting in October 1933, Bohr only discussed measurements in relation to the Pauli-Heisenberg field theory and Dirac's relativistic theory of the electron; see Bohr, 'Solvay Conference', NBA (ref. 16).

[62] Bohr and Rosenfeld, 'Question' (ref. 17), 360.

[63] This is evident from a letter from Pauli to Heisenberg, 18 January 1933, *PSC*, vol. 2, 148–51; *see also* Jørgen Kalckar, 'Introduction', *BCW*, vol. 7, 3–51, on 11.

von Neumann's book, *Matematische Grundlagen der Quantenmechanik* (Berlin, 1932), for an elucidation of quantum measurements.[64] In contrast, Bohr never recognised von Neumann's axiomatic approach, which nevertheless came to serve as a starting point for all the later discussions of *the* so-called measurement problem.[65] During 1933 Bohr and Rosenfeld's paper went through numerous proofs before it was finally published at the end of the year. A letter from Pauli to Peierls in May 1933 indicates that after Bohr had resumed contact with Pauli, Pauli began to look at the Landau-Peierls paper with new eyes, realising for example that, 'for the field measurement the negative energy states play no role [because the exchange relations] do not contain the charge, mass, or dimensions of the probe body'.[66] However, only in early 1934, probably after he had studied the Bohr-Rosenfeld paper and after it had been discussed at the Solvay Conference in October 1933, did Pauli arrive at a 'complete agreement with all of Bohr's assertions about *measurements*'.[67] From then on he became a source of great encouragement to Bohr and Rosenfeld in their continued work on measurability.[68] Pauli had returned to the fold.

It ought to be noted that despite their controversy, Bohr immediately developed warm and close friendships with both Landau and Peierls, which only became stronger over time. Bohr excused them in a letter to George Gamow: 'I hope it will be a comfort to Landau and Peierls that the stupidity they have committed in this respect is no worse than those which we all, including Heisenberg and Pauli, have been guilty of in this controversial subject'.[69] In proposing Landau for the Nobel Prize in 1962, Bohr referred to Landau and Peierls' work in the following manner: 'By illuminating the known paradoxes of the radiation phenomena in a new way, Landau and Peierls' work gave rise to a resumed examination and clarification of the problems with measuring electromagnetic field quantities'.[70] However, Landau and Peierls apparently never reconciled

---

[64] Pauli, *Die allgemeinen Prinzipien* (ref. 10), 143–54; *see also* Max Jammer, *The Philosophy of Quantum Mechanics: The Interpretations of Quantum Mechanics in Historical Perspective* (New York: Wiley, 1974), 487.

[65] Rosenfeld, 'Report', RP (ref. 17).

[66] Pauli to Peierls, 22 May 1933, *PSC*, vol. 2, 163–65, on 165; English translation is quoted in Miller, *Early Quantum Electrodynamics* (ref. 14), 42.

[67] Pauli to Heisenberg, 23 February 1934, *PSC*, vol. 2, 300 (emphasis in the original).

[68] Niels Bohr and Léon Rosenfeld, 'Field and Charge Measurements in Quantum Electrodynamics', *Physical Review* 78 (1950): 794–98; *SP*, 400–12, on 412; Léon Rosenfeld, 'Quantum Electrodynamics' (ref. 16), 424.

[69] Bohr to George Gamow, 21 January 1933, *BCW*, vol. 9, 571; quoted in Pais, *Niels Bohr's Times* (ref. 1), 361.

[70] Bohr, Aage Bohr, Ben Mottelson, Christian Møller and Rosenfeld to Kungliga Vetenskapsakademiens Nobelkommitté, 30 January 1962, carbon copy, BSC.

themselves to Bohr's 'clarification' of their paper. When their diverging ideas were discussed at the Solvay meeting in 1933, Peierls, the only one of the two to be present, seems to have accepted that their requirements of repeatibility and predictability were asking too much of a measurement in Bohr's sense of the word.[71] However, Peierls did not acknowledge Bohr and Rosenfeld's distinction between *idealised* thought experiments used to examine the measurement process logically in a classical setting to gain a better *understanding* of the statements of quantum theory and *real* experiments used to test the quantitative predictions of the theory in the laboratory.[72] In comparing Bohr and Rosenfeld's analysis with how an experimental physicist would go about the problem, he later insisted the operations involved in Bohr and Rosenfeld's measurement 'look quite unlike any kind of measurement that an experimentalist would design'.[73] He continued, 'Bohr and Rosenfeld's arguments are based on the fundamental laws of the quantum theory of the electromagnetic field, without regard to the sort of small test bodies or particles the physicist actually has at his disposal'.[74] Landau's objections went along a somewhat different path although intimately connected with Peierls'. In particular, Landau continued to argue that point interaction was possible as the limit of 'smeared' interaction when the smearing radius tends toward zero.[75]

Bohr and Rosenfeld's paper at the time was generally seen as having proved the consistency of the uncertainty relations of the electromagnetic field quantities with the possibility of measuring these quantities in an idealised measurement.[76] The paper had therefore secured the foundations of quantum electrodynamics,

---

[71] Rosenfeld, 'Ehrenfest', RP (ref. 43); Bohr, 'Solvay Conference', NBA (ref. 16).

[72] Bohr and Rosenfeld, 'Field and Charge Measurements' (ref. 68), 411.

[73] Rudolf Peierls, 'Field Theory since Maxwell', in *Clerk Maxwell and Modern Science: Six Commemorative Lectures*, ed. C. Domb (London: Athlone, 1963), 26–42, on 36. See also Rudolf Peierls, *Bird of Passage: Recollections of a Physicist* (Princeton: Princeton University Press, 1985), 66–67; Rudolf Peierls, preface to *Landau: A Great Physicist and Teacher* by Anna Livanova (Oxford: Pergamon Press, 1980), viii; Rudolf Peierls, 'Some Recollections of Bohr', in *Niels Bohr: A Centenary Volume*, ed. A.P. French and P.J. Kennedy (Cambridge, MA: Harvard University Press, 1985), 227–31, on 228–29.

[74] Peierls, 'Field Theory' (ref. 73), 36.

[75] Léon Rosenfeld, 'Nogle minder om Niels Bohr', in *Niels Bohr: Et mindeskrift* (Copenhagen: Jul. Gjellerups Boghandel, 1963), 65–75, on 72–73; E.M. Lifshitz, 'Lev Davidovich Landau (1908–68)', in *Landau: The Physicist and the Man: Recollections of L. D. Landau*, ed. I.M. Khalatnikov (Oxford: Pergamon Press, 1989), 7–27, on 16–17; Lev Landau, 'Fundamental Problems', in *Theoretical Physics in the Twentieth Century: A Memorial Volume to Wolfgang Pauli*, ed. Markus Fierz and Victor F. Weisskopf (New York: Interscience, 1960), 245–48; Lev Landau, 'On the Quantum Theory of Fields', in *Niels Bohr and the Development of Physics*, ed. W. Pauli (London: Pergamon Press, 1955), 52–69, on 53.

[76] See, for example, Walter Heitler, *The Quantum Theory of Radiation* (Oxford: Clarendon, 1936), 81.

resulting in a revival of the physicists' faith in the theory after it had been shattered by Landau and Peierls' paper. This understanding of the Bohr-Rosenfeld paper may also reflect what the theoretical physicists craved — a way out of the crisis they found themselves in; Bohr and Rosenfeld had assured physicists that the elaboration of quantum electrodynamics could be continued without anybody having to worry about its foundations. Thus, if the physicists craved for crisis in the Weimar period, as suggested by Paul Forman in 1971, this was not the case in the 1930s.[77]

Furthermore, this understanding of the Bohr-Rosenfeld paper as having secured the foundations of quantum electrodynamics may also explain why the second quantum generation did not occupy themselves with epistemology — Bohr and Rosenfeld had said what needed to be said about that topic.[78]

Concluding Remarks

Previously it has been suggested the difference between German and British scientific culture could partly explain the development of the two competing research programs in quantum electrodynamics.[79] The present study instead draws attention to a cultural divide between different generations of physicists; that is, a generational barrier concerning scientific values or ideologies in theoretical physics. The 'old' guard comprising Bohr, Born, Einstein, Ehrenfest, de Broglie, etc., as well as the young quantum physicists Heisenberg, Pauli and Jordan were, besides developing mathematical descriptions of physical phenomena, busy drawing epistemological implications of their findings. They faced a new generation, including Dirac, Rosenfeld, Landau, Peierls, Heitler, Hans Bethe, Christian Møller, Hendrik B.G. Casimir, Robert Oppenheimer, Max Delbrück, Enrico Fermi and others, who either had difficulty in grasping the subtleties of the older generation's critical reflections or were too impatient to care about them. This new generation of physicists was more solidly anchored in the 'beauty' and consistency of the mathematical formulation, i.e., mathematical elegance and simplicity, and whether the theory dealt with observables and could stand the test of its experimental predictions. Members of this new generation of physicists were the ones

---

[77] Paul Forman, 'Weimar Culture, Causality, and Quantum Theory, 1918–1927: Adaptation by German Physicists and Mathematicians to a Hostile Intellectual Environment', *HSPS* 3 (1971): 1–115, on 58–63, 109.

[78] Pais, *Niels Bohr's Times* (ref. 1), 362.

[79] Cini, 'Cultural Traditions' (ref. 14); Paul Forman, 'The Reception of an Acausal Quantum Mechanics in Germany and Britain', in *The Reception of Unconventional Science*, AAAS Selected Symposia Series, ed. Seymour H. Mauskopf (Boulder: Westview, 1978), 11–50.

who eventually authored the first extremely influential textbooks on quantum mechanics with only sporadic treatments of epistemological issues if any, viz., Dirac, Heitler and Landau.[80]

Given there existed such a barrier between scientific ideologies of the first and second quantum generation, the Bohr-Rosenfeld versus Landau-Peierls controversy nevertheless seems to display an element of the young generation's eager aspiration to qualify themselves in the eyes of their older peers with respect to contributing to interpretation. Indeed, concrete opportunity and impetus to transgress the divide between these generations arose for many of the younger physicists when they visited Bohr's institute in Copenhagen, but apart from Heisenberg and Pauli, probably only Rosenfeld really experienced this transformation; Rosenfeld became Bohr's closest collaborator from then on.[81] Landau and Peierls, on the other hand, seem to have failed miserably in that regard, at least in Bohr's eyes.[82] As David Kaiser has suggested, however, while Bohr and Rosenfeld may have come out as winners of the 'small war' with Landau and Peierls over the interpretation of measurement in quantum electrodynamics, in the long run Bohr and Rosenfeld lost the game about what theoretical physics should be about.

---

[80] Karl Hall, 'Think Less about Foundations: A Short Course on Landau and Lifshitz's Course of Theoretical Physics', in *Pedagogy and the Practice of Science: Historical and Contemporary Perspectives*, ed. David Kaiser (Cambridge, MA: MIT Press, 2005), 253–86; Laurie M. Brown, 'Paul A. M. Dirac's *The Principles of Quantum Mechanics*', *Physics in Perspective* 8, no. 4 (2006): 381–407. Also, it is important to mention Enrico Fermi's famous and influential review article, 'Quantum Theory of Radiation' (ref. 6) in this connection. See Schweber, 'Fermi and Quantum Electrodynamics' (ref. 6). As for the influence of Heitler's textbook, *Quantum Theory* (ref. 76), see Schweber, *QED* (ref. 14), 82. To be sure, Heitler included a thorough review of the Bohr-Rosenfeld paper. He stated the importance of this paper in line with the aforementioned general reception of it, i.e., as having secured the consistency of quantum electrodynamics. See Heitler, *Quantum Theory* (ref. 76), 56–81.

[81] Anja Skaar Jacobsen, 'Léon Rosenfeld's Marxist Defense of Complementarity', *HSPS* 37, suppl. (2007): 3–34.

[82] Ehrenfest, on the other hand, recognised in Landau a 'Talmudist', a metaphor for a physicist who disregarded the phenomenological approach and took an interest in foundations. Ehrenfest wrote to Abram Fedorovich Ioffe, in a letter dated 6 January 1933: 'One should admit that something of a talmudist is present in his [Landau's] style of thinking (I can say the same about Einstein and myself). In any case, it is more prominent in his talks than in his thinking. A couple of heated discussions about his unfounded paradoxical judgments convinced me that he is a *clear* and *graphic* thinker — especially in classical physics'. Cited in Gennady E. Gorelik, *Matvei Petrovich Bronstein and Soviet Theoretical Physics in the Thirties*, trans. Valentina M. Levina (Basel: Birkhäuser, 1994), 57. However, Landau soon distanced himself from foundational issues. The reason for that might be found in the dangerous risks faced by advocates of philosophical positions opposing the official Party ideological line in the Soviet Union in the 1930s. See Hall, 'Think Less' (ref. 80), 269–71.

Instead the new generation of physicists set that agenda, focusing on mathematical techniques and calculations devoid of deeper epistemological reflection. Because of the political development in Europe during the 1930s, further developments took place predominantly in an American context.[83]

Acknowledgments

I would like to thank the Carlsberg Foundation for supporting this research and the Niels Bohr Archive, Copenhagen, for making it possible. The inspiration to write this paper came from the conference *The Cultural Alchemy of the Exact Sciences: Revisiting the Forman Thesis* in Vancouver, British Columbia, in March 2007, and from the discussions in the Quantum History Reading Group at the Max Planck Institute for the History of Science in Berlin during the winter 2008–09. I want to thank Jørgen Kalckar, Finn Aaserud, Donald Salisbury, Christoph Lehner, Skúli Sigurdsson, Silvan S. Schweber and Luis J. Boya for their critical comments and valuable suggestions for improvements on a previous version of this paper. For permission to consult and quote correspondence I am indebted to the Niels Bohr Archive, Académie des sciences-Institut de France and Fondation Louis de Broglie, and the Special Collections and Western Manuscripts, Bodleian Library, Oxford University.

---

[83] David Kaiser, comments on 'Interpreting Quantum Mechanics: A Century of Debate', History of Science Society Annual Meeting, 1–4 Nov 2007, Arlington, VA; Silvan S. Schweber, 'The Empiricist Temper Regnant: Theoretical Physics in the United States, 1920–1950', *Historical Studies in the Physical and Biological Sciences* 17, no. 1 (1986): 55–98. *See also* David Kaiser, *Drawing Theories Apart: The Dispersion of Feynman Diagrams in Postwar Physics* (Chicago: The University of Chicago Press, 2005); and Kragh, *Quantum Generations* (ref. 1), 249–56.

Causality in Physics and in the History of Physics: A Comparison of Bohm's and Forman's Papers

*Olival Freire Jr.**

David Bohm's 1952 proposal of a causal interpretation of quantum mechanics and Paul Forman's 1971 thesis that Weimar's cultural climate was a major factor in dispensing with causality in the birth of quantum mechanics are both milestones in physics and the history of physics in the twentieth century.[1] The very existence of Bohm's physics had already corroborated some of Forman's claims on the role played by contextual factors in the history of physics. Indeed, Bohm continued to work on what physicists had abandoned in the Weimar Republic of the 1920s: a causal interpretation of the quantum mathematical formalism, internally consistent and compatible with the experimental results from the non-relativistic domain of atomic phenomena. However, as at that time there were no new experimental or theoretical results able to justify Bohm's quest for a causal interpretation, one has to take into account both the larger context in which Bohm's work was embedded and his idiosyncratic approach to quantum mechanics, supporting Forman's claims to a certain extent. Despite the fact that

---

* Universidade Federal da Bahia, Instituto de Fisica, Salvador, Brazil 40210-340; freirejr@ufba.br.
  The following abbreviations are used: DB, David Bohm Papers, Birkbeck College, London; *HSPS, Historical Studies in the Physical Sciences* (later, *Historical Studies in the Physical and Biological Sciences*).

[1] David Bohm, 'A Suggested Interpretation of the Quantum Theory in Terms of "Hidden Variables"', parts 1 and 2, *Physical Review* 85, no. 2 (1952): 166–79, 180–93; Paul Forman, 'Weimar Culture, Causality, and Quantum Theory, 1918–1927: Adaptation by German Physicists and Mathematicians to a Hostile Intellectual Environment', *HSPS* 3 (1971): 1–115.

both papers dealt with the role of causality in quantum physics, literature about this relationship is scarce.[2]

Less known — except among experts in the foundations of quantum mechanics — is the fact that Bohm did not persevere in his causal interpretation of quantum theory. While always looking for alternatives to standard quantum mechanics, he shifted emphasis from the recovery of determinism toward the primacy of realism through an 'ontological interpretation' of this physical theory.[3] Although such a move was more evident in the 1980s, the abandoning of the quest for a causal interpretation had already taken place some time in the late 1950s and the early 1960s. Once again, an explanation for this move requires an account of its context and Bohm's personal circumstances. A major reason for this was his relationship with Marxism, in particular his rupture with it in 1956–57. His philosophical investigations, which led him to study Hegel, and his scientific work drawing him toward deterministic chaos also played their roles.

If Forman's thesis is loosely considered as *extrascientific reasons shaping philosophical choices in science*, the story of Bohm's quest corroborates it. Indeed, in 1971, the physics historian Forman argued that the adoption of a non-causal quantum mechanics was *determined* more by the intellectual accommodation of the German physicists to the dominant cultural climate in the Weimar Republic than by any intrinsic scientific reason. According to Forman, 'the movement to dispense with causality in physics, which sprang up so suddenly and blossomed so luxuriantly in Germany after 1918, was primarily an effort by German physicists to adapt the content of their science to the values of their intellectual environment'. This environment is described by Forman in the following terms: 'in the aftermath of Germany's defeat the dominant intellectual tendency in the Weimar academic world was a neo-romantic, existentialist "philosophy of

---

[2] Max Jammer earlier noted the relationship without further development; see his *The Philosophy of Quantum Mechanics: The Interpretations of Quantum Mechanics in Historical Perspective* (New York: Wiley, 1974), 250. In 1993, Paul Feyerabend, in private communication, noted 'there is nothing like [Forman's analysis] for the Fifties and later, neither for inside science, nor for the political + intellectual ambiance', and added that an exception was Cushing's studies; see Paul Feyerabend to David Peat, 30 August 1993, DB, Folder A21. However, while Cushing used Forman's thesis for claiming historical contingency in the victory of complementarity in 1927, he failed to see that the rise of Bohm's causal interpretation was also the fruit of historical, albeit different, contingencies; see James Cushing, *Quantum Mechanics: Historical Contingency and the Copenhagen Interpretation* (Chicago: University of Chicago Press, 1994), xi.

[3] David Bohm and Basil Hiley, *The Undivided Universe: An Ontological Interpretation of Quantum Theory* (London: Routledge, 1993). While causality and determinism are different philosophical concepts, physicists, when dealing with quantum mechanics, do not usually make such a distinction, and for this reason I will use both terms interchangeably in this paper.

life", revelling in crises and characterised by antagonism toward analytical rationality generally and toward the exact sciences and their technical applications particularly'.[4] In 'Kausalität, Anschaulichkeit, and Individualität', for instance, Forman restated his previous claim, adding neither loss of visualisability nor loss of individuality was considered a lesson from quantum mechanics for the same reasons the German physicists abandoned causality: the dominant intellectual climate accepted both requirements as central to its values while rejecting causality. However, this time Forman seemed to relate the broader context mainly to the speedy acceptance of the abandoning of determinism than to the very production of an acausal quantum mechanics. He claimed, 'thus we should not be surprised that once a non-deterministic theory of atomic processes was at hand, German physicists were disposed to view it and represent it in public as providing the liberation from causality so generally desired'.[5] Thus, the roots of the concepts of the strong and weak Forman thesis — now current in the literature — may be found in Forman's own works.[6]

This paper mainly argues that Bohm's story corroborates Forman's thesis, but it also provides an exploratory review of the reception of Forman's thesis among scholars working on interpretational issues of quantum mechanics and suggests a resonance between Bohm's and Forman's works. It is organised as follows: the opening section immediately below presents a brief review of the history of Bohm's ideas on determinism in quantum physics and the context of their inception and reception, the middle section discusses when and why Bohm gave up the causal interpretation, and the final section includes conjecture on the reception of Forman's thesis among physicists and makes concluding remarks.

Bohm's Quest for a Causal Quantum Mechanics

Until 1927, physicists who shared determinism or causality as a cognitive value intrinsic to physical sciences — a value previously dominant in the history of

---

[4] Forman, 'Weimar Culture' (ref. 1), 4–7.

[5] The follow-up papers by Paul Forman include: 'The Reception of an Acausal Quantum Mechanics in Germany and Britain', in *The Reception of Unconventional Science*, ed. S.H. Mauskopf (Boulder: Westview Press, 1979), 11–50; 'Kausalität, Anschaulichkeit, and Individualität, or How Cultural Values Prescribed the Character and the Lessons Ascribed to Quantum Mechanics', in *Society and Knowledge: Contemporary Perspectives in the Sociology of Knowledge and Science*, 2nd edn., ed. Nico Stehr and Volker Meja (New Brunswick, NJ: Transaction Books, 2005), 357–71, on 362.

[6] For more on the concepts of weak and strong Forman thesis, see Cushing, *Quantum Mechanics* (ref. 2), 100; and Helge Kragh, *Quantum Generations: A History of Physics in the Twentieth Century* (Princeton: Princeton University Press, 1999), 153–54.

physics — hoped that the then current construction of the new quantum theory would preserve this value when the huge task was accomplished. In the fall of 1927 the advocates of a causal quantum mechanics heard its swan song at the Solvay International Physics Council, held in Brussels. It was here that the French physicist Louis de Broglie presented his 'pilot wave' theory — an attempt to interpret quantum formalism in a causal fashion. De Broglie's proposal was criticised by Wolfgang Pauli and poorly received by most of the audience. After leaving the Solvay council, de Broglie abandoned his pilot wave approach and adopted the complementarity view.[7] While I do not intend to discuss this story here, nor the enduring support complementarity received from physicists, it is worth pointing out that it is this preamble to the causal interpretation which explains why Max Jammer called David Bohm's works in the early 1950s 'the revival of hidden variables'.[8] As we know, for the de Broglian pilot wave as well as for Bohm's models, quantum systems such as electrons are represented by particles with well-defined positions in the space-time arena and guided by the wave function from Schrödinger's equation. In contrast, standard quantum mechanics does not use this representation; instead it takes the wave function as the most complete description of these systems.[9] Therefore one can say that a 'real' particle position is hidden in standard quantum mechanics.[10]

It is well documented that Bohm independently resumed de Broglie's work from the point where the latter had abandoned it. Indeed, both Pauli and Einstein warned Bohm that the preprint Bohm had sent to them was a revival of de Broglie's 1927 work. Pauli further summarised his earlier criticisms of it, saying that such an approach worked well for elastic collisions but not for inelastic scattering and reaffirming his claim that in the inelastic scattering of particles by a rotator such an approach would not give stationary states before and after the

---

[7] Louis de Broglie, 'Personal Memories on the Beginning of Wave Mechanics', in *Physics and Microphysics* (New York: Pantheon Books, 1955), on 142–85.

[8] Max Jammer, *Philosophy* (ref. 2), 278.

[9] In this paper I loosely use the terms standard, complementarity, usual or Copenhagen as equivalent labels for the interpretation of quantum mechanics associated with Niels Bohr, Werner Heisenberg, Max Born and Wolfgang Pauli. A short presentation of this common and dominant view can be found in Silvan S. Schweber, review of *Meeting the Universe Halfway: Quantum Physics and the Entanglement of Matter and Meaning*, by Karen Barad, *Isis* 99, no. 4 (2008): 879–82. For a critical appreciation of the origins of the term 'Copenhagen interpretation', see Don Howard, 'Who Invented the "Copenhagen Interpretation"? A Study in Mythology', *Philosophy of Science* 71 (2004): 669–82; and Kristian Camilleri, 'Constructing the Myth of the Copenhagen Interpretation', *Perspectives on Science* 17, no. 1 (2009): 26–57.

[10] Bohm, 'Suggested Interpretation' (ref. 1).

scattering. In reacting to Pauli's criticisms, Bohm's 1952 paper may thus be seen as a resumption of the 1927 interrupted debates at the Solvay council. As Bohm replied to Pauli,[11]

> In the second version of the paper, these objections are all answered in detail .... The second version differs considerably from the first version .... The essential new point that I have added is to show in detail (especially by working out the theory of measurement in paper II) that his interpretation leads to all of the results of the usual interpretation. Section 7 of paper I is also new [transitions between stationary states — the Franck-Hertz experiment], and gives a similar treatment to the more restricted problem of the interaction of two particles, showing that after the interaction is over, the hydrogen atom is left in a definite "quantum state" while the outgoing scattered particle has a corresponding definite value for its energy.

If, from the conceptual point of view, the 1952 controversy was just a continuation of that of 1927, what else was new? My claim is that the different setting was the cultural context in which Bohm moved to his causal interpretation and the context in which he received minimal support to defend it. Despite this minimal support, it was more than that obtained by de Broglie in 1927. Bohm's works were developed in a time in which the monocracy of the Copenhagen school, in the words of Max Jammer,[12] began to be challenged. Bohm's decision to be a critic of the complementarity interpretation was motivated by two other contemporary critical trends: Einstein's enduring and renewed stand considering quantum theory an incomplete physical theory and Soviet Marxist criticisms that portrayed the complementarity view as an idealistic trend in science. In addition to these criticisms, the 1950s was a decade marked by many momentous events: Schrödinger's resuming of his criticism of the quantum 'jumps'; Everett's thesis on the relative state interpretation of quantum mechanics as an alternative approach to the standard one; the birth of the stochastic interpretation and the first voices to question the existence of the measurement problem as a flaw in the foundations of quantum

---

[11] Olival Freire, 'Science and Exile: David Bohm, the Cold War, and a New Interpretation of Quantum Mechanics', *HSPS* 36, no. 1 (2003): 1–34, on 7–10. Bohm's letters are: David Bohm to Wolfgang Pauli, July 1951; Bohm to Pauli, summer 1951; Bohm to Pauli, October 1951 and Bohm to Pauli, 20 November 1951; these letters appear in Wolfgang Pauli, *Wissenschaftlicher Briefwechsel mit Bohr, Einstein, Heisenberg u. A.*, vol. 4, part 1, ed. Karl von Meyenn (Berlin: Springer-Verlag, 1996), 343–46, 389–94 and 429–62.
[12] Jammer, *Philosophy* (ref. 2), 250.

mechanics.[13] Despite Bohm's poor reception, the controversy that arose from it was the beginning of what in hindsight we may refer to as the 'twilight of Copenhagen monocracy'.[14]

In the 1950s, however, the paper to stir up the most controversy was written by Bohm on the interpretation of quantum mechanics. In proposing an alternative interpretation of quantum mechanics, Bohm was challenging the widely accepted idea of quantum theory as a complete physical theory. This was because his proposal provides the same results for non-relativistic physical processes as the usual mathematical formalism interpreted from the complementarity point of view. As a matter of fact, not only did he suggest a new interpretation, which he called hidden variable interpretation, but he also added something to the quantum formalism in order to build a model of particles with well-defined trajectories — something that was not allowed in standard quantum theory. Although Bohm's proposal gathered adherents, such as Louis de Broglie (who reconverted to the causal interpretation), Jean-Pierre Vigier and, for a brief period, Mario Bunge, the causal interpretation by no means became widely accepted. In fact, it was severely criticised by Pauli, Rosenfeld, Heisenberg and Born, and viewed with scepticism by several other scientists. The way in which Bohm's causal interpretation was received among physicists and the fact that that interpretation led to no different prediction from those of usual quantum mechanics reinforced an early idea that the interpretation of quantum physics was a philosophical subject, not a subject for professional physicists.[15] Labelling it a philosophical controversy was initially an expression of professional bias against critics of the standard interpretation of quantum mechanics or even against the acknowledgment of the existence of unsolved problems in the foundations of such a theory. It survived in common discourse on the subject as late as 1974 when Max Jammer entitled his book *The Philosophy of Quantum Mechanics*.

Bohm's proposal touched upon a number of different philosophical and conceptual issues, such as the dispute over realism versus instrumentalism and positivism and the measurement problem in quantum mechanics. However, none of

---

[13] Olival Freire, 'The Historical Roots of "Foundations of Quantum Physics" as a Field of Research, 1950–1970', *Foundations of Physics* 34, no. 11 (2004): 1741–60; Stefano Osnaghi, Fabio Freitas and Olival Freire, 'The Origin of the Everettian Heresy', *Studies in History and Philosophy of Modern Physics* 40 (2009): 97–123. Marxists, however, were not unanimous in criticising the complementarity view; see Anja Skaar Jacobsen, 'Léon Rosenfeld's Marxist Defense of Complementarity', *HSPS* 37, Suppl. (2007): 3–34.

[14] Osnaghi *et al.*, 'Origin' (ref. 13).

[15] Also influential was the fact that Bohm and his collaborators did not get a satisfactory relativist treatment for the causal interpretation at the time; see Freire, 'Science and Exile' (ref. 11).

these resonated as intensely and acutely as the resurgence of causality and determinism. In fact, Bohm opened his 1952 paper arguing that his interpretation 'provides a broader conceptual framework than the usual interpretation, because it makes possible a precise and continuous description of all processes, even at the atomic level'. Bohm and his collaborators supported this emphasis by choosing causal interpretation as their label for their approach. Indeed, Bohm did not use the term 'causal interpretation' in his initial 1952 papers. He used it in his subsequent paper, while reacting to criticisms by Otto Halpern. Since then both critics and supporters have emphasised the philosophically minded *causal interpretation* over the philosophically neutral while technically accurate *hidden variable interpretation*.[16]

While the context of the debates on Bohm's causal interpretation has been well charted, historians have yet to come to grips with the context of its origins in part due to the paucity of documentary material on his switch from the 1951 *Quantum Theory* textbook — which philosophically was a blend of realism with Bohr's complementarity — to the 1952 papers on the causal interpretation. However, a number of testimonies support the claim that at least three different factors played roles in this switch: the influence of criticism from Soviet physicists, given further weight by Bohm's commitment to Marxism; his discussions with Einstein after his textbook came out and the dramatic changes coming from his rupture with American academia as a consequence of being a victim of McCarthyism. Christian Forstner made use of a Fleckian framework to argue forcibly that the last factor was decisive in Bohm's switch. While limitations of space here preclude a discussion of the documentary evidence to support this, Forstner rightly points out the role that Bohm's commitment to Marxism may have played a part in his research. Alexei Kojevnikov had already remarked upon its influence in Bohm's previous works on plasma physics.[17] In general, while discussing how influential Bohm's Marxism was on his switch toward the causal interpretation, there is no doubt that the influence of Marxism in the Cold War was effective in supporting the causal interpretation, and that such support was influential on Bohm himself, albeit weaker than Bohm had hoped for.[18] A fragment of

---

[16] David Bohm, 'Reply to a Criticism of a Causal Re-Interpretation of the Quantum Theory', *Physical Review* 87, no. 2 (1952): 389–90; O. Halpern, 'A Proposed Re-Interpretation of Quantum Mechanics', *Physical Review* 87, no. 2 (1952): 389.

[17] Christian Forstner, 'The Early History of David Bohm's Quantum Mechanics through the Perspective of Ludwik Fleck's Thought-Collectives', *Minerva* 46 (2008): 215–29; Anja Skaar Jacobsen, review of *Quantenmechanik im Kalten Krieg: David Bohm und Richard Feynman*, by C. Forstner, *Centaurus* (2009); Alexei Kojevnikov, 'David Bohm and Collective Movements', *HSPS* 33, no. 1 (2002): 161–92.

[18] Freire, 'Science and Exile' (ref. 11).

a letter he sent in 1955 to the American physicist Melba Phillips brings together these three threads:[19]

> At times I feel discouraged about the state of the world. A thing that particularly strikes home to me is the report I got from Burhop (confirmed by others) on Russian physicists. Apparently, they are all busy on doing calculations on electrodynamics according to Feynman, Dyson, et al. Their orientation is determined strongly by the older men, such as Fock and Landau, who in addition to their training, are influenced by the fear of a sort of "Lysenko affair" in physics. The typical physicist appears to be uninterested in philosophical problems. He has not thought much about problems such as the re-interpretation of qu. mchs, but tends to like the word of the "big-shots" that ideas on this such as mine are "mechanistic". Actually, the standard procedure is just to label such a point of view, and then most people accept the label without even bothering to read about such questions. There are some philosophers in Moscow who criticized the usual interpretation, but they haven't had much influence on the physicists. All in all, the situation in Soviet physics doesn't look very different from that in Western physics. It is disappointing that a society that is oriented in a new direction is still unable to have any great influence on the way in which people work and think.

When and why Bohm Abandoned the Primacy of Causality

It is beyond the scope of this paper to chart Bohm's changing ideas on the interpretation of quantum mechanics; however, as the physicists who work on the foundations of this theory know, while de Broglie, Vigier and their Parisian group continued the causal program, Bohm abandoned his own approach. Indeed, in his most comprehensive work on the subject, he clearly stated, 'the question of determinism is therefore a secondary one, while the primary question is whether we can have an adequate conception of the reality of a quantum system, be this causal or be it stochastic or be it of any other nature'.[20] The remaining question is thus to discover when and why he gave up primacy for causality in his research. Bohm himself argued that his abandonment of the causal interpretation was caused by the limited response to these ideas and 'because [he] did not see clearly, at the time, how to proceed further'. Then, his 'interests began to turn in other directions.

---

[19] Bohm to Melba Phillips, 18 March 1955, DB, Folder C49.
[20] Bohm and Hiley, *Undivided Universe* (ref. 3), 2.

During the 1960s, [he] began to direct [his] attention toward *order* ...'[21] To assess Bohm's shift of interest, one needs to understand its broader context. As far as chronology is concerned, my point is that Bohm no longer gave precedence to causality some time between 1956, after the conclusion of his book *Causality and Chance*, and 1960.[22] In addition to his becoming discouraged, at least two other reasons for his shift should be considered, both external to his investigations on quantum mechanics: his ideological and political rupture with Marxism and the research that brought him closer to what we now call deterministic chaos.

The best evidence of how and when Bohm shifted his focus from causal laws can be found in the correspondence he exchanged with the American artist Charles Biederman, now being edited by the Finnish philosopher Paavo Pylkkänen. As remarked by Pylkkänen, 'here we have Bohm, who is internationally known as a defender of a deterministic interpretation of the quantum theory, and thus for many a defender of strict determinism in nature, arguing strongly for the objective existence of properties such as contingency, chance, determinism, etc. Of course, Bohm does this already in *Causality and Chance*, but here the point is made more vividly, given that Bohm is defending the role of indeterminism rather than questioning it, as he most famously did in his 1952 papers'.[23]

From this extensive correspondence — over 4,000 pages between March 1960 and April 1969 — I shall select fragments from a few letters to provide the reader with an idea of the issues at stake. In his very first letter Biederman was clear-cut in his defence of determinism: 'To explain my interest in your book [*Causality and Chance*]. To put it briefly, the notion of indeterminism has always seemed contrary to experience, which, even after reading your very fine book, I cannot accept even as an eventually limiting case'. And yet, 'I sympathize with your belief that a deeper penetration will reveal a nature of causality. But there is the possibility that this will also dispel the basis for the present "lawless" view of nature and, rather than make it a limited case, will dispense with it entirely'. Bohm's answer to Biederman is that time implies a certain ambiguity. 'Thus, there is some ambiguity in past and future. We experience this ambiguity in certain ways directly. For when we try to say "now", we find that by the time we have said it, the time that we meant is already past, and no longer "now"'. He continues, citing an example closer to physics, 'and if we try to do it with clocks, so as to be more precise, quantum theory implies that a similar ambiguity would arise

---

[21] David Bohm, 'Hidden Variables and the Implicate Order', in *Quantum Implications: Essays in Honour of David Bohm*, ed. B.J. Hiley and F. David Peat (London: Routledge, 1987), 33–45, on 40.
[22] David Bohm, *Causality and Chance in Modern Physics* (London: Routledge, 1957).
[23] David Bohm and Charles Biederman, *Bohm-Biederman Correspondence: Creativity and Science*, vol. 1, ed. Paavo Pylkkänen (London: Routledge, 1999), xix.

because of the quantal structure of matter. In fact, there is no known way to make an unambiguous distinction between past and future'. Thus, 'it becomes impossible that the past shall completely determine the future, if only because there is no way to say unambiguously what the past really was until we know its future'. As Biederman might have compared that letter with the book which was the catalyst of their correspondence, Bohm anticipated such a question, 'as you may perhaps have noticed, my ideas on determinism and indeterminism have developed since I wrote *Causality and Chance*, although what I now think about these questions was, to a considerable extent, implicit in the point of view expressed in the book'. His conclusion, in short, is that 'neither determinism nor indeterminism (causality or chance) is absolute. Rather, each is just the opposite side of the whole picture', and that 'in the question of determinism vs. indeterminism, there is as I have said, a necessary complementary relation of the two ideas'.[24]

Bohm's reference to *Causality and Chance* deserves some attention. The philosophical convictions he held while writing this book weakened the prominence he attributed to causal laws in science, as he concluded that causal and probabilistic laws should be accorded the same philosophical status. Also noteworthy is the fact that these philosophical studies were motivated, at least partially, by his ideological commitment to Marxism. Indeed, it was the Brazilian physicist Mario Schönberg, his colleague at the Universidade de São Paulo in Brazil at the time (where Bohm worked for three years to escape from McCarthyism in the United States), who 'first pointed Bohm in the direction of the philosopher G.W.F. Hegel, saying that Lenin had suggested that all good Communists read the German philosopher'.[25] For our purposes, however, the most meaningful remark in his letter to Biederman was the comment that 'my ideas on determinism and indeterminism have developed since I wrote *Causality and Chance*'. This can be seen as a clue to the kind of change Bohm experienced after writing the book and before the first letter from Biederman. In that period he left Brazil for Israel, visited Paris and Bristol, and eventually settled in London. My belief is that the singular most relevant change he experienced during that period was his break with Marxism.

As we have seen in the letter to Melba Philips, as late as 18 March 1955, Bohm still had expectations that Soviet and Marxist physicists would support his causal interpretation in a stronger manner. Then, in late 1956 or early 1957, a crisis

---

[24] Charles Biederman to Bohm, 6 March 1960; Bohm to Biederman, 24 April 1960; both in *Bohm-Biederman* (ref. 23), 3–4, 8–19.

[25] For discussions between Bohm and Schönberg, see David Peat, *Infinite Potential: The Life and Times of David Bohm* (New York: Addison-Wesley, 1997), 155–57, quote on 155; and Freire, 'Science and Exile' (ref. 11), 19.

point in his commitment to Marxism was reached, triggered by Khrushchev's report on Stalin's crimes and by the invasion of Hungary by Soviet troops.[26] Bohm's crisis in Paris was witnessed by the physicist Jan Meyer and is well recorded in two long letters to Melba Phillips.[27] How dramatic Bohm's involvement was with these critical events may be seen from the following fragments:

> It is clear from the above that what is needed in the left-wing movement today is a certain measure of disengagement from Russia. Russia has made an enormous number of errors. ... This raises the question of the probable future of the C.P.'s [Communist Parties] throughout the world. ... As soon as a man opposed the direction of the C.P. he became a traitor guilty of the most heinous crimes. Confessions were manufactured and extorted on a large scale. The truth had nothing to do with the case; what was published was only what would be convenient for the interests of the gov't. This was a direct perversion of the principle that dialectical materialism should be scientific and objective. Perhaps some people said that false confessions served the interests of a "larger truth". Similarly, *Humanité* [the official newspaper of the French Communist Party] still publishes lies about Hungary; quite cynically since the truth is evident. It is clear also that the Russian gov't publishes whatever it thinks is convenient about world affairs. Perhaps they have already ceased to lie consciously, and they may be only deceiving themselves.

After Bohm's break with Marxism he made few references to its ideas; however, one of them, again in a letter to Biederman, regards determinism in society: 'For they [Marxists] felt that by studying the evolutionary process of the past, they could pick out the main direction in which history was moving. They became so attached to their theories that they were unable to review their own role objectively, or to admit new and unexpected developments not fitting into these theories'.[28] How much Marx's historical materialism depends on adopting determinism in history is debatable, however. As the historian Eric Hobsbawm remarked, at least two features of Marxism should not be abandoned unless one

---

[26] For an account of those events, see John L. Gaddis, *The Cold War: A New History* (New York: Penguin Books, 2005), 83–194. For more on the relationship between those events and the idea of crisis in Marxism, see Jack Lindsay, *The Crisis in Marxism* (Bradford-on-Avon, England: Moonraker Press, 1981), 6–8.

[27] Jan Meyer, conversation with Olival Freire, 30 January 1997; Bohm to Phillips, undated, DB, Folder C49. This rupture is noted by Kojevnikov, 'David Bohm' (ref. 17), 191; Peat, *Infinite Potential* (ref. 25), 178; and Freire, *David Bohm* (ref. 15).

[28] Bohm to Biederman, 2 February 1961, *Bohm-Biederman* (ref. 23), 95.

gives up historical materialism as a way to change the world: a) the triumph of socialism is the logical end of all historical evolution until the present, and b) socialism marks the end of prehistory as it cannot and will not be an antagonistic society.[29] For the purpose of our analysis, it is enough to note that Bohm's rupture with Marxism destroyed his general belief in determinism as a feature of science and history.

As to the second reason of his abandonment of causality — Bohm's interest in deterministic chaos — I need to make a bolder conjecture in light of the meagre documentary evidence. Bohm's reflections on the role of statistics in physics led him to the emerging studies on dynamical systems, but unfortunately he did not exploit the far-reaching implications of his own insights.[30] For instance, in the mid-1950s while in Brazil, he worked with W. Schützer on statistics in physics and its relationship to the theory of probability. They dealt with systems in which the relationship between 'the long-run properties of the orbit in phase space are essentially independent of the initial conditions', but they also approached systems in which instabilities precluded this characteristic. Bohm then arrived at a definition remarkably close to the current definition of chaotic systems, a definition he unfortunately neither published in the paper nor developed further: 'I am working on another line, which can perhaps be called "The Causal Interpretation of the Theory of Probability" .... I have defined the concept of "chaos", which is qualitatively speaking, a lack of order as regularity in the objects or events under consideration. A chaotic distribution of objects or events generally results from an unstable process, such as collisions of molecules, in which the final orbit is very sensitively dependent on the initial one. Thus chaos comes from the very form of the causal laws'.[31]

Historical comparisons between the causal interpretation of quantum mechanics and the study of dynamical systems in classical physics were initially made by Max Jammer, who noted as a 'strange coincidence in the history of modern physics', that 'at the time when hidden variable theories were proposed to reinstate the determinism of classical physics in quantum mechanics, claims were voiced that classical physics is indeterministic, perhaps even more so than quantum mechanics'. While Jammer presents the new ideas relating classical

---

[29] Eric Hobsbawm, *On History*, chap. 11 (London: Weidenfeld and Nicolson, 1997).

[30] On the history of dynamical systems, see David Aubin and Amy Dalmedico, 'Writing the History of Dynamical Systems and Chaos: *Longue Durée* and Revolution, Disciplines and Cultures', *Historia Mathematica* 29, no. 3 (2002): 273–339.

[31] Bohm to Phillips, [1953?], DB, Folder C47; David Bohm and Walter Schützer, 'The General Statistical Problem in Physics and the Theory of Probability', *Supplemento del Nuovo Cimento* 2, no. 4 (1955): 1004–47.

physics to indeterminism, technically one may still speak of deterministic chaos. However, this term is misleading as the main appeal of determinism in classical mechanics is predictability of paths, which disappears in the study of non-linear dynamical systems. It is worth emphasising that determinism was rooted in modern science through Newtonian mechanics, where Newton's second law plus initial conditions and the definition of the relevant force, for instance gravity, provides in principle the future trajectories of systems such as planets. Similar predictability is no longer available in non-linear dynamical systems. Indeed, sensitivity to initial conditions can lead these systems ultimately into a very broad range of different paths. Jammer concluded without further elaboration that the 'two opposing tendencies developed independently of each other at the same time'.[32] He did not remark that Bohm, the champion of the first (deterministic) tendency, also approached the second, and this may have contributed to Bohm's abandonment of the first.[33]

Conclusion

This paper claims that Forman's 1971 thesis and Bohm's 1952 causal interpretation resonate. It has analysed how the context of the post-World War II period encouraged the resurgence of a causal quantum mechanics, a choice physicists had abandoned in 1927. Additionally, it has been argued that Bohm's later abandonment of a priority for causal descriptions in science was strongly conditioned by an extrascientific factor. Thus, my account of Bohm's story is akin to Forman's approach to the history of physics. However, the claim that there is a resonance between Bohm's and Forman's works requires an examination of the influences they might have had on each other.

Forman was familiar with Bohm's paper but was not particularly influenced by it. However, both Forman and Bohm were influenced by Einstein's enduring criticism of quantum mechanics, which Forman read as a quest for determinism. 'What did impress me was Einstein's attachment to the goal of causal description', Forman recently recalled. In a follow-up paper to his original thesis, Forman had already suggested such an influence when he removed his physics historian hat to

---

[32] Jammer, *Philosophy* (ref. 2), 277.
[33] Since the early 1990s, studies have appeared connecting Bohm's causal interpretation to quantum chaos. However, Bohm did not work out the connection between his causal interpretation and the concept of chaos in studying the role of probabilities in physics. For a review of the connections between the causal interpretation and quantum chaos, see James Cushing and Gary Bowman, 'Bohmian Mechanics and Chaos', in *From Physics to Philosophy*, ed. Jeremy Butterfield and Constantine Pagonis (Cambridge: Cambridge University Press, 1999), 90–107.

align Einstein with his physical assessments of the foundations of quantum mechanics: 'As Einstein repeatedly but vainly emphasized, it [quantum mechanics] cannot be regarded as a complete description of an independently subsisting microscopic world'. And '... they slur over the fact that quantum mechanics is a deterministic theory of probabilities'.[34] Bohm apparently was unaware of Forman's paper, as Basil Hiley, his longtime collaborator, remembers: 'I can't recall any discussion Bohm and I had about Paul Forman's ideas'. However, the person Bohm most influenced among physicists was maybe John Bell, the physicist who would put Bohm's and Einstein's worries about causality into the physics mainstream through the technical concept of locality and the theorem now known as Bell's theorem. Bell was the most influential physicist to promote the Forman thesis among fellow physicists, as we will now see.[35]

To conclude, we move to another aspect related to the resonance between Bohm's and Forman's work, that of the reception of the Forman thesis among physicists, and in particular its reading and citing by physicists and philosophers who worked on the foundations of quantum mechanics. If there indeed was a resonance between Bohm's causal interpretation and Forman's thesis, one should expect the latter to be praised by supporters of Bohm's approach, or at least by critics of standard quantum mechanics, and ignored by those who support the complementarity interpretation. In fact, so far as I was able to ascertain, no physicist who supports the standard interpretation of quantum mechanics cites Forman's paper, according to the almost 200 citations registered on the *ISI — Web of Science* database (accessed 8 March 2009). In contrast, one may find a meaningful number of what I have called quantum dissidents among the citations. By using this term I want to name not only those who are critics of the usual interpretation of quantum mechanics but also those who fought against the dominant attitude among physicists at the time according to which foundational issues had already been solved by the founding fathers of the discipline.[36] Indeed, in addition to James Cushing, previously referred to, one may point to H. Pierre Noyes as the

---

[34] Paul Forman, e-mail message to Olival Freire, 6 March 2009; Forman, '*Kausalität*' (ref. 5), 361. However, later analysis of Einstein's stands has attenuated the role of determinism in his criticisms of quantum mechanics. See Michel Paty, 'The Nature of Einstein's Objections to the Copenhagen Interpretation of Quantum Mechanics', *Foundations of Physics* 25, no. 1 (1995): 183–204.

[35] Basil Hiley, e-mail message to Olival Freire, 8 March 2009. On the role of Bell's theorem and its connection to Bohm's work, see Olival Freire, 'Philosophy Enters the Optics Laboratory: Bell's Theorem and its First Experimental Tests (1965–1982)', *Studies in History and Philosophy of Modern Physics* 37 (2006): 577–616.

[36] Olival Freire, 'Quantum Dissidents: Research on the Foundations of Quantum Theory circa 1970', *Studies in History and Philosophy of Modern Physics* 40, no. 4 (2009): 280–89.

earliest to make a citation of Forman's paper, followed by John Bell, who cited Forman twice, as the most notable physicist, and Franco Selleri who quoted it at greatest length. Forman's thesis also appears outside the context of the quantum controversy in Kurt Gottfried's paper, or in technical papers in specialised journals, such as an article by Trevor Marshall and Emilio Santos.[37] Thus, at least among practitioners of physics, the story about Forman's thesis seems to be linked either to quantum dissidents or Bohm's causal interpretation supporters. This paper has aimed to make this intertwining between physics and the history of physics explicit.

Acknowledgments

I am thankful to Alexei Kojevnikov and Cathryn Carson for their invitation to the Forman conference; CNPq for supporting this research; the Office for History of Science and Technology, University of California, Berkeley, for hosting me while I was working on this paper; Birkbeck College, London, for hosting me while I consulted the David Bohm Papers and Denise Key for helping me in the English editing of this paper.

---

[37] H. Pierre Noyes, 'A Democritean Approach to Elementary Particle Physics', SLAC–PUB–1956 (Menlo Park, CA: Stanford Linear Accelerator Center, 1977); available online at: http://www.slac.stanford.edu/cgi-wrap/getdoc/slac-pub-1956.pdf (accessed 14 August 2009); J.S. Bell, 'Bertlmann's Socks and the Nature of Reality', *Speakable and Unspeakable in Quantum Mechanics* (Cambridge: Cambridge University Press, 1981), 139–58; J.S. Bell, 'On the Impossible Pilot Wave', *Foundations of Physics* 12 (1982): 989–99; Franco Selleri, *Quantum Paradoxes and Physical Reality* (Dordrecht: Kluwer, 1990), 28–32; Kurt Gottfried, 'Was Sokal's Hoax Justified?' *Physics Today* 50, no. 1 (1997): 61–62 and Trevor Marshall and Emilio Santos, 'Stochastic Optics: A Reaffirmation of the Wave Nature of Light', *Foundations of Physics* 18, no. 2 (1988): 185–223.

Part III

SCIENCE AND CULTURE: CROSS-DISCIPLINARY DEBATES

Forman Reformed, Again[†]

M. Norton Wise[*]

How, I have been asking myself for many years, could one of my most admired historians of science be so misguided? Paul Forman's work has always stood as an exemplar for meticulous research as well as for a wonderfully polemical style that puts his arguments in unambiguous, if controversial, form. How then could such a magnificent contribution to the cultural history of science as Forman's 'Weimar Culture, Causality, and Quantum Theory' be so deeply flawed? And how might we fix it?

First, a preliminary: Forman's argument for the impact of Weimar culture on physicists' expression of an acausal ideology for physics depends empirically on his identification of a number of individuals who underwent a conversion from causal to acausal explanation and who stand in for a 'tide' of conversions in the period 1918–27. Although I am sceptical both of the tide and of what seem to me overly cynical interpretations of a number of the individual cases, I will not be particularly concerned here with contesting Forman's data, which remains extremely impressive. Instead, I want to critique his model. It is a model of sociological causation that plays out under the term 'influence'. In reflecting on this model, on what would constitute evidence for it and on the intensity of negative responses it generates among scientists and others who are its objects of explanation,

---

[*] Department of History, UCLA, 6265 Bunche Hall, Los Angeles, CA 90095-1473; nortonw@history.ucla.edu.

[†] This discussion derives from a commentary titled 'Forman Reformulated' (shortened to 'Forman Reformed' at Forman's suggestion), presented at a workshop organised by Sam Schweber at Brandeis University in 1988. The short paper has been widely read but never published. I revise it somewhat here but without attempting to take into account the large subsequent literature, which would be essential to a full-scale evaluation.

I have come to refer to it as 'the historical dynamics of influence'.¹ Individual action and belief, in this model, take place 'under the influence' of outside agents. Just as for astrological influences, however, it is impossible to say how or on whom the emanations will, or will not act — they are a very slippery sort of causal agent and we do not need them for cultural history.² The following critique aims ultimately at suggesting an alternative model that would use the same empirical material in a different interpretive framework.

Causal Influence in History of Ideas

My own interest in the historical dynamics of influence may be placed within the history and sociology of science. Most familiarly, consider Thomas Kuhn's *Structure of Scientific Revolutions*. As is well known, Kuhn situated himself problematically between two traditions, one of which he improved, the history of ideas, and the other of which he ultimately disapproved, the sociology of scientific knowledge. In his 'Preface', the history of ideas appears under Arthur Lovejoy's *Great Chain of Being* (1936) and the sociology of knowledge under Ludwig Fleck's *Genesis and Development of a Scientific Fact* (1935). Lovejoy presents the lineage from Plato to the nineteenth century of a set of elementary ideas and the men who thought them. Fleck presents more nearly the social structure of a community of practices and people who stabilised the Wasserman test for syphilis. Although the two traditions were supposed to fit together in Kuhn's *Structure*, he famously failed to merge them in the term 'paradigm'. The relation of individual thought to community practice remained problematic. One reason for the difficulty — one that infects the subsequent literature as well — is apparent in the way the term 'influence' gets transferred from the influence of ideas to social

---

[1] My discussion of influence here is taken in part from M. Norton Wise, 'Kultur als Ressource: Die Rhetorik des Einflusses und die Kommunikationsprobleme zwischen Natur- und Humanwissenschaftlern', in *Wissenschaftsfeinde?: "Science Wars" und die Provokation der Wissenschaftsforschung*, ed. Michael Scharping (Münster: Westfälisches Dampfboot, 2001), 63–88.

[2] Two caveats are necessary. First, historians very often say that one person was influenced by another or by his or her culture when they actually only want to identify sources of inspiration, techniques, modes of expression or other resources that were important for what the person did. I am concerned here only with causal influence. Second, in the natural sciences, influence almost always implies a causal force. Crucially, however, one must be able to specify when such causal influence will act and when not. Coercive situations in history may also be of this kind, but when describing such cases, we should use a stronger term than influence. If we must use influence talk, as perhaps is necessary in describing how funding by the National Science Foundation influences the direction of research, we should be careful to specify what exactly we mean by the term and how the influence acts.

(or cultural) influence, such that ideas and society become alternative sources of individual action.³

In locating himself historiographically in *Structure* Kuhn cited several authors as 'particularly influential': Alexandre Koyré, Emile Meyerson, Hélène Metzger and Anneliese Maier, among whom Koyré certainly loomed largest. 'Their works', he wrote, 'together with A. O. Lovejoy's *Great Chain of Being*, have been second only to primary source materials in shaping my conception of what the history of scientific ideas can be'. Thus Kuhn placed his own work in the tradition of the history of ideas. More particularly, he cited these authors for the anti-whig perspective that he always represented. They had shown 'what it was like to think scientifically in a period when the canons of scientific thought were very different from those current today'.⁴ Like his histories, Kuhn's autobiographical narrative is one about thinking, about how his own ideas about paradigms were shaped by the influence of earlier thinkers. Friends and colleagues also enter as 'influences' on 'my ideas'. And yet the regular repetition of 'my ideas' will not allow us to suppose his ideas were theirs. Here again is the problem of influence in the history of ideas. What is the content of the emanation called 'influence'? Even Koyré observed, 'we ourselves determine the influences we are submitting to'.⁵

The problem continues in the alternate sociological path that Kuhn found in Ludwig Fleck while working out his *Structure*. Fleck's *Scientific Fact*, 'an essay that anticipates many of my own ideas ... made me realize that those ideas might

---

³ In what follows I will treat 'cultural' as a subset of 'social', since this was common practice for scholars in the 1970s who were working to integrate intellectual history into social history, as exemplified in Paul Forman's treatment of cultural history as sociological, as well as in studies of the sociology of knowledge done by such authors as Barry Barnes and David Bloor. More recently, many historians have preferred to separate 'the cultural', as the realm of ideology and representation, from the more material relations and interests of 'the social'. See Paul Forman's own reflections in 'The Primacy of Science in Modernity, of Technology in Postmodernity, and of Ideology in the History of Technology', *History and Technology* 23 (2007), 1–152, esp. 68 and references therein. Forman is convinced the cultural takes priority over the social in shaping the unity of historical eras, 'integrating the outlooks of actors in diverse social situations', and that to reject such unifying cultural constructs is tantamount to 'rejecting history as a scholarly discipline'. (68) To me, this conception of the cultural is too monolithic. I take my cue from complex social and cultural formations like the city of Los Angeles, which is as famous (perhaps infamous) for its lack of a centre as for the diversity of cultures that meet and interact there. The challenge for the historian is to understand how such interactions produce dynamical stability in the absence of any unifying idea or ideology that defines the whole culture.

⁴ Thomas S. Kuhn, *The Structure of Scientific Revolutions*, 2nd ed. (Chicago: University of Chicago Press, 1970), vi.

⁵ Alexandre Koyré, *From the Closed World to the Infinite Universe* (Baltimore: Johns Hopkins University Press, 1957), 5–6.

require to be set in the sociology of the scientific community'.[6] The scientific community thereby becomes the putative location for the characteristic modes of thought of its members, their paradigms. As has often been observed, however, *Structure* contains nothing particularly social or sociological (despite some remarks about textbooks, journals and professional associations). The community of practitioners is unified by its paradigms, which are ideas carried by its exemplary problem solutions and inaccessible to outsiders.

Here lies the great blind spot in Kuhn's conception of the social. With scientists isolated in their paradigmatic disciplinary thoughts, everything else becomes an external influence on those thoughts, of the same kind as and in competition with internal intellectual influence. 'I have said nothing about the role of technological advance or of external social, economic, and intellectual conditions in the development of the sciences'.[7] Clearly in conflict with himself about this limitation, he first stated in the text that such external 'influence' would not modify his main conclusions; then he added a footnote stating it was only with respect to 'this essay' that he took 'the role of external factors to be minor', citing his papers on external intellectual and economic conditions in the development of energy physics. But even these papers, including the classic paper on 'Energy Conservation as an Example of Simultaneous Discovery', focused on the *ideas* of energy and their mathematical formulation. For example, it is the concept of work employed by engineers that figures in the discussion, not the practices of engineers, nor the work done by steam engines in producing commodities, nor the nature of economic exchange relations.[8] And in fact, Kuhn never again returned to the analysis of 'external factors' even in their intellectualised form. External social influence always stood in tension with internal intellectual influence.

The Social Turn

One of the main contributors to this continuing tension was the course taken by sociological studies of science in the late 1960s and 70s, for they were presented in direct opposition to the history of ideas. But the opposition took the form of replacing the old dynamics of influence with a new dynamics of influence. For the influence of great books and great men in the near or distant past, it substituted the

---

[6] Kuhn, *Structure* (ref. 4), vii.
[7] Ibid., x.
[8] Thomas S. Kuhn, 'Energy Conservation as an Example of Simultaneous Discovery', in *Critical Problems in the History of Science*, ed. M. Clagett (Madison: University of Wisconsin Press, 1959), 321–56; reprinted in Thomas S. Kuhn, *The Essential Tension: Selected Studies in Scientific Tradition and Change* (Chicago: University of Chicago Press, 1977), 66–104; concept of work on 83–94.

influence of present social forces. The internal-external struggle of the 1970s was a contest of influences.

No doubt we are most familiar with the social turn through SSK, the Sociology of Scientific Knowledge, particularly in the form of the 'strong programme' of Barnes, Bloor and the Edinburgh School. Citing Kuhn's paradigms as a major motivation, this hyperscientific school sought a deterministic causal account of scientific belief. It was laid out particularly clearly by Barry Barnes in a chapter of his *Scientific Knowledge and Sociological Theory* (1974) entitled 'Belief, Action and Determinism: The Causal Explanation of Scientific Change'. With clear reference to Kuhn's normal and revolutionary science, Barnes emphasised normality and deviation from normality within any given culture. As he put it:

> the alternative suggested here would make the sociological study of beliefs more consistently deterministic than before. On the one hand, beliefs representing departures from normality are held to be in need of causal explanation. On the other hand, beliefs which are bound up in normal practice, although in a sense unproblematic, must be accounted for by some theory of socialization which will give a deterministic account of their acquisition. And the stability of these same beliefs will depend on the stability of the constellation of normal *causal influences* which serves to sustain them as unproblematic, institutionalized features of the overall social landscape.[9]

I call attention to the way in which 'influence' is rendered as 'causal influence' here and throughout the text. That rendering follows from the definition of science as theory-dominated and deterministic that Barnes espoused for scientific sociology: 'Indeed, given that science is constitutively theoretical, and always proceeds by metaphorical redescription in deterministic terms, it could be said that causal explanation was constitutive to "the idea of a social science"'.[10]

Barnes' methodological pursuit of deterministic causal influences proceeded through the metaphor or model of 'man as a programmed system', like a computer, whereby 'to hold a system of beliefs corresponds to being programmed in a certain way' and therefore to carrying out determined actions in response to a given situation.[11] Within the subculture of a scientific specialty, the program that practitioners acquire through professional training and socialisation into a subculture is

---

[9] Barry Barnes, *Scientific Knowledge and Sociological Theory* (London: Routledge, 1974), 70 (emphasis added).
[10] Ibid., 76.
[11] Ibid., 80.

incorporated as a sub-routine into the larger program that they acquire from their general culture. Thus cultural influences divide into 'internal' and 'external' causes according to whether they reside in the sub-program or the more general program. But we should note that in this sociological rendering both influences are external in that they are causes emanating from outside the individual who becomes programmed by them. Acknowledging some of the limitations of the metaphor, notably for creative science, Barnes proceeded to give an extended discussion of internal versus external cultural influences, depending largely on whether practitioners would consider them as legitimately part of their subculture or not. Paul Forman figures prominently as representative of external causal influences.[12] [The most important feature of Barnes' discussion for my purposes (below) is that he thoroughly conflates cultural resources with cultural causes.]

Similar remarks apply to David Bloor's *Knowledge and Social Imagery* (1976). Again, 'social' includes both internal and external social relations, but both act from outside individuals to determine their action. Even more than Barnes, Bloor represented social influences as like forces acting on individuals, including an analogy to the parallelogram of forces from elementary mechanics. Although the 'strong programme' generated much valuable discussion, as well as polemics, it is worth noting no satisfactory example of a causal deterministic account of scientific activity has yet been given. By far the most dramatic candidate, for Bloor as for Barnes, was Forman's 1971 analysis.[13] This leads me to state the obvious point that the Forman thesis was very much a part of the social turn, especially in its causal claims, which motivates my critique.

Forman began his classic account by urging that 'the historian ... must insist upon a causal analysis' and that one could choose either 'psychological' (e.g. Kuhn) or 'sociological' (e.g. Barnes and Bloor) causation. Choosing the sociological option as most fruitful, he himself would treat 'mental posture as socially determined response to the immediate intellectual environment and current experiences'.[14] He then argued extrinsic cultural influences led Weimar physicists to abandon their traditional principles. Capitulating to a hostile intellectual environment, they sought an acausal quantum mechanics.

It should be apparent that Forman, as a rebellious student of Kuhn, formulated the relation of culture and science in a manner quite incompatible with Kuhn's

---

[12] Ibid., 109–11, 116, 141–42.

[13] David Bloor, *Knowledge and Social Imagery* (London: Routledge and Kegan Paul, 1976), 4, diagram of parallelogram of forces on 27.

[14] Paul Forman, 'Weimar Culture, Causality, and Quantum Theory, 1918–1927: Adaptation by German Physicists and Mathematicians to a Hostile Intellectual Environment', *HSPS* 3 (1971): 1–115, on 3.

lifelong stress on shifts in individual thinking in response to technical problems internal to the discipline. Forman was now dealing with 'mental posture' and cultural influences turned the hinge of mental posture, setting in motion a search for a new kind of physics, independent in the first instance of any pressing technical problems. In rejecting the individualistic emphasis then current in history of science and continuing in Kuhn's work, Forman made the characteristic switch of influences of the social turn, replacing the intellectual influences of the history of ideas with the social influences of the sociology of knowledge, as a kind of mirror image and in causal deterministic form. It is the causal form, I argue, rather than the cultural content, that constitutes the problem. The historical dynamics of influence is no more adequate for sociological history than it is for the history of ideas. To put the point polemically, why in the world, if we cannot give a causal account of the formation of snowflakes, would we imagine that we could give a causal account of the formation of thoughts?[15]

Forman's Model: Characteristics and Critique

The Forman thesis rests on a capitulation model. It develops in three stages: *hostility* of the intellectual culture, prompts ideological *accommodation* and *capitulation* by physicists, who therefore suddenly *restructure* their physical concepts in culturally acceptable ways, but thereby violate the integrity of their discipline.

Intellectual Hostility. In Weimar culture the concept of causality very often stood for mechanical determinism, rigidity and death, all running counter to the widespread sympathy for *Lebensphilosophie* and Spenglerism. This point agrees in general with many other studies by intellectual and cultural historians of Weimar. Since no one has seriously challenged it, I will take it as basically true. Intellectual attitudes hostile at least to mechanist explanation, and hostile to science *to the degree that it was identified with mechanism*, were widespread. These attitudes attended hostility to materialist technological society and ambivalence to liberal rationalist government. 'Republican Automatons', by George Grosz, may serve to capture the mood (Figure 1).

Ideological Accommodation and Capitulation. To the hostile culture, numerous theoretical physicists and mathematicians accommodated themselves socially and attitudinally, without any good reasons grounded in their physics for doing so. Even some who opposed the anti-scientific sentiments of their culture capitulated

---

[15] For the latest on the complexity of understanding snowflakes, their uniqueness and their (rare) sixfold symmetry, see the simulations of Janko Gravner and David Griffeath, 'Modeling Snow-Crystal Growth: A Three-Dimensional Mesoscopic Approach', *Physical Review E* 79 (2009): 1–18; the online publication contains full-colour images.

Fig. 1: George Grosz, *Republican Automatons* (*Republikanische Automaten*), 1920.
Source: Museum of Modern Art, New York.

to its power. This core argument for the entire thesis rests on identifying a wide variety of attitudes with vitalism, read as *Lebensphilosophie* and Spenglerism.

Sudden Restructuring of Knowledge. In order to adapt the internal content of their discipline to their new attitudes, physicists suddenly sought an acausal understanding of phenomena for which previously they would have sought causal explanations. At this point, if not already in their ideological accommodation, physicists violated the integrity of their discipline. Capitulating to the hostile environment, they forsook their most fundamental principles.

I will respond to the following features of this model: *capitulation, external forces, suddenness, vitalism, disciplinary integrity*.

Capitulation. Forman's sociological analysis relies on a negative social model. Physicists are not treated as participants in the culture, or as a subculture nested within it, but stand largely outside it as a separate group. Better, they are surrounded by it without being within it or of it. Even where 'participation' is acknowledged, it shades off into increasingly externalising metaphors, moving from 'adaptation' to 'accommodation' to 'capitulation'.[16] The result is a re-creation of the internal/external dichotomy. Because the physicists are not of the society they must violate their own internal integrity in order to accommodate themselves to a critical external world. They cannot share its critical perspective and adapt it to their own professional aims. In this sense, the capitulation model is not social enough; it does not portray physicists as interactive social beings embedded in their culture.

Causation by External Forces. Forman's image of social causation answers to his belief that the historian 'must insist on a causal analysis'. That is, historians must do what Weimar physicists did not do — maintain the nerve to demand rigorous causal explanations. The capitulation model, however, looks more like causation within an older Newtonian mechanics, with society acting like an external force on an isolated body of physicists, than like the field theories and extremum principles of the late nineteenth century, which made the behaviour of any part of a system a product of its (local) participation in the system as a whole. Although Forman claims to be employing a field approach, and his analysis would be much improved by it, he actually treats cultural hostility as an external force.[17] As a causal agent for transforming physical theory, however, external hostility has serious limitations. It may offer motivation but it cannot offer the positive resources necessary for generating a new conceptual apparatus, one that does not already exist either inside or outside of physics. Acausality, as the mere negation of deterministic mechanical causality, could not — and did not — solve problems in physics. And the positive concept of statistical causality, or probabilistic causality, which did solve problems, did not derive from hostility to science in general, although it certainly involved severe doubts about the adequacy of mechanistic explanation and also had strong resonance with the broader culture, as Forman has shown.

Suddenness. The capitulation model requires that the transition within physics from deterministic causality to statistical causality be quite sudden and

---

[16] Forman, 'Weimar Culture' (ref. 14), 5, 46, 55, 62.

[17] Ibid. Forman says that he has 'sought a model in which certain "field variables" and their derivatives at a certain place and time are regarded as evoking corresponding attitudes'. (p. 4) A field model, however, would require understanding how any particular object introduced into the field interacts with and contributes to it, for it is this contribution that explains the moving force the object experiences. Magnetic 'influence', for example, depends on whether the object is ferromagnetic, paramagnetic, diamagnetic or nonmagnetic, without even considering the nuances of more complex behaviours. One might imagine explaining individual susceptibilities in terms of yet other 'influences' of family, education, intellectual heritage, etc., but this only drives the same problem of specifying causal influence into the traditional realm of history of ideas.

without good internal reasons in physics.[18] But the transition was by no means so sudden as the capitulation model requires and there were plenty of good internal reasons. Most obviously, the problems of giving a causal account of the Second Law of Thermodynamics were widely discussed from the late nineteenth century. That is, the inability of atomistic mechanics to account for directionality of development in time, without appeal to additional assumptions, was widely regarded as a symbol of its problematic status and of the need for a new theoretical foundation for causal laws.[19] Forman is well aware of this background but still considers the many expressions of acausality that he has identified in the period 1919–26 as different in tone and quality, and as constituting a 'tide' of conversions. I am not convinced of the tide or of the causal connection of individual cases to Weimar culture. My scepticism begins with his inclusion of Franz Exner (albeit ambivalently) as an early convert in 1919 and then Schroedinger as a later convert in 1922. No serious connection to Weimar culture is made in either case, and Schroedinger's conversion is merely asserted. Much more satisfying is Deborah Coen's recent cultural analysis of the Viennese research group around Exner — including Schweidler, Przibram, Ehrenhaft, Kohlrausch, Smoluchowski and Felix Exner, as well as Schroedinger — who for two decades had been systematically studying fluctuation phenomena in diverse fields and the fundamental role of probability and statistics in nature.[20] A similar problem is that Forman does not include the major role of Niels Bohr in Copenhagen, whose famous postulates of stationary states and acausal transitions in the atom had developed into a sophisticated formulation of statistical causality by 1918, one studied by everyone in the field and having nothing in the first instance to do with Weimar culture,[21] although it

---

[18] Forman has most recently reiterated his conviction that the 'great size and suddenness of the change' required the 'invocation of social pressure to explain ideological reorientation' in Weimar in 'Primacy of Science' (ref. 3), 126–27, n417.

[19] Lorraine J. Daston, Michael Heidelberger and Lorenz Krüger, eds., *The Probabilistic Revolution, 1800–1930: Dynamics of Scientific Development*, vol. 1, *Ideas in History* (Cambridge, MA; MIT Press, 1986). The essays collected in this work make the point over a much wider domain of science and society, including papers by Daston, Heidelberger, Ian Hacking, Andreas Kamlah, Eberhard Knobloch, Theodore Porter and M. Norton Wise.

[20] Deborah R. Coen, *Vienna in the Age of Uncertainty* (Chicago: University of Chicago Press, 2007), esp. chap. 8. *See also* Michael Stöltzner, *Vienna Indeterminism: Causality, Realism, and the Two Strands of Boltzmann's Legacy* (PhD dissertation, University of Bielefeld, 2003).

[21] For the post-1927 period, however, see John Heilbron's sarcastic portrayal of Bohr as 'guru to his group' of young associates (Wolfgang Pauli, Werner Heisenberg, Pascal Jordan), locating their attachments to Bohr and to the principle of complementarity in their psychological needs, in J.L. Heilbron, 'The Earliest Missionaries of the Copenhagen Spirit', *Revue d'histoire des sciences* 33 (1985): 193–230, quote on 223.

had a great deal to do with the local culture in which Bohr worked. Less problematic than the Exner and Bohr stories is the attribution of conversion to Richard von Mises, with emphasis on its 'suddenness' between two inconsistent printings of a lecture in 1920 and 1921. At first glance this does indeed look sudden but a closer look at von Mises' trajectory by Michael Heidelberger has shown that his probabilistic view of hydrodynamics, elasticity of solids and Brownian motion was grounded in work from 1919 that refers back to Gustav Fechner's indeterministic *Kollectivmasslehre* (1897), which had become a standard source on probability and statistics in Germany.[22] More generally, the suddenness thesis would require careful documentation of the intellectual histories of the various actors to see how they came to their perspectives on acausality.

Monolithic Vitalism. Forman's capitulation model rests in its causal structure on associating all sympathies for holism and all perceptions of cultural crisis with *Lebensphilosophie* and Spenglerism; in short, with vitalism. Vitalism becomes the identifying mark and causal force of Weimar culture. To identify with it was to abandon rigorous science. As an identifier, however, it is far too monolithic. It makes every holism into a form of quasi-religious mysticism. To put some long-term historical perspective on the problem, such an indictment would reduce to anti-science all ideas of emergent properties developed in the sciences of complexity of the last 40 years. Similarly, it would set the romantic Alexander von Humboldt, who led the rejuvenation of German science in the 1830s, against the currents of true science. Useful antidotes to this representation are widely available.[23]

---

[22] Michael Heidelberger, 'Fechner's Indeterminism: From Freedom to Laws of Chance', in Daston, Heidelberger and Krüger, *Probabibilistic Revolution* (ref. 19), 117–56, esp. 144–47; and Michael Heidelberger, *Nature from Within: Gustav Theodor Fechner and His Psychophysical Worldview*, trans. Cynthia Klohr (Pittsburgh: University of Pittsburgh Press, 2004).

[23] Timothy Lenoir, *The Strategy of Life: Teleology and Mechanics in Nineteenth Century German Biology* (Dordrecht: Reidel, 1982); Anne Harrington, *Reenchanted Science: Holism in German Culture from Wilhelm II to Hitler* (Princeton: Princeton University Press, 1996); Michael Dettelbach, *Romanticism and Administration: Mining, Galvanism, and Oversight in Alexander von Humboldt's Global Physics* (PhD dissertation, Cambridge University, 1992); Michael Dettelbach, 'Global Physics and Aesthetic Empire: Humboldt's Physical Portrait of the Tropics', in *Visions of Empire: Voyages, Botany, and Representations of Nature*, ed. D.P. Miller and P.H. Reill (Cambridge: Cambridge University Press, 1996), 258–92; Jan Faye, *Niels Bohr: His Heritage and Legacy: An Anti-Realist View of Quantum Mechanics* (Dordrecht: Kluwer, 1991); and M. Norton Wise, ed., *Growing Explanations: Historical Perspectives on Recent Science* (Durham, NC: Duke University Press, 2004).

Disciplinary Integrity. Forman makes disciplinary identity the touchstone of his diagnosis.[24] Such a primary identity is necessary within the capitulation model, for it gives to physicists the isolation required if they are to be acted on by an external force which causes them to sacrifice their integrity by questioning causality or pursuing acausal explanation. But there are several grounds on which to question this primary identity. Most apparent from a sociological perspective is the status of German academics as civil servants and especially their prized role as *Kulturträger* (which Forman thoroughly exploits in his many citations of public lectures). No doubt cultural identity melds with disciplinary identity everywhere — compare national and regional identity — but their interrelation was particularly strong in Germany. We need to know more about how the relation was experienced, whether as tension or mutual reinforcement, or both.

A second worry concerns the age and status of theoretical physics as a discipline. Physics as a whole had barely attained that status by the mid-nineteenth century and the characteristic features of theoretical physics remained in flux at least through the 1920s. Only then did the 'physics of principles' espoused by Planck and Einstein, on whom Forman ultimately relies to define disciplinary integrity,[25] acquire a dominant position, as opposed in particular to Sommerfeld's 'physics of problems', and only then did the ideological split of theoretical physics from applied physics and technological development become such a strong ideology.[26]

---

[24] The priority of disciplinary identity, along with causal cultural history, continues in Forman's analysis in Forman, 'Primacy of Science' (ref. 3). In an important footnote, Forman makes an impassioned plea for disciplinary science, with its traditional value of basic research in pursuit of natural laws, in the face of 'the anti-disciplinary effects of post-modern values', which are producing 'a chaos of purposes and practices in which only unlovely characters can thrive' (p. 127–420). Forman's target is the primacy of technology and means-ends rationality in post-modern culture. I share his concerns so far as they relate to the rampant commercialisation of academic research and its corrosive effect on pursuit of the public good in the university, but as an active proponent of cross-disciplinary education and research, I do not agree that a loss of 'disciplinarity' is the problem. Furthermore, as an analyst of the fundamental role of technologies in modern science, I would respond to the view that 'if the fact of scientific laws is not regarded as a greater miracle than the fact that the machine works, then it is "curtains" for the scientific enterprise', by saying that these are often much the same miracle. Consider, for example, the Nobel Prizes in Physics awarded in 2007 for discovery of the giant magnetoresistance/ hard disc and in 2009 for fibre optics (e.g. broadband phone lines) and charge-coupled devices (e.g. digital cameras).

[25] Forman, 'Weimar Culture' (ref. 14), 91–96.

[26] Suman Seth, '*Mystik* and *Technik*: Arnold Sommerfeld and Early-Weimar Quantum Theory', *Berichte zur Wissenschaftsgeschichte* 31, no. 4 (2008): 331–52; Seth deals at full length with the 'problems' vs. 'principles' approaches to theory in Suman Seth, *Crafting the Quantum: Arnold Sommerfeld and the Practice of Theory, 1890–1926* (Cambridge, MA: MIT Press, 2010).

How much weight ought we to give to disciplinary identity compared with natural-scientific, academic or broader cultural identity?

Thirdly, we need to know more about causality as an historical object, about the specific role causality played in this contest to define theoretical physics. How, and for whom did it come to be such a crucial principle that to question it was to violate disciplinary integrity? The relatively small group of physicists who protested against relaxing causality may suggest that it did not carry such a defining status, rather than that Weimar culture swamped the critical faculties of the non-protesters.

An Alternative Model

In seeking to invert these points of criticism, I begin from the premise that, in general, the social and political context of science acts productively. A cultural history of physics, therefore, ought to offer a better understanding of the physics, of the conditions under which problems are analysed, experiments done and theory generated. Such conditions are conditions of possibility, not causes.

Participation vs. Capitulation. With respect to Weimar physicists, I would argue that Paul Forman should have contented himself with treating them as participants in their culture, rather than as capitulators to it. If we are primarily interested in their attempts to solve problems in physics, they are better understood as drawing on their culture for relevant resources, rather than being influenced by it, since it is only by taking up and adapting those resources for their own needs that the culture becomes effective. 'Influence' typically connotes the action of a foreign agent, as in 'under the influence'.[27] But in using such language we tend to forget that we choose our influences, or at least they depend on our susceptibilities: e.g. the different children in a given family and community, even twins, experience very different 'influences' and have very different memories of shared events. Thus, the primary task for an influence account would be to understand how it is that certain people take on a supposed influence and others do not under very similar circumstances. But that would be a story of individual motivations and choices within a given context rather than a cultural influence story, and it would place susceptibility and agency in the individual rather than the culture, reversing the causal arrows of the influence story. And with that, the supposed influences would become resources available to be taken up by particular people for particular purposes, or not, depending on how the individual participated in the culture. That is the transformation of models I would advocate. The influence

---

[27] Forman, 'Weimar Culture' (ref. 14), 3, 76.

story would become a resource story and the capitulation story a participation story.

Under such a redescription, we might attempt to understand how it happened that acausal quantum mechanics developed originally in Central Europe and not, for example, in Britain. One answer could be that concepts like statistical causality were widely available and acceptable in Central Europe along with a supporting structure of meaning that filled the concept with rich content. One of its specific origins, I have argued elsewhere,[28] lay in an insistence in the late nineteenth century — in Central Europe but not Britain — on differentiating psychical from physical causality in the interpretation of social statistics. The history of the generation of the concept, then, and the resources that generation brought with it, help us to understand both how it happens that so many quantum physicists were willing to discard earlier notions of causation and how they happened to be so concerned with psychology (e.g. Bohr, Jordan, Pauli).

Methodologically speaking, the strategy of interpreting quantum mechanics in acausal and psychological terms cannot be attributed simply to the crisis of Weimar culture, however real the crisis and however important it became for the individual physicists who experienced it. Their interpretations drew on and developed strategies well-established in Central Europe for approaching phenomena that seemed to violate the categories of mechanics.

Nevertheless, Forman has shown convincingly that a considerable number of physicists did see their interests, their personal and cultural interests, to lie in the search for an acausal account of quantum phenomena. I would put this, however, in a positive sense of interests rather than a crass sacrifice of integrity. I would emphasise the utility of cultural resources. Utilities include the structure of motivations and rewards: personal support; social acceptance; intellectual prestige; professional rewards like position, salary and power; and indeed, coherence between scientific and political beliefs. There is nothing necessarily crass about pursuing one's personal and political goals simultaneously with, and through, the search for scientifically valid explanations of the world, even if that pursuit involves rejecting presently accepted canons of explanation. Crassness and sacrifice of integrity enter when valid knowledge is turned to immoral ends or when validity itself is sacrificed to other interests, such as the reward structure of society. The Forman thesis locates the turn to acausal explanation in such a

---

[28] M. Norton Wise, 'How Do Sums Count?: On the Cultural Origins of Statistical Causality', in Daston, Heidelberger and Krüger, *Probabilistic Revolution* (ref. 19), 395–425. For interpretations of social statistics in Germany, see the following essays in *Probabilistic Revolution*: Theodore Porter, 'Lawless Society: Social Science and the Reinterpretation of Statistics in Germany, 1850–1880', 351–75; and Ian Hacking, 'Prussian Numbers 1860–1882', 377–94.

sacrifice of physics to social utilities and there I depart from it. This is not to say that individuals did not make immoral choices. My immediate example for such a person would be Pascual Jordan, who turned his holistic quantum mechanics into Nazi propaganda through his *Führerprinzip*.[29] But the fault lay more nearly in his lack of humanistic integrity than his lack of disciplinary integrity.

Localist vs. Internalist History. The problems of a discipline are usually posed and solved internally, in terms of the subjects and languages that define the specialised research community. But in general this is better described as localisation within, rather than separation from, the larger society, thereby avoiding the internalist/externalist trap. On the one hand, concepts of wide significance, such as statistical causality, obtain concreteness, point and development with reference to localised and narrowly specified subjects. On the other hand, the larger sphere supplies a context of meaning which makes the concepts available as a resource in the first place. A localist account, as opposed to an internalist one, does not close its eyes to the constructive role of the society, which penetrates into and between the particular problems, methods and accomplishments of physics. The task, however, is not to derive the physics from the society in which it participates, but to understand the physics within the society and, in part, to draw the physics out of the society in terms of the resources available to its practitioners.[30]

In another dimension, a localist account carves up the ground differently. It looks not only at research specialities and disciplines embedded in a larger cultural context but at geographically local contexts cutting across disciplines. For the story of statistical causality, with respect to Niels Bohr, for example, the social and intellectual life of the University of Copenhagen and of the friends of Bohr and his family are pertinent. Purely at the intellectual level, one quickly discovers Bohr espoused views closely related to those of others in his circle of acquaintances, including: his father, the professor of physiology, his physics professor Christiansen and his philosophy professor Höffding.[31] Their variable positions and perspectives formed a local nexus of concepts and explanatory strategies, a nexus

---

[29] M. Norton Wise, 'Pascual Jordan: Quantum Mechanics, Psychology, National Socialism', in *Science, Technology, and National Socialism*, ed. M. Walker and M. Renneberg (Cambridge: Cambridge University Press, 1994), 224–54; reprinted in Peter Galison, Michael Gordin and David Kaiser, eds., *Science and Society: The History of Modern Physical Science in the Twentieth Century*, vol. 4 (New York: Routledge, 2001), 332–69.

[30] Coen, *Vienna* (ref. 21). Coen richly develops such a perspective for the Exner clan's statistical interpretation of natural law and causality, integrating their family life with their various scientific pursuits and their liberal politics in Vienna, such that their statistical interpretation was 'a resource for liberal self-fashioning' (p. 298) and vice versa.

[31] Faye, *Niels Bohr* (ref. 23), chaps. 1–4; Wise, 'How Do Sums Count?' (ref. 28), 406–20.

that provided resources and utilities for Bohr's own explanatory strategy in quantum mechanics.

A Web of Local Nodes. There is a much-discussed problem with localist history. It begs for an account of how the local ties into larger formations. My reinterpretation of the Forman thesis begins from the broad-scale view that there can be no question, for Central Europe, of the continuity from the late nineteenth century in attempts to find relief from the causal inadequacies of mechanics, especially when supposed to operate on the individual elements of collective objects (or statistical ensembles). If we conceive these partially interactive attempts as forming something like a web, then the question becomes how and when people in particular locations were tied into the web of interactions and how they used its resources. Exner and Bohr serve as examples. They may be seen as accessing through local nodes a cultural web extending through Central Europe to which they contributed new linkages and new perspectives on physics. Other physicists accessed this web at different nodes at different times and also added to it in diverse ways. If this metaphor of a continually developing web of more local cultural formations has validity, we might replace the 'suddenness' of Forman's thesis with something like the following view. After World War I the problems of physics became more tightly enmeshed in the cultural web, largely as a result of the resources offered for reconceptualising long-standing difficulties in physics, but certainly also as a support for cultural renewal in a condition perceived by many in Weimar Germany as one of crisis. The crisis, however, should not be overestimated. Linkages between statistical physics and other cultural domains had been well established before the dramatic stretching and straining of the postwar years. Furthermore, the prior existence of the cultural web helps to explain where some of the resources came from for articulating an acausal quantum mechanics, why the articulation was relatively restricted to Central Europe until the late twenties and why considerable reinterpretation attended its dissemination in quite different cultural conditions, as in Britain and the United States.

Conclusion

In proposing a model in which resources and participation replace influences and capitulation, I am attempting to find a more satisfying picture of how individuals relate to the environment in which they are embedded. It is a picture in which an entity like Weimar culture represents a more or less stable dynamic of interaction and exchange rather than a monolithic object that acts on its own. It is interactions all the way down. Individuals enter into this dynamic as participants. They are not autonomous agents who can simply do as they please, for their action is constrained by what is possible in their local culture and the larger web with which it interacts. But they do have a great deal of freedom to work as they see fit on the challenges and opportunities, problems and solutions that sustain them, intellectually and materially.

In this position individuals make moral and ethical choices. One of the problems of the influence model is that it tends to relieve individuals of responsibility for their own action and places that responsibility on the culture that acts on them. Certainly that was not Paul Forman's intention. Nevertheless, the model allows that interpretation. I want to see a model in which individuals are fully responsible for their choice of 'influences', or rather resources.

Acknowledgments

Elaine Wise has contributed comments and references for the discussion of causal influence for twenty years. So too have many colleagues and graduate students at colloquia and in seminars. I am grateful to all of them.

From *Kosmos* to *Koralle*:
On the Culture of Science Reading in
Imperial and Weimar Germany[†]

Arne Schirrmacher[*]

Forman on Science Communication

Paul Forman's influential studies on the history of physics and its historiography, which appeared almost 40 years ago in the leading English-language journals on the history of science, were a major breakthrough because they presented an interpretation, soon to be known as the 'Forman thesis', claiming that scientists in the Weimar Republic had, in a certain sense, adapted or anticipated scientific content on the basis of societal conditions. More important, his contributions laid the foundation for an entire set of innovative research perspectives in the history of science. Forman examined the doublet riddle in atomic physics just prior to the formulation of quantum mechanics; questioned the established historiography on the discovery of the diffraction of X-rays by crystals; published a book-length article in 1971 entitled 'Weimar Culture, Causality, and Quantum Theory, 1918–1927: Adaptation by German Physicists and Mathematicians to a Hostile Intellectual Environment', in which he vigorously posed the question of the contextual dependence of so-called hard sciences such as physics and mathematics;

---

[*] Max-Planck-Institut für Wissenschaftsgeschichte, Boltzmannstr. 22, 14195 Berlin; aschirrmacher@mpiwg-berlin.mpg.de.

[†] This article is based on a paper presented at the conference, 'The Cultural Alchemy of the Exact Sciences: Revisiting the Forman Thesis', Vancouver, British Columbia, 23–25 March 2007. A slightly longer German version appeared as 'Kosmos, Koralle und Kultur-Milieu: Zur Bedeutung der populären Wissenschaftsvermittlung im späten Kaiserreich und in der Weimarer Republik', *Berichte zur Wissenschaftsgeschichte* 31 (2008): 353–371.

The following abbreviation is used: *HSPS, Historical Studies in the Physical Sciences*.

and attempted to specify this in subsequent publications.[1] The diversity of groundbreaking ideas attributable to Forman is most clearly evident in his 1967 unpublished dissertation, 'The Environment and Practice of Atomic Physics in Weimar, Germany'.[2]

Several passages from Forman's wide-ranging studies attracted particular interest and were compiled and generalised into theses. They became the starting point for a long debate within history of science.[3] Essentially this debate revolved around the question of the possible existence and the mechanism of a correlation between a Weimar cultural milieu — however defined — and science, particularly physics. In the revised and pointed wording Forman used in 1984, this meant:

> We are therefore led to look for special circumstances, unique, or largely unique, to the environment of German-speaking Central European physicists, which may have called forth their misuses of quantum mechanics.

One such 'circumstance' was indeed an 'antagonism toward physical science, and more particularly toward theoretical physicists' because their objectives were diametrically opposed to *Lebensphilosophie*. Therefore, the physicist was held responsible for modern technology and its alienating impact on life. According to Forman, both were to lead to social 'despisement', attempts at self-rehabilitation and finally a kind of public relations by physicists:

> The German physicist, especially the theoretical physicist, who before and during World War I enjoyed so much esteem, now, in the Weimar period, found himself despised. Gone were those values and attitudes which had ensured him an honored place in German society and in the eyes of his university, displaced

---

[1] See the following publications by Paul Forman: 'The Doublet Riddle and Atomic Physics circa 1924', *Isis* 59 (1968): 156–74; 'The Discovery of the Diffraction of X-rays by Crystals: A Critique of the Myths', *Archive for the History of Exact Sciences* 6 (1969): 38–71; 'Alfred Landé and the Anomalous Zeeman Effect, 1919–1921', *HSPS* 2 (1970): 153–261; 'Weimar Culture, Causality, and Quantum Theory, 1918–1927: Adaptation by German Physicists and Mathematicians to a Hostile Intellectual Environment', *HSPS* 3 (1971): 1–115; 'Scientific Internationalism and the Weimar Physicists: The Ideology and its Manipulation in Germany after World War I', *Isis* 64 (1973): 151–80; 'The Financial Support and Political Alignment of Physicists in Weimar Germany', *Minerva* 12 (1974): 39–66 and 'The Reception of an Acausal Quantum Mechanics in Germany and Britain', in *The Reception of Unconventional Science*, ed. Seymour H. Mauskopf (Boulder, CO: Westview Press, 1979), 11–50.

[2] Paul Forman, 'The Environment and Practice of Atomic Physics in Weimar, Germany: A Study in the History of Science' (PhD dissertation, University of California, Berkeley, 1967).

[3] For commentary on this debate and a collection of critical papers, see Karl von Meyenn, ed., *Quantenmechanik und Weimarer Republik* (Braunschweig and Wiesbaden: Vieweg, 1994).

by others distinctly hostile to the theoretical physicist's enterprise. In an effort to realign their discipline with dominant cultural values, individual German physicists had been delivering manifestos against causality even before the discovery of quantum mechanics ... Thus we should not be surprised that once a nondeterministic theory of atomic processes was at hand, German physicists were disposed to view it and represent it in public as providing that liberation from causality so generally desired.[4]

Yet how widespread was the 'hostility' against physics and mathematics? Where and for which societal groups was the 'despisement' of the theoretical physicist discernible? Where did general 'cultural values and attitudes' evolve and how widespread were they among the general populace? What did scientists present to the public and with what intention?

What is missing most in the discussions on the 'Forman thesis' is a clear identification of the counterpart of the physicists, without which it is difficult to assess the pervasiveness, not only of 'despisement', but also of a negative intellectual and political attitude in society toward theoretical physics. Reference is made here not only to a 'hostile cultural environment' or specifically to a 'cultural milieu'. In addition, 'social', 'intellectual' or 'antagonistic' 'milieus' or 'environments' play a role. Social pressure, antagonistic societal forces, etc., are discussed and physicists described as searching for a place within the 'broader society'.[5] In any case, they 'accommodated' to 'values of the Weimar intellectual milieu'.[6] These formulations implicitly suggest a public discourse, and Forman also hints at a communication between science and the public at the end of the quote cited above. However, most of the publications popularising natural science in the public at large are not among the sources considered by Forman and his commentators.[7]

---

[4] Paul Forman, '*Kausalität, Anschaulichkeit*, and *Individualität*, or How Cultural Values Prescribed the Character and the Lessons Ascribed to Quantum Mechanics', in *Society and Knowledge: Contemporary Perspectives in the Sociology of Knowledge and Science*, ed. Nico Stehr and Volker Meja (New Brunswick, NJ: Transaction Books, 1984), 333–47, quotes on 337–38.

[5] *See also* the discussion in Meyenn, *Quantenmechanik* (ref. 3), and in the contributions to this volume.

[6] Forman, 'Weimar Culture' (ref. 1), 7.

[7] Sources such as academic speeches in university settings (rectorial addresses, anniversary celebrations, etc.), general lectures held at large scholarly meetings (e.g. *Naturforschertagungen* — the annual meetings of the Gesellschaft Deutscher Naturforscher und Ärzte) and even science communication in the journal *Die Naturwissenschaften* are considered 'inter-specialist communication' in the literature on the popularisation of science. See Joske Bunders and Richard Whitley, 'Popularisation within the Sciences: The Purposes and Consequences of Inter-Specialist Communication', in *Expository Science: Forms and Functions of Popularisation*, ed. Terry Shinn and Richard Whitley (Dordrecht: Kluwer, 1985), 61–78.

Originally, Forman pursued a premise that has not yet been challenged seriously: Notwithstanding debates about a general 'Weimar culture' that was generally felt, at least by physicists, to be a 'hostile intellectual environment', Forman maintained '[the only relevant question is] what image the educated public held of the physical scientist and his world view'.[8] Yet this view can be challenged by pointing out that cultural and social developments, especially during the Weimar period, cannot be understood merely as a history of the high culture embraced by the educated classes. Instead, the period witnessed the breakthrough of a politically and socially influential mass culture, which also expressed a particular interest in the scientific and technological dimensions of modernity.[9] Evident here is a lack of clarity with respect to determining the relevant social groups. Therefore, we should ask whether it is wise to focus only on the 'intellectual environment of the early Weimar period' when dealing with 'extraordinarily heavy social pressures', 'which the German academic environment could and did exert upon the individual scholar or scientist placed within it'. Likewise, we should ask whether the convictions expressed were those of an 'academically educated general audience' or of a culturally influential segment of the population that can be quantified by the number of people who read Spengler's *Untergang des Abendlandes* (ca. 100,000) or scientific journals and popular science magazines (whose regular circulation surpassed the number of people holding a degree of higher education). Also we should question whether cultural attitudes of the Weimar period were decisively influenced by the entire public, not just the 'educated middle class'.[10] Moreover, Forman based his ideas in 1971 on a model of description that postulated a basic interchange between science and the public: the scientists' motivation was said to lie in the achieved 'prestige' whereby they attempted 'to alter the public image of science as to bring this image back into consonance with the public's altered values'.[11]

On the basis of my study of periodicals popularising science for a large, interested public — primarily magazines but also book series, newspapers and radio broadcasts — I would like to show in this essay that it is hardly possible to speak of disdain toward physicists or hostility toward exact natural science during the

---

[8] Forman, 'Weimar Culture' (ref. 1), 8, 19.
[9] Peter Gay, *Weimar Culture: The Outsider as Insider* (New York: Harper and Row, 1970); Detlev Peukert, *The Weimar Republic: The Crisis of Classical Modernity* (New York: Hill and Wang, 1992).
[10] Forman, 'Weimar Culture' (ref. 1), 6, 31, 36. Based on *Sperlings Zeitschriften-Adressbuch* (Leipzig: Börsenverein, 1902–1947), the number of journal issues published in the mid-1920s on popular science was around 300,000 to 500,000. The number of readers at this time can be roughly estimated by keeping in mind the number of issues shared within families, libraries, cafes, etc.
[11] Forman, 'Weimar Culture' (ref. 1), 6.

Weimar Republic. At best, such tendencies can be found within the scientific public (*Fachöffentlichkeit*) — which is itself a part of science — and therefore took place in a framework barely extended beyond that of scientists and intellectuals. My argument is presented in two steps: first, I discuss how the rough distinction between science and the public should be further differentiated, using an analysis of magazine publishing (*Zeitschriftenwesen*), to expose that distinction as multi-layered; over the course of my discussion I present the relevant popular science magazines and outline the discourse on atomic physics in them. Second, I consider the degree to which the political and economic crises of the Weimar Republic were reflected in atomic physics and science communications.

The *Zeitschriftswesen* of Science: The Case of Atomic Physics

Popular science as part of the science environment

Because the history of science concentrates on sources close to science and scientists, the general societal attitude toward natural science and technology — be it interest or disinterest, hope or mistrust, awe or hostility — usually eludes it. The availability of public opinion polls or other primary sources that describe general societal attitudes toward natural science for the period under study here is limited at best.[12] A more promising path of inquiry follows the discourse found in popular science media, which had to find their audiences in the general marketplace. Magazines such as *Umschau*, *Kosmos* and *Koralle* had little chance of success if they did not appeal to the interests and views of their readership. They were forced to search for subscribers outside the realm of subsidised science and to convince their readers at the newsstand from issue to issue. Consequently, the topics selected by their publishers and editors became reflections of general attitudes and the popular science periodicals became suitable sources, which sometimes — if only to a limited degree — also contain the direct reactions of their public in the form of letters to the editor, question boxes and reader surveys.[13]

---

[12] Occasional calls for letters from the readers of popular science magazines and listeners of broadcast programs provide some small insight. Public opinion polling on scientific and technological topics started with the British social-research organisation, Mass Observation, founded in 1937. In 1940, Mass Observation began to focus on topics like the 'treatment of science', 'everyday feelings about science' and the question, 'Where is science taking us?' See Peter Broks, *Understanding Popular Science* (Maidenhead: Open University, 2006), 71–72, 75–76.

[13] The journal *Umschau*, which claimed in its subtitle to be a 'Review of the Progress and Movements across the Entire Spectrum of Science, Technology, Literature, and Art' ('Übersicht über die Fortschritte und Bewegungen auf dem Gesamtgebiet der Wissenschaft, Technik, Literatur und Kunst'), conducted in its very first year of publication a call for answers to the question, 'Was ist

In my examination of popular science literature, I proceed from the premise of several notions and findings that can be traced back to Forman's 1967 dissertation. His use of the term 'environment' as the general area in which science is conducted encompassed a far broader scope than the later term, 'cultural milieu'. As a passive agent, 'environment' includes that which more recently has been called 'resources', thus giving it an active component because resources can be 'mobilised' — they are institutional and disciplinary, economic and instrumental, and cognitive. According to Mitchell Ash, for example, shifts in scientific paradigms can be understood from this perspective as the 'reorganizing and newly organizing of resource ensembles'.[14] Norton Wise once attempted to reform Forman in this sense by arguing that the public discourse should be understood as a 'productive agent'.[15]

In the history of science, magazine and journal publishing is considered part of the 'environment' of atomic physics, for it has proven to be an important arena for shaping attitudes toward science. Forman pointed out and described connections between political factions among physicists and the various journals and magazines, and what impact these political or ideological interpretations had on the style of the respective publication process (editorial procedure, manuscript selection). For example, he contrasted the 'editor pope' of the conservative *Annalen der Physik* and the 'editor impotent' of the liberal *Zeitschrift für Physik*. Although Forman did not study the popular journals reaching out to a broader audience and only included *Die Naturwissenschaften*, which strove to inform scientists about the advances being made in their respective neighbouring disciplines, already he could recognise several essential points. Two observations are of particular interest here. First, his insight that the 'totality of physical journals' formed a system in which each individual journal assumed a specific function.[16] This discovery of a system can be applied to publications geared to a

---

Bildung?' ('What is education?'), the purpose of which was not least to discover their readers' interests. As part of its new, more welcoming appeal, in 1901 *Naturwissenschaftliche Wochenschrift* introduced the column 'Readers' Questions'. In 1925, *Umschau* needed to place the great number of questions it received in a supplementary section; by 1928, for example, the number of questions reached more than 900.

[14] Mitchell G. Ash, 'Wissenschaft und Politik als Ressourcen füreinander: Programmatische Überlegungen am Beispiel Deutschlands', in *Wissenschaften und Wissenschaftspolitik. Bestandsaufnahmen zu Formationen, Brüchen und Kontinuitäten im Deutschland des 20. Jahrhunderts*, ed. Rüdiger vom Bruch and Brigitte Kaderas (Stuttgart: Steiner, 2002), 32–51, on 32.

[15] M. Norton Wise, 'Forman Reformed' (unpublished manuscript), as presented at Brandeis University in 1988; revised and published as 'Forman Reformed, Again' in this volume.

[16] Forman, 'Environment and Practice' (ref. 2), Section I.4.e, 171–205, on 171, 180.

broader readership. Second, Forman uncovered an obvious contradiction, which he left unresolved:

> Despite the 'unpropitiousness of circumstances' — namely the inflation and the antagonism toward the physical sciences and mathematics — semi-professional journals of general scientific content prospered.[17]

I would like to place this 'curious fact', as Forman called it, at the heart of my argument: Why was a society that appeared to physicists to be a 'hostile intellectual environment', and to which they were supposedly compelled to adapt their science, so interested, even enthusiastic, to read about science and its progress? At any rate, the journal *Kosmos* reached the highest circulation of its nearly century-long existence in 1925 with more than 200,000 copies. In 1922, it had already surpassed its prewar circulation and at least five additional popular science journals of various orientations achieved circulations of several tens of thousands in the second half of the 1920s. Thus the interest in natural science, as documented by such demand, was twice if not three times as high as prior to World War I.[18]

Establishing differentiations within the public and science

There are indeed clear indications that science communication increasingly opened up to a mass public and that the scientific community communicated not only with an 'intellectual milieu'. Granted, not everyone was inevitably affected by this science boom. It is necessary to distinguish between the various levels within the scientific community and the public.[19] For the purpose of this study, it is sufficient to identify the 'science' of a discipline (for example, physics) as including all subdisciplines (theoretical and experimental physics) or even special areas (atomic physics, etc.) and to label it 'research science' *(Fachwissenschaft)*. This label distinguishes science from the more general and supra-disciplinary scientific community — or scientific public *(Fachöffentlichkeit)*, for short. The existence of this scientific public was most evident at large conventions such as the *Naturforscherversammlungen*, but also in the academic and university institutions that sponsored events, often with an interdisciplinary focus.

---

[17] Ibid., 176.
[18] *Sperlings Zeitschriften-Adressbuch* (ref. 10).
[19] For a more detailed account, see Arne Schirrmacher, 'Nach der Popularisierung: Zur Relation von Wissenschaft und Öffentlichkeit im 20. Jahrhundert', *Geschichte und Gesellschaft* 34 (2008): 73–95, on 84–88.

Table 1: A differentiation of groups within science and the public and their respective publications.

Research science/discipline	Annalen der Physik, Zeitschrift für Physik	~ 100
Scientific public	Die Naturwissenschaften	~ 1,000
Attentive public	Umschau, Technische Monatshefte ... Kosmos	~ 10,000
(Occasionally) interested public	Koralle, Wissen und Fortschritt, Urania / Radio	~ 100,000
Broad public	Berliner Illustirte, B.Z. am Mittag	~ 1,000,000

While Forman studied the discourse on this level, I would like to examine the public beyond the community of scientists, using the reach of various publications to make distinctions about those who read them. The market for popular science journals and magazines can be roughly grouped according to their circulation, which can be seen at the same time as an indication of the systematic character of science communication (Table 1).

Whereas a scholarly journal during the first third of the twentieth century usually reached several hundred scientists within a single discipline, the weekly circulation of the journal *Die Naturwissenschaften*, which was established in 1913 and addressed an interdisciplinary audience, reached between 2,000 and 3,000, numbers that essentially remained unchanged for decades. In addition, the weekly magazines *Naturwissenschaftliche Wochenschrift* and *Umschau* also achieved circulations of 15,000 and 20,000 respectively in the Weimar period. *Umschau* explicitly set out to treat equally all-important developments within natural science and technology (and beyond), whereby scientific terminology was to be avoided whenever possible and findings classified in a framework according to a certain worldview as well as graphically illustrated. In this way, an interested and educated public, i.e., an 'attentive public', was addressed. Although not scientists themselves, the members of this science-interested public were therefore regularly informed of the latest progress.

The magazine *Kosmos* also served essentially the same readership. Compared to the previous types of popular science magazines, however, it was exceptional in two regards. For one, a subscription was included in the membership to the *Gesellschaft der Naturfreunde*; also, its circulation reached extraordinarily large numbers. This can be attributed in part to its appearance only on a monthly basis and the correspondingly low annual cost. More important, however, was the special focus of the journal on nature and geography, topics attractive beyond a pure scientific interest. Furthermore, technology was almost completely excluded.[20]

---

[20] In addition to *Kosmos*, a bi-weekly journal entitled *Natur* began publication in 1910. It was a conceptual copy of *Kosmos*, started by authors who were disenchanted with the journal. One of the only evident differences between *Kosmos* and *Natur* was that the latter had a stronger orientation toward *Lebensphilosophie*. See Andreas Daum, *Wissenschaftspopularisierung im 19. Jahrhundert:*

In addition to this established public of popular science regulars, there were those who showed an occasional interest in science topics. This 'occasionally interested public' would be likely to visit museums of science and technology and purchase popular scientific magazines sold at newsstands. For this already rather large segment of the public, the magazines had to resort to more generic, general language, completely forego the use of scientific formulae and use illustrations designed specifically for their readership (particularly for the cover) to capture their readers' interest. Typical of this sort were *Koralle*, *Technik für Alle* and *Wissen und Fortschritt*, which together sold more than 100,000 copies per month in the years after 1927. Although the Ullstein publication *Koralle* adopted a more modern and in particular a more attractively illustrated format than the rather traditional *Kosmos*, the former did not outsell the latter until after 1933, when the scope of topics was expanded to include art, sports, popular culture and even humour, with the space dedicated to science and technology correspondingly reduced. With a circulation of 800,000, *Koralle* was transformed from a popular science magazine into a mass medium in National Socialist Germany.[21]

As a rule, the members of the 'broad public' did not demonstrate any specific interest in natural science that would have been reflected in the reading of specialised periodicals. They turned their attention to stories on natural science topics in cases where they were linked to attractions outside the realm of science, such as sensations, records or genius. This segment of the public was thus only reached through mass media, namely newspapers and illustrated magazines.

By thus expanding our view and differentiating among the various publics associated with science, we have gained greater leeway, compared to Forman's analysis, to discuss the various levels that defined a general intellectual environment, which physicists then perceived as an influential 'cultural milieu' and may have caused them to react. At the same time, this fundamental transformation of science communication, starting at the turn of the century, illustrates a system of communication and a dynamic that resulted in the increasing exposure of broader segments of the public to science. It is also characterised by the emerging mass media in the years from the turn of the century to about 1930. Corresponding to

---

Bürgerliche Kultur, naturwissenschaftliche Bildung und die deutsche Öffentlichkeit 1848–1914 (Munich: Oldenbourg, 1998), 371–75; Arne Schirrmacher, 'Der lange Weg zum neuen Bild des Atom: Zum Vermittlungssystem der Naturwissenschaften zwischen Jahrhundertwende und Weimarer Republik', in *Wissenschaft und Öffentlichkeit als Ressourcen füreinander: Studien zur Wissenschaftsgeschichte im 20. Jahrhundert*, ed. Sybilla Nikolow and Arne Schirrmacher (Frankfurt and New York: Campus, 2007), 39–73, on 55–57.

[21] *125 Jahre Ullstein. Presse- und Verlagsgeschichte im Zeichen der Eule* (Berlin: Axel Springer, 2002), 64–65.

this trend was the impartial inclusion of theories, sometimes contradictory, within a single issue and the emphasis placed on a superficial presentation of the science without any effort to provide a depth of analysis. This trend may appear to be a sign of both democratisation and trivialisation. However, it was countered by the publisher Springer Verlag, which intended once again to give research scientists a say in disseminating information about recent findings in natural science by way of the journal *Die Naturwissenschaften* as well as the book series *Verständliche Wissenschaft*, launched in 1927. In this way, scientific experts could assume the task of conveying science news, rather than leaving it to professional communicators, who at most merely availed themselves of sources available at the level of the scientific public.

Atomic physics in the communication system

In order to demonstrate that negotiation processes between the various levels of science and the public actually took place,[22] I will briefly sketch the revolutionary development of the concept of the atom. Starting in the early twentieth century, this concept changed from the early modern representation of a globular atom to the planetary and disembodied atom — as depicted by Rutherford, Bohr and Sommerfeld — that became widely established during the years upon which Forman based his study.

The iconography of the globular atom disappeared around the turn of the century when radioactivity and especially Philipp Lenard's experimental findings on the emptiness of atoms were discussed. Lenard's article in the *Annalen der Physik* in 1903 is an instructive exemplar of the different ways in which the various popular science periodicals — most of which have already been mentioned above — reported on recent discoveries of scientific research, with regard both to the time lag between their story and the publication of the related article and to the manner in which they treated the story.[23] Month after month, more and more popular articles appeared on this new insight into matter and its strange emptiness. Whereas the review journals ran articles on Lenard's findings within a couple of months, half a year passed before *Naturwissenschaftliche Wochenschrift*, a periodical aimed at the attentive public, took up the topic. However, the findings were presented here in a broader framework that placed them in the context of the historical and philosophical discussion of matter.

---

[22] Ludwik Fleck, *Genesis and Development of a Scientific Fact*, chap. 4 (Chicago: University of Chicago Press, 1979).

[23] Philipp Lenard, 'Über die Absorption von Kathodenstrahlen verschiedener Geschwindigkeit', *Annalen der Physik* 12 (1903): 714–44.

Noteworthy is that the article in *Naturwissenschaftliche Wochenschrift* not only reported the findings available in Lenard's original work, such as his conclusion that the atom is empty to the same degree as the universe is empty, but also published subsequent conclusions derived from his study but not included in it, for example that

> indeed each cathode ray particle contained within a force field [will] orbit rapidly around the positive point or describe paths, the knowledge thereof is to be expected from a yet-to-be-found solution to the three-body problem, which takes into consideration not only attracting forces but also repelling ones.[24]

In this way, readers of *Naturwissenschaftliche Wochenschrift* were introduced already in May 1904 to a concept of the atom that anticipated essential characteristics of the later (scientific) description of the atom, meaning not only the planetary structure of the 1911 Rutherford model but also the problem of determining the orbits of electrons analogous to the mechanical many-body problem, a task taken on 20 years later by Max Born and Werner Heisenberg. Independent of the article in *Naturwissenschaftliche Wochenschrift*, a planetary analogy appeared in *Kosmos* in 1904, in which experimental findings on radium and cathode rays were harmoniously combined with speculative public knowledge on matter. In this way knowledge emerged that had been developed *with* the public.

How could the modern atom have been known before science *discovered* it? We can solve this apparent puzzle at least partially by noting that, on one hand, the planetary atom was introduced into the discourse at the level of the scientific public as early as 1901 by Jean Perrin in an article for an upscale popular science journal, the French *Revue scientifique*.[25] As a result, the scientific discussion appeared in magazines aimed at an interested and educated public as a pre-discourse to the later considerations in professional scientific journals. On the other hand, there is a simple and plausible reason why the planetary model of the atom was so popular in magazines such as *Kosmos* and *Natur*: many of the published authors had been recruited from the field of astronomy, a discipline that had long been attractive to the lay public and produced professional popularisers from early on.[26] Thus, it may be attributed to a *déformation professionelle* that, for

---

[24] August Becker, 'Über die Konstitution der Materie', *Naturwissenschaftliche Wochenschrift*, n.s., 3 (1904): 532.

[25] Jean Perrin, 'Les hypothèses moléculaires', *Revue Scientifique* 15 (1901): 449–61. For more on the classification of journals, see Dominique Pestre, 'Les revues de vulgarisation scientifique en France 1918–1940: Un panorama', *Cahier d'histoire et de philosophie des sciences* 24 (1988): 71–81.

[26] On the large percentage of astronomers, see Daum, *Wissenschaftspopularisierung* (ref. 20), 385–90.

astronomers, the microscopic world could be described by applying structures and notions from astronomy.

The next stage in this interaction between science and the public is the establishment of the planetary model of the atom in science, or more precisely, in the discipline and on the research level and its communication with the scientific public. This can only be treated briefly here. First, scientists such as Rutherford and Bohr only needed to follow up on, rather than invent, the planetary analogy for the atom because it was already being widely discussed. Hence this was by no means a move to adapt to a hostile 'environment' of popular views in the Formanian sense. Rather, the articles appearing in *Naturwissenchaften* from 1913 to 1918 show a peculiar reserve, but not hostility, toward the model from within science.

The first mention of an atomic model that came close to Bohr's model as presented in *Naturwissenschaften* can be found in the publication in 1913 of the habilitation lecture by Kasimir Fajans, a young assistant for physical chemistry from Karlsruhe, who would soon become a colleague of Arnold Sommerfeld in Munich and eventually rise to the position of Ordinarius (full professorship) there in 1925. He considered such a theory in 1913 as only one among many speculations about the atom. At the time, Fajans chose to support the ideas of J.W. Nicholson and told his colleagues in physics, chemistry and biology that all matter consisted of hydrogen, 'Coronium', 'Nebulium', and 'Protofluor'.[27] Further articles followed by one of Lenard's students in that same year, and in 1916 Ernst Gehrcke explained to *Naturwissenschaften* readers that the drawbacks to Bohr's theory suggested his own idea about rings of 'aether vacuum' were just as tenable.[28]

So we see that the public apparently had no problem with the planetary atom. Instead, the physicists themselves created confusion or developed hostility toward the evidently so attractive new atom in the debate within their scientific public. Not until Sommerfeld's influential, semi-popular book *Atombau und Spektrallinien* (1919) did scientists publicly embrace the modern atom and authorise its validity. In 1918, Sommerfeld and Fajans described to the Deutsches Museum how a wood-and-wire model of the hydrogen atom was to be made and helped in planning a special exhibition room called 'Constitution of the Elements', devoted to the topic.[29]

---

[27] Kasimir Fajans, 'Die neueren Vorstellungen von der Struktur der Atome', *Naturwissenschaften* 1 (1913): 237–41, on 240.

[28] Heinrich Baerwald, 'Über die Förderung unserer Kenntnis vom Bau des Atoms durch die Erforschung der positiven Strahlen', *Naturwissenschaften* 1 (1913): 355–59, 384–88; Ernst Gehrcke, 'Atommodelle und Serienspektren', *Naturwissenschaften* 4 (1916): 586–90.

[29] For more detail, see Schirrmacher, 'Der lange Weg' (ref. 20).

In sum, it can be said of the scientific and public discourse about the atom from the turn of the century, through the war years and well into the Weimar period, that physicists were apparently able to integrate findings, ideas and even speculations from the public debate into their own scientific research as a productive resource. In the end, the acceptance of the new atom was — as Ludwik Fleck explained — fostered by this mutual 'genesis and development of a scientific fact' by science in conjunction with the public.[30]

This successful process of working out notions and concepts is, however, only part of the story. Just as professional science was toning down its criticism of the new model and as journals, books and exhibits were beginning to use the scientifically authorised models and illustrations held back for some time, quantum mechanics refuted the accepted and popular image of the planetary atom, categorically rejecting the idea of determinable orbits. The popular science magazines of the Weimar period refused to represent the new quantum-mechanics atom pictorially (not until 1936 did the first such illustrations appear in *Kosmos*). The major success of constructing the atom in conjunction with the public made it nearly impossible to revoke this planetary image; to this day we tend to favour the planetary image, if for reasons that are, strictly speaking, not scientific.[31]

Was the Weimar Crisis One of Atomic Physics in Particular, of Science in General or of Science Communication?

Since to this point we have considered magazines already in publication in Imperial Germany, I will now focus on the changes occurring during the Weimar period and analyse the meaning and role of crises. 'Crisis' became a key term both in the description of the times by Weimar contemporaries and later in the historiography on the first German republic, usually linked to a general discussion on modernity.[32] I will deal strictly with the question concerning possible crises in science communication and limit my observations to the discourse on crisis as related to the natural sciences. The question is this: Was the publication of science periodicals affected by a real crisis or was a crisis conjured up and used to reshape this field of publishing? In other words, what role did the diagnosis of crises play in science communication?

---

[30] Fleck, *Genesis and Development* (ref. 22). Due to space limitations, I cannot go into further detail here on the elaborations in this book about the theory of thought styles and its application to 'Zeitschriftenwissenschaft'; see Schirrmacher, 'Nach der Popularisierung' (ref. 19).

[31] Usually educational reasons are cited to explain why this model is used, despite the fact that the model's explanatory ability hasn't been weighed against its perpetuation of certain misconceptions.

[32] Relevant here is Peukert, *Weimar Republic* (ref. 9).

Undoubtedly, the dramatic inflation of the currency took its toll. Long-established magazines such as *Naturwissenschaftliche Wochenschrift* and *Prometheus* ceased publication; officially they were merged into *Umschau*, but no evidence of continuity can be found.[33] Magazines propagating a stronger ideological direction, for example *Neue Weltanschauung* of the Monist League, disappeared without a successor. A new popular science magazine typical of the Weimar era was *Koralle*, published by Ullstein Verlag, starting in 1925. It engaged several big names in the natural sciences as authors, including Erwin Schrödinger, who wrote on atomic and quantum physics. *Koralle* would remain the exception in this regard: in most cases, professional communicators handled the topic of physics. Two years after *Koralle* began publication, the monthly journal *Wissen und Fortschritt* was founded, which presented itself from the very first issue as a German defender against rampant Americanism in science and technology. The Franck'sche Verlagshandlung brought out *Technik für Alle*, a magazine dedicated to technology that served as a complementary publication to its own journal *Kosmos*, which for the most part ignored anything technological. *Technik für Alle* actually appeared first in 1916, but it was not until the Weimar Republic that its circulation expanded considerably. Not to be forgotten in this overview, *Urania* played a singular role as a decidedly Marxist magazine, a role that would be revived in the GDR.[34]

For a detailed analysis of these magazines it would be necessary to consider the new link between the publications and politics, their ownership by publishing conglomerates and the ways in which the relationship between science and the public changed in general. Space limitations prevent me from presenting this analysis and likewise I cannot delve into the impact of colour printing, large formats and the effect of competition from the medium of radio — all factors that changed the framework for science communication. Instead I will focus on the phenomena and perceptions of crises, since Forman considered this as constituting the characteristic trait of the Weimar 'environment' for modern physics and for

---

[33] The goal of such a merger might have been to make use of the existing subscriber pool. This was just about the same reason that *Naturwissenschaften* declared itself the successor of *Naturwissenschaftlichen Rundschau* and continued for a while to make formal mentions of its publisher. See Heinz Sarkowski, *Der Springer Verlag: Stationen seiner Geschichte, Vol. 1: 1842–1945* (Berlin: Springer, 1992), 192.

[34] See editorial, *Wissen und Fortschritt* 1 (1927): 1. *Technische Monatshefte*, established in 1910, was the predecessor of *Technik für Alle*, which appeared in 1916. On *Urania*, see Nick Hopwood, 'Producing a Socialist Popular Science in the Weimar Republic', *History Workshop Journal* 41 (1996): 117–53; and Thomas Schmidt-Lux, 'Wissenschaftliche Weltanschauung und Religion: Ein Beitrag zur Theorie des Säkularisierungsprozesses und seiner institutionellen Akteure am Beispiel der URANIA' (PhD dissertation, Leipzig University, 2006).

science, by and large. Further, I ask whether the readers of *Kosmos* and *Koralle* sensed such a crisis and whether it was then a crisis of science in general or modern physics in particular, or rather a problem of science communication.

First of all, a crisis debate did indeed occur in these periodicals, but it had no apparent influence on the attention paid to quantum and atomic physics. Even though much was subject to change in magazines geared to an interested public — such as an increasing emphasis on spectacular photography, cover illustrations and reports on psychological and supernatural phenomena — neither *Kosmos* nor *Koralle* hesitated to report on quantum mechanics, Schrödinger's equation, the laws of probability and causality, wave–particle duality or even matter waves.[35] The respective articles demonstrate a strikingly high level of quality when it comes to explaining, illustrating and presenting in an understandable manner such difficult topics. A general hostility toward these topics is not found in *Kosmos*, *Koralle* or *Wissen und Fortschritt*.

Still, we discover a great deal that fits Forman's findings. In 1927, a lead story in *Kosmos* mentioned Oswald Spengler and called for establishing new ties between physics and philosophy. In addition, pseudoscience flourished in the form of the *Welteislehre* (cosmic ice theory), divining rods and supposedly new radiation phenomena. However, at the same time there were critical analyses and discussions with readers on these topics. In 1931, *Kosmos* asked its readers to send in any scientific hoaxes found in newspapers and popular magazines. The following year, *Koralle* ran a long article on astrology, palm-reading and other similar topics under the title 'Irrwege schlechter Zeiten' ('Errant Ways of Bad Times') and thereby directly established the connection between flourishing pseudoscience and economically and politically troubled times.[36] Also revealing is the treatment accorded to cosmic ice theory, which was criticised factually in an article appearing in *Kosmos* in 1925, whereas *Koralle* resorted to subtle sarcasm in treating the subject a few years later.[37]

---

[35] See the following articles written by Erwin Schrödinger: 'Das Rätsel des Lichts', *Koralle* 5 (1929): 294–98; 'Das Gesetz der Zufälle', *Koralle* 5 (1929): 417–18; 'Was ist eigentliche Elektrizität?' *Koralle* 6 (1930), 110–12; 'Wissenschaft–Kunst–Spiel', *Koralle* 6 (1930): 404–05, 425–26. *See also* the following references to Schrödinger in *Koralle*: a mention of Schrödinger equations and the Pauli principle in 'Kleinste Welten', *Koralle* 5 (1929/30): 331 and an article on matter waves by Alfred Stern, 'Die Materie als Analogon des Lichtes', *Koralle* 7 (1931/1932): 474–76, which closes with: 'Corpuscles and waves certainly do belong together — as two mutually complementary sides of a physical reality,' on 476. (My translation.)

[36] Fritz Wassmann, 'Irrwege schlechter Zeiten: Astrologie, Chiromantie, Kartomantie, Siderismus', *Koralle* 8 (1932/33): 60–64.

[37] D. Grebe, 'Welteislehre und Himmelsforschung', *Kosmos* 22 (1925): 289–92; Hermann von Socher, 'Welteislehre: Astronomie der Nichtastronomen', *Koralle* 8 (1932/33): 302–04.

Especially in the case of atomic physics, we find a plurality of explanations rather than any particular hostility toward the field. In a 1929 issue of *Koralle*, the interested reader could read Schrödinger's description of quantum theory in an article written on the occasion of his appointment to succeed Max Planck at Berlin University. He availed himself of the story on the riddle of light to enlighten his 'gentle female reader' about spectroscopy (the *Leuchtreklame*, that is, the illuminated 'advertising' that every atom makes) and about quanta (as small as 'inflation Reichmarks') and he admitted to her that he himself did not even really know whether the quantum hypothesis was the genuine expression for the phenomena. 'As soon as we really know what we mean', he wrote, 'we will also be able to describe it to the layperson interested in natural science'.[38]

Here we have an example of how the discourse between scientific experts and an interested or even more general public could lead to a certain casual and flirtatious form of communication between science and the public. Contrary to Schrödinger's language adaptation, by way of a more familiar and lighthearted tone, the explanation offered by the professional communicator of science Paul Kirchberger on the quantisation of energy states in an atom provided the readers of his 1928 article in *Kosmos* precisely that which Schrödinger's article did not — namely, a graphic explanation. Even when Kirchberger indicated in his text that no illustration could claim to explain the solution of the quantum problem, since only mathematical theory could do this when empirically confirmed, his drawings are nevertheless good examples of a visualisation of Schrödinger's 'quantization as eigenvalue problem' (*Quantisierung als Eigenwertproblem*) from 1926 (Fig. 1).[39]

Kirchberger's was not the only such article to appear in *Kosmos*. There were others, just as there were in *Koralle*, where Werner Leszynski often reported on atomic physics and contributions by professional scientists were the exception. About two months before Schrödinger's article appeared, an engineer reported extensively in *Kosmos* on the atomic theory of Johannes Stark. In this article, the interior structure of the atom was depicted with colourful ball-and-stick models (Fig. 2). Whereas today Stark's ominous axial theory of the atom represents no more than a footnote in the history of atomic physics, it then attracted notable attention, as evidenced by the fact that one of the few and expensive colour pages in *Kosmos* was used to present it.[40]

Perhaps the most obvious interpretation for this plurality of offered explanations is the thesis that a democratisation of science communication took place, which both expanded the circle of those able to participate in the deliberation of

---

[38] Schrödinger, 'Das Rätsel' (ref. 35), 298.
[39] Paul Kirchberger, 'Ein Fortschritt in der Atomtheorie', *Kosmos* 25 (1928): 109–12.
[40] H. Schütze, 'Wie sieht das Atom aus?' *Kosmos* 26 (1929): 281–83, colour plate on 280.

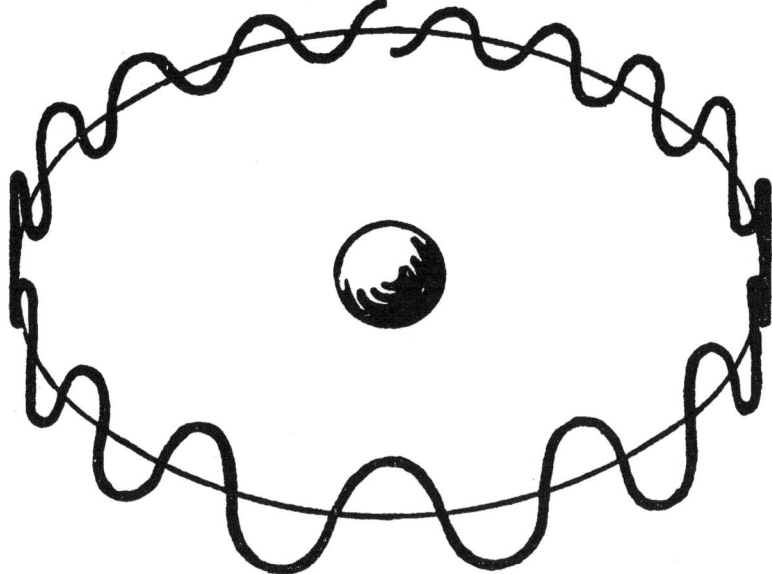

Abb. 2. Das neue Modell des Wasserstoffatoms; eine un=
mögliche Wellenbahn. Dort, wo die Welle mit dem
Wellental begann, kommt nach Umlauf der Welle ein Wel=
lenberg an. Ergebnis: Aufhebung der Welle; sie kann
sich nicht schließen, weil die Welle nicht mit einer Tal=Phase
anlangt

Fig. 1: Drawing originally designed as a visualisation of possible energy eigenstates in an article on quantum mechanics in *Kosmos*.

Source: Kirchberger, 'Ein Fortschritt' (ref. 39), 111.

scientific propositions and established a new culture of critique: no longer did the status of unassailability exist in the form that the scientific community of Imperial Germany was able to count on. There had been a considerable expansion of the spectrum of topics and of the societal sectors from which authors claiming to work scientifically were recruited. At the same time, every author, regardless of topic, had to be prepared generally to face criticism and if necessary to endure, even as soon as the next journal issue, the presentation of a completely different theory on the same topic.[41] Here is precisely where I concur with Forman's evaluation of

---

[41] Additionally, one often finds in science and technology journals a tendency to correct the occasional sensationalist newspaper story and put it in context. One such episode occurred in October 1929 when a German newspaper reported it had been said at a conference of the American Chemical Society that the hydrogen atom could explode. See the short notice, 'Irrtum aus Amerika', *Koralle* 5 (1929/30): 331.

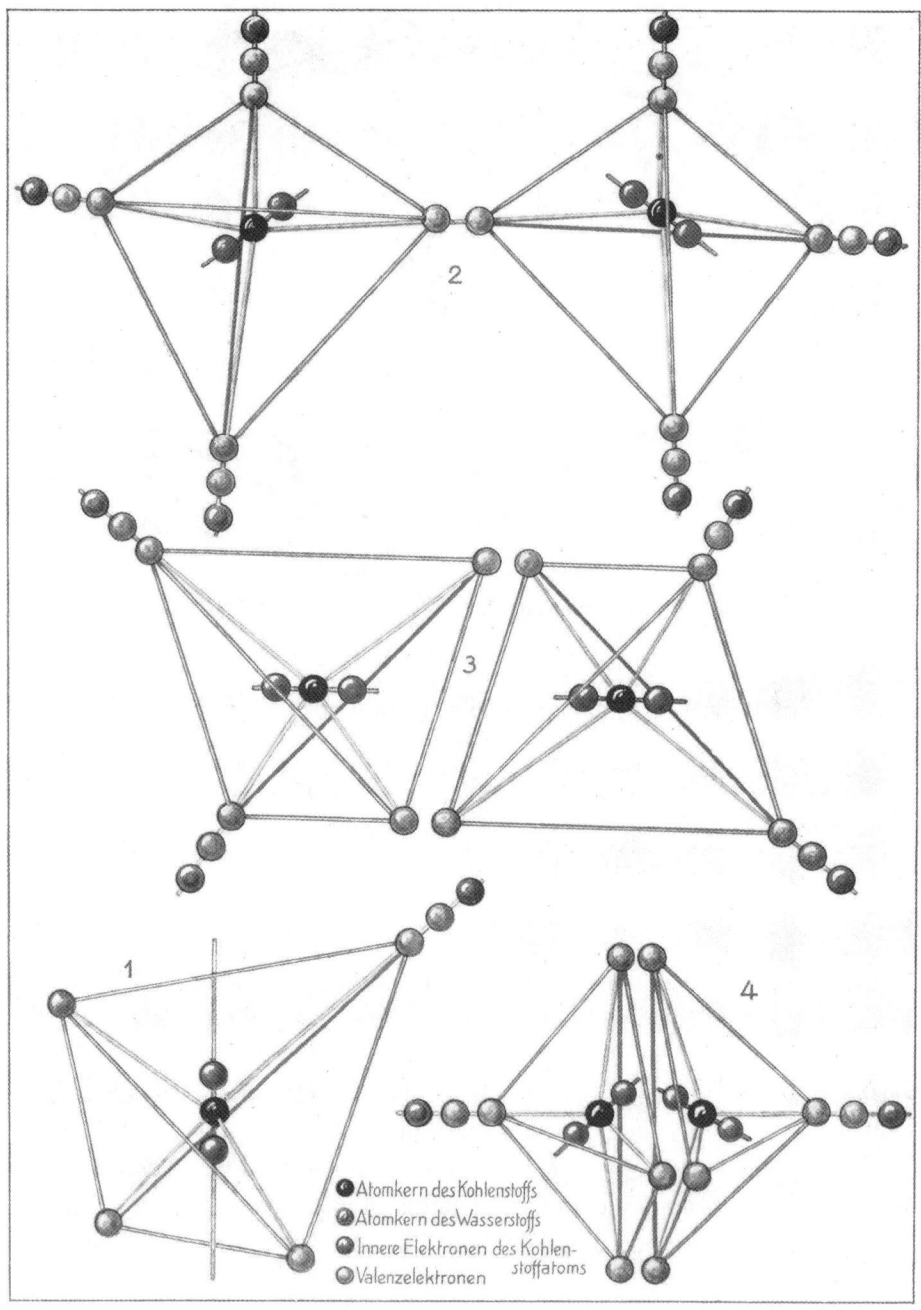

Fig. 2: Elaborate (colour) illustration published in *Kosmos* of a scientifically dubious alternative theory on the structure of the atom.

Source: Schütze, 'Wie sieht das Atom aus?' (ref. 40), 280.

physics journals during the Weimar period. What was true for *Kosmos* and *Koralle* is analogous to what happened in *Zeitschrift für Physik*, the leading periodical for atomic physics that was also publishing a great number of stories of rather dubious content at the time.[42] Evidently it was an expression of Weimar liberal democracy that publishers and editors opted for greater diversity in the system of science communication instead of stipulating strict criteria with an aim to provide a clearer demarcation between the scientific and the unscientific. For popular magazines this policy paid off. To offer a wide range of topics — be it on the progress in quantum mechanics or a criticism of science in general — corresponded to the interest of the publishing business to tap into as large an audience of readers as possible. In this respect, the changes in the way natural science was communicated also reflect a change in the public's interest in natural science and technology, as evidenced by the expanding market for such information.

Having thus extended the range of material labelled science communication, in comparison to Forman, who mainly focused on the criticism originating from the scientific public, the crisis discourse now includes the debate that took place at the level of the educated and interested public. Starting in August 1930, a series of 12 articles appeared in *Koralle* on the crisis of science, in which a number of well-known intellectuals attempted to answer the simple question: 'Why do we pursue science today at all?' The first in the series was by Adolf von Harnack, the great representative of the meritocratic model, which stipulated that the right to be publicly heard should be linked to scientific achievements; it was to be his last publication. Other answers were proffered by a sociologist, political geographer and medical doctor, who in turn were followed by the graphologist Ludwig Klages, the economist Werner Sombart, the physicist Erwin Schrödinger and the biologist Jakob von Uexküll. An engineer-turned-writer and a physician also contributed to the series, which was wrapped up by an historian and a chemist.

Particularly instructive is Sombart's response, which was immediately picked up by Schrödinger. Sombart points out at the very beginning that two things should not be confused: crises in scientific fields that are not destructive but rather act as necessary agents to open up a vital area of research and crises that endanger entire scientific projects. It was in this context that Schrödinger wrote his often-reprinted article 'Science, Art and Play', which is perhaps more well known under the title 'Science as Culture'. Schrödinger resolved the problem in a similar

---

[42] *See also* Dieter Hoffmann, '"You Can't Say to Anyone to Their Face: Your Paper is Rubbish": Max Planck as Editor of the *Annalen der Physik*', *Annalen der Physik* 17 (2008): 237–301. In contrast, it is more difficult to classify the respective political orientations of the journals. *Kosmos* geared itself to a conservative, nature-loving public, whereas *Koralle* set a pointedly progressive tone and described the potential to emancipate from nature through natural science and technology.

fashion by arguing that all of the (productive) crises possibly occurring in the various fields of science were nowhere near becoming a general crisis in science — i.e., a '*Wissenschaftsdämmerung*' (twilight of science). He contended science was a universal human need, like art and play, and could not be defined solely by aims and purposes. Since science was culture first and foremost, its cultural value had to be deemed higher than its utilitarian value. Following the demise of an authoritarian state, a new Zeitgeist would forge ahead and lead to crises in science. However, 'there is never any need to oppose the assaults of the spirit of the age [*Zeitgeist*]; that which is fit to live will successfully resist'.[43] Science would thus prove itself more immune to the public than the discourse among the scientists themselves sometimes seemed to suggest.

If we turn our attention away from the intellectual discourse within science and focus instead on the more general discourse of popularised science, the emerging picture of science communication in the Weimar Republic is dominated neither by hostility toward the natural sciences nor does it indicate any contempt of scientists. Instead of a destructiveness due to external factors in general acting on science and in particular during the Weimar period, we find the public to be a productive resource for science, both in sharing the assets of public knowledge and providing the opportunity to draw on the public interest in science.

---

[43] Schrödinger, 'Wissenschaft–Kunst–Spiel' (ref. 35); the English translation appears as 'Science, Art and Play', *The Philosopher* 13 (1935): 11–18, quote on 18. *See also* Erwin Schrödinger, 'Science as Culture', *The International Forum*, 1 January 1931, 10–11.

Living Ambiguity: Speculative Bodies of Science in Weimar Culture

Cornelius Borck*

Writing on the post-revolutionary Moscow he had visited in the winter of 1926–27, Walter Benjamin observed such a trip opened up a 'new optic', a clearer and more conscious appreciation of Europe's historical situation. For him, the Soviet revolution had resulted in a radical experimentalisation, where 'every thought, every day, and every life is stretched out as if on a laboratory table'.[1] Weimar Germany, in contrast, was not the product of a successful revolution but of a lost war; and retrospectively, after the ensuing catastrophe, it has shrunk to the short-lived and problematic interlude between the Kaiserreich and the Nazi regime.[2] The result of a failed attempt at revolution, and destined to a fractured course of economic, political and social instability, Weimar Germany accepted Oswald Spengler's *Decline of the West* quite naturally as its bestseller. And yet, the very same Weimar Germany, with its nostalgic longing for bygone greatness and romantic harmony, saw the birth of abstract modernism in the arts and sciences, forged the Bauhaus, enjoyed a thriving cultural life in its metropolis, suffered from the radicalisation of politics, went through hyperinflation, had a burgeoning film industry, engaged in American-style rationalisation and discussed the value of racial purity.[3] Weimar Germany differed from revolutionary Russia — so

---

* Königstr. 42, 23552 Lübeck, Germany; borck@imgwf.uni-luebeck.de.
[1] Walter Benjamin, 'Moskau', in *Gesammelte Schriften*, vol. 4, ed. Tillman Rexroth (Frankfurt am Main: Suhrkamp, 1972), 316–48, on 325 (my translation).
[2] S. William Halperin, *Germany Tried Democracy: A Political History of the Reich from 1918 to 1933* (New York: Norton, 1965).
[3] Daniel Peukert, *The Weimar Republic: The Crisis of Classical Modernity* (New York: Hill and Wang, 1992). A kaleidoscope of the divergent strands of Weimar culture that serves as a valuable introduction is Anton Kaes, Martin Jay and Edward Dimendberg, eds., *The Weimar Republic Sourcebook* (Berkeley: University of California Press, 1994).

one could argue in the vein of Benjamin's insight — in that it was not a single or singular experiment in revolution which pervaded the last corners of society and subverted every life and moment, but rather a short-lived, conflict-ridden and incoherent period amounting to an ongoing experimentalisation of life and rationality that dissolved and disintegrated pre-established norms and disciplines. And yet, at the same time, a divergent productivity of opposing extremes provided the crucial ingredient for the Weimar Republic to become a period of remarkable cultural innovation.

It is the lasting charm of Paul Forman's (first) thesis that it directly addressed this peak of intellectual innovation in the midst of a society far better characterised by antagonism than a striving for the modernist project it nonetheless generated.[4] Taking the bull by the horns, Forman famously argued that the answer to this apparent paradox lay in the antagonism itself. In the style of an almost Hegelian resolution, he suggested the opposition against modern, scientific rationalism was of such a scale and nature that it forced scientists to actively detach themselves from the existing standards of scientific rationality in order to survive in this arena of hostility. The notorious 'adaptation to a hostile intellectual environment', with which Forman succinctly summarised his first masterpiece in its subtitle, has meanwhile, however, turned into a conundrum almost as difficult to tackle as the problem he once aimed to address. Certainly, Forman's case study marshalled the end of a purely internalist history of science, according to which the truth of a theory ultimately guided and determined the path of history. But what exactly did the kind of externalism that Forman argued for entail?

What could possibly be meant by a conscious adaptation to irrationalism?[5] Certainly, human beings frequently behaved irrationally and continue to do so, if judged by external observers or asked retrospectively; but in the present tense of action, irrationalism is necessarily opaque to an actor. In the actor's perspective, every activity follows some ratiocination. What exactly was the mechanism that Forman had in mind when he spoke of 'quasi religious conversions' or of an intellectual 'capitulation' on behalf of the scientists in Weimar Germany?[6] This choice

---

[4] Paul Forman, 'Weimar Culture, Causality, and Quantum Theory, 1918–1927: Adaptation by German Physicists and Mathematicians to a Hostile Intellectual Environment', *Historical Studies in the Physical Sciences* 3 (1971): 1–115.

[5] This review of some aspects of the critical discussion kindled by Forman's thesis follows John Hendry, 'Weimar Culture and Quantum Causality', *History of Science* 18 (1980): 155–80; P. Kraft and P. Kroes, 'Adaptation of Scientific Knowledge to an Intellectual Environment', *Centaurus* 27 (1984): 76–99 and Steven Shapin, 'Discipline and Bounding: The History and Sociology of Science as Seen through the Externalism-Internalism Debate', *History of Science* 30 (1992): 333–69.

[6] Forman, 'Weimar Culture' (ref. 4), 80, 86.

of words points to activities decisively outside the repertoire of both scientific practice and logical reasoning. Does he assume a scientist to embrace irrationality because of extra-rational experiences, by way of a rational decision-making or simply because of overwhelming social pressures that 'virtually dictated'[7] to dispense with causality? Each of these 'mechanisms' would entail distinct psychological dispositions and give rise to different sets of cognitive, social and material consequences. In addition and beyond its questionable nature, how can the assumed psychodynamic mechanism be applied to scientists as a social group; does this not imply some concept of a collective consciousness and a somewhat coherent group intentionality? Finally, there is the epistemological question: what can be gained by such an explanation? Scientists, just like other human beings, act and behave with many psychologically well-known (ir)regularities. But how illuminative and relevant is it to link the path of scientific development to motives, beliefs and assumptions? Such explanations would immediately generate more questions than they would answer, such as how a particular hostility was perceived, how it intervened into laboratory routines and how it translated into new practices, alliances or conceptualisations. Looked at in light of this critique, the Forman thesis is so suggestive precisely because it 'blackboxes' the nitty-gritty of scientific work and confines the navigation through a complex cultural context inside an easier-to-handle container of antagonism. From the beginning to the end, a clear demarcation line separates science from the hostile environment. Forman's externalism is as absolute as his scientific idealism; the externalist explanations stay away from proper science, leaving it absolutely unaffected, avoiding it like a taboo. His famous paper demonstrates how some work in the humanities remains to be stimulating because its engaging inconclusiveness provokes debate.

Forman's thesis comes across as so awkwardly remote today that it calls out for contextualisation to be rescued by the cultural history of science.[8] Almost four decades after its publication, Forman's magic formula may indeed shed more light on the history and philosophy of science in the Cold War era than on the period it addressed. With now almost the same historical distance to the publication of the thesis as between it and the Weimar Republic, a similar historiographical scrutiny can be applied to both. A proper contextualisation of quantum physics as well as of Forman's theorising may reveal how local material, social and intellectual interactions paved the way for the quantum leap from the history of scientific theory

---

[7] Ibid., 7.

[8] At the conference from which this edited volume originated, M. Norton Wise intervened by asking, 'How can such a potentially fantastic contribution to cultural history of science be so deeply and seriously flawed?' Compare with his unpublished, but widely circulated manuscript, 'Forman Reformed' (2007), revised and published in this volume as 'Forman Reformed, Again'.

to externalism. His efforts to provide a causal explanation for acausality may seem perplexing today, but this very determination is probably indicative of the rigid rationalism of the history of science prevalent at the time Forman wrote his thesis. Part and parcel of this rationalism in the history of science was a strict hierarchy of the disciplines according to mathematical formalisation, with physics and the theoretical sciences ranking at the top. This prioritising of formal and mathematical sciences over qualitative and descriptive forms of knowledge also informed the Forman thesis as it exaggerated the contrast between quantum theory and Weimar Germany to a rather strict opposition between scientific rationality, conceived of as coherent, abstract and mathematical, and more speculative and allegedly irrational styles of thought, assumed to be predominant in Weimar culture and exemplified by *Lebensphilosophie*.

There is no question that Forman rightly identified the intellectual milieu as the crucial ingredient in his reconstruction of the history of science during the Weimar period.[9] The debate, however, centres around the issue of whether anti-intellectualism and irrationality are the appropriate and useful general descriptors for this milieu. And here we arrive at the starting point of this paper. It agrees with the Forman thesis that the cultural context is of central importance for scientific debates in a specific period and for 'ways of knowing' in general, to use John Pickstone's phrase.[10] But instead of distancing this context to a hostile milieu best characterised by externalising it into a monolithic block, the paper intends to open the 'black box' of irrationalism and 'socio-political pressure'[11] by asking where and how this particular cultural context furnished opportunities, prompted responses or provided intellectual means that articulated in specific ways with scientific questions or debates. This paper thus aims to transform Forman's important insight about the immersion of science in its context into a culturally grounded theory, providing a working framework for rich and meaningful reconstructions of scientific developments. It focuses on examples from the life sciences — which may seem unfair, since Forman explicitly and exquisitely addressed quantum

---

[9] In addition to the literature on Weimar culture physics and natural sciences referenced in the other contributions to this volume, see, for the life sciences, Jonathan Harwood, 'Weimar Culture and Biological Theory', *History of Science* 34 (1996): 347–77; Anne Harrington, *Reenchanted Science: Holism in German Culture from Wilhelm II to Hitler* (Princeton: Princeton University Press, 1996); Mitchell G. Ash, *Gestalt Psychology in German Culture, 1890–1967: Holism and the Quest for Objectivity* (Cambridge: Cambridge University Press, 1998) and *Greater than the Parts: Holism in Biomedicine, 1920–1950*, ed. Christopher Lawrence and George Weisz (New York: Oxford University Press, 1998).

[10] John V. Pickstone, *Ways of Knowing: A New History of Science, Technology and Medicine* (Manchester: Manchester University Press, 2000).

[11] Forman, 'Weimar Culture' (ref. 4), 110.

theory, not biology or the human sciences. But the purpose here is not to accuse Forman of shortcomings concerning sciences on which he had not set his sights. Quite the opposite; with the intention to correct for the bias toward physico-mathematical sciences alluded to above, this paper explores examples from the life sciences with an aim to arrive at a more detailed analysis of the articulation of science and society by way of examining a field where scientific rationality cannot be conflated with a mathematical apparatus.

Seeking out substantial and valid reconstructions that extend beyond the field of rigid scientific logic, this paper reverses, in fact, the logic of Forman's argumentation. Forman generously dismissed biology as too weak a case to prove his thesis: 'As today with the "ecology" fad, so also in the Weimar period it was the biologist who could most easily adapt his ideology and values to those of the intellectual milieu. Life, that central symbol, was his own subject. Paraphrasing a spokesman for the discipline, of all natural sciences biology … brings us to the edge of the irrational'.[12] Whereas Forman disregards biology to such a degree that its dubious status almost merits putting 'biology' in quotes, 'life sciences' is understood here to cover a vastly heterogeneous terrain of scientific activities. There may be doubts as to whether some of these activities can legitimately be claimed a 'science' or whether they rather represent defective variants, as I am sure Forman would argue. However, their more questionable nature — as in the case of electrophrenology or telepathic energy transmission explored here — turns them into particularly significant examples for my purposes. The intent of this paper is to demonstrate how the reconstruction of rational argumentation can be extended into domains judged irrational or speculative by later and more stringent standards. The ambiguities of the Weimar period certainly included an uncertainty about the proper limits of science because its most respected branches underwent massive transformations, many new applied forms of knowledge aspired to the status of science and more esoteric styles of reasoning sailed favourably for a while under its banner.[13]

Arguing with the epistemic relevance and significance of particulars for a philosophically informed history of science, this paper has no intention of

---

[12] Ibid., 40, referring to the biology pedagogue Philipp Depdolla, who was quick to embrace eugenics and the NS Rassenlehre. Cf. Philipp Depdolla, *Erblehre, Rasse, Bevölkerungspolitik, vornehmlich für den Unterricht in höheren Schulen bestimmt* (Berlin: Alfred Metzner Verlag, 1934).

[13] Using Ian Hacking's notion, I want to allude to his different forms of reasoning without grouping them into particular styles, as defined by him; cf. Ian Hacking, 'Styles of Scientific Reasoning', in *Post-Analytic Philosophy*, ed. John Rajchman and Cornel West (New York: Columbia University Press, 1985), 145–65.

commenting on the life and human sciences of the Weimar period in their entirety, or of making claims about their most important fields or developments. It confines itself to very few examples, deliberately taken from the fringes of the field, so that they inherently promise by their sociocultural position to elucidate the intrinsic ambiguities of scientific reasoning if understood as a historical process — e.g., as materialisations of 'how men actually think'.[14] Obviously, this paper must confine itself to sketchy outlines of a small number of examples. Beyond obvious differences, the three fields I want to briefly explore here have in common that they blur the strict dualism of progressive rationalism versus irrational romanticism. Comparing and contrasting some of the arguments around mind, brain and personality, the body and industrialisation and finally, science and epistemology during the Weimar Republic may be helpful toward arriving at a richer picture of the complex intertwining of reason and rationality, sentiment and sensibility, in this very short but innovative period.

The Birth of the Electric Brain

The first case I want to discuss is the entanglement of electrophysiology with psychodiagnostics in Weimar Germany.[15] In the spring of 1926, Emil Ludwig, prolific author of profitable biographies, claimed to have witnessed a very special scientific breakthrough. According to his report in the *Berliner Illustrirte Zeitung*, Weimar Germany's most widely read weekly, a little gadget, an electrical stimulation and measuring apparatus had produced a near-perfect personality profile of him simply by applying weak electrical test currents to a sequence of diagnostic points scattered about his head.[16] In the cold light of later-day neuroscience, this was just a revival of Franz Joseph Gall's outdated phrenology in the guise of modern-looking machinery and hence trumpeted as a scientific achievement, thanks to the then-recent availability of electrical gadgets. Zachar Bissky, the propagator of this new electric 'diagnoscopy' (as it was called), did not hesitate to

---

[14] Karl Mannheim, *Ideology and Utopia: An Introduction to the Sociology of Knowledge* (London: Routledge & Kegan Paul, 1936), 1. The entire sentence reads, 'This book is concerned with the problem of how men actually think. The aim of these studies is to investigate not how thinking appears in textbooks of logic, but how it really functions in public life and in politics as an instrument of collective action'. As I will argue later, it is hardly a coincidence that the sociology of scientific knowledge emerged within the cultural context of Weimar Germany.

[15] For a more detailed account of this episode, see Cornelius Borck, 'Electricity as a Medium of Psychic Life: Electrotechnical Adventures into Psychodiagnosis in Weimar Germany', *Science in Context* 14 (2001): 565–90.

[16] Emil Ludwig, 'Die Durchleuchtung der Seele', *Berliner Illustrirte Zeitung* 35, no. 10 (1926): 299–306.

make the link to phrenology explicit in his advertising material, where he spun yarns about how the device resonated with the brain's intrinsic frequencies and thus opened the most fantastic avenues for diagnostic as well as therapeutic interventions.[17] And indeed, diagnoscopy enjoyed a stellar, though short-lived, heyday.

Phrenological personality profiling by means of a surprisingly simple technology enrolled quite a remarkable list of supporters and thus forced major players into action. According to several critical scientific reports, diagnoscopy was first used for personality testing within the Swiss postal services and clock-manufacturing industry. After some positive reports and a conference on the new method in Karlsruhe in 1925, several German institutions became interested, including technical universities, police departments, prison administrations, large industries such as the Gelsenkirchener Bergwerks-AG and small enterprises recommending diagnoscopy for marriage counselling.[18] At many of these places the interest in the apparatus may have been little more than curiosity; however, diagnoscopy held out a more specific promise for rational modernisers. And this is what makes the case interesting with regard to Forman's thesis. Fritz Giese and Robert Werner Schulte, young and aspiring experimental psychologists, embarked on a scientific investigation of diagnoscopy.[19] Envisioning a bright future for their discipline in the form of the applied science of 'psychotechnics', these two psychologists hoped for a leading role in the rationalisation of life and labour on a national scale — which Giese in particular advertised as an 'Americanization'.[20] As part of their relentless initiatives to rationalise and optimise all occupations, professions and affairs, they identified diagnoscopy as an individually beneficial form of counselling, praising it as an efficient means for Germany's occupational modernisation. It is obviously difficult to accept that academically trained scientists such as Giese and Schulte truly 'believed' in diagnoscopy but that is not the point here. Typical for the style of thought of scientific management as

---

[17] Zachar Bissky, *Die Diagnoskopie: Eine neue Methode zur medizinischen, psychologischen und forensischen Diagnostik* (Karlsruhe/Berlin-Charlottenburg: Bios-Institut für praktische Menschenkunde, 1925).

[18] Fritz K. Walter, 'Über die Elektrodiagnose seelischer Eigenschaften ("Diagnoskopie") nach Bissky: Eine kritische Besprechung', *Jahrbuch der Charakterologie* 4 (1927): 297–324.

[19] According to historical sources, the two psychologists worked independently on diagnoscopy. *Cf.* Robert Werner Schulte, 'Über Elektrodiagnose seelischer Eigenschaften', *Psychologie und Medizin* 1 (1925), 62–94; Fritz Giese, 'Elektrodiagnostik des Characters', in *Bericht über den neunten Kongreß für experimentelle Psychologie in München vom 21–25. April 1925* (Jena: Gustav Fischer, 1925), 162.

[20] Fritz Giese, *Girlkultur: Vergleiche zwischen amerikanischem und europäischem Rhythmus und Lebensgefühl* (Munich: Delphin-Verlag, 1925).

exemplified in their approach toward the problem of rationalisation, the very question of a proper and appropriate scientific theory could be bypassed in light of convincing empirical evidence. Independent of one another, Giese and Schulte straightforwardly tested the experimental method with positive results. This was enough to get the ball rolling; because of the successful enlisting of professional psychologists into its camp, diagnoscopy provoked additional players into action and various congresses of psychotherapy and psychology established commissions to critically evaluate diagnoscopy. Eventually, eminent scientists such as Oscar Vogt, director of the Kaiser Wilhelm Institute for Brain Research, and Georg von Arco, German pioneer of wireless telegraphy and radio technology, forged counterstrategies on how to secretly obtain one of Bissky's machines in order to expose his method as a fraud. But on the side of the critics, there was hardly less ambiguity about Zeitgeist and modernity. The next project Arco undertook, for example, was a large-scale experiment on telepathy by means of radio and broadcasting, driven as much by his personal curiosity into mental energy as by his critical agenda.

These debates having set the scene, just a few years later a stern believer in telepathy managed to assemble the first brain-wave recorder.[21] Hans Berger did not have to adapt to a hostile anti-intellectual milieu; on the contrary, he embodied this milieu to a remarkable degree. In fact, with his *Weltanschauung*, Berger would make a perfect spokesperson for Forman's Weimar Zeitgeist. Deeply conservative and with a strong nostalgia for the alleged lost harmony of nature and culture, Berger was convinced of the true existence of mental energy and psychic forces as natural processes. Since a peculiar near-death experience early in his life, coupled with a moment of telepathic communication with his sister, he believed in psychophysical dualism and searched for experimental ways to objectively and scientifically prove the reality of psychic energy.[22] This was the guiding idea that connected the many experimental projects utilising various instrumental approaches, which he embarked upon in the course of his life. And yet, socially,

---

[21] The first report on the human electroencephalogram is Hans Berger's 'Über das menschliche Elektrenkephalogramm', *Archiv für Psychiatrie* 87 (1929): 527–70, though his 'Das Elektrenkephalogramm des Menschen', *Medizinische Welt* 4 (1930): 91–113, had a bigger public impact. For a more extended version of this history, see Cornelius Borck, *Hirnströme: Eine Kulturgeschichte der Elektroenzephalographie* (Göttingen: Wallstein, 2005).

[22] This became immediately evident when David Millett and I worked on Berger's personal diaries and notes, which are now located in the archives of the Ernst-Haeckel-Haus in Jena. See David Millett, 'Hans Berger from Psychic Energy to the EEG', *Perspectives in Biology and Medicine* 44 (2001): 522–42; here he takes up an interpretation from Martin Schrenk, 'Hans Bergers Idee von der "psychischen Energie"', *Nervenarzt* 41 (1970): 263–79.

he was not at all an outsider. Trained as a physician with a specialisation in modern methods of experimental physiology and brain research, he earned a professorship and directed an important psychiatric clinic, was elected dean and rector of Jena University and served actively in academic societies.

His professional activities, his experimental pursuits and his personal beliefs may appear disconnected in retrospect, but in combination they formed the fertile environment that enabled electroencephalography to materialise, without question one of the most important brain research methods of the twentieth century.[23] Without his speculative inclinations, Berger would have been unlikely to record brain waves, his most important scientific success. Only his deep convictions kept him going, against all odds and manner of mishaps, to try and try again, and to follow paths abandoned by others. To cut a convoluted and complicated story short, he pursued a project that most of his colleagues in electrophysiology regarded as impossible and futile: the capturing of minute electrical activity from the myriads of neurons inside the head. And yet, he found a surprisingly regular, large and stable brain rhythm. This was not mere serendipity since the biggest obstacle he had to overcome was his own scepticism. Initially he disbelieved his recordings, and through a difficult and circuitous process, struggled to convince himself of his findings because the data conflicted with his expectations.[24] Berger's story not only demonstrates how speculative thinking may yield significant scientific achievements, it reveals how the very perseverance of outdated and esoteric beliefs can result in innovative experimental procedures when such beliefs translate into intellectual resources. Every step of his winding path of discovery was the result of a hypercritical analysis of the experimental data, adhering to a sceptical routine of reflection and corroboration. The case of Berger thus exemplifies an internal rationality of speculative and allegedly irrational beliefs.

Cool Harmony and New Objectivity in Fritz Kahn's Bodyworlds

The second example comes from a very different trend in Weimar culture, the *Anschaulichkeit* of the modern, rational and industrialised world, as exemplified by the aesthetic movement of factual realism or new objectivity. The *Neue Sachlichkeit* emerged as the label for a distinctive style in the visual arts after a

---

[23] Cornelius Borck, 'Writing Brains: Tracing the Psyche with the Graphical Method', *History of Psychology* 8 (2005): 79–94.

[24] Cornelius Borck, 'Recording the Brain at Work: The Visible, the Readable, and the Invisible in Electroencephalography', *Journal of the History of the Neurosciences* 17 (2008): 367–79.

seminal exhibition in Mannheim in 1925.[25] Carl Grossberg and Christian Schad count among the more prominent artists associated with this movement and exemplify its almost sterile realism, focusing with the same radical detachment on machines and bodies, and portraying contemporary culture with a heightened awareness of social issues. More than just a new style in the arts that spread from painting to film, architecture and photography, the *Neue Sachlichkeit* resonated in Weimar Germany as an attitude or habitus. The revolutionary visual language of photographers such as August Sander and Albert Renger-Patzsch helped to illustrate how *Neue Sachlichkeit* also encompassed a cultural trend. These photographers applied a cool atmosphere with abstract precision to nonetheless highly idealised images of object types. In a certain way, the new objectivity in photography aimed at a form of realism that did not betray an aesthetic idealism. *Die Welt ist schön* [The World is Beautiful] was the ostentatious title of Renger-Patzsch's famous collection of 100 photographs that jumped from rather abstract images of natural objects to detailed studies of massive industrial buildings and the endless sequentiality of modern production spaces — thus claiming an extension of traditional norms of beauty into modernism and the modern industrialised world.[26] Behind this aestheticisation of industrial architecture stood a larger 'culture of distance', a strategy of cool conduct as the paradoxical combination of instrumental rationality with *Lebensphilosophie*, as Helmuth Lethen has shown.[27]

A particularly telling example of this blend of rationalism and romanticism, Fritz Kahn's modern bodyworlds, *Das Leben des Menschen* [The Life of Man], presents a human biology for general audiences. Kahn, a doctor with a practice in Berlin, started a second career, reporting regularly on medical topics in the *Berliner Illustrirte* and the *Frankfurter Zeitung*. The popular textbook *Das Leben des Menschen* was his magnum opus and appeared in five volumes between 1924 and 1931 for the *Kosmos Gesellschaft der Naturfreunde* and its more than 200,000 members.[28] Kahn's recipe for success was the combination of a lucid writing style with intriguing visuals that portrayed the body and its functions as machine ensembles. In addition to the over 1,500 images in the five volumes, the set also

---

[25] Hans-Jürgen Buderer and Manfred Fath, eds., *Neue Sachlichkeit: Bilder auf der Suche nach der Wirklichkeit — figurative Malerei der zwanziger Jahre* (Munich: Prestel, 1994).

[26] Albert Renger-Patzsch, *Die Welt ist schön: Einhundert photographische Aufnahmen* (Munich: Kurt Wolff, 1928).

[27] Helmut Lethen, *Cool Conduct: The Culture of Distance in Weimar Germany* (Berkeley: University of California Press, 2002).

[28] On *Kosmos*, see Arne Schirrmacher, 'Kosmos, Koralle und Kultur-Milieu', *Berichte zur Wissenschaftsgeschichte* 31 (2008): 353–71.

included an oversized poster, summarising the book by showing *Man as Industrial Palace* — thus encapsulating Kahn's explanatory strategy. The poster showed the human body as an assembly line in an industrialised workspace, packed with technology and populated, according to the division of labour, by anonymous workers and white-collar members of the modern society. The image did not simply compare the body to an industrial plant metaphorically — it directly represented every organ and bodily system with a technical analogue topographically proportionate in size and location. The heart was the typical pump, the lungs a ventilation apparatus, the brain a combination of switchboard and calculation technology, and so on.[29] In this way, Kahn appeared to regard technological and mechanistic models as appropriate vehicles through which to propagate a modernist view of science and medicine to larger audiences.

Does Kahn's success demonstrate how a biological modernism thrived side by side with Weimar Germany's purportedly organic holism? Certainly, Kahn was a moderniser, but his modernism represented more of a blending with holism than a contrast to it. Precisely this combination makes Kahn such an apt example for the discussion here. On the one hand, he employed a modernist technological style to describe the functionality of the human body in terms of industrial production and to provide performance indicators as though for a company's business report. On the other, he engaged in different narratives, for example, when he poetically described the processes of infection and repair as the fairy tale of the life cycle of a lymphocyte. In the midst of a society where the traditional doctrine of marriage clashed with the political agenda of Magnus Hirschfeld's Institute for Sex Research, Kahn explained the details of sexual arousal in terms of hydraulic difficulties and disturbances of the male sexual apparatus, but offered at the same time humanist encouragement for romantic love and frank advice about successful lifelong relationships. Thus, he blended explicit information and humanist enlightenment into a particularly successful holistic modernism; in addition to his several books, he acted as an advertising consultant for various companies. In light of this career in the Weimar Republic, Kahn's latitude in narratives and visualisation styles can be seen as indicative of the ambivalences so typical of the era's Zeitgeist.

As decisively modern as most of Kahn's visualisations are, his multi-faceted book is more accurately viewed as an amalgamation of opposing trends rather than a depiction of the clash of extremes allegedly so typical for Weimar Germany. This characterisation extends to his personal life. After a childhood in New York and

---

[29] Cornelius Borck, 'Communicating the Modern Body: Fritz Kahn's Popular Images of Human Physiology as an Industrialized World', *Canadian Journal of Communication* 32 (2007): 495–520.

Hoboken, he grew up in a Jewish family with Zionist leanings in Berlin, where he also finished his medical training shortly before World War I. Following the war, he divided his life between a medical practice in Berlin-Charlottenburg, extended geographical adventures to the Sahara and the Norwegian ice, and an increasingly successful career as a popular science author before he had to leave Germany and eventually emigrated to the United States.[30] Because of his success in Weimar Germany, some of his publications had already been translated and printed in the US, which allowed him to continue his career with a similar success after he arrived. Commenting on *Man in Structure and Function*, an abridged version of *Das Leben des Menschen*, a more critical reviewer observed, however: 'The most noteworthy feature of the publication consists in its 461 well reproduced illustrations, many of which are highly imaginative and most are instructive to a rare degree. At times, however, the desire for originality has gone to almost absurd extremes'.[31] It may well be that American audiences were less prepared to accept and acknowledge Kahn's style of visualisation but the critique of the reviewer captures succinctly that distinctive characteristic of Weimar German culture: the abundance of absurd extremes.

In a certain way, images like *Man as Industrial Palace* formed part of a larger strategy to 'organicize' alien technology through comparison to the body, as, for example, McLuhan would later argue with regard to communications technology.[32] It should be noted, however, that Kahn's images turned this strategy upside down. Where McLuhan recognised familiar biological principles in modern communications technology, Kahn familiarised the body's concealed organic interior by way of common gadgetry, as if some techno-literacy would bear the potential to reconnect with the body's machinery in new ways. *Man as Industrial Palace* depicted a technological paradise that measured biological bodies in terms of technical artefacts. The images extended the common comparison of bodies and machines to a manifest amalgamation of nature and technology, visualising a constructivism in which technological civilisation and experimental science intervened in the biological nature of human bodies. And yet, Kahn's colonisation of the human body by machines did not result in the dystopia of, for example, Fritz Lang's *Metropolis*. Quite the contrary, Kahn's

---

[30] For biographical information on Kahn and an introduction to his style of visualisation, see Fritz Kahn, *Man Machine: Maschine Mensch*, ed. Uta von Debschitz and Thilo von Debschitz (Vienna: Springer, 2009).

[31] Fritz Kahn, 'Review of Man in Structure and Function', *Quarterly Review of Biology* 18 (1943): 385–86.

[32] Marshall McLuhan, 'The Gadget Lover: Narcissus as Narcosis', in *Understanding Media: The Extensions of Man* (Cambridge, MA: MIT Press, 1994), 41–47.

images reflect a harmony of perfection where every machine and each individual human being has found their place. In contrast to the industrial realities of the times, Kahn's machines did not leak or produce waste while the workers and operators diligently did their jobs. What Kahn created with his popular images was a romantic utopianism of industrialisation, where human ingenuity in instrument making and machine building finally arrives at a stage at which machines epitomise the complexity of the human body. Thanks to technological progress, human beings can and will understand their nature; the process of technological civilisation ultimately arrives at an enculturation of nature into technology. Kahn thus extended the classical enlightenment program of cognitive self-reflexivity and moral autonomy to the body; technological advances enable a radically new form of 'know thyself', the technological explanation of bodily processes. And in all this, there was no apparent danger to undermine human freedom and liberty.

Beyond the combining of divergent trends to form a humanist utopia of technology, Kahn's visualisations also unfold as an historical epistemology of biological knowledge in the making. How much can an image explain that goes no further than introducing various miniature operators into its world of automata and machines, who do exactly what the technology is said to do? And how far does the rather disingenuous alignment of cognition with something as dull and limited as the control of a switchboard go? Precisely because Kahn took his mission to educate the public on the human body so seriously, he arrived at visualisations that clearly reveal their own limitations. The many little assistants in Kahn's images do their work so diligently and so smoothly, right out in the open of the image, it seems as if the limits of this visualisation strategy are put on display as well. The images visualise knowledge together with the questions that come with it. In the explicitness of their technological style they trace the optical unconscious of the machine philosophy. The ambiguity of their iconography provides a key to how nature is constituted in the contractions of the social and the scientific that is culture.

The Emergence of Historical Epistemology from the Debates on Science

The 1920s saw an escalation of crisis rhetoric in the writing about science. Physicists debated the general epistemological implications of theoretical physics, while psychologists discussed the fragmentation of their discipline into ever more specialisations, biologists argued over the lack of a coherent framework and the crisis talk in medicine often comprised a holistic critique of an allegedly inhumane techno-medicine.[33] Right in the middle of society, a wide-ranging debate about the meaning of recent science and on the implications of scientific research for the most fundamental questions was taking place. It was a debate of a highly ambivalent nature, driven as much by sincere concern as by a 'craving for crisis', as Forman put it.[34] Quite in contrast, however, to the prevailing rhetorical function that Forman attributed to this debate, most contemporary authors agreed there was no crisis of scientific productivity in the Weimar period but rather a crisis that resulted *from* scientific productivity, as Edmund Husserl argued in his unfinished *Crisis of European Sciences*.[35] And this is the last path I will pursue in this exploration of ambiguity in Weimar culture.

Among the many writings on crisis, the leading science journal *Naturwissenschaften* published a paper by the neo-Kantian philosopher Kurt Riezler. In 'The Crisis of Reality', Riezler argued that the very progress of the sciences undermined their progress towards 'absolute reality'.[36] Today his paper and its concerns are largely forgotten. When arguing that the diversification of knowledge resulting from the progress of the sciences along divergent trajectories had led to an epistemological crisis threatening the intelligibility of absolute reality, Rietzler probably expressed the consensus of the Zeitgeist, even though he shared little with *Lebensphilosophie* and was not exactly modern either.[37] His intellectual position is of no particular importance to my argument here, but Riezler's paper inspired a then little-known Polish microbiologist to an epistemological intervention that appeared a few months later in *Naturwissenschaften*. Discarding Riezler's layered model of reality culminating in an absolute truth, Ludwik Fleck responded with a critical analysis of the socio-epistemic assumptions implied in Riezler's account of the crisis.[38] Fleck's

---

[33] In addition to the references provided by Forman for physics and mathematics, *cf.* Hans Driesch, *Grundprobleme der Psychologie: Ihre Krisis in der Gegenwart* (Leipzig: Reinicke, 1926); Karl Bühler, *Die Krise der Psychologie* (Jena: Fischer, 1927); Oswald Bumke, *Eine Krisis der Medizin* (Munich: Huber, 1929); Pascal Jordan, 'Die Quantenmechanik und die Grundprobleme der Biologie und Psychologie', *Naturwissenschaften* 20 (1932): 815–21; Carsten Timmermann, 'Constitutional Medicine, Neoromanticism, and the Politics of Antimechanism in Interwar Germany', *Bulletin of the History of Medicine* 75 (2001): 717–39.

[34] Forman, 'Weimar Culture' (ref. 4), 58–63.

[35] Edmund Husserl, *The Crisis of European Sciences and Transcendental Philosophy* (Evanston, IL: Northwestern University Press, 1970).

[36] Kurt Riezler, 'Die Krise der Wirklichkeit', *Naturwissenschaften* 16 (1928): 705–12.

[37] Forman introduced Riezler as an exception to the prevailing *Lebensphilosophie*; Forman, 'Weimar Culture' (ref. 4), 111.

[38] Ludwik Fleck, 'On the Crisis of "Reality"', in *Cognition and Fact: Materials on Ludwik Fleck*, ed. R.S. Cohen and T. Schnelle (Dordrecht: Reidel, 1986), 47–57; *see also* Cornelius Borck, 'Message in a Bottle from "The Crisis of Reality": On Ludwik Fleck's Interventions for an Open Epistemology,' *Studies in History and Philosophy of Biological and Biomedical Sciences* 35 (2004): 447–64.

response to Rietzler undercut the whole discussion about crisis in arguing for a social explanation of the intellectual functioning of a primarily collective cognising body. This short text, Fleck's first piece of epistemological writing in German, already entailed an outline of his radical epistemology soon to follow. Since every act of knowing necessarily relates to, and connects with, pre-existing knowledge, tradition and education, for Fleck, epistemology had to start with the social and historical context: 'To observe, to cognize is always to test and thus literally to change the object of investigation. This is the day-to-day praxis of science. Here the social and the historical-traditional element is predominant'.[39] In its concrete materialisations as well as in its theoretical content, science is moulded and shaped by its history — and this history is as much the product of ongoing interactions in a sociocultural context, of which the sciences are always part, as it is the result of the sciences' active interrelationship with their objects.

The recent discussions in physics served Fleck as a welcome example for illustrating his idea of the eminent role of the social, and of tradition and education, in epistemology that he had arrived at from his own familiarity with developments in medicine and bacteriology. The sciences differed always and everywhere in their styles of thought, but across their very heterogeneity the epistemologist can trace previous and ongoing interactions: '[Science] is an eternal, synthetic rather than analytic, never-ending labour — eternal because it resembles that of a river that is cutting its own bed'.[40] To Fleck, the 'crisis of reality' provided the test case for the endless chain of epistemological ruptures he was to describe in his coming monograph.[41] A critical study of the history of science, he would argue, reveals a continually changing and shifting enterprise, a permanent and infinite reconstruction of bits and pieces of concepts and theories. The reflexive socio-historical analysis of such changes offered a new epistemological foundation for the sciences, according to him, if and only if the problematic ideal of an approximation of the sciences toward absolute truth was to be abandoned. Fleck replaced it by a normative principle, the democratic competition between scientific concepts and different styles of thought: 'Natural science is the art of shaping a democratic reality and being directed by it — thus being reshaped by it'.[42]

The times to come clearly were 'a hostile intellectual milieu,' leaving Fleck's epistemology little chance to flourish and, indeed, directly threatening his life. The rise of National Socialism and the murderous extensions of the Nazi regime all

---

[39] Fleck, 'Crisis of "Reality"' (ref. 38), 53.
[40] Ibid., 54.
[41] Ludwik Fleck, *Genesis and Development of a Scientific Fact* (Chicago: University of Chicago Press, 1979).
[42] Fleck, 'Crisis of "Reality"' (ref. 38), 54.

over Europe almost extinguished the Fleck family. Ludwik Fleck survived — together with his wife and son alone — because of his expert knowledge on typhus that proved indispensable for the Nazis when they ordered him to continue with bacteriological work within the concentration camp system. The following Cold War period, with its military enforcement of the antagonism of political systems, hardly proved more advantageous and beneficial for his ideas of a free development of the sciences by way of competition under democratic ideals. Fleck's very last paper, an urgent appeal for democracy as the necessary basis for science and humanity that he wrote after his emigration to Israel and while suffering from a fatally weakening heart, was rejected by the journal *Science*,[43] thus testifying directly to the conclusions of Forman's second seminal study on the adversarial effects of the Cold War ideology on science and the distorting consequences of massive military funding for scientific appliances.[44] The form of socio-cultural analysis of science Fleck propagated with his *Genesis and Development of a Scientific Fact* hence required a much prolonged time before it finally won recognition — first by means of a famous reference in passing to 'an almost unknown monograph' by Forman's teacher.[45] The lively and controversial exchanges about the history and philosophy of science had not yet begun when Fleck died; in addition to the crisis of reality, it took the *Structure of Scientific Revolutions* and Forman's insistence on external influences on science to rediscover Fleck's finegrained and dynamic theory of social-material-cognitive interactions in the history of science.

Epistemologies of Life and Science

Given the centrality of *Lebensphilosophie* in the Forman thesis, should one not expect an even more immediate impact of such an intellectual climate on the life sciences than on quantum physics? The broad field of the sciences dealing with living beings may have lacked in their plurality the same disciplinary coherence as physics and did not undergo a similar revolution during this period, but various strands within the life sciences come to mind that seem to testify clearly to a trend toward more speculative ways of thinking: the various forms of 'holism' in biology, medicine, psychology and philosophy, from Jakob von Uexküll's *Umweltlehre* and Ludwig von Bertalanffy's systems biology, via gestalt psychology or Victor von

---

[43] Fleck, 'Crisis in Science', in *Cognition and Fact* (ref. 38), 153–58.
[44] Paul Forman, 'Behind Quantum Electronics: National Security as Basis for Physical Research in the United States, 1940–1960', *HSPS* 18, no. 1 (1985): 149–229.
[45] Thomas S. Kuhn, *The Structure of Scientific Revolutions*, 2nd ed. (Chicago: University of Chicago Press, 1962), vi.

Weizsäcker's philosophy of illness, to Kurt Goldstein's theory of the organism.[46] Or one could think of the special resonances of psychoanalysis and psychotherapy in Weimar Germany from Karl Abraham via Johannes Heinrich Schultz to Wilhelm Reich and Georg Groddeck.[47] Last but not least, the renaissance of physiognomic approaches could be mentioned, from the interest in graphology, shared by scholars as different as Emil Kraepelin and Ludwig Klages, to the implementation of racialised science long before the rise of National Socialism.[48] Given their relevance during the Weimar period, these scientific trends seem to invite an extension of the Forman thesis to the life and human sciences: if the developments in physics resulted from a hostile intellectual environment, a wholesale embrace of the Zeitgeist manifested in the life sciences in the form of a direct translation of *Lebensphilosophie* into science. Indeed, physiognomy, Gestalt psychology and holistic biology shared a disregard for mathematical formalisation even in relation to the standards of their disciplinary fields and a tendency toward *Anschaulichkeit*.[49] Judged by the orthodoxy of the philosophy of science, these movements lacked scientific rigour and their status as science was questioned, most famously in Karl Popper's attack on psychoanalysis.[50]

However, these and other human sciences do not simply lack theory or abstraction but rather specialise in a different type of theorising and reasoning. And these types of reasoning describe vital trends in their respective domains of knowledge, far too important to be ignored for an alleged lack of scientific standards or disciplinary rigour as measured by idealisations which are themselves questionable. Without physiognomic approaches, gestalt theory or holistic

---

[46] Jakob von Uexküll, *Umwelt und Innenwelt der Tiere* (Berlin: Springer, 1921); Ludwig von Bertalanffy, *Das Gefüge des Lebens* (Leipzig: Teubner, 1937); Ludwig von Bertalanffy, 'An Outline of General System Theory', *British Journal for the Philosophy of Science* 1 (1950): 134–65; Viktor von Weizsäcker, *Ärztliche Fragen: Vorlesungen über allgemeine Therapie* (Leipzig: Thieme, 1934); Kurt Goldstein, *The Organism: A Holistic Approach to Biology Derived from Pathological Data in Man* (New York: American Book Company, 1939).

[47] Ulfried Geuter, *The Professionalization of Psychology in Nazi Germany* (Cambridge: Cambridge University Press, 1992); Geoffrey Cocks, *Treating Mind and Body: Essays in the History of Science, Professions, and Society under Extreme Conditions* (New Brunswick, NJ: Transaction Books, 1998).

[48] Armin Schäfer, 'Lebendes Dispositiv: Hand beim Schreiben', in *Psychographien*, ed. Cornelius Borck and Armin Schäfer (Zürich: Diaphanes, 2005), 241–65; Hans-Walter Schmuhl, ed., *Rassenforschung an Kaiser-Wilhelm-Instituten vor und nach 1933* (Göttingen: Wallstein, 2003).

[49] Paul Forman, '*Kausalität, Anschaulichkeit, and Individualität*, or How Cultural Values Prescribed the Character and the Lessons Ascribed to Quantum Mechanics', in *Society and Knowledge: Contemporary Perspectives in the Sociology of Knowledge and Science*, ed. Nico Stehr and Volker Meja (New Brunswick, NJ: Transaction Books, 1984), 333–47.

[50] Karl R. Popper, *Conjectures and Refutations: The Growth of Scientific Knowledge* (London: Routledge, 1963).

biology, any history of twentieth-century life and human sciences would draw not just an incomplete picture, but a grossly misleading one too. Not the least because of Forman's stimulating paper, there is little need to continue the debate over the proper role of theory in science, but much opportunity to build on his work on science in culture. Precisely the omnipresence of *Lebensphilosophie* in Weimar Germany questions its usefulness as an explanatory category for the history of the life sciences. In fact, the Zeitgeist, the *Lebensphilosophie*, can be linked to so many different intellectual projects and scientific programs that it loses much of its explanatory power. Rather than searching for promising applications of the Forman thesis in the life sciences, the very ease of its applicability should call for more specific ways to contextualise scientific knowledge.

By way of conclusion, three issues shall be commented on briefly. The first is the insufficiency of *Lebensphilosophie* to account for the ambivalences that characterised the intellectual culture in Weimar Germany where, for example, the nostalgic longing for the alleged lost richness of the Abendland's culture went hand in hand with a passion for America, scientific management, Taylorisation and *Neue Sachlichkeit*. Bissky, Giese and Schulte used the modernity of an electrical gadget to promise extravagant scientific advances, from esoteric psychology to efficiency management. Berger lamented the loss of natural-cultural harmony — and contracted Siemens to build an ultra high frequency brain-wave recorder. The English title of Spengler's *Untergang des Abendlandes* may have been *The Decline of the West*, but in the burgeoning culture of the 'golden twenties', the declining Abendland appears to have only accelerated the public esteem for the West — for example, the Tiller Girls who attracted large audiences in the mid-1920s in Berlin, or in the form of Kahn's iconising integration of American architecture into his visual explanations of organic functions. This is certainly no argument against the importance of Spengler's book for the intellectual life in Weimar Germany, but *Lebensphilosophie* suggests an overly homogeneous image of the Weimar period. While some lamented the decline of traditional culture in the face of what they deemed 'mere civilization',[51] others fought for the political and social liberation from the restricting norms of conservative traditions; and frequently such divisions went through one and the same project, such as in the case of Kahn's technological utopia of organic knowledge. In the midst of an increasing radicalisation of politics in Weimar Germany, opposing groups and tendencies bound together nonetheless; labour unions, political institutions and industrial leaders, for example, orchestrated jointly the streamlining of mass

---

[51] Georg Simmel, 'Der Begriff und die Tragödie der Kultur', *Gesamtausgabe Bd. 12: Aufsätze und Abhandlungen 1909–1918* (Frankfurt am Main: Suhrkamp, 2001), 194–223.

fabrication that also offered a reintegration of crippled soldiers into the workforce — hoping for a future of rationalisation that would be as bright as it would be humane.[52]

The second issue is the intrinsic but problematic vagueness of *Lebensphilosophie*. Since 1971, the clear-cut opposition between progressive rationalism and various forms of romantic speculation, intuition and irrationalism that Forman suggested (and indeed required for his construction of anti-intellectual milieu) has been intensively problematised in many historical and epistemological directions. A much-increased interest in the intellectual sophistication of the Romantic period, for example, corresponds with compelling analyses of the far more convoluted paths toward modernism around the turn of the century.[53] Many divergent projects, programs and ideas displayed some affiliation with Zeitgeist arguments and often they shared an investment in an opposition to what they regarded as narrow concepts of rationality. With the term 'irrationality', however, this entire debate is effectively shunned by declaring any reasoning that deviates from the path ultimately confirmed by empirical science to be spurious. Forman's analysis thus can only insufficiently capture what was at stake in the debates about *Lebensphilosophie* that roiled less about demarcating the limits of rationality *per se* than its limits vis-à-vis notions of allegedly higher forms of reasonability. It is amazing at what great length Forman deals with many examples of such forms of reasoning, but only for the purpose of demonstrating their defective nature because everything else would endanger his concept of external causation. This is Forman's Manichaeism. He does not stop at describing what he assumes to be failures of ratiocination but construes out of these a guilty form of rationality defect. In order to safeguard a particular view of science and rationality, he has to foreshorten the range of historical analysis. Of course, many of the arguments and forms of evidence with which actors came forward in defence of esoteric or strange concepts now and retrospectively appear as problematic to a degree of disbelief, as for example the massive wave of interest in extrasensory perception and telepathy that arose after the invention of telecommunications media at the turn of

---

[52] Peter Hinrichs and Lothar Peter, *Industrieller Friede? Arbeitswissenschaft, Rationalisierung and Arbeiterbewegung in der Weimarer Republik* (Cologne: Pahl-Rugenstein, 1976); Jennifer Karns Alexander, *The Mantra of Efficiency: From Waterwheel to Social Control* (Baltimore: Johns Hopkins University Press, 2008).

[53] Andrew Cunningham and Nicholas Jardine, eds., *Romanticism and the Sciences* (Cambridge: Cambridge University Press, 1990); Bernd Apke and Veit Loers, eds., *Okkultismus und Avantgarde: von Munch bis Mondrian, 1900–1915* (Ostfildern: Edition Tertium, 1995); Robert J. Richards, *The Romantic Conception of Life: Science and Philosophy in the Age of Goethe* (Chicago: University of Chicago Press, 2002).

the century.⁵⁴ But would it not be much more rewarding and productive if the historical analysis could reach out to what were in their time acceptable forms of argumentation? Already the extended debates in Weimar Germany on culture and civilisation indicate to what an astonishing degree rational argumentation was involved on the side of an alleged irrational Zeitgeist. Even the Frankfurt school's famous distinction between instrumental rationality and reason, for example, must be regarded as an offspring of this debate — with further internal ramifications between Horkheimer's and Adorno's Hegelian critique of instrumental rationality and Benjamin's speculative historical dialectics. Bertalanffy and Goldstein argued about the shortcomings of reductionist biological explanations, but does that mean their conceptualisation of scientific reasoning was more closely related with Spengler or Klages — and, if so, what would it mean?

Finally, allegedly progressive trends and *Lebensphilosophie* typically intertwined insolubly in Weimar culture, as has been stressed over and again in this paper. No neat distinction can be drawn between the two, nor a clear coupling of progressive modernism versus romantic irrationality. The Bauhaus, for example — that icon of progressive modernism already mentioned by Forman as a site of ambivalence — is just the most famous example.⁵⁵ Often it has been pointed out how surprisingly open the Weimar Bauhaus was toward esoteric philosophies such as Kandinsky's interest in theosophy and Klee's in occultism, or Johannes Itten's leanings toward mysticism. And yet, this ambivalence was more than the conflict between opposing camps or an internal resistance against Gropius steering the Bauhaus toward rationalism. With this demarcation, Paul Forman misses probably the most important engine of the Weimar Bauhaus dynamics. The same Itten of the esoteric Weltanschauung who rallied a vegan sect in Weimar also conceived the famous Bauhaus Vorkurs. At the Bauhaus, abstraction emerged as modernity's dominant style directly from the very radicalness of Itten's esoteric views.⁵⁶ And because of these roots, the Bauhaus's modernist slogan 'form follows function' generated new forms of *Anschaulichkeit* that integrated abstraction and rationality to aesthetic harmony. The interwar years were the formative period for high modernism as they witnessed a general polarisation of the political debate and a broad popularisation of esoteric, more speculative forms of *Weltanschauung*. Science

---

[54] Clément Chéroux, *The Perfect Medium: Photography and the Occult* (New Haven, CT: Yale University Press, 2005); Corey Keller, ed., *Brought to Light: Photography and the Invisible, 1840–1900* (San Francisco: San Francisco Museum of Modern Art, 2008).

[55] Forman, 'Weimar Culture' (ref. 4), 22: 'When one looks at the manifestos of this movement, one can be struck by its ambivalence'.

[56] Timothy O. Benson, ed., *Expressionist Utopias: Paradise, Metropolis, Architectural Fantasy* (Berkeley: University of California Press, 2001).

took new and unexpected turns not because of adaptation, conversion or capitulation but because of the unusually large number of diverging opportunities for interaction, exchange, interrelation and articulation in and across this densely woven web of culture.

In short, the very heterogeneity and ambiguity of the many faces of Weimar, the many differences in their historical coexistence, appear to be the first and foremost characteristic of this intriguing period. A heightened ambiguity characterised the Weimar cultural context — and the intellectual witnesses were acutely aware of it when they established new forms of critical analysis and media theory that still inspire present-day cultural analysis, including science studies. The perplexing complexity of the period is probably also the main reason for postmodernity's intensified interest in it.[57] Alas, Paul has been there already, demonstrating his intellectual sovereignty;[58] he sees his criticisms of a highly problematic intellectual *habitus* confirmed.[59] This, however, is not a question of historical evidence but an argument about philosophical attitudes. I introduced Fleck as the final example of the epistemological richness of the Weimar period because this also offered a reflexive turn in the debate about acceptable and legitimate forms of reasoning. After the Holocaust, Fleck questioned whether science needed a more rigorous defence than his earlier call for an open and dynamic democracy. The result was Fleck's most skilfully written and rhetorically crafted text that invited the reader to participate in a dialogue. One could describe this writing as an intellectual form of *Anschaulichkeit*. Certainly there are many different voices that should be listened to in such a dialogue, including radical pleas for rigid standards of scientific work, especially when scholars out of passion submit, first and foremost, themselves to these standards. Such an open intellectual arena builds the laboratory of historical epistemology and continues to change our understanding of science — just as the experimentalisation of science and life in Weimar Germany created a laboratory of ambiguity.

---

[57] See, for example, the over 40 volumes of the *Weimar and Now: German Cultural Criticism* series (Berkeley: University of California Press) begun in 1991, not coincidentally after the end of the Cold War.

[58] Paul Forman, 'Independence, Not Transcendence, for the Historian of Science', *Isis* 82 (1991): 71–86.

[59] Paul Forman, 'The Primacy of Science in Modernity, of Technology in Postmodernity, and of Ideology in the History of Technology', *History and Technology* 23 (2007): 1–152.

Science and Politics: Pathology in Weimar Germany (1918–33)

*Cay-Rüdiger Prüll**

As the title suggests, my remarks here will relate directly to Paul Forman's thesis developed in his article published in 1971.[1] This relation is not by chance: the idea that the cultural milieu influences or shapes the performance of scientific work and its results turned out to be one of the most productive ideas in modern historiography of the history of medicine over the last four decades. This paper will analyse two main issues: to what extent Forman's ideas on the relationship of science and politics, which he investigated for physics in Weimar Germany, can still be adapted to current views on the subject, and how Forman's contribution can be combined with new perspectives developed since 1971.

My intent is to pursue these issues first by way of the history of the medical discipline of pathology in Weimar Germany, specifically the branch of constitutional (holistic) pathology and its relation to right-wing political discourse at the time. Although constitutionalism had been a broad movement since at least the last decade of the nineteenth century, spanning several political ideologies,[2] the juxtaposition of constitutionalism and a right-wing mentality serves well to reveal those conditions that forged a connection between pathology and politics in the period

---

* Institut für Geschichte und Ethik der Medizin, Ruprecht-Karls-Universität Heidelberg, Im Neuenheimer Feld 327, 69120 Heidelberg, Germany; pruell@uni-heidelberg.de.
 The following abbreviations are used: DFG, Deutsche Forschungsgemeinschaft (German Research Association); EAHMH, European Association for the History of Medicine and Health; *HSPS*, *Historical Studies in the Physical Sciences*.
[1] Paul Forman, 'Weimar Culture, Causality, and Quantum Theory, 1918–1927: Adaptation by German Physicists and Mathematicians to a Hostile Intellectual Environment', *HSPS* 3 (1971): 1–115.
[2] See, for example, Carsten Timmermann, 'Constitutional Medicine, Neoromanticism, and the Politics of Antimechanism in Interwar Germany', *Bulletin of the History of Medicine* 75, no. 4 (2001): 717–39.

from 1918 to 1933. For the purposes of this analysis, I do not restrict 'politics' to indicate solely a membership in a particular political party or the holding of a specific political title. Moreover, I see political activities as the implementation of specific theories with regard to how life is ordered and how problems are solved in a society.[3] Right-wing constitutional pathology in Weimar Germany will serve as the second avenue through which to discuss and re-evaluate Forman's ideas.

This paper is divided into four parts. First, I explain what pathology is and investigate pathology and its political environment in the period between approximately 1850 and 1918. Then I turn to the Weimar Republic, but instead of proceeding with an overview, I illustrate my ideas through two individual case studies, based on the activities of the pathologists Ludwig Aschoff (1866–1942) and Otto Lubarsch (1860–1933). Both were prominent scientists in their discipline in the first half of the twentieth century. Finally, I analyse my findings in order to discuss Forman and his 1971 paper.

The Field of Pathology and the Political Culture in Germany, 1850–1918

Pathology developed as a discipline of scientific medicine around 1850. In its German version, pathology was carried out primarily in the hospital morgue. There, the pathologist performed clinical dissections, i.e., autopsies, on deceased patients with an aim to determine the process of the disease as well as the cause of death. The field was institutionalised as an independent discipline with chairs and institutes within the hierarchically organised university system of Wilhelmine Germany. This strengthened the authority of the pathologists and helped lead to the establishment of their field as a core discipline of scientific medicine in the last decades of the nineteenth century.[4] Both strategy and theory of pathology were significantly influenced by the Berlin pathologist Rudolf Virchow (1821–1902). According to Virchow, a wide range of human diseases could be detected by routine, controlled autopsy procedures in combination with laboratory work and clinical knowledge. The theory behind this approach to medical research was

---

[3] Volker Sellin, 'Politik', in *Geschichtliche Grundbegriffe*, vol. 4, ed. Otto Brunner, Werner Conze and Reinhart Koselleck (Stuttgart: Klett-Cotta, 1978), 789–874, esp. 873–74.

[4] See an overview on the history of pathology in Russell C. Maulitz, 'The Pathological Tradition', in *Companion Encyclopedia of the History of Medicine*, vol. 1, ed. William F. Bynum and Roy Porter (London: Routledge, 1993), 169–91; Roger Cooter, 'The Dead Body', in *Medicine in the Twentieth Century*, ed. Roger Cooter and John Pickstone (Amsterdam: Harwood, 2000), 469–85; Cay-Rüdiger Prüll, *Medizin am Toten oder am Lebenden?: Pathologie in Berlin und in London, 1900–1945* (Basel, Switzerland: Schwabe, 2003), esp. 15–49.

based on 'cellular pathology', which claimed the cell is the smallest unit of life, all cells stem from cells and the cell is the seat of all diseases. In total, Virchow's concept was unequivocally morphological, oriented toward investigating the static condition of human tissues and organs and preserving these specimens in pathology museums.[5]

Although Virchow himself was liberal and a critic of the German monarchy, his pathology was a good fit with the politics of Imperial Germany in the second half of the nineteenth century. It boosted the reputation of German science and helped the government bear the setback posed by the late reunification and the late (and marginal) acquisitions of colonial possessions. Many of Virchow's followers and pupils had a pan-Germanic bias and saw morphological pathology, or pathological anatomy, as a specifically German science, which contributed to the worldwide distribution of German cultural achievements.[6]

In addition to making an impact outside of Germany, the Virchowian system was also adapted to solve contemporary socio-political problems within Germany itself. But at the turn of the century, German society, unsettled by the rapid speed of industrialisation and social changes, began to gravitate to a great extent toward alternative medicine. In the context of scientific medicine, Virchowian pathology was attacked as materialistic, focusing on tissues and organs and neglecting the whole organism and the patient as such.[7] Many pathologists answered these attacks by developing a new concept known as constitutional pathology. Based on cellular pathology, the main focus of constitutional pathology was an analysis of the interrelationship between the person and his/her environment from a morphological perspective. Pathologists were interested in human defence mechanisms against artificial and natural enemies, the hereditary precondition of the human body and the options of the human organism with regard to continued survival.[8] Constitutional pathology was strengthened after 1900 by the hygiene movement,

---

[5] For Virchow, see Prüll, *Medizin* (ref. 4), 161–65; Erwin H. Ackerknecht, *Rudolf Virchow: Doctor, Statesman, Anthropologist* (Madison: University of Wisconsin Press, 1953); Constantin Goschler, Rudolf Virchow (Cologne, Germany: Böhlau, 2002) and Cay-Rüdiger Prüll, 'Rudolf Virchow (1821–1902)', in *Theater der Natur und Kunst/Theatrum Naturae et Artis: Wunderkammern des Wissens. Essays*, ed. Horst Bredekamp, Jochen Brüning and Cornelia Weber (Berlin: Henschel/Humboldt University of Berlin, 2000), 208–12.

[6] Cay-Rüdiger Prüll, 'Otto Lubarsch (1860–1933) und die Pathologie an der Berliner Charité von 1917 bis 1928', *Sudhoffs Archiv* 81, no. 2 (1997): 193–210; Prüll, *Medizin* (ref. 4), 337–67.

[7] Eva-Maria Klasen, 'Die Diskussion über eine "Krise" der Medizin in Deutschland zwischen 1925 and 1935' (PhD dissertation, Johannes Gutenberg University Mainz, 1984).

[8] Dietrich von Engelhardt, 'Kausalität und Konditionalität in der modernen Medizin', in *Pathogenese: Grundzüge und Perspektiven einer Theoretischen Pathologie*, ed. Heinrich Schipperges (Berlin: Springer, 1985) 32–85.

which promoted a person's responsibility to lead a healthy life and to take preventive measures, above all against infectious diseases. Also, constitutionalist thinking was significantly supported by the crisis of World War I. Competition between the nations was fuelled during this 'war of cultures', and in Germany a keen need arose to secure the health and wellbeing of its people. Between 1914 and 1918, Ludwig Aschoff, a professor at the University of Freiburg, promoted the discipline of 'war pathology', aimed specifically at investigating the constitution of the German body through the dissection of German soldiers killed in action. Pathology as a German science now served the needs of the German nation in a period of crisis and this form of constitutionalism allowed the integration of social Darwinist, racial hygienist and overtly racist ideas.[9] In 1916, Otto Lubarsch, a professor at the University of Kiel and serving in the army as an advising pathologist, attacked France and Britain as 'racial disgraces', having sent against Germany 'the yellow, brown and black mob and, if possible, the white scum of the earth'.[10] The ideas of those pathologists reinforced tendencies already prevalent in other disciplines, such as psychiatry.[11]

All in all, German pathology from 1850 to 1918 was very much entangled in the German political system and the needs and fears of its people. In particular, constitutional pathology was both co-founder and fruit of a specific Zeitgeist. Since this branch of pathology prevailed during the years of the Weimar Republic, I discuss Ludwig Aschoff and Otto Lubarsch in order to investigate the relationship of constitutional pathology and right-wing politics during this period.

Constitutional Pathology and the Apolitical Professor — Ludwig Aschoff

Ludwig Aschoff, sometimes called the Virchow of the first half of the twentieth century, was born in Berlin in 1866, the son of a physician. He too studied

---

[9] Cay-Rüdiger Prüll, 'Die Sektion als letzter Dienst am Vaterland: Die Deutsche "Kriegspathologie" im Ersten Weltkrieg', in *Die Medizin und der Erste Weltkrieg*, vol. 3, ed. Wolfgang U. Eckart and Christoph Gradmann (Pfaffenweiler, Germany: Centaurus, 1996), 155–82; Cay-Rüdiger Prüll, 'Holism and German Pathology (1914–1933)', in *Greater than the Parts: Holism in Biomedicine, 1920–1950*, ed. Christopher Lawrence and George Weisz (Oxford: Oxford University Press, 1998), 46–67; Cay-Rüdiger Prüll, 'Pathology at War 1914–1918 — Germany and Britain in Comparison', in *Medicine and Modern Warfare*, ed. Roger Cooter, Mark Harrison and Steve Sturdy (Amsterdam: Rodopi, 1999), 131–61.
[10] Otto Lubarsch, *Wissenschaft und Volkstum* (Kiel: Lipsius & Tischer, 1916), 9–10.
[11] Peter Riedesser and Axel Verderber, *Maschinengewehre hinter der Front: zur Geschichte der Deutschen Militärpsychiatrie* (Frankfurt: Fischer, 1996); Paul Lerner, *Hysterical Men: War, Psychiatry, and the Politics of Trauma in Germany, 1890–1930* (Ithaca, NY: Cornell University Press, 2003).

medicine, and upon graduating became assistant to Friedrich-Daniel von Recklinghausen (1833–1910) and Johannes Orth (1847–1923), who had been assistants to Virchow. In 1903, he became a professor of pathology in Marburg, and then in 1906 held the same position in Freiburg, where he retired in 1936.[12]

Aschoff, a child of Wilhelmine Germany, developed an ambivalent attitude toward the social and political conditions in the German Empire. On the one hand he grew up during the glorification of Bismarck and the unification of Germany, but on the other he was also surrounded by prejudice against certain social groups that appeared to endanger the unity of those who considered themselves truly German. Presumably during his youth in Berlin, Aschoff developed an aversion toward the Jews from Eastern Europe, who were seen as invaders and enemies of German culture. Furthermore, around the turn of the century, he belonged to a group of intellectuals who suspected society was fragmenting. They believed there was an ongoing decline in standards among the representatives of intellectual education and that true (scientific) values and virtues were giving way to materialistic ideas in an industrialised world dominated by the struggle between labourers and industrialists. Around 1900, Aschoff developed his ideal of a unified nation, a 'Volksgemeinschaft', where the constitutional health of the German people would be supported and maintained by intellectual bourgeois leaders (whom Aschoff called 'Führer').[13]

Aschoff's ideal was the source of constitutional 'war pathology' between 1914 and 1918, and after the German defeat in 1918 he continued to promote the use of constitutional pathology in the Weimar Republic. His nationwide efforts were supported by discussions, which began around 1925, about a 'crisis' in medicine. Criticism of modernity and modern life after the war defeat also extended to issues about technical and medical progress, the Janus-faced character revealed by the devastation caused by industrial warfare and its failure to produce a German victory. This kind of scepticism promoted a certain acceptance of 'irrational' beliefs in approaches based on intuition and *Lebensgefühl*. Holistic attitudes, such as constitutionalism, seemed to accommodate these less rational approaches, as well as serving the needs of postwar medicine in light of its concern with improving the biological quality of the German race.[13]

For this reason Aschoff concentrated on publishing the results of his constitutional investigations carried out during World War I. Together with several

---

[12] For more on Aschoff's life, see Cay-Rüdiger Prüll, 'Aschoff, Karl Albert Ludwig', in *Dictionary of Medical Biography*, vol. 1, ed. William F. Bynum and Helen Bynum (Westport, CT: Greenwood Press, 2007), 132–35; Cay-Rüdiger Prüll, 'Pathologie und Politik — Ludwig Aschoff (1866–1942) und Deutschlands Weg ins Dritte Reich', *History and Philosophy of Life Sciences* 19 (1997): 331–68.
[13] Ibid., 335–338; Klasen, 'Diskussion' (ref. 7).

colleagues and above all, his Berlin colleague Walther Koch (1880–1962), he had already initiated a series of forums in 1920, entitled 'Publications from the Fields of War Pathology and Constitutional Pathology',[14] in which pathologists discussed problems of war-related constitutional pathology. Papers appeared in this series that directly or indirectly related to the idea of racial hygiene, e.g., the constitutional deficiencies of soldiers infected with tetanus or the improved racial quality of schoolchildren between 1880 and 1921. Remarkably, Aschoff and his colleague Koch managed to preserve the discipline of war pathology in the Weimar period by shifting to occupational pathology in 1930. The 'war battle' was equated to the 'occupational battle' because soldiers as well as labourers were seen as the pillars of a constitutionally healthy society. In 1931, Aschoff and Koch renamed the series 'Publications on Occupational and Constitutional Pathology'. Koch summarised these events when he wrote: 'Occupational pathology is actually the war pathology of peace'.[15] Similar to the shift in publication emphasis was a shift in the direction of research with regard to the huge collection of war pathology specimens in Berlin.[16]

Eventually, Aschoff's approach spread beyond his colleagues to his students. One example is the introduction of sports medicine to the medical faculty of Freiburg University in 1924. Together with the pan-Germanic professor of internal medicine, Oscar de la Camp (1871–1925), Aschoff researched to what extent the human body could adapt to extraordinary physical stress. Within this research program, the heart of the sportsman and the shift from regular to pathological conditions were essential topics. In 1924, Aschoff and de la Camp founded an Institute of Sports Medicine, which was incorporated into the Clinic of Internal Medicine and handed over to an associate professor, Hermann Rautmann (1885–1956). Already in 1921, Rautmann had been entrusted with a lectureship on 'Leibesübungen und Körperentwicklung' (Physical Exercises and Bodily Development). In the summer term of 1925, the lecture was renamed 'Klinische Constitutions- und Vererbungslehre' (Clinical Constitutional and Hereditary Science). And in the winter term of 1926/27 the title of the lecture was 'Klinische

---

[14] Ludwig Aschoff, Maximilian Borst, Martin Benno Schmidt, *et al.* eds., *Veröffentlichungen aus der Kriegs- und Konstitutionspathologie*, vol. 1 (Jena: Gustav Fischer, 1920).

[15] Prüll, 'Holism' (ref. 9), 50–55, on 53–54; *see also* Ludwig Aschoff, Maximilian Borst, Martin Benno Schmidt and Ludwig Pick eds. under the direction of Walter Koch, *Veröffentlichungen aus der Gewerbe- und Konstitutionspathologie*, vol. 7 (Jena: Gustav Fischer, 1931).

[16] Another example of Aschoff's work on constitutional pathology is his interest in the so-called *Völkerpathologie* to correlate specific living conditions of people with specific diseases. See Susan Gross Solomon and Jochen Richter, ed., *Ludwig Aschoff: Vergleichende Völkerpathologie oder Rassenpathologie — Tagebuch einer Reise durch Russland und Transkaukasien* (Pfaffenweiler, Germany: Centaurus, 1998).

Constitutions- und Vererbungslehre mit besonderer Berücksichtigung der Leibesübungen' (Clinical Constitutional and Hereditary Science with Special Consideration of Physical Exercises). Therewith constitutional pathology as well as racial hygiene was taught to students with the strong support of the Freiburg University medical faculty and Rautmann received several travel and research grants to further his research in this area.

But for Aschoff and de la Camp, constitutional pathology via sports and physical education also had a sociopolitical context. At a time when sports were becoming an integral part of leisure time and similarly, sports heroes the subjects of admiration, the two researchers were able to persuade the university senate to make physical exercise compulsory for all students in their first and second semesters (1924/25). In Aschoff's view, physical exercise strengthened the will of the individual and the unity of the group, and would eventually lead to the development of a people's community (*Volksgemeinschaft*). Therefore, he also promoted physical exercise among the general public. The pan-Germanic aims behind the program became clear in April 1933, when the 77-year-old Aschoff himself participated in 'military sports exercises', which had been introduced in the very same year.[17]

Aschoff also propagated constitutional morphological pathology at the international level. To compensate for his country's defeat he advanced his field as a specifically German contribution to the major successes of Western European culture. During his travel activities, for example to Japan in 1924, Aschoff pronounced his views and in this manner fought against the exclusion of German scientists from international meetings after 1918.[18]

In summary, the pathology promoted by Aschoff in Weimar times was deeply influenced by the nationalistic worldview of Imperial Germany and by a corresponding branch of pathology, namely constitutional pathology. Remarkably, Aschoff was not fully aware of the impact of his political activities, seeing himself as an advocate of true independent science and a servant of state interests independent from specific political agendas. In the mid-1920s he withdrew any credit he had previously given to the Weimar government, believing they had failed to exert adequate national influence to create a *Volksgemeinschaft*.

---

[17] For more on the significance of sports in Weimar culture, see Christiane Eisenberg, 'Der Sportler', in *Der Mensch des 20. Jahrhunderts*, ed. Ute Frevert and Heinz-Gerhard Haupt (Frankfurt/New York: Campus, 1999), 87–112, esp. 99–104. For more on the introduction of sports medicine to the University of Freiburg and racial hygiene, see Cay-Rüdiger Prüll, 'Zur Ambivalenz medizinischen Fortschritts: Neue Techniken und der Einzug völkischen Gedankengutes in die Medizin', in *Von der badischen Landesuniversität zur Hochschule des 21. Jahrhunderts*, vol. 3, ed. Bernd Martin (Freiburg: Alber, 2007) 243–58, esp. 254–55.

[18] See Prüll, 'Pathologie und Politik' (ref. 12), 347–51.

Constitutional Pathology and the Lay Politician — Otto Lubarsch

Like Aschoff, Otto Lubarsch was deeply rooted in Wilhelmine Germany. He was born in 1866, the son of a businessman in Berlin. Though the family was Jewish, Lubarsch's father did not choose a Jewish upbringing for his children, favouring instead the assimilation of Christian culture and the social values of Imperial Germany. Otto Lubarsch consequently converted to Protestantism and in 1892 married a woman of the gentry (Margerete Freiin von Hanstein). He became an unreserved supporter of the Wilhelmine Empire and in 1890 was a founding member of the Pan-German League (*Alldeutscher Verband*). Educated in the Virchowian tradition, Lubarsch worked as a pathologist at different hospitals and institutions before being appointed to a professorship at Düsseldorf Medical Academy in 1907. In 1914 he was a professor in Kiel, and finally in 1917, he occupied the famous Virchowian Chair of Pathology of the Charité Hospital in Berlin. He died in 1933, shortly after the seizure of power by the National Socialists.[19]

As a broad-minded and flexible thinker in the area of scientific pathology, Lubarsch reconsidered Virchowian pathology and became an early adherent of holistic attitudes. He consequently supported constitutional pathology with an aim of integrating all aspects of life when reconstructing the cause and development of disease, taking this approach as early as the turn of the century, but more intensively — and not by chance — in World War I, especially around 1915 and continuing through the 1920s.[20] Lubarsch also felt the need for reform in an age of crisis of scientific medicine. In 1917, he wrote that many unsolved questions in the area of constitutional pathology would need to be clarified.[21] He discussed related questions in different scientific papers until the end of his life, and finally in his autobiography. In this way, his ideas and attitudes were accessible to his colleagues.

After 1918, Lubarsch's interest in holistic thinking in medicine was coupled with and supported by an ambivalent, if not hostile, attitude toward industrialisation and modernisation.[22] Health and illness were closely related to man and his social and political environment. In Lubarsch's view, the modern lifestyle had fostered an increase in hazardous and health-damaging habits such as smoking and

---

[19] For more on Lubarsch's life, see Robert Rössle, *Otto Lubarsch, 4 January 1860–1 April 1933*, in *Verhandlungen der Deutschen pathologischen Gesellschaft* 27 (1934): 341–49.

[20] Ibid., 345–47.

[21] Otto Lubarsch, 'Über Aufgaben und Ziele der pathologischen Forschung und Lehre', *Deutsche Medizinische Wochenschrift* 43 (1917): 1377–80, esp. 1379.

[22] See the remarks of Otto Lubarsch, which were made in a meeting at the Berlin-Brandenburg physician chamber and quoted in his 'Um die Freiheit des Aerztestandes', *Aerztliches Vereinsblatt* 53 (1924): 373–76.

drinking. He associated the young democracy with these tendencies and did not keep his opinions from his students. In 1927, Lubarsch autopsied a patient and told his student audience: 'I do not hesitate to tell you the name of the deceased. It is Iwan Kutisker, a Jew from the East, who was sentenced to five years in prison. At the age of 20 he infected himself with syphilis and he smoked, according to the habits of Eastern Jews, 30 to 40 cigarettes every day'. Having disclosed his patient's identity made Lubarsch's remarks a scandal and the minister of education withdrew his licence to lecture, starting a discussion that would last until Lubarsch's retirement one year later, in 1928.[23]

Despite restrictions placed on his ability to lecture, Lubarsch could still spread his ideas to the general public. As the poisons of civilisation were thought to damage the hereditary material of mankind, he fought aggressively against the consumption of alcohol. He was closely connected to the German Society against the Abuse of Spirits (Verein gegen den Mißbrauch geistiger Getränke e.V), which Lubarsch provided with specimens of human organs that revealed the devastating morphological results of alcohol abuse. The Society presented the specimens in its 'Anti-alcoholic Exhibition'.[24]

Lubarsch's constitutional pathology also reached the international stage through its influence on his disciple Max Kuczynski (1890–1967) during the time of the Weimar Republic. The latter wished to develop an ethnic pathology, the purpose of which would be to examine the cause and development of diseases in different regions of the world in connection with the respective geographic and cultural environments. Kuczynski shared with Lubarsch the holistic view of medicine in general and of pathology in particular, along with a certain nationalism. But in contrast to Lubarsch, Kuczynski refused to pursue speculations about racial hygiene and after emigrating from Nazi Germany in 1936, he continued his work in Peru.[25]

Similar to Aschoff, Lubarsch saw morphological pathology as a German science. At an early stage of his career, from 1899 to 1904, when he was a pathologist in Posen, Eastern Prussia, Lubarsch used his lectures to educate Polish and German general practitioners about pan-Germanic ideas, as though he hoped to build a bulwark along the Empire's eastern frontier. During the Great War, his Institute in Berlin performed war-related research, and after 1918 promoted morphological

---

[23] Prüll, *Medizin* (ref. 4), 366–67.
[24] Ibid., 182–83.
[25] Cay-Rüdiger Prüll, 'Wie Laboratorien an den Amazonas kamen oder die "Ethnopathologie" des Max Kuczynski', *Praxis* 96 (2007): 1787–90. For more on Kuczynski's work in Peru, see Michael Knipper, 'Antropología y "crisis de la medicina": el patólogo M. Kuczynski-Godard (1890–1967) y las poblaciones nativas en Asia Central y Perú', *Dynamis. Acta Hispanica ad Medicinae Scientiarumque Historiam Illustrandam* 29 (2009): 97–121.

pathology as a truly German development in an effort to improve the reputation of German science internationally. This policy was directed against the alliance of enemy states (*Feindverbundstaaten*) in the West, as Lubarsch tried to provide support to the Soviet Union, which he regarded as the other loser of World War I. Against this background, Kuczynski performed research on typhoid fever, prevalent above all in Eastern Europe.[26]

It is not surprising that Lubarsch, from the beginning of the Weimar Republic, supported the anti-democratic and anti-Semitic German Nationalistic People's Party (*Deutschnationale Volkspartei*) and was a founding member of the party's association of university teachers. His official political activity corresponded with the general alignment of his pathology and also his constitutional pathology. In contrast to Aschoff's belief that science could maintain its distance from politics, Lubarsch saw himself as a lay politician, making no effort to draw a clear-cut distinction between scientific and political matters.[27]

Conclusion

In this final section I discuss the examples described above in light of Forman's main ideas on science, especially physics, and their relation to the intellectual context in Weimar Germany. Basically, I agree with Forman that scientists are influenced by their environment and react or adapt to the various challenges posed by society. And I believe my example of right-wing constitutional pathology shows Forman was also correct when he pointed out the life sciences were more readily adapted to the new holistic ideas after 1918 than physics and chemistry. Already the boundaries between the biological and sociopolitical realms had started to erode before 1900 with the acceptance of Darwinism, social Darwinism and racial hygiene.

There were some specific differences between physics and constitutional pathology in Weimar Germany, due to the individual characters of the disciplines, but limitations of space prevent a fuller discussion here.[28] Nonetheless, the examples presented above point to three key areas where Forman's thesis has to be modified and extended due to new, fruitful research in the field of history of science and medicine within the last 30 years.

---

[26] Prüll, *Medizin* (ref. 4), 361–63.

[27] Ibid., 358–67.

[28] In my view, there are no specific correlations between new acausalist attitudes with modernism, and causalist notions with conservative notions, as claimed by Forman. Indeed, in at least the case of constitutional pathology, it seems acausalist notions actually often correlate with conservative political ideas.

First, it is commonly agreed among historians of science that scientific knowledge is not 'objective'. In contrast, the basic ideas and the process of acquisition of knowledge are moulded by the biographical experiences, education and environment of the researcher, and further the theory must be taken seriously that scientific knowledge is 'socially constructed' to best serve the needs of the particular society. This understanding goes beyond Forman, who more or less implicitly supported the demarcation line between internal causal 'exact' science and external philosophical speculations.[29]

Second, much more attention has been paid in the last decades to analysing the nineteenth-century roots of the sociopolitical standing of science in the Weimar Republic because the 'wholeness hunger' made itself felt much earlier than 1918. So we must consider, more carefully than Forman did, the development of science and medicine within the respective cultural setting starting from about 1850 onward to achieve a fuller understanding of Weimar-era science and medicine. Based on the long history of constitutionalism, Aschoff and Lubarsch did not see criticism of causality as a major problem.

Third, new research in the social history of science has changed our attitudes toward the so-called 'internalist' and 'externalist' views on the performance of research. Forman focused very much on the *reaction* of physics to a hostile environment, whereas more recent research considers scientists also as *active* creators of their societal and cultural environment. The internalist-externalist boundaries have been blurred by microsociological historical research, showing the multidirectional transfer of ideas and notions.[30] Taking into account ideas on networking and the flow of information and communication, our examples of constitutional pathology in Weimar Germany show Aschoff and Lubarsch as vectors communicating and negotiating specific scientific theories among different social agents — colleagues, students, the public. Science is embedded in society. The scientist is not only scientist but also citizen. The example of Aschoff illustrates the impact of the Zeitgeist on both this allegedly apolitical scientist and his colleagues at the time. The example of Lubarsch portrays the frank expression of (scientific) thoughts based on involvement in a political party.

---

[29] It is not possible here to give a full account of the literature written within the realm of science studies. A good overview can be found in Timothy Lenoire's *Instituting Science: The Cultural Production of Scientific Disciplines* (Palo Alto, CA: Stanford University Press, 1997). For constructivism, see Jan Golinski, *Making Natural Knowledge: Constructivism and the History of Science* (Cambridge: Cambridge University Press, 1998).

[30] See Karin Knorr-Cetina, *The Manufacture of Knowledge: An Essay on the Constructivist and Contextual Nature of Science* (Oxford: Pergamon Press, 1981).

Papers such as the one by Forman of 1971 have inspired important research into the social history of science and medicine over the last four decades.[31] Subsequent work on the differing theories of scientific practice has led to a deepened understanding of science in the context of cultural practices, an issue that nonetheless remains a challenge for future research activities.

Acknowledgments

I would like to thank the participants of the conference on the work of Paul Forman, held at the University of British Columbia, Vancouver, Canada, 23–25 March 2007, for fruitful comments on my paper. Furthermore, I would like to thank Helmuth Trischler, Cathryn Carson and Alexei Kojevnikov for their support, as well as Lindsay Crawford for her editorial work.

---

[31] See, for example, Helmuth Trischler, Cathryn Carson and Alexei Kojevnikov, 'Beyond Weimar Culture: Die Bedeutung der Forman-These für eine Wissenschaftsgeschichte in kulturhistorischer Perspektive', *Berichte zur Wissenschaftsgeschichte* 31 (2008): 305–10.

Jordan alias Domeier: Science and Cultural Politics in Late Weimar Conservatism

Richard H. Beyler*

A spectre is haunting the 'Forman thesis': the possibility that exponents of scientific rationality might abet, knowingly or unknowingly, the worst varieties of cultural and political unreason. Forman's Weimar physicists were allowing themselves 'to sacrifice physics, indeed the scientific enterprise, to [a] *Zeitgeist*' of anti-intellectualism. As Forman confesses, he had 'not been able to … maintain a perfectly neutral stance' toward his historical characters.[1] The particular power of the 1971 paper is due not only to the status of the advent of quantum theory as an epochal event in the history of science, but also to the connotations of the specific

---

* History Department, Portland State University, P.O. Box 751, Portland, OR 97207–0751; beylerr@pdx.edu.

The following abbreviations are used: DHV, Deutschnationaler Handlungsgehilfen-Verband (German National Union of Commercial Employees); *DV*, *Deutsches Volkstum*; *HSPS*, *Historical Studies in the Physical Sciences* (later, *Historical Studies in the Physical and Biological Sciences*); HVA, Hanseatische Verlagsanstalt (Hanseatic Publishing House).

[1] Paul Forman, 'Weimar Culture, Causality, and Quantum Theory, 1918–1927: Adaptation by German Physicists and Mathematicians to a Hostile Intellectual Environment', *HSPS* 3 (1971): 1–115, on 112–13; the argument is extended temporally and conceptually in Paul Forman, '*Kausalität, Anschaulichkeit* and *Individualität*, or How Cultural Values Prescribed the Character and Lessons Ascribed to Quantum Mechanics', in *Society and Knowledge: Contemporary Perspectives in the Sociology of Knowledge and Science*, ed. Nico Stehr and Volker Meja (New Brunswick, NJ: Transaction Books, 1984), 333–47. A vigorous advocacy that intellectual disciplines should remain unbeholden to outside influences appears in several of Forman's other writings, e.g., 'Behind Quantum Electronics: National Security as Basis for Physical Research in the United States, 1940–1960', *HSPS* 18, no. 1 (1987): 149–220; 'Independence, Not Transcendence, for the Historian of Science', *Isis* 82 (1991): 71–86; 'The Primacy of Science in Modernity, of Technology in Postmodernity, and of Ideology in the History of Technology', *History and Technology* 23 (2007): 1–152. I have experienced at least a little of this haunting; in my very first graduate seminar I wrote a review essay on the 'Forman thesis' and ever since I have been thinking about many of the issues raised by Forman.

historical context: the Weimar Republic as a brilliant but troubled era leading, somehow, to the catastrophic National Socialist era that followed.

To be fair to Forman, he is too sophisticated an historian to espouse the simplistic view of the Weimar as nothing but a precursor to Nazism. He does, however, suggest to his readers this catastrophic denouement with the third of his categories of 'intellectual-political orientation' among physicists. After those who were both politically progressive and culturally modernist, and then, secondly, the 'principled political conservatives' with classical literary and artistic preferences, Forman comes, finally, to the 'outright reactionaries', including several individuals later known as outspoken National Socialists. Forman finds the cultural-political progressives (with some prominent exceptions, such as Albert Einstein) were inclined to boldly abandon causality in physics, while the political and aesthetic conservatives tended to be conservative on the causality issue.[2] While the 'progressives' lose Forman's sympathy for their willingness to abandon classical standards to an anti-rationalist *Zeitgeist*, those at the opposite end of the physics spectrum were — according to Forman's alignment model — susceptible to the most extreme forms of political reaction.

A physicist whose work combined both objects of Forman's antipathy was Pascual Jordan. Jordan — the beginnings of whose career lay chronologically just beyond the scope of Forman's article — was one of the architects of quantum mechanics and a consistently strong advocate of the principle of acausality. He later earned notoriety for an apparent alignment with National Socialism, manifested in Nazi Party and SA membership, as well as in numerous books and articles during the years 1933–45, which sought to associate modern science, especially quantum physics and biophysics, with one or more strains within NS ideology. In apologias after 1945, Jordan rationalised his actions as working for moderation within the Nazi system and defending respectable science from the 'Aryan physics' movement, which included some of Forman's reactionaries. However, his often-controversial associations with right-wing causes continued throughout the postwar period.[3]

---

[2] Forman, 'Weimar Culture' (ref. 1), 113.

[3] The balance of this essay relies extensively on Richard H. Beyler, 'From Positivism to Organicism: Pascual Jordan's Interpretations of Modern Physics in Cultural Context' (PhD dissertation, Harvard University, 1994), esp. 207–24; and expands on Richard H. Beyler's 'Exporting the Quantum to Biology: Jordan's Biophysical Initiatives', in *Pascual Jordan (1902–1980): Symposium zum 100. Geburtstag des Physikers*, preprint no. 328, ed. Jürgen Ehlers, Dieter Hoffmann and Jürgen Renn (Berlin: Max-Planck-Institut für Wissenschaftsgeschichte, 2007), 69–82, on 71–72. For more on Jordan's career in its political relations, see M. Norton Wise, 'Pascual Jordan: Quantum Mechanics, Psychology, and National Socialism', in *Science, Technology, and National Socialism*, ed. Monika Renneberg and Mark Walker (Cambridge: Cambridge University Press, 1994), 224–54;

Other historians have since endeavoured to interpret the Jordan case in ways indebted to, but somewhat diverging from, Forman's alignment model. John Heilbron includes Jordan among the 'missionaries of the Copenhagen spirit', but one whose 'extravagances' in extending the scope of complementarity, as well as his Nazi sympathies, alarmed Niels Bohr and other Copenhagen interpretation adherents. Where Forman disapproves of the capitulation of culturally weak physicists to external pressures, Heilbron critiques ambitious or 'imperialistic' physicists attempting to draw sweeping conclusions from the mathematical formalisms of quantum theory.[4]

Norton Wise analyses Jordan's description of modern quantum physics as *anschaulich* (perceptible) in the context of Nazi ideology, writing that Jordan, in making such linkages, produces 'shifts of emphasis', 'transforms meanings' and thereby 'coopt[s] ideas typically located in a liberal socialist context, often also Jewish, for the National Socialist cause'. In short, Jordan committed 'violent reversals' of the work of Niels Bohr and others. For Wise, Jordan's commitment to acausalism was hardly a capitulation per the Forman thesis, but rather an advocacy; it was a long-term project, rather than a product of the immediate postwar environment; finally, the case exhibited the danger of lumping together under a single category of 'vitalism' a diverse range of theories and associations.[5] However much it may 'stick in the craw', Wise writes, Jordan's case shows that it is a 'myth that the acquisition of fundamental knowledge had to cease when scientists embraced Hitler'.[6] Wise thus conjures up the Formanian spectre of an alliance between science and unreason — it could indeed happen — but also defends most of Heilbron's 'Copenhagen missionaries' against collective guilt.

For Heilbron and Wise, Jordan thus represents a kind of limiting case within their categories of analysis. In comparison to most other prominent quantum theorists, Jordan's linkage of physics and cultural politics does appear

---

Dieter Hoffmann, *Pascual Jordan im Dritten Reich: Schlaglichter*, preprint no. 248 (Berlin: Max-Planck-Institut für Wissenschaftsgeschichte, 2003); Dieter Hoffmann and Mark Walker, 'Der gute Nazi: Pascual Jordan und das Dritte Reich', in *Pascual Jordan (1902–1980)*, ed. Ehlers *et al.*, 83–112; Richard H. Beyler, 'Pascual Jordan: Freedom vs. Materialism', in *Eminent Lives in Twentieth Century Science & Religion*, ed. Nicolaas A. Rupke (Frankfurt: Peter Lang, 2007), 157–76; Arne Schirrmacher, 'Physik und Politik in der frühen Bundesrepublik Deutschland: Max Born, Werner Heisenberg und Pascual Jordan als politische Grenzgänger', *Berichte zur Wissenschaftsgeschichte* 30 (2007): 13–31.

[4] J.L. Heilbron, 'The Earliest Missionaries of the Copenhagen Spirit', *Revue d'histoire des sciences* 38 (1985): 195–230, on 222; also in E. Ullmann-Margalit, ed., *Science in Reflection: The Israel Colloquium* (Dordrecht: Kluwer Academic, 1988), 201–33.

[5] Wise, 'Pascual Jordan' (ref. 3): 227–28.

[6] Ibid., 244.

anomalous. But the second component of the linkage was also relevant: Jordan's speculations were squarely within a loosely delimited but nonetheless distinctive discourse in German conservatism. Moreover, Jordan's cultural-political agenda emerged before 1933 (and it continued well after 1945). So while particular attention may — deservedly — focus on Jordan because of his actions during the Nazi regime, those later events cannot serve as a causal account of the early development of his cultural-political views. Nor should they necessarily serve as a correlative account, though this is partly a question of 'lumping' or 'splitting' in the history of German conservatism: while there were obviously similarities and synergies between National Socialism and various other strains of Weimar conservatism, and although the Nazis probably could not have come to power without the support or cooptation of many non-Nazi conservatives, it is also possible to discern non-trivial differences among the various, sometimes cooperating and sometimes competing, camps within the German Right in the late Weimar era.[7]

Along with many other academics and white-collar workers, Pascual Jordan was a member of the Deutschnationale Volkspartei (DNVP, German National People's Party), which was, apart from the Nazis, the party furthest to the right on the Weimar political spectrum.[8] More specifically, Jordan's cultural-political sympathies were expressed in a series of essays he wrote under the pseudonym Ernst Domeier for the periodical *Deutsches Volkstum: Monatsschrift für das deutsche*

---

[7] Within the voluminous literature on German conservatism, some authors emphasise continuities among various segments of the right wing, while others emphasise the non-Nazi or even anti-Nazi traits of the varieties of conservatism. Overviews of Weimar conservatism, with a diversity of historiographical viewpoints, include Armin Mohler, *Die konservative Revolution in Deuschland 1918–1932: Grundriss ihrer Weltanschauungen* (Stuttgart: Friedrich Vorwerk Verlag, 1950); Kurt Sontheimer, *Antidemokratisches Denken in der Weimarer Republik: Die politischen Ideen des deutschen Nationalismus zwischen 1918 und 1933* (Munich: Nymphenburger Verlagshandlung, 1964); Kurt Lenk, *Deutscher Konservatismus* (Frankfurt: Campus Verlag, 1989); Jost Hermand, *Old Dreams of a New Reich: Volkish Utopias and National Socialism*, trans. Paul Levesque (Bloomington: Indiana University Press, 1992); Stefan Breuer, *Anatomie der konservativen Revolution* (Darmstadt: Wissenschaftliche Buchgesellschaft, 1993); various essays in Larry Eugene Jones and James Retallack, eds., *Between Reform, Reaction, and Resistance: Studies in the History of German Conservatism from 1789 to 1945* (Providence: Berg, 1993); Axel Schildt, chap. 5 in *Konservatismus in Deutschland: Von den Anfängen im 18. Jahrhundert bis zur Gegenwart* (Munich: C.H. Beck, 1998); Thomas Rohkrämer, *A Single Communal Faith? The German Right from Conservatism to National Socialism*, Monographs in German History, vol. 20 (New York: Berghahn Books, 2007).

[8] Pascual Jordan, questionnaire, 27 August 1934, Archiv der Humboldt-Universität Berlin, Folder Personalkte-UK-Pers., J69, p. 1; also available at Rostock Universitätsarchiv, Folder Phil. Fakultät/Lehrstuhl Theoretische Physik/Berufungsakte Jordan.

*Geistesleben.*⁹ The journal was published by the Hanseatische Verlagsanstalt (HVA, Hanseatic Publishing House). The HVA, in turn, was associated with the Deutschnationaler Handlungsgehilfen-Verband (DHV, German National Union of Commercial Employees), which was in 1931 the largest non-Marxist labour organisation in Germany. The DHV proposed to be a bulwark against the proletarianisation of the middle classes and a defender of traditional values. Moreover, as historian Gary Stark notes, the term *deutschnational* was often a code word for 'anti-Semitic', and Jews were indeed excluded from the union.[10] The HVA was formed in 1920 to bring under one management various publishing and literature distribution enterprises of the DHV, but soon grew into a full-scale book and magazine publisher and became essentially autonomous from the union itself. Among HVA's authors were Arthur Moeller van den Bruck, populariser of the term 'Third Reich'; 'friend-or-foe' political theorist Carl Schmitt; aesthete of 'total mobilization' Ernst Jünger; nationalist sociologist Hans Freyer and Oswald Spengler, patron of Forman's Weimar *Zeitgeist*.[11]

Deutsches Volkstum began publication in 1917, but the journal's rise to prominence dates from the appointment of Wilhelm Stapel as editor in 1919.[12] English

---

[9] Throughout this essay, for the sake of simplicity and convenience, I will discuss these articles in the text and cite them in the footnotes using Jordan's pseudonym, Ernst Domeier. The identification of Domeier with Jordan rests on three independent pieces of evidence: (1) At least one of the Domeier essays is very reminiscent of passages in a subsequent HVA publication appearing under Jordan's own name (*cf.* Ernst Domeier, 'Wohin treibt die Technik?' *DV* 15 (1933): 41–42, and Pascual Jordan, *Physikalisches Denken in der neuen Zeit* (Hamburg: Hanseatische Verlagsanstalt, 1935), 49); (2) In later correspondence with Vieweg Verlag, Jordan discussed the possibility of including one of the Domeier essays as a chapter in his 1947 book *Physik im Vordringen*, though nothing came of the plan (see Pascual Jordan to Vieweg Verlag, 12 February 1945, Jordan file, Vieweg-Archiv, Vieweg + Teubner, Wiesbaden, Germany; and (3) Jordan mentions the essays in later correspondence with Henning Stapel, the son of the journal's editor, Wilhelm Stapel, after the latter's death (see Pascual Jordan to Henning Stapel, 30 June 1973, Folder 782; and Henning Stapel to Pascual Jordan, 7 July 1973, Folder 609, both in Staatsbibliothek zu Berlin, Hanschriftenabteilung, Nachlass Jordan).

[10] On the origins of the HVA and its associations with the DHV, see Gary Stark, *Entrepreneurs of Ideology: Neoconservative Publishers in Germany, 1890–1933* (Chapel Hill: University of North Carolina Press, 1981), 22–32, esp. 24.

[11] On the subsequent history of the HVA, see Stark, *Entrepreneurs* (ref. 10); Siegfried Lokatis, *Hanseatische Verlagsanstalt: Politisches Buchmarketing im "Dritten Reich"* (Frankfurt: Buchhändler-Vereinigung, 1992); Ascan Gossler, *Publizistik und konservative Revolution: Das "Deutsche Volkstum" als Organ des Rechstintellecktalismus 1918–1933*, Uni Press Hochschulschriften, vol. 122 (Münster: LIT Verlag, 2001).

[12] On Stapel's career, see Heinrich Keßler, *Wilhelm Stapel als politischer Publizist: Ein Beitrag zur Geschichte des konservativen Nationalismus zwischen den beiden Weltkriegen* (Nuremberg: Lorenz Spindler Verlag, 1967). *See also* Stark, *Entrepreneurs* (ref. 10); Lokatis, *Hanseatische* (ref. 11) and Gossler, *Publizistik* (ref. 11).

renderings of *Volkstum* range from 'folk customs' to 'nationality' to 'national characteristics'. In Stapel's parlance, the term referred to a national character, tradition, or 'nomos' which had an essential reality transcending any mere aggregation of individuals and their (merely) utilitarian interests; it likewise transcended any particular state or political manifestation. In any event, parliamentary democracy, in which national decisions resulted from the summing up of individual preferences and sordid bargaining among interest groups, was for Stapel not conformable to *Volkstum*.[13] Stapel was particularly concerned with the role of religion in this context — not so much as a source of correct doctrine or moral guidance, but rather as a carrier of the national tradition. He was thus heir to a tradition of militant Erastian Lutheranism, its antipathy directed now not toward Catholicism (as in the days of Luther) but materialism, socialism and liberalism.[14]

Stapel directed his animus against Judaism too. If national character was something that transcended the rights and interests of individuals and constituent groups, and if the Christian church alongside the state was the bearer of the German national character, then Jews could not fully partake in it. Negative comment on alleged Jewish cultural and political over-representation in Weimar Germany was common in the pages of *Deutsches Volkstum*, as well as in some of Stapel's other writings. It should be noted that Stapel couched his anti-Semitism in what he thought of as cultural, not biological, terms; the Jews were even deserving of some respect, in Stapel's view, though they could never contribute to reviving German nationhood.[15] This is not to say that biologically rooted anti-Semitism was unknown or unwelcome to Stapel. For example, his review of Philipp Lenard's *Grosse Naturforscher* mildly objected to Lenard having given such short shrift to philosophy and the other humanistic disciplines, but left uncriticised Lenard's overtly anti-Semitic interpretations of history.[16]

---

[13] See, for example, the following works of Wilhelm Stapel: 'Demokratie und Diktatur', *DV* 10 (1923): 387–91; *Die Fiktionen der Weimarer Verfassung: Versuch einer Unterscheidung der formalen und der funktionalen Demokratie* (Hamburg: Hanseatische Verlangsanstalt, 1928); 'Volk und Staat', *DV* 10 (1928): 421–28; and 'Versuch einer Metaphysik des Staates', *DV* 13 (1931): 409–19. For further discussion, see Keßler, *Wilhelm Stapel* (ref. 12), chaps. 2 and 3; Gossler, *Publizistik* (ref. 11), 141–56.

[14] The salient text is Stapel's (probably) best-known work, *Der christliche Staatsmann: Eine Theologie des Nationalismus* (Hamburg: Hanseatische Verlagsanstalt, 1932); Keßler, *Wilhelm Stapel* (ref. 12), 159–76; Gossler, *Publizistik* (ref. 11), 122–29; Roland Kurz, *Nationalprotestantisches Denken in der Weimarer Republik: Voraussetzungen und Ausprägungen des Protestantismus nach dem Ersten Weltkrieg in seiner Begegnung mit Volk und Nation*, Die Lutherische Kirche: Geschichte und Gestalten, vol. 24 (Gütersloh: Gütersloher Verlagshaus, 2007), 221–86.

[15] See Keßler, *Wilhelm Stapel* (ref. 12), 148–59; Gossler, *Publizistik* (ref. 11), 129–40; Kurz, *Nationalprotestantisches Denken* (ref. 14), 287–94.

[16] Wilhelm Stapel, review of *Grosse Naturforscher* by Philipp Lenard, *DV* 13 (1931): 662–63.

Though Stapel evidently longed for some kind of overthrow of the liberal parliamentary regime, it was unclear what form he thought the new order should take. According to biographer Heinrich Keßler, Stapel favored an authoritarian, clerically inclined regime, to which the Heinrich Brüning government of 1930 was the closest approximation. All of the possible right-wing coalitions, however, were unstable and within a year or so Stapel began to consider the National Socialists as the only party capable of 'breaking through' the republican impasse. Stapel expected the Nazis would fade after an initial success and foresaw a chance thereafter for more responsible conservative forces to construct a new political order.[17]

He was not alone in asserting, after the fact, that he had underestimated the virulence of Nazism and overestimated his own ability to deal with it.[18] After January 1933, he soon found himself meeting with disapproval from the new regime, or rather from powerful factions within the Nazi polycracy. Stapel's culturally based anti-Semitism was insufficiently radical and biological for the Nazis. Furthermore, his insistence on a 'Christian' foundation for German *Volkstum* conflicted with the neo-paganism favoured by official Party ideologue Alfred Rosenberg and by the SS.[19] The Propaganda Ministry pressured Stapel out of the

---

[17] Stapel's opinions on Weimar politics are reviewed in Keßler, *Wilhelm Stapel* (ref. 12), 109–76; Stark, *Entrepreneurs* (ref. 10), 186–89, 224.

[18] This sentence is in some respects a quick synopsis of the history of non-Nazi German conservatism in the 1930s (see ref. 7). It is worth considering, as some historians have done, the extent to which the 'underestimation' trope became self-serving after 1945. Numerous examples can be adduced. On Martin Heidegger, see Arnold Davidson, 'Questions Concerning Heidegger: Opening the Debate', *Critical Inquiry* 15 (1989): 407–26; Victor Farías, *Heidegger and Nazism*, trans. Paul Burrell and Gabriel R. Ricci (Philadelphia: Temple University Press, 1989); Philippe Lacoue-Labarth, *Heidegger, Art, and Politics: The Fiction of the Political*, trans. Chris Turner (Oxford: B. Blackwell, 1990); Richard Wolin, *The Politics of Being: The Political Thought of Martin Heidegger* (New York: Columbia University Press, 1990); Michael E. Zimmerman, *Heidegger's Confrontation with Modernity: Technology, Politics, Art* (Bloomington: Indiana University Press, 1990); Tom Rockmore, *On Heidegger's Nazism and Philosophy* (Berkeley: University of California Press, 1992); Hans Sluga, *Heidegger's Crisis: Philosophy and Politics in Nazi Germany* (Cambridge, MA: Harvard University Press, 1993). On Carl Schmitt, see Jerry Z. Muller, 'Enttäuschung und Zweideutigkeit: Zur Geschichte rechter Sozialwissenschaftler im 'Dritten Reich'', *Geschichte und Gesellschaft* 12 (1986): 289–316; John McCormick, *Carl Schmitt's Critique of Liberalism: Against Politics as Technology* (Cambridge: Cambridge University Press, 1997). On Hans Freyer, see Jerry Z. Muller, *The Other God That Failed: Hans Freyer and the Deradicalization of German Conservatism* (Princeton: Princeton University Press, 1987). On Bernhard Bavink, see Klaus Hentschel, 'Bernhard Bavink (1879–1947): Der Weg eines Naturphilosophen vom deutschnationalen Sympathisanten der NS-Bewegung bis zum unbequemen Non-Konformisten', *Sudhoffs Archiv* 77 (1993): 1–32.

[19] Wilhelm Stapel, 'Der Neocalvinismus und die Politik', *DV* 14 (1932): 395–400, draws the parallel that Calvinism is to democracy as Lutheranism is to monarchy or authoritarianism.

editorship of *Deutsches Volkstum* in 1938; thereafter the journal became a strictly Party-line organ.[20]

To characterise *Deutsches Volkstum* under Stapel's editorship as anti-intellectual, vitalist, or occultist would be at least an oversimplification. Mysticism was at odds with religious tradition and scientifically suspect.[21] The overall tenor of the magazine was one of elevated, self-consciously elite scholarship (*Wissenschaft*). To be sure, this scholarship (*Wissenschaft*) was placed into the service of the essentialisation of German culture and, not infrequently, the cultural exclusion of German Jews. Soon after the Nazi takeover, co-editor Albrecht Erich Günther explicated the role he saw for intellectuals in the new order: the National Socialist state defined itself as a 'blood community of German national comrades', a culturally and racially specific entity in contrast to the purportedly internationalist, universalist empires of liberalism and Bolshevism. Proceeding from this premise, the Nazi state would take steps to 'remove' (*beseitigen*) 'national alien or national enemy' (*volksfremden oder volksfeindlichen*) scientists — Jews above all because of their status as a people without a true national identity (*Zwischenvolk*).[22] Beyond this, the new German state was to be avowedly totalitarian, dissolving the supposed divisions among private, professional and public functions. Therefore, the new regime needed scholars who, in their scholarship as well as in their non-professional lives, were 'carriers of the new will to live of the nation' (*Träger des neuen Lebenswillens der Nation*). Neither the specific content of the science nor whether it attempted to create a 'scientific' National Socialism was as important as the act of ideological commitment itself.[23]

This is the specific intellectual context in which Jordan's writings as 'Ernst Domeier' appeared from 1930 to 1933: ten articles, as well as several contributions to the monthly section of observational or anecdotal snippets, the 'Beobachter'.[24]

---

[20] See Keßler, *Wilhelm Stapel* (ref. 12), 177–224, 125, 277–78; Stark, *Entrepreneurs* (ref. 10), 228–31. For background on the HVA as a whole during the Third Reich, see Lokatis, *Hanseatische* (ref. 11).

[21] Felix Krueger, 'Okkultismus und Wissenschaft', *DV* 11 (1929): 517–28.

[22] Albrecht Erich Günther, 'Gibt es eine nationalsozialistische Wissenschaft?' *DV* 15 (1933): 761–67, on 762–64.

[23] Günther, 'Gibt es' (ref. 22): 765–66. Günther cited as examples biologists Jacob von Uexküll and Edgar Dacqué, anthropologist Bruno Gutmann, political scientist Carl Schmitt and philosopher Alfred Baeumler.

[24] It would be plausible to assume a connection between Jordan and Stapel began during the former's stint in Hamburg from 1928 to 1929; however, there is no direct evidence of this. The first letter between them in the Jordan papers (see Staaatsbibliothek zu Berlin, Hanschriftenabteilung, Nachlass Jordan, Folder 610) dates from 1932; however, it should be noted the archival holdings prior to 1945 are incomplete.

Jordan's reasons for using a pseudonym remain obscure, though the circumstances provide some clues. He became an *Extraordinarius* (roughly equivalent to associate professor) in Rostock in 1929 — on the one hand, an advancement to this rank at a relatively young age; on the other hand, a move to a comparatively obscure university. On the personal front, Jordan married Hertha Stahn in 1930; their first of two sons was born the following year. Throughout his subsequent career, alongside strictly scientific work, Jordan was a prolific author of popularisations of science in long and short forms, as well as many books, essays and articles of a cultural-political character. A biographical article by one of his students suggests this vigorous literary productivity was spurred, in part, by financial concerns — and at the time of the Domeier articles, Jordan had just started a young family.[25] Yet persisting themes across five decades suggest these works were also the product of conviction. From all of these circumstances, one can sketch the picture of a man at the beginning of a potentially promising though still tenuous career, capable of writing for a popular audience and desiring to put that skill to practical use, but possessing controversial opinions that ranged beyond his strictly professional duties. In such a position a pseudonym seems not unexpected, at least until the Nazi takeover dramatically changed the context of political controversy. The Domeier appellation ceased after 1933; thereafter, Jordan was evidently willing to publish similar sentiments under his own name.

The Domeier articles expressed a distrust of democratisation and, conversely, an unapologetically elitist conception of culture and science; they resented the decline of German national power; they bemoaned the erosion of traditional social values and artistic tastes. At first glance these seem little more than variations on themes promoted by many of the HVA's authors, but there were also distinctive emphases: a confidence in the scientific method when properly understood and a fascination with the powerful potentialities of technology.

Domeier's first contribution, 'Überschulung und Bürokratie', appeared in March 1930 in a special issue devoted to pedagogy. According to this article, the growth of the number of students in German education, ascribed to a longer average period of schooling, had created 'our "school misery"'. Employers had higher expectations for the educational level of employees, even though the studies were frequently irrelevant to the work required. The extended schooling only brought grief to students and placed a financial burden on their families, contributing in turn to a 'world record' decline in the birth rate. Furthermore, the extension of schooling meant that more of the experience of youth was placed under the control

---

[25] Engelbert Schücking, 'Jordan, Pauli, Politics, Brecht, and a Variable Gravitational Constant', *Physics Today* 52, no. 10 (1999): 26–31, on 28.

of the educational bureaucracy.[26] This last point particularly echoed the theme of the issue's preceding article, in which Stapel complained that state bureaucracy had taken control of education away from the family, part of a creeping socialism in which private concerns gradually became state concerns.[27]

Domeier's analysis of the root cause of the supposed educational bloat was quite original, however. It was connected to Jordan's later theories on the relation between quantum physics and psychoanalysis, first published in 1932; the *Deutsches Volkstum* article shows that (as Domeier) he was already contemplating this psychological problem by 1930.[28] According to this analysis, the push to extend the length of schooling was a compensation for feelings of powerlessness at both the personal and national levels:

> Since 1918 we are an unfree people. Foreign powers essentially regulate our taxes and fiscal management, our police and military. Foreign troops have conducted maneuvers on German soil. There is hardly any really incisive question in politics and economics on which we could decide independently from foreign orders … Instead of a subject we are an object of world politics; and the tone of diplomatic intercourse of foreign, especially small, states with Germany is adjusted accordingly.[29]

Domeier thus lamented a situation in which the exercise of the subjective national will was thwarted by alien forces; the feeling of internal conflict must be especially acute for civil servants, nominally those who were supposed to be the executors of the national will:

> All those who have anything to do with the government of our Reich … must be inwardly filled with feelings which result with lawlike necessity from the pressing sensation of powerless dependence on foreign powers …. One can avoid the

---

[26] Ernst Domeier, 'Überschulung und Bürokratie', *DV* 12 (1930): 175–79.

[27] Wilhelm Stapel, 'Soziologische Einordnung der Schule', *DV* 12 (1930): 169–75, on 174–75.

[28] For Pascual Jordan on connections between quantum complementarity and psychoanalysis, see his 'Quantenmechanik und die Grundprobleme der Biologie und Psychologie', *Naturwissenschaften* 20 (1932): 815–21; 'Quantenphysikalische Bemerkungen zur Biolgie und Psychologie', *Erkenntnis* 4 (1934): 215–52, esp. 246–48; 'Positivistische Bemerkungen über die parapsychischen Erscheinungen', *Zentralblatt für Psychotherapie und ihre Grenzgebiete* 9 (1936): 3–17; *Anschauliche Quantentheorie: Eine Einführung in die moderne Auffassung der Quantenerscheinungen* (Berlin: Julius Springer, 1936), 313–19; *Verdrängung und Komplementarität: Eine philosophische Untersuchung* (Hamburg: Stromverlag, 1947). For discussion, see Heilbron, 'Earliest Missionaries' (ref. 4); and Wise, 'Pascual Jordan' (ref. 3).

[29] Domeier, 'Überschulung und Bürokratie' (ref. 26), 179.

psychological damages ... by taking up the reality into the consciousness. But it is exactly the method of the ruling parties to avoid reality and to flee into a haze of do-good domestic- and foreign-political fantasies.[30]

The psychotherapeutic treatment of frustrations and anxieties by bringing their repressed, unconscious source into the realm of consciousness was exactly the phenomenon Jordan later analogised to quantum complementarity. (Jordan himself had undergone psychoanalysis to treat his recurring stutter.)

Rather than striking out at the foreign nations holding Germany down, the externally cowed Weimar regime had chosen to repress the unpleasant fact of external constraint by turning its aggressions inward:

[T]he domestic-political imperialism of the current bureaucracy reveals itself to be a psychologically natural phenomenon .... [I]nferiority complexes create the necessity of a compensation at any price .... Bureaucratic pomposity and bureaucrats' terrorizing of [their] own Volk comrades produce the needed compensation, the needed release, and make the recovery of the damaged self-image possible.[31] The busybody extension of schooling was an outlet for the frustrations of weakling bureaucrats. More broadly, the Weimar Republic as a whole was the project of weaklings and defeatists.[32]

But in what sense did schooling more children longer constitute 'terrorizing' the population? A follow-up article by Domeier provided an answer: overschooling fomented social conflict. Certain interest groups were carrying on the class struggle through concealed means, such as the Social Democrats' efforts toward equalising the duration and quality of schooling for all children. Domeier took particular umbrage at the efforts of Carl Heinrich Becker, formerly Prussian minister of education, to lower the barriers among the various tracks (classical, modern, technical, vocational) in German secondary education so as to allow more 'worker students' to attend German universities. Avowedly for Becker, such

---

[30] Ibid.

[31] Ibid.

[32] This is the gist of an anecdote Domeier relates in the magazine's 'Beobachter' section in his contribution 'Prädestination', *DV* 12 (1930): 402. An (unnamed) acquaintance from secondary schooldays, who had then been the target of much bullying, ended up attending the same university as Jordan, where the unnamed student served as the head of the Republican Student League, an organisation in support of the Weimar regime. For more on how Domeier saw the parliament as weak and indecisive, see 'Parlamentarismus' (in the 'Beobachter' section) in 'Prädestination', on 561–62.

reforms were needed so that education could serve a democratic policy.[33] Domeier's conception of education, particularly higher education, was precisely anti-democratic: 'The flooding of the universities with students of proletarian character would simultaneously effect the downfall of conservative academic traditions and the suppression (and economic annihilation) of the current academy.'[34] For Domeier, it was a 'natural fact' that there were differences in intellectual capacity among the classes, so that some would benefit from advanced studies while others would not; in contrast, socialist and democratic efforts to equalise differences were artificial machinations:

> [I]t cannot be called desirable that regularly a considerable portion of the student body in higher education should come from the lower social classes. For no political will can remove facts of biology from the world ... [I]f one is of the opinion that it should be the duty of the state to equalize mistakes of the dear Lord and [amend] "social injustices," then it is consistent to help bring about the state of the structureless mass by the leveling of differences in education ...[35]

This analysis of social crisis as a failure to acknowledge natural hierarchy was later mirrored in Jordan's analysis of the functioning of living organisms: hierarchical power structures were literally vital; the 'structureless mass', on the other hand, literally the state of death.[36]

One might be inclined to think that out of professional self-interest, a university professor would favour a growth in the size of the universities. More important, however, to Domeier was the preservation of distinct social classes and the preservation of scholarship as an avowedly *elite* vocation. Numerical growth, in this view, would only weaken academia's ability to serve as a haven for individual accomplishment in the face of mass culture, which was the domain of 'surrogatism' — i.e., satisfaction with pale imitations or 'residuals' (*Eliminate*) of real culture.

---

[33] On Becker and Weimar educational reform, see Erich Wende, *C. H. Becker, Mensch und Politiker: Ein biographischer Beitrag zur Kulturgeschichte der Weimarer Republik* (Stuttgart: Deutsche Verlags-Anstalt, 1959), esp. 99–105, 138–34, 235–51; Fritz K. Ringer, *The Decline of the German Mandarins: The German Academic Community, 1890–1933* (Cambridge, MA: Harvard University Press, 1969), 69–80, 269–82, 404–18; Forman, 'Weimar Culture' (ref. 1): 23–24.

[34] Ernst Domeier, 'Überschulung und Klassenkampf', *DV* 12 (1930): 271–76, on 272n; *cf.* Wilhelm Stapel, 'Die Widerherstellung der Universität', *DV* 14 (1932): 488–96.

[35] Domeier, 'Überschulung und Klassenkampf' (ref. 34): 275.

[36] For a psychoanalytic interpretation of fears among right-wing radicals of the 'structureless mass' in Weimar Germany, see Klaus Theweleit, *Male Fantasies*, 2 vols. trans., Stephen Conway *et al.*, Theory and History of Literature, vols. 22–23 (Minneapolis: University of Minnesota Press, 1987–89).

Sports, for example, became a surrogate for health and sports heroes commercially safe (and valuable) figures who were at the same time regarded by the masses as (originally) one of their own.[37] From Domeier's perspective, this was not rationality conceding power to the *Zeitgeist*, but just the opposite: a defence of the unique status of scientific inquiry opposed to entropic forces of irrationalism. It was precisely the organicist model of society that brought Domeier to this conclusion.

Such sentiments were by no means unique to Domeier. In 1927 and 1928, the League of German Universities, for example, officially expressed its 'grave concern [over] the ever greater expansion of the circle of those admitted to the universities' and the consequent dilution of standards. Many, if not most members of the professoriate, according to Fritz Ringer's classic study of the German humanistic professoriate, saw educational reforms as symptomatic of a dangerous spread of social democracy.[38]

In several short writings on art and architecture criticism, Domeier expressed further dismay at the dilution of aesthetic and social values. He professed a dislike of much of contemporary art criticism, particularly Kurt Pfister, on the grounds that it was too self-consciously ingenious, asserted too much of the critic's personality and showed disrespect to the genius of the artists themselves.[39] Respect for role differentiation was again the watchword. In 'Gulliver besucht die Marxisten', Domeier mocked avant-garde architectural experiments with cheaper materials, built-in furniture and 'cellular' *(zellenmäßig)* design.[40] He directed particular scorn at the Reichsforschungsgesellschaft für Wirtschaftlichkeit im Bau- und Wohnungswesen (Reich Research Society for Economy in Construction and Housing), established in 1927 to advance efficient construction techniques. Among the projects completed under its aegis were the Haselhorst settlement in Berlin-Spandau, a Walter Gropius housing project in Dessau and Ernst May's Praunheim project near Frankfurt.[41] Domeier's rhetoric drew upon the reputation of May and other associated architects

---

[37] Ernst Domeier, 'Surrogatismus', *DV* 14 (1932): 431–37, on 433, 435–36. The term *Eliminate* was borrowed from Hans Blüher; see his *Traktat über die Heilkunde: Insbesondere die Neurosenlehre* (Jena: Eugen Diederichs, 1928), 24–47.

[38] *Mitteilungen des Verbandes der deutschen Hochschulen*, vol. 8 (1928): 45, quoted in Ringer, *Decline* (ref. 33), 79. *See also* Ringer, *Decline* (ref. 33), 282–95. Another example of the defence of the elite university from *DV* is Gerhard Günther's 'Die kleine Universität', *DV* 12 (1930): 577–91.

[39] Ernst Domeier, 'Wider den Geistreichigkeit in der Kunstschriftstellerei', *DV* 12 (1930): 757–66; *see also* Ernst Domeier, 'Charlatanerie', *DV* 14 (1932): 74–77. The former article is the one Jordan considered reprinting in a subsequent book (see ref. 9).

[40] Ernst Domeier, 'Gulliver besucht die Marxisten', *DV* 13 (1931): 107–13.

[41] For more on the Reich Research Society for Economy in Construction and Housing and Ernst May, see Barbara Miller Lane, *Architecture and Politics in Germany, 1918–1945* (Cambridge, MA: Harvard University Press, 1968), 90–103, 122–23.

as left-wingers. In 1930, May took up an urban design assignment from the Soviet government, which provoked his critics to complain about the 'Bolshevization' of architecture under his influence.[42] Domeier did not mention this specifically, but by juxtaposing Germany and the Soviets he strongly implied a similar criticism. What he found objectionable, in both the USSR and Germany, was the imposition of centrally planned, technical designs upon a sanctified domestic sphere.

For all his distaste for modern architecture, art and art criticism, Domeier was hardly an anti-modernist in general. On the contrary, he celebrated the achievements of *modern* science and the current power and enormous future potential of technology. But his technoscientific modernism was not a worldview of progress and social improvement; modern science and technology revealed, rather, just where progressivism, in both its liberal and socialist variants, had reached its limits and failed. Readers of the Forman thesis might be inclined to see here the influence of Oswald Spengler, but Domeier's explicitly avowed resource was more the history of religion — consonant with the overall religious tenor of *Deutsches Volkstum* — and the philosophy of Schopenhauer. His 1930 article, 'Religion, Moralismus, und soziale Gesinnung', called for a 'counterattack against the Enlightenment's critique of religion'.[43] Religious impulses, he suggested, were simply an empirical fact of history for all cultures and as such needed no further justification. In higher cultures, it was manifested as 'turning away from the world' seen as 'the scene of eternally peaceless, sinful, vain and futile striving ….' This attitude, he pointed out, was essentially the same as Schopenhauer's.[44] Therefore, in this view, it was empirically mistaken to place morality, or doing service to humanity and society, at the centre of the religious impulse. Pacifism, as an attempt to avoid violent conflict, was antithetical to this understanding of religion, as was a concern with 'social needs'.[45] After quoting Schopenhauer again on the renunciation of life, he concluded:

> A deep-rooted irreligiosity is therefore hidden in those who wish to help in that they recommend birth control, want to "do away" with war, and overall set aside the cruelty of existence, in order to provide humanity a painless life in undisturbed enjoyment. They do not realize that in pain and passion is concealed the directing care of the Father, who calls his children home into his kingdom.[46]

In removing moral concerns and specific theological doctrines as the central tenets of religion, and in particular in asserting his contempt for pacifist inclinations,

---

[42] See Lane, *Architecture* (ref. 41), 103, 164–65.
[43] Ernst Domeier, 'Religion, Moralismus, und soziale Gesinnung', *DV* 12 (1930): 905–12, on 906–7.
[44] Ibid., 907–8.
[45] Ibid., 910–11.
[46] Ibid., 912.

Domeier's views resonated with Stapel's. Domeier went even further, however, in his denigration of the desire for material wellbeing, peace and social equity, as we have already seen in his rejection of meliorism and social reform in fields such as education and housing.

It should therefore be no surprise that he posited violence and power as the essence of scientific and technical knowledge. Amid the world unemployment crisis, he noted with irony the notion that technologies had often been promoted as beneficial 'labor-saving' devices — both liberalism and socialism had mistakenly believed the essence of technology was to provide convenience or pleasure.[47] If this were true, then John Maynard Keynes' solution to the crisis, i.e. encouraging consumption, might work. But 'cigarettes, movies, radio, automobiles, lipstick, silk stockings, *Uhu* magazines, and newspapers', far from being 'healthy and necessary', were in fact manifestations of a 'life-devouring power'.[48] A contempt for mass culture was again apparent.

Domeier saw technological improvement as an inevitable necessity, but did not regard this inevitability with optimisim:

> [I]n full purity and force the inner nature of technology appears in the warlike weapon. Armored cruisers, submarines, machine guns, airplanes, gas bombs and heavy artillery: those are things ... which mercilessly reveal the demonic in technology ... Only in the service of war can technology unfold its true nature without deception; only in the technical weapon is the complete agreement between the destructive power of technology and the pursued ends achieved: the demonic of technology is mastered and exceeded by the demonic of war.[49]

His governing premise was that life itself demanded violence, national survival required military preparedness, and in turn the constant forward development of technology. Domeier cited as inspiration for this view Ernst Jünger's semi-autobiographical essays in *Das abenteuerliche Herz*, which (as in much of his other writings from this period) expressed fascination with the power of modern technology — not excluding its role in conflict — as a nearly autonomous entity possessing its own demonic force.[50]

---

[47] Ernst Domeier, 'Technik, Arbeitslosigkeit und Krieg', *DV* 13 (1931): 838–46, on 838.
[48] Ibid., 840.
[49] Ibid., 845.
[50] Ibid., 845, quote on 846. He refers to Ernst Jünger, *Das abenteuerliche Herz: Aufzeichungen bei Tag und Nacht* (Berlin: Frundsberg-Verlag, 1929). Recall that Jordan's bureaucrats were also bound by psychological law to express aggression, either in foreign affairs or in domestic 'imperialism' and 'terrorism'.

Despite some remarkable parallels, the match between Domeier's and Jünger's views was not exact; if anything, the former went even further than Jünger in his apocalyptic vision of modern technology. In a commentary on a lengthy review by Albrecht E. Günther of Jünger's book *Der Arbeiter*, Domeier — who confessed he had not read the book, but only Günther's article — criticised the notion that technical knowledge could ever come to a state of 'perfection'.[51] Such a state implied stagnation, whereas in reality technology (and the science behind it) experienced continuous innovation. Moreover, atomic physics had recently discovered energy sources whose technical exploitation (the article takes for granted that eventually this will occur) will result in energy sources of literally unimaginable abundance: the power plants at Niagara Falls and the Dnjeprostroi Dam look like 'playthings', but likewise weaponry will see an 'immeasurable increase in its degree of effectiveness'.[52] The optimistic vision of progress reaching perfection was an 'avoidance' (*Ausweichen*) of the 'eschatological character of the technical future'.[53]

If science likewise could not hope to achieve a state of perfection, or at least some version of Enlightenment progress, what was left to motivate the scientist's activity? Domeier's answer came in a 1933 article on 'The Future of Science', which deserves a careful exegesis. He begins with an acknowledgment of the current climate of contemporary science reminiscent of the Forman thesis: university youth had grown impatient with the 'liberal spirit' of science and 'more and more often treated the sanctified "objectivity" of science with open contempt'.[54] Without necessarily applauding this state of affairs, Domeier first accepts that prior triumphant achievements of science coincided with the liberal political culture of the Enlightenment and the nineteenth-century faith in secular progress. Ultimately, however, he concludes the liberal understanding of scientific objectivity was an attempt to avoid conflict, to absolve individuals from personal decision and to quell the 'boldness and daring demanded by religious belief'.[55]

The 'spiritual desires of our times', Domeier contended, regarded such a liberal understanding with scepticism. One manifestation, he noted, was the cultural relativisation of science as found in — to take a prominent example — the work

---

[51] Ernst Domeier, 'Wohin' (ref. 9): 41–42, on 41; cf. Albrecht Erich Günther, 'Die Gestalt des Arbeiters: Zu Ernst Jüngers neuem Buch', review of *Der Arbeiter: Herrschaft und Gestalt*, by Ernst Jünger, *DV* 14 (1932): 777–90.

[52] Domeier, 'Wohin' (ref. 9): 42. The same argument (with similar turns of phrase) was later repeated in Jordan, *Physikalisches Denken* (ref. 9), 49.

[53] Domeier, 'Wohin' (ref. 9): 43.

[54] Ernst Domeier, 'Die Zukunft der Wissenschaft', *DV* 15 (1933): 320–25, on 320.

[55] Ibid., 321.

of Oswald Spengler. Spengler's argument was not that the results of mathematics, say, were valid only in particular times and places, but that the 'meaning and attractive enticement' embedded in the theories, which drove the development of the field, was culturally variable.[56] It was therefore necessary, according to Domeier, to find a new motivation for the 'formative power' of science in an era that was secularised but also disallowed any 'avoidance' of personal commitment — evidently a euphemism for the friend-or-foe importunity of National Socialism. Moreover, science (especially physics) had begun to renounce the erstwhile ambition of 'understanding' phenomena in favour of a more modest but clearer goal: complete, simple descriptions of experienced observations.[57] In other words, Domeier here offered a short exposition (without using the term) of Jordan's favoured 'radical positivist' epistemology.

After such renunciations, two possibilities remained to provide the drive (*Antrieb*) for scientific research. The first was beauty, an aesthetic pleasure of discovery that was almost impossible to convey to a non-specialist. The second was the 'will to power', a pursuit of knowledge that was 'most sublime, finest and yet of almost brutal vitality'. This will to power had always existed in science, but in the immediate historical circumstances was pushed 'nakedly' to the forefront.[58] Standing at the threshold of the Nazi takeover, Domeier thus perceived liberal visions of impersonal objectivity and secular progress were — for better or worse — defunct. In their place were visions of personal voluntarism — decisions valued merely for the sake of decisiveness, rather than for their moral content; the unapologetic exercise of power for its own sake and the inevitability of violence. Whether this world was to be encountered with dismay or relish was beside the point. What was pertinent was that modern science and technology had developed in just such a way as to be congruent with, and useful to this new post-liberal, post-progressive culture. In raising this possibility, Jordan writing as Domeier did not act idiosyncratically, but availed himself of the conceptual resources — vocabulary, targets of animosity, aspirations — of a specific sector of the German right-wing intelligentsia (*Bildungsbürgertum*). For Jordan/Domeier, this was not a hostile but a congenial intellectual environment.

---

[56] Ibid., 322.
[57] Ibid., 324.
[58] Ibid., 324–25.

The Causality Debates of the Interwar Years and Their Preconditions: Revisiting the Forman Thesis from a Broader Perspective

*Michael Stöltzner**

This paper will unavoidably resemble more of a sketch than a full-fledged drawing, as within these space limitations I attempt to convey the central argument of my 400-page PhD dissertation and the papers ensuing therefrom.[1] My intent in these works was to demonstrate how the causality debates in Weimar Germany and interwar Austria were an integral part of a more extended causality debate that emerged from two different readings of Boltzmann's legacy, statistical mechanics, at the end of the nineteenth century and only came to a close in the late 1930s when the philosophical debates surrounding quantum mechanics abated. Viewed within the context of this lengthier, and conceptually more profound, debate, it is true that the years 1918–27 were a time of turmoil, ranging from cultural and political upheaval to changes even in philosophical terminology. But in contrast to Forman, I assert that most scientists seriously examining and writing on causality

---

* Department of Philosophy, University of South Carolina, Columbia, SC 29208; stoeltzn@sc.edu.
[1] Michael Stöltzner, *Vienna Indeterminism: Causality, Realism and the Two Strands of Boltzmann's Legacy* (PhD dissertation, University of Bielefeld, 2003); available online at: http://bieson.ub.uni-bielefeld.de/volltexte/2005/694/ (accessed 9 December 2009); Michael Stöltzner, 'Vienna Indeterminism: Mach, Boltzmann, Exner', *Synthese* 119 (1999): 85–111; Michael Stöltzner, 'Die Kausalitätsdebatte in den *Naturwissenschaften*: Zu einem Milieuproblem in Formans These', in *Wissensgesellschaft: Transformationen im Verhältnis von Wissenschaft und Alltag*, ed. H. Franz, W. Kogge, T. Möller, T. Wilholt; available online at: http://bieson.ub.uni-bielefeld.de/volltexte/2002/90/html/Inhalt_Wissensgesellschaft.pdf (accessed 9 December 2009); Michael Stöltzner, 'Vienna Indeterminism II: From Exner to Frank and von Mises', in *Logical Empiricism: Historical and Contemporary Perspectives*, ed. P. Parrini, W. Salmon and M. Salmon (Pittsburgh: University of Pittsburgh Press, 2003), 194–229.

had already taken their general philosophical stand long before.² Rather than an adaptation to a hostile intellectual milieu that, in 1927, boosted the participants' fitness in the struggle for cultural survival, we find a complex but continuous debate about the philosophical consequences of modern physics that, after 1913, mainly appeared in the pages of one of the leading scientific journals of the German-speaking world, Arnold Berliner's *Die Naturwissenschaften*.

Let me be direct, as Forman himself was when he spoke of his protagonists as 'converts to acausality' who made 'quasi-religious confessions to the [anti-scientific] milieu' (dominated by *Lebensphilosophie* and Oswald Spengler's prophecy of a cultural decline), and who advocated, as did (allegedly) Reichenbach, 'an existentialist philosophy, disguised as logical empiricism', and published books, as did (allegedly) Frank, that contained 'largely blather'.³

To my mind, Forman's thesis misses core aspects of the causality debate from 1918 to 1927, and it does so because its author followed a methodology that, albeit fruitful in many other domains, is unable to assess the interactions between philosophical and scientific commitments in the historical context of the Weimar Germany and interwar Austria. More generally, in order to appraise the historical dynamics of a philosophical concept, such as causality, any broader historical and sociological approach must be reconciled with the recently sharpened methods of the history of philosophy of science.

My own picture of the causality debate is a complex one and it involves philosophical as well as historical, ideological and sociological motives. Philosophically, the debate about causality and determinism was hardly separable from the quest for a proper interpretation of physical probability, the conception of microphysical reality and the more general debates about the effects of modern physics on the conception of nature. Some of these lines reached back into the 1890s, when the old mechanistic worldview broke into pieces and they concerned statistical mechanics, relativity theory and quantum physics alike. Many critics, among them John Hendry, have noted that Forman's main philosophical misconception was to adopt, as did Spengler, a conception of causality intimately linked

---

² Paul Forman, 'Weimar Culture, Causality, and Quantum Theory, 1918–1927: Adaption by German Physicists and Mathematicians to a Hostile Intellectual Environment', *HSPS* 3 (1971): 1–115.

³ Forman, 'Weimar Culture' (ref. 2), quotes on 90, 68n158. Forman's criticisms were directed, respectively, at Hans Reichenbach's 'Die Kausalstruktur der Welt und der Unterschied von Vergangenheit und Zukunft', *Sitzungsberichte der Bayerischen Akademie der Wissenschaften, mathematisch-naturwissenschaftliche Abteilung* (1925): 133–75; English translation in his *Selected Writings, 1909–1953*, vol. 2 (Dordrecht: Reidel, 1978), 81–119; and Philipp Frank's *Das Kausalgesetz und seine Grenzen* (Vienna: Springer, 1932); translated into English by Marie Neurath and Robert S. Cohen as *The Law of Causality and Its Limits*, ed. Robert S. Cohen (Dordrecht: Kluwer, 1998).

to the strict determinism that predated statistical mechanics and characterised the mechanistic world-view.[4] Already in 1872, Emil du Bois-Reymond had touted the idea that firmly sticking to this explanatory ideal forced scientists to forgo a full understanding of the essence of matter and force. In the Weimar days, hardly any scientist professed adherence to du Bois-Reymond's *Ignorabimus,* but many considered it the basic misconception responsible for the diagnosis of crisis in modern physics.[5] It is true that Ludwig Boltzmann, in his battles with energeticism, portrayed himself as the last exponent of the mechanistic worldview. But on the other hand, in developing step by step the statistical interpretation of the second law of thermodynamics, he introduced a notion entirely alien to the mechanistic picture that Du Bois-Reymond had built upon, to wit, an objective physical probability that could no longer be understood as a degree of ignorance.

The debate as to whether physical science required a causal foundation at all began with Franz Serafin Exner's rectorial address of 1908, during which he launched a local tradition of empiricist indeterminism that I call Vienna Indeterminism. Among its main advocates, I count, besides Exner, his former assistant Erwin Schrödinger and the Logical Empiricists Philipp Frank and Richard von Mises. Exner argued chance is the basis of all natural laws and the apparent determinism in the macroscopic domain emerged only as the thermodynamic limit of very many random events.[6] Max Planck, in a rectorial address of 1914 that equally sung the praises of Boltzmann, fiercely objected to Exner's indeterminism and insisted on a deterministic foundation of all natural laws including the probabilistic ones.[7] The debate, as I view it, ended at the 1936 Copenhagen Congress for the Unity of Science when Frank and Moritz Schlick, Planck's former student and a resolute critic of any indeterminacy in principle, joined arms to combat the increasing number of metaphysical misinterpretations of quantum mechanics.

My main sociological point is that in the German-speaking world the causality debates took place among 'physicist-philosophers', a distinct type of scientist-thinker more widespread in Germany than in other countries, and whose historical prototypes could be traced back to such influential figures as Hermann

---

[4] John Hendry, 'Weimar Culture and Quantum Causality', *History of Science* 18 (1980): 155–80.
[5] Emil du Bois-Reymond, *Über die Grenzen des Naturerkennens* (Leipzig: Veit, 1872). For the historical reception of this speech, see Kurt Bayertz, Myriam Gerhard and Walter Jaeschke, ed., *Weltanschauung, Philosophie und Naturwissenschaft im 19. Jahrhundert,* vol. 3, *Der Ignorabimus-Streit* (Hamburg: Meiner, 2007).
[6] Franz S. Exner, *Über Gesetze in Naturwissenschaft und Humanistik* (Wien-Leipzig: Alfred Hölder, 1909).
[7] Max Planck, *Dynamische und statistische Gesetzmäßigkeit* (Leipzig: Barth, 1914).

von Helmholtz and Ernst Mach. There were well-defined steps of how a physicist could come to achieve this physicist-philosopher status. Most important, the philosophical ambitions of these men were only loosely, if at all embedded into the then current philosophical schools, among them neo-Kantianism and *Existenzphilosophie*.

Since the philosophical convictions of physicist-philosophers were not forced into a coherent philosophical system, there was ample leeway to simultaneously participate in different thought collectives. The notion of a thought collective, which we owe to Ludwik Fleck, provides an important perspective from which to understand the nine years at the heart of the Forman thesis.[8] Yes, there were those who defended scientific modernism and technological progress while simultaneously diagnosing a cultural crisis. Richard von Mises, for one, as late as 1951 remained sympathetic to Spenglerian ideas.[9] And there were others, such as his fellow physicist-philosophers from the Vienna Circle, who at the 1929 Congress of the German Physical Society went public with the claim that the achievements of modern science demanded an entirely new style of scientific philosophising and, more broadly, a new scientific world conception. No wonder that Richard von Mises disliked the manifesto, but he and Frank, in the opening session of the same congress, appeared almost as intellectual twins in their plea for abandoning the old triad of 'school philosophy', the categories of space, time and causality, replacing them with more suitable notions.[10] This message was well understood by those who rejected the neopositivist assault on metaphysics, among them the third speaker of the Prague opening session, physicist-philosopher Arnold Sommerfeld.[11]

---

[8] Ludwik Fleck, *Entstehung und Entwicklung einer wissenschaftlichen Tatsache: Einführung in die Lehre vom Denkstil und Denkkollektiv* (Basel: Schwabe and Co., 1935); translated into English as *The Genesis and Development of a Scientific Fact*, ed. T.J. Trenn and R.K. Merton (Chicago: University of Chicago Press, 1979). Fleck was no stranger to the readers of *Die Naturwissenschaften*; *cf.* Ludwik Fleck, 'Zur Krise der "Wirklichkeit"', *Die Naturwissenschaften* 17 (1929): 425–30.

[9] See Richard von Mises, *Positivism: A Study in Human Understanding* (New York: Dover Publications, 1968), 388n3.

[10] Philipp Frank, 'Was bedeuten die gegenwärtigen physikalischen Theorien für die allgemeine Erkenntnislehre?' *Die Naturwissenschaften* 17 (1929): 971–77, 987–94; translated into English as, 'Physical Theories of the Twentieth Century and School Philosophy', in *Modern Science and Its Philosophy* (New York: Collier Books, 1961), 96–125. Richard von Mises, 'Über kausale und statistische Gesetzmäßigkeit in der Physik', *Die Naturwissenschaften* 18 (1930): 145–53; also appears in *Erkenntnis* 1 (1930): 189–210.

[11] Arnold Sommerfeld, 'Einige grundsätzliche Bemerkungen zur Wellenmechanik', *Physikalische Zeitschrift* 30 (1929): 866–70.

My paper is organised as follows. First, I briefly sketch the causality debate. Second, I discuss the physicist-philosopher model and the extent to which *Die Naturwissenschaften* provided a forum for discourse during the interwar years. To conclude, I place the Forman thesis itself in an historical context and indicate in what way new methodological insights in the history of philosophy of science may prove helpful in understanding the causality debate.

Vienna Indeterminism and the Causality Debate

Rather than representing merely a 'subterranean anticausality current', as Forman put it, Exner's inaugural speech of 1908 made a great stir and triggered a polemic with Planck that continued the earlier Mach-Planck controversy.[12] In an interview with T.S. Kuhn, Frank recalls this speech was pretty influential for the then-younger generation of Vienna physicists, who, 'strange as it was ... were all followers of Mach *and* Boltzmann.'[13] Not that Exner in 1908 would have continued Mach's scepticism about atoms and the energeticist interpretation of thermodynamics. To the contrary, he closely followed the brand of empiricism that Boltzmann had developed from the late 1880s on in order to employ Mach's antimetaphysics against his primary opponent, Wilhelm Ostwald's energeticism. Boltzmann's consistent empiricism had important consequences for the basic principles of physical science. In the last years of his life, and especially in his 1903–06 lectures on natural philosophy, Boltzmann contemplated even the law of energy conservation was only statistically valid — an idea that would gain momentum only much later in the 1924 Bohr-Kramers-Slater quantum theory — and that the entropy of a system might be described by a nowhere differentiable function.[14]

Exner added an important dimension to Boltzmann's late indeterminism. While Boltzmann devoted surprisingly little attention to the interpretation of probability, Exner brought the relative frequency interpretation, or the *Kollektivmaßlehre* (developed by Gustav Theodor Fechner), to bear on the kinetic theory of gases.[15] On this basis, he could simultaneously claim that (i) in physics 'we observe regularities which are brought out exclusively by chance' but whose

---

[12] Forman, 'Weimar Culture' (ref. 2), 67.

[13] Quoted from John Blackmore, R. Itagaki and S. Tanaka, ed., *Ernst Mach's Vienna 1895–1930: Or Phenomenalism as Philosophy of Science* (Dordrecht: Kluwer, 2001), on 63.

[14] Ludwig Boltzmann, 'Über die sogenannte H-Kurve', *Mathematische Annalen* 50 (1898): 325–32.

[15] Gustav Theodor Fechner, *Kollektivmaßlehre*, ed. Gottlob Friedrich Lipps (Leipzig: W. Engelmann, 1897). For a scholarly analysis, see Michael Heidelberger, *Nature from Within: Gustav Theodor Fechner's Psychophysical Worldview* (Pittsburgh, PA: University of Pittsburgh Press, 2004).

probability is so high 'that it equals certainty for human conceptions', while (ii) in the domain of the humanities and the descriptive sciences 'the random single events succeed one another too slowly [such that] there can be no talk about a law'.[16] Still, or so he would claim in an unpublished manuscript, even the evolution of culture was shaped by the second law of thermodynamics by virtue of which culture and science necessarily advance and spread despite the death of the individual cultural organism.[17] Thus read Exner's own resolution of the problems raised by Spengler's cultural morphology. A first reaction to Spengler can already be found in the preface to the second edition of *Lectures on the Physical Foundations of Natural Science*.[18]

Let me turn to the opposite side in the first phase of the causality debate. Planck, in his famous Leiden lecture of 1908 that had started the polemics with Mach, vigorously defended what he took to be Boltzmann's legacy against Mach's anti-realism. In 1914, he now felt obliged to save Boltzmann's legacy from Exner's over-interpretation. Planck stressed 'the fundamental importance of performing an exact and fundamental separation between ... the *dynamical*, strictly causal, and the merely *statistical* type of lawfulness for understanding the essence of all scientific knowledge'.[19] This distinction finds its expression in the sharp contrast between reversible processes subsumed under a dynamical law and irreversible processes governed by the second law of thermodynamics. 'This dualism ... to some may appear unsatisfactory, and one has already attempted to remove it — as it does not work out otherwise — by denying absolute certainty and impossibility at all and admitting only higher or lower degrees of probability .... But such a view should very soon turn out to be a fatal and shortsighted mistake'.[20] This was an obvious allusion to Exner, who responded to Planck in a separate chapter of his 1919 *Lectures on the Physical Foundations of Natural Science*.[21]

If we look at Planck's argument, we come to realise that the principal difference between what I call Vienna Indeterminism and the Berlin reading of Boltzmann resided in the relationship between causality and physical ontology.

---

[16] Exner, *Über Gesetze* (ref. 6), 13, 16, 14. (My translation.)
[17] Franz S. Exner, *Vom Chaos zur Gegenwart*, 1923, unpublished mimeographed typescript, Vienna University Library and Austrian Central Library of Physics, Vienna.
[18] Franz S. Exner, *Vorlesungen über die physikalischen Grundlagen der Naturwissenschaften*, 2nd edn. (Wien-Leipzig: Franz Deuticke, 1922). A detailed analysis of this response can be found in Erwin N. Hiebert, 'Common Frontiers of the Exact Sciences and the Humanities', *Physics in Perspective* 2 (2000): 6–29. For an analysis of Exner's manuscript, see Michael Stöltzner, 'Franz Serafin Exner's Indeterminist Theory of Culture', *Physics in Perspective* 4 (2002): 267–319.
[19] Planck, *Dynamische* (ref. 7), 10. (My translation.)
[20] Ibid., 23.
[21] Exner, *Vorlesungen* (ref. 18), lecture 94.

Either (a) one followed — as did the Berliners — Kant by claiming that to stand in a causal relationship was a condition of the possibility of the reality of a physical object (Kant called this 'empirical realism'), or (b) one agreed with Mach that causality consisted in functional dependencies between the determining elements and that physical ontology was about 'facts' (*Tatsachen*) consisting in stable complexes of such dependencies. To those standing in the Kantian tradition, the latter stance fell short of the aims of scientific inquiry. Those standing in the Hume-Mach tradition, however, had more leeway in searching for an ontology suitable for a new scientific theory. Notice this difference in ontology extends across a longer historical time span than the debate I am focusing on — it reached back to Mach's works of the 1870s and 1880s, and ended only when philosophers of science abandoned the ideal of descriptivism after 1945 and started once again to seek explanations.

Based upon this basic distinction between two notions of causality, Vienna Indeterminism — as touted by Exner in 1908 — can be characterised by the following three tenets: (1) The highly *improbable events* admitted by Boltzmann's statistical derivation of the second law of thermodynamics exist. (2) In a radically empiricist perspective, the burden of proof rests with the determinist, who must provide a sufficiently specific theory of microphenomena before claiming victory over a merely statistical theory. Even worse, assuming a deterministic microtheory without cogent reasons would lead to a 'duplication of natural law [that] closely resembles the animistic duplication of natural *objects*' — as Schrödinger put it in his 1922 Zurich inaugural address.[22] (3) The only way to arrive at an empirical notion of objective probability is by way of the limit of relative frequencies. It is meaningful to assume the existence of statistical collectives (*Kollektivgegenstände*) and relate them to experience even though the limit is only obtained for infinitely large collectives. In 1912 and 1919, von Mises provided a rigorous mathematical framework for the relative frequency interpretation.[23]

---

[22] Erwin Schrödinger, 'Was ist ein Naturgesetz?' *Die Naturwissenschaften* 17 (1929): 9–11, on 11. (My translation.) Schrödinger's 1922 address was not published until 1929. In a short preface, he relates: 'The subsequent rise and development of quantum mechanics has brought Exner's sphere of ideas into the focus of scientific interest, by the way, without his name ever being mentioned.' (p. 9) And thus he describes Exner's teaching: 'Within the past four or five decades physical research has clearly and definitely shown that *chance* is the common root of all the strict regularity that has been observed, at least in the overwhelming majority of natural processes, the regularity and invariability of which have led to the establishment of the postulate of universal causality'. (Ibid.)

[23] Richard von Mises, 'Über die Grundbegriffe der Kollektivmaßlehre', *Jahresbericht der Deutschen Mathematiker-Vereinigung* 21 (1912): 9–20; and Richard von Mises, 'Grundlagen der Wahrscheinlichkeitsrechnung', *Mathematische Zeitschrift* 5 (1919): 52–99.

Let me add that one has to distinguish two approaches to the issue of realism within the younger generation of Vienna Indeterminists. While Frank and von Mises elaborated the conventionalist picture and took theories as purely symbolic entities coordinated to experience, Schrödinger never abandoned Boltzmann's conception of theories as universal pictures and, accordingly, considered it a major shortcoming of quantum mechanics in the Copenhagen interpretation that the domain of validity of some of its basic concepts was inherently restricted.

In 1914, Planck rejected all three tenets of Vienna Indeterminism, but in the 1920s he reconciled himself with the highly improbable events (1). But even after the advent of quantum mechanics, Planck still cherished the hope for a deterministic reformulation of atomic physics. His former student Schlick gradually approached Vienna Indeterminism as far as the burden of proof (2) was concerned, but Schlick never accepted the relative frequency interpretation (3). In 1925, at a time when the Vienna Circle had begun to assemble around him, Schlick still held that 'only in the utmost case of emergency will the scientist or philosopher decide to postulate purely statistical micro-laws, since the scope of such an assumption would be enormous: The principle of causality would be abandoned, ... and hence the possibility of exhaustive knowledge would have to be renounced.'[24] After the 'emergency' had occurred in the form of quantum mechanics, Schlick presented an entirely new theory of causality in which the verificationist criterion of meaning blocked the assumption of a microworld that was deterministic but unobservable in principle.[25] But Schlick still insisted on separating all statistical regularity (*Gesetzmäßigkeit*) into strict law and pure randomness, such that there were strictly speaking no 'statistical laws' — a thesis that surprised physicists as diverse as Einstein and Heisenberg. The reason for Schlick's insistence was that, instead of subscribing to the relative frequency interpretation, as almost all physicists had done, he remained committed to Johannes von Kries's *Spielraumtheorie* of probability, in which objective randomness is integrated into a deterministic Kantian universe.[26]

While Schlick had to openly revoke his earlier theory of causality in the face of quantum mechanics, the Vienna Indeterminists could feel themselves

---

[24] Moritz Schlick, 'Naturphilosophie', in *Lehrbuch der Philosophie: Die Philosophie in ihren Einzelgebieten*, ed. Max Dessoir (Berlin: Ullstein, 1925), 397–492, on 461; the English translation appears in Henk Mulder and Barbara F.B. van de Velde-Schlick, ed., *Philosophical Papers*, vol. 2 (Dordrecht: Reidel, 1979), 1–90, on 61.

[25] Moritz Schlick, 'Die Kausalität in der gegenwärtigen Physik', *Die Naturwissenschaften* 19 (1931): 145–62; the English translation appears in *Philosophical Papers*, vol. 2 (ref. 24), 176–209.

[26] Johannes von Kries, *Prinzipien der Wahrscheinlichkeitsrechnung* (Freiburg im Breisgau: Mohr, 1886), 2nd ed. 1927.

confirmed. It is important for the sociological coherence of the latter tradition that they typically combined such a declaration with an explicit reference to Exner's being the first one to have contemplated the probabilistic nature of natural laws, while Schlick held that Exner's works contained nothing beyond the traditional philosophical criticism against determinism.[27]

This confrontation provides a framework in which other alleged 'converts to acausality' can be integrated. Here are just two examples. Walter Nernst, for one, had not forgotten the days he had worked with Boltzmann in Graz. 'Among all laws [of physics] the thermodynamical ones occupy a distinctive position because unlike all others they are not just of a special kind, but applicable to any process one can imagine'.[28] In the same vein, Exner had, in 1908, argued the second law of thermodynamics is the basic principle of nature. If one related all physical laws to the second law, so Nernst continued, this would not reduce their significance or rank; 'it would however put an end to the logical overuse of the laws of nature'.[29]

In a review of Exner's *Lectures,* Hans Reichenbach endorsed 'that Exner unequivocally advocates the objective meaning of the probabilistic laws in which he rightly conceives a very general regularity of nature'.[30] As did Exner, Reichenbach held that the basic laws of nature were of a statistical kind. But he did so for reasons that contradicted the radical empiricism of the Vienna Indeterminists. In his PhD dissertation of 1915 and in a series of papers ensuing from it in the early 1920s, he argued the principle of causality must be supplemented with a principle of lawful distribution (later called the principle of the probability function) that guarantees future empirical findings do not constantly change the form of the law. Initially Reichenbach considered both principles as synthetic *a priori*. After 1923 he allowed the principle of causality could be empirically false and developed the second principle into a probability logic that represented a precondition of any empirical science. Due to the unavoidability of measurement errors the connection between our experiences and probability theory was of a more basic kind than the one between any other theory and our experiences. This thesis of Reichenbach became a source of conflict with Frank and von Mises.[31]

---

[27] *Cf.* Schrödinger, 'Naturgesetz' (ref. 22), preface; and Frank, *Kausalgesetz* (ref. 3).
[28] Walter Nernst, 'Zum Gültigkeitsbereich der Naturgesetze', *Die Naturwissenschaften* 10 (1922): 489–95, on 492. (My translation.)
[29] Ibid., 493.
[30] Hans Reichenbach, review of *Vorlesungen über die physikalischen Grundlagen der Naturwissenschaften*, by Franz Exner, *Die Naturwissenschaften* 9 (1921), 414–15, on 415. (My translation.)
[31] Hans Reichenbach, 'Kausalität und Wahrscheinlichkeit', *Erkenntnis* 1 (1930): 158–88; partially translated into English in: Reichenbach, *Selected Writings*, vol. 2 (ref. 3), 333–44. For further details, see Michael Stöltzner, 'The Logical Empiricists', in *The Oxford Handbook of Causation*, ed. Helen Beebee, Christopher Hitchcock and Peter Menzies (Oxford: Oxford University Press, 2009), 108–27.

I am afraid I must leave this discussion, admittedly brief, about the continuity of the causality debate across the breakdown of the two empires in 1918 and the quantum revolution in 1926. Let me just add that the problem of causality remained a subject of philosophical dispute even at a time when most physicists had accepted quantum mechanics and Born's statistical interpretation of the wave function, and it did so even within the narrower circle of the Logical Empiricists.

The Physicist-Philosophers and Their Main Forum

Let me now turn to the sociological context of the causality debates. I have already emphasised the importance of the role-model of physicist-philosopher to my argument. Now I describe how a physicist laid claim to this status, even though he might have remarked to his colleagues, cautiously or with a wink, that this represented only his Sunday activity.

Forman was right to assume 'that institutions of German academic life provided frequent occasions for addresses before university convocations', and that this indicates 'the extraordinarily heavy social pressure which the German academic environment could and did exert upon the individual scholar or scientist placed within it'.[32] Both in the Wilhelmine Empire and the Weimar Republic, the main duty of a physicist elected rector, dean of the *Philosophische Fakultät* or secretary of an academy was to build a bridge to his colleagues from the humanities. They demanded something more profound than just popularising the most recent scientific achievements and emphasising their importance to technology and the state.

Academic customs thus set the stage for the physicist-philosophers. The most influential role models were Hermann von Helmholtz and Ernst Mach, who led the way out of the older *Naturphilosophie* and toward a new style of discussing the foundations of science as a philosophical problem. For this reason, many physicist-philosophers following in their footsteps remained critical of excessive speculation and eschewed entering into popular discourse. Quite a few of them, accordingly, regarded Wilhelm Ostwald's monistic sermons and Ernst Haeckel's writings with suspicion.

The publication of the academic addresses followed a typical pattern. Initially, they came out as separate booklets and were republished in one or two journals of the learned societies. At a certain point, a physicist would then assemble a collection of those academic speeches into a book that bore a title such as 'Popular

---

[32] Forman, 'Weimar Culture' (ref. 2), 6–7.

Writings' or 'Physical Panoramas'.³³ The publication of such a book testified to the author's new status as a physicist-philosopher. There were also a few journals that combined publications of scientists who took a philosophical approach with papers penned by guilded philosophers positively disposed toward the sciences, most prominently among them the *Vierteljahrsschrift für wissenschaftliche Philosophie und Soziologie* and Ostwald's *Annalen der Naturphilosophie*. These two journals ceased publication in 1916 and 1921 respectively.

In 1913, the media landscape for the physicist-philosopher underwent a significant change when a large percentage of the above-mentioned academic addresses were published by the newly-founded *Die Naturwissenschaften*. Modelled after the British *Nature,* the 'scientific weekly for the progresses of science, medicine, and technology' — thus went *Die Naturwissenschaften*'s subtitle — strove to 'follow the major developments within the whole of natural science and present them in a generally comprehensible and captivating form'.³⁴ The journal not only emphasised the unity of the sciences in a time of rapidly progressing specialisation, but also — and this was among its most distinctive features as compared to *Nature* — the philosophical and cultural context of the sciences. To a large extent, this orientation was the product of the singular nature of Arnold Berliner, who ran the journal from 1913 until he was forced out in 1935 under the Nuremberg laws. Berliner was both one of the early technical physicists, having worked as a factory director, and a '*Kulturmensch*', who venerated Goethe and was among Gustav Mahler's closest friends. From the recollections of Wilhelm Westphal, we learn that to the younger Berlin physicists, Berliner was an intellectual father figure not unlike Exner's place of honour within his circle.³⁵

The impressive number of philosophical articles solicited by Berliner can be divided into two groups. On the one hand, he ran a kind of education program by publishing survey papers on Goethe, Kant, Schopenhauer and the like. On the other, he published papers in which scientists reflected on the conceptual and philosophical foundations of their most recent achievements, among them most of the papers weighing in on the causality debate and many papers and reviews penned by Logical Empiricists. Conversely, until 1930 those Logical Empiricists who had a science background published roughly a third of their papers in *Die Naturwissenschaften*.³⁶

---

³³ Ludwig Boltzmann, *Populäre Schriften* (Leipzig: J.A. Barth, 1905); Max Planck, *Physikalische Rundblicke* (Leipzig: J.A. Barth, 1922).
³⁴ 'Zur Einführung', *Die Naturwissenschaften* 1, no. 1 (1913): 1. (My translation.)
³⁵ Wilhelm Westphal, 'Arnold Berliner zum Gedächtnis', *Physikalische Blätter* 8 (1952): 121; *see also* Max von Laue, 'Arnold Berliner (26.12.1862–22.3.1942)', *Die Naturwissenschaften* 33 (1946): 257–58.
³⁶ For more detailed figures, see Stöltzner, 'Kausalität' (ref. 1).

This proves that, among the readers of this journal, neopositivism was by far less fringy than Forman assumed.

Berliner's journal unequivocally took sides in two debates of great importance to science as a whole in the early years of the Weimar Republic. First, in the struggles about relativity theory that had intensified after 1918, *Die Naturwissenschaften* became an important fixture in the 'defense belt' (*Verteidigergürtel*) around Einstein.[37] This debate shaped the philosophical understanding of modern physics, be it relativistic or quantum, and prompted physicist-philosophers to take a stand against public dismissals of the unintuitive new physics.

Second, Berliner's journal devoted two papers to a severe criticism of Spengler's views on biology and physics. One of the critics, the applied mathematician Paul Riebesell, had already participated in the relativity debates.[38] As did the Vienna Indeterminists, Riebesell accepted statistical laws as genuine laws. This even permitted him to turn the tables on Spengler. 'Science — not the philosophy of nature — will now as before stick to the principle of causality and will approach precisely Spengler's problem of the predetermination of history with its new methods. For, by means of statistical laws — which Spengler incidentally does not recognize as mathematical laws — one has already successfully analyzed those mass phenomena, which historical questions are all about'.[39] Thus Riebesell drew terminological consequences from the present state of physics and the social sciences, but not in an act of adaptation. In a certain sense, the break between Spengler's cultural morphology and the quantitative social science mentioned by Riebesell was even more radical than the one between Spengler's concept of causality and statistical mechanics that represented the basis of Exner's criticism.

In evaluating the sociological impact of this confutation of Spengler for the German scientific community, we have to consider that *Die Naturwissenschaften* was closely connected to two leading research organisations of the German-speaking world, the *Deutsche Gesellschaft der Naturforscher und Ärzte* and the *Kaiser-Wilhelm-Gesellschaft*. By virtue of this authoritative character, I have considered *Die Naturwissenschaften* as a submilieu that provided scientists with a pro-scientific cultural identity more specific than just being *Bildungsbürger*, such

---

[37] Klaus Hentschel, *Interpretationen und Fehlinterpretationen der speziellen und der allgemeinen Relativitätstheorie durch Zeitgenossen Albert Einsteins* (Basel: Birkhäuser, 1990).

[38] Paul Riebesell, 'Die Beweise für die Relativitätstheorie', *Die Naturwissenschaften* 4 (1916): 98–101. *See also* Paul Riebesell, 'Die neueren Ergebnisse der theoretischen Physik und ihre Beziehungen zur Mathematik', *Die Naturwissenschaften* 6 (1918): 61–65.

[39] Paul Riebesell, 'Die Mathematik und die Naturwissenschaften in Spenglers "Untergang des Abendlandes"', *Die Naturwissenschaften* 8 (1920): 507–09, on 508. (My translation.)

that they did not face the general milieu directly without a stabilising group identity.[40] Today, however, I believe that the concept of milieu — even in the operationalist sense of Fleck — is too unspecific and ultimately forces us to accept the alternative that Forman posed at the beginning of his study, to wit, retrenchment versus adaptation.[41] To my taste, all this sounds too passive for physicist-philosophers. Moreover, after 1900 there no longer existed a homogeneous intellectual milieu that could, as a whole, change under the influence of the lost war. As I will outline in the final section, one has to take a multi-layered approach instead. Thus I would now say that *Die Naturwissenschaften* simply provided a forum for those who endorsed scientific modernism, which could mean different things in different disciplines, and considered science an integral part of the general culture, toward which scientists held different attitudes. Much of this orientation of *Die Naturwissenschaften*, especially in the fields of physics and philosophy, was due to the unique position of Arnold Berliner within the Berlin scientific elite.

A few notable exceptions notwithstanding, independence from ruling philosophical schools was a major characteristic of physicist-philosophers. This permitted them to form strategic alliances. Let me first provide an abstract characterisation of this notion. A strategic alliance is formed if there is a set of basic philosophical convictions that a group of physicist-philosophers considers as central in order to further their philosophical agenda within a particular intellectual, social or disciplinary context. In this case, they confine their disagreements to internal discussions, even though in retrospect these may appear substantial. Strategic alliances dissolve and their members regroup, as the convictions considered pivotal within the respective context undergo changes. It is important to stress the philosophical ambitions of a physicist-philosopher are not limited to forming a given strategic alliance. Or to cast it into Fleckian terms, the members of a strategic alliance are typically members of different thought collectives.

It seems to me that both Logical Empiricism, at least initially, and the Göttingen-Copenhagen interpretation of quantum mechanics represented strategic alliances of this kind. When the Logical Empiricists joined up with the latter, the tradition of Vienna Indeterminism ended in the mid 1930s because this move alienated Schrödinger from Frank and von Mises. The reasons were at least twofold. For one, the Logical Empiricists decided to combat the 'metaphysical misinterpretations of quantum theory' by developing an empiricist reading of Bohr's complementarity. To do so they invoked a verificationist criterion of meaning that, in its language-oriented version, was unacceptable to Schrödinger, who sought

---

[40] Stöltzner, 'Kausalität' (ref. 1); Stöltzner, *Vienna Indeterminism* (ref. 1).
[41] Fleck, *Entstehung* (ref. 8).

universally valid pictures rather than concepts with a limited domain of applicability. More generally, after the EPR paper of 1935 and Schrödinger's cat paper, the discussions about the interpretation of quantum mechanics shifted from causality and indeterminism to questions of reality in the atomic domain.[42]

But there was also a sociological element. By organising specific meetings, through the foundation of their own journal *Erkenntnis*, during their search for international allies and by way of a debate on whether their distinctive method consisted of the logic of science (Carnap), scientific philosophy (Reichenbach), or encyclopedism (Neurath), Logical Empiricists after 1930 accomplished the basic steps in establishing a new scientific discipline. Through this process of discipline formation, the role model of physicist-philosopher, albeit still extant, lost its pervasiveness in the German-speaking world even before many physicist-philosophers were forced into emigration after 1933. The further course of the new discipline 'philosophy of science' was not to be without implications for how the Forman thesis was cast.

Contextualising the Forman Thesis

Here I first want to show that the Forman thesis is a child of its era in more ways than one; that is, not solely by its setting out to demonstrate a strong, or even causal, influence of social factors on the conceptual structure of empirical science — a perspective that would prove most influential during the 1970s and beyond. Forman's treatment of the relationship between physical theory and cultural milieu was also deeply informed by what was common to both Rudolf Carnap's philosophy of science and Thomas S. Kuhn's revolutionist perspective on the historiography of science. As did most members of their respective disciplines, both focused on theory and in doing so treated a scientific theory as a single and largely homogeneous entity. This, so I claim, prevented Forman and others from conceiving the causality debate in its larger historical context and its proper discursive mode, and led them to disregard the philosophical continuities beyond how the protagonists of his study, after 1918, used philosophical keywords such as 'causality', 'crisis' and 'Spengler'.

In the famous debates with W.V. Quine, Carnap argued that questions about the existence of scientific objects were only meaningful once a linguistic

---

[42] Philipp Frank, 'Philosophische Deutungen und Mißdeutungen der Quantentheorie', *Erkenntnis* 6 (1937): 303–17; Moritz Schlick, 'Quantentheorie und Erkennbarkeit der Natur', *Erkenntnis* 6 (1937): 317–26; Erwin Schrödinger, 'Die gegenwärtige Situation in der Quantenmechanik', *Die Naturwissenschaften* 23 (1935): 807–12, 823–28, 844–49; Albert Einstein, Brian Podolsky and Nathan Rosen, 'Can Quantum-Mechanical Description of Physical Reality Be Considered Complete?' *Physical Review* 47 (1935): 777–80.

framework had been specified.⁴³ While Carnap continued to hold that the choice of a framework was guided by pragmatic concerns, Kuhn's *Structure of Scientific Revolutions* pointed out that at certain moments in history, scientific revolutions overthrew an old framework, as well as the paradigm formulated within it, and instituted a new one.⁴⁴ The main point of Kuhn's argument, at least in its original form, was that the old and new frameworks were incommensurable, such that there was no rational bridge from one to the other. Kuhn's book has often been understood as the final blow to a philosophy of science in the Vienna Circle tradition even though it received Carnap's endorsement. For both agreed that a transition between two linguistic frameworks could only be rationally justified if there existed a suitable meta-framework in which both paradigms could be formulated.⁴⁵ The main difference between the men was that Carnap denied strong versions of the idea that observation is essentially theory-laden, such that brute facts always provided a bridge between two paradigms even if there was no theoretical bridge between old and new. Both their consensus and disagreement hence concerned the relationship between one or two theories and (unstructured or theory-laden) empirical data.

The subsequent debate among philosophers centred around whether the history of science could be rationally reconstructed — as Lakatos and Popper held — or whether it was essentially contingent — as Feyerabend came to radicalise Kuhn's analysis. This debate was still in its early phase when Forman's paper came out. Seen from the perspective of the frontlines of the Lakatos-Kuhn-Popper-Feyerabend debates of the 1970s, we find an interesting ambiguity in Forman's thesis. On the one hand, by claiming a causal influence of the postwar milieu he argued in favour of historical contingency in the style of Feyerabend. On the other hand, Forman held that after 1927 there was sufficient reason to abandon causality and thus assumed, in contrast, the rationality of scientific development. The second point shows that Forman was not a social constructivist.⁴⁶

---

⁴³ Rudolf Carnap, 'Empiricism, Semantics, and Ontology', *Revue internationale de philosophie* 4 (1950): 20–40.

⁴⁴ Thomas S. Kuhn, *The Structure of Scientific Revolutions* (Chicago: University of Chicago Press, 1962).

⁴⁵ *Cf.* Thomas S. Kuhn, 'What are Scientific Revolutions?' *The Road Since Structure: Philosophical Essays, 1970–1993* (Chicago: The University of Chicago Press, 2002), 14n2.

⁴⁶ Still, in a later paper he tracked the anti-causal rhetoric beyond 1927 and supplemented his considerations with two other keywords of the Weimar 'milieu'. *Cf.* Paul Forman, '*Kausalität, Anschaulichkeit*, and *Individualität*, or How Cultural Values Prescribed the Character and the Lessons Ascribed to Quantum Mechanics', in *Society and Knowledge: Contemporary Perspectives in the Sociology of Knowledge and Science*, ed. Nico Stehr and Volker Meja (New Brunswick, NJ: Transaction Books, 1984), 333–47.

In her *Quantum Dialogue,* Mara Beller has made the case that philosophical orientations of quantum physicists were 'local and provisional'. Neither they nor the mileu could exert a deeper, let alone a causal, influence.[47] In her account, it was mainly Bohr's power politics that changed a fruitful continuous dialogue into a revolutionary narrative. In the form of de Broglie's pilot wave theory, she stressed, the possibility of a causal quantum mechanics had always existed. I doubt the radical juxtaposition between dialogical emergence and rhetorical consolidation — the two parts of her book — does justice to the rather stable philosophical convictions of the protagonists of her narrative because it downgrades them to justificatory rhetoric.

Again motivated by the fact that de Broglie-Bohm theory has the same empirical content as the dominating Copenhagen interpretation, James T. Cushing has rightly interpreted quantum mechanics as a case of Duhemian underdetermination.[48] But he additionally construed a counterfactual history showing how the causal picture could have prevailed, thus filing an equal rights claim for alternative interpretations of quantum mechanics. The move was justly criticised by Beller and Forman among others.[49] Once again, we find the above-mentioned confrontation between one or two theories and empirical data. No wonder Cushing stressed that philosophical motives were of little importance, apart perhaps from positivism's role in justifying the Copenhagen dogma. Both Beller and Cushing identify philosophy with academic philosophy and thus miss the peculiar role of physicist-philosopher that lies at the heart of my reconstruction of the causality debate.

It seems to me, in contrast to Forman, Beller and Cushing, that the causality debates among German physicist-philosophers can only be assessed by assuming a multi-layered structure of beliefs and attitudes encompassing general philosophical principles, mathematical formalisms, specific theories, personal research agendas and cultural self-identities that evolved and changed on different time scales. Some of these were instantiated by Fleckian thought collectives, while others are situated on the timescale of *longue durée*. Let me integrate my narrative into this picture.[50]

---

[47] Mara Beller, *Quantum Dialogue: The Making of a Revolution* (Chicago: University of Chicago Press, 1999).

[48] James T. Cushing, *Quantum Mechanics: Historical Contingency and the Copenhagen Hegemony* (Chicago: University of Chicago Press, 1994).

[49] Beller, *Quantum Dialogue* (ref. 47); Paul Forman, review of *Quantum Mechanics: Historical Contingency and the Copenhagen Hegemony*, by James T. Cushing, *Science* 267, no. 5205 (1995): 1844.

[50] *Cf.* Michael Stöltzner, 'Gangarten des Rationalen: Zu den Zeitstrukturen der Quantenrevolution', *Berichte zur Wissenschaftsgeschichte* 32 (2009): 176–92.

I have claimed that the causality debate extended across roughly three decades, from Exner's 1908 inaugural address until the discussions ensuing from the 1935 EPR paper. There was one thought collective — Vienna Indeterminism — whose members were not forced to change their philosophical principles: not in 1918, when a deep political crisis began; nor in the face of the growing problems in atomic physics that started around 1920; nor after the recognition of the strange features of the new quantum theory that Born's interpretation of the wave function brought to light. In Schlick's case we have seen that for those who, unlike Schrödinger, fully endorsed the new quantum mechanics, the philosophical reorientation was eased by the fact that indeterminism had already been a widely discussed option. If we look at physicist-philosophers such as Frank, von Mises and Schlick, we see the orientation at a Machian or Kantian conception of causality, which stretches over an even longer period, proved to be an enormously stable philosophical disposition. Schlick, in particular, needed much longer to abandon the Kantian category of causality than to abandon the Kantian conceptions of time and space in the context of relativity theory.

I have remained largely silent about the role of mathematics within my narrative, now taken as one level within the aforementioned multi-layered structure. Doing so demonstrates that I am not writing the kind of 'winners' history' rightly criticised by Beller. Despite the important role of probability for the causality debate, the crucial breakthrough in the field came only in 1932 with Kolmogorov's axiomatisation that, by virtue of its abstract nature, avoided the problems that plagued Fechner's and von Mises's statistical collectives.[51] This achievement, however, would subsequently only make clear that the problem of quantum probabilities was not fully resolved and a full-blown frequentism as advocated by the Vienna Indeterminists does not do the job.

It remains to be seen in further historical examples whether such higher-level principles, philosophical and mathematical ones, may exert such a strong force on concept formation that they can be considered historically relativised constitutive *a priori* — as Michael Friedman has suggested in order to save the Kuhnian insight from social constructivism — or whether they only mediate across fractures in the conceptual development that occur on another level.[52] The latter claim, it seems to me, can be reconciled with Fleck's insight that a single scientist may simultaneously belong to different thought collectives, which may be located on various levels in Friedman's sense. Such was the tack taken in the present analysis

---

[51] See Thomas Hochkirchen, *Die Axiomatisierung der Wahrscheinlichkeitsrechnung und ihre Kontexte: von Hilberts sechstem Problem zu Kolmogoroffs Grundbegriffen* (Göttingen: Vandenhoeck & Ruprecht, 1999).

[52] Michael Friedman, *Dynamics of Reason* (Stanford, CA: CSLI Publications, 2001).

because it permitted me to incorporate other thinkers, such as Reichenbach and Nernst, who were neither part of Vienna Indeterminism nor shared Planck's insistence on determinism or Schlick's separation between lawfulness and randomness.

Finally, it is important to note that the existence of local traditions, such as Vienna Indeterminism, does not contradict the historical integrity of the causality debate described in the present paper insofar as this took place within a single but multifarious German-speaking scientific culture. For this reason, I am happy to observe that Forman no longer considers the Vienna tradition as just a 'subterranean anticausality current'[53] but rather as part of a broader Austro-Hungarian tradition in which positivist tendencies were pivotal and had roots extending back long before 1918.[54] I hope to have also shown that the stance of the Weimar participants in the debate, both thematically and socially, was not primarily a product of the fall of the Wilhelmine Empire but shaped by earlier philosophical commitments defined in a struggle over the philosophical interpretation of Boltzmann's statistical mechanics.

Acknowledgments

I am greatly indebted to Veronika Hofer for many discussions about historical methodology, including intellectual milieus and Fleckian thought collectives, and a critical reading of the present paper. This paper has emerged from a talk at the conference 'The Cultural Alchemy of the Exact Sciences: Revisiting the Forman Thesis' held at the University of British Columbia, Vancouver, BC, Canada, in March 2007 and was revised for presentation at the HQ1 conference at the Max Planck Institute for the History of Science in Berlin. I thank Paul Forman for his critical comments on an earlier version of this paper; also the organisers and participants of both conferences for their very helpful questions and remarks — among them my Vancouver commentator Alan Richardson — and Cathryn Carson and Lindsay Crawford for manifold suggestions on how to improve the clarity of my writing.

---

[53] Forman, 'Weimar Culture' (ref. 2), 67.
[54] Paul Forman, 'Die Naturforscherversammlung in Nauheim im September 1920', in *Physiker zwischen Autonomie und Anpassung: Die Deutsche Physikalische Gesellschaft im Dritten Reich*, ed. Dieter Hoffmann and Mark Walker (Weinheim: VCH-Wiley, 2007), 29–58, on 40.

Modern or Anti-modern Science? Weimar Culture, Natural Science and the Heidegger-Heisenberg Exchange

*Cathryn Carson**

Weimar Culture and Scientific Rationality

Paul Forman's 'Weimar Culture' is a *generative* work: each time I go back to it, I find something new. It has accompanied my own thinking ever since I first read it, weaving in and out of my work in patterns of which I cannot pretend even to have always been aware. This essay takes off from one line of argument in 'Weimar Culture', following it in a direction marked out between intellectual history and the history of the philosophy of science. While Forman's work treats of physicists retailing philosophical themes, the traffic in the opposite direction holds its own interest, particularly when it involves a philosopher not commonly thought to have cared much for science.

In Forman's essay, Weimar's perceived crisis of science (*Krisis der Wissenschaft*) signals above all a tide of anti-rationalism and hostility to science: the antagonism of the era's tone-setting intellectuals, as the author cogently puts it, 'toward analytical rationality generally and toward the exact sciences and their technical applications particularly'.[1] Counterpoised to scientific rationality in Forman's account, and in every sense its antagonist, stands a widespread

---

* Office for History of Science and Technology, 543 Stephens Hall, University of California, Berkeley, CA 94720-2550; clcarson@berkeley.edu.
    The following abbreviation is used: *GA*, Martin Heidegger, *Gesamtausgabe*.

[1] Paul Forman, 'Weimar Culture, Causality, and Quantum Theory, 1918–1927: Adaptation by German Physicists and Mathematicians to a Hostile Intellectual Environment', *Historical Studies in the Physical Sciences* 3 (1971): 1–115, on 4.

allegiance to *Lebensphilosophie* ('life philosophy'). In the resonant phrasing of 'Weimar Culture', *Lebensphilosophie* amalgamated in one reactionary package the 'rejection of reason as an epistemological instrument because inseparable from positivism-mechanism-materialism, and because, as fundamentally disintegrative, incapable of satisfying the "hunger for wholeness"; glorification of "life," intuition, unmediated and unanalyzed experience, with the immediate apprehension of values, and not the dissection of causal nexus, as the proper object of scholarly or scientific activity'.[2] Ostwald Spengler's *Decline of the West*, historicising science and proclaiming the artificiality of causality, serves as a canonical text of the movement.

Yet Spengler is clearly one figure among many amid the multiple branching points in the reaction against modern scientific reason. Indeed, commenting on the contemporary fascination with Spengler's ideas, Theodor Adorno offered his own assessment: 'When the second volume of the *Decline* appeared, in 1922, its reception did not even remotely approach that of the first ... The laymen who had read Spengler as they had read Nietzsche and Schopenhauer before him had in the meantime become estranged from philosophy'. Adorno continued, more pointedly, 'The professional philosophers were soon to flock to Heidegger, whose work was to give their irritation more dignified and refined expression'.[3] When Adorno highlighted the philosopher Martin Heidegger, in fact, he was only speaking the obvious. Heidegger would become a chief spokesman of anti-modernist intellectuals and (Adorno would surely also have noted) one of the most visible and influential articulators of a critique of scientific reason as well.[4]

Heidegger's writings may have been hard to understand, but *Being and Time* (*Sein und Zeit*) of 1927 instantly captured the spotlight and was widely read in an existentialist vein. In Weimar he provoked a broad spectrum of reactions, which

---

[2] Ibid., 16. The *lebensphilosophisch* animus can be described as a reaction against modernity, even though in this study of Weimar that theme of Forman's later work does not yet play a large role. *Cf.* Paul Forman, 'The Primacy of Science in Modernity, of Technology in Postmodernity, and of Ideology in the History of Technology', *History and Technology* 23 (2007): 1–153; and Forman's introductory talk on 'Weimar Culture, Causality, and Modernity' at the conference 'The Cultural Alchemy of the Exact Sciences: Revisiting the Forman Thesis', University of British Columbia, Vancouver, 2007.

[3] Theodor W. Adorno, 'Spengler after the Decline', trans. Samuel M. Weber, in *Prisms* (Cambridge, MA: MIT Press, 1981), 50–72, on 53.

[4] *Cf.* Theodor W. Adorno, *The Jargon of Authenticity*, trans. Knut Tarnowski and Frederic Will (Evanston, IL: Northwestern University Press, 1973); Max Horkheimer and Theodor W. Adorno, *Dialectic of Enlightenment*, trans. John Cumming (New York: Continuum, 1999).

are captured especially trenchantly in the chapter on 'The Hunger for Wholeness' in Peter Gay's classic study:

> What Heidegger did was to give philosophical seriousness, professorial respectability, to the love affair with unreason and death that dominated so many Germans in this hard time. Thus Heidegger aroused in his readers obscure feelings of assent, of rightness; the technical meaning Heidegger gave his terms, and the abstract questions he was asking, disappeared before the resonances they awakened. Their general purport seemed plain enough: man is thrown into the world, lost and afraid; he must learn to face nothingness and death. Reason and intellect are hopelessly inadequate guides to the secret of being; had Heidegger not said that thinking is the mortal enemy of understanding? ... Whatever the precise philosophical import of *Sein und Zeit* and of the writings that surrounded it, Heidegger's work amounted to a denigration of Weimar, that creature of reason, and an exaltation of movements like that of the Nazis, who thought with their blood, worshipped the charismatic leader, praised and practiced murder, and hoped to stamp out reason — forever — in the drunken embrace of that life which is death.[5]

When Weimar's original intellectual historians were writing, the 'destruction of reason' motif was strong. Since then, academic assessments have largely backed away from the notion, and with it from Gay's sort of univocal account. While historians of science have been preoccupied with the complex affiliations of intellectual projects such as holism, cultural historians have highlighted the multi-valence and ambiguity of modernist and antimodernist movements at large.[6] Yet there has been considerably less effort, for good reason, to recapture any kind of 'modernist' Heidegger; Heidegger made it quite clear that he did not belong in that camp. For intellectual historians and historically minded philosophers, the

---

[5] Peter Gay, *Weimar Culture: The Outsider as Insider* (New York: Harper & Row, 1968), 82. In his own 'Weimar Culture', Forman notes Heidegger in passing: Forman, 'Weimar Culture', 34. For his sustained engagement see Forman, 'Primacy' (ref. 2), esp. 7–10.

[6] E.g., Jonathan Harwood, *Styles of Scientific Thought: The German Genetics Community 1900–1933* (Chicago: University of Chicago Press, 1993); Mitchell Ash, *Gestalt Psychology in German Culture, 1890-1967: Holism and the Quest for Objectivity* (Cambridge: Cambridge University Press, 1995); Anne Harrington, *Reenchanted Science: Holism in German Culture from Wilhelm II to Hitler* (Princeton: Princeton University Press, 1996), each taking note of Forman's work. For examples from cultural history, see Martin H. Geyer, *Verkehrte Welt: Revolution, Inflation und Moderne, München 1914–1924* (Göttingen: Vandenhoeck & Ruprecht, 1998); Suzanne Marchand and David Lindenfeld, eds., *Germany at the Fin de Siècle: Culture, Politics, and Ideas* (Baton Rouge: Louisiana State University Press, 2004). In 1971, along with Peter Gay, Forman called on Georg Lukács, Kurt Sontheimer and Fritz Ringer.

main question in this line has been what forward-looking notions can be taken from him at all. At the end of the day, his philosophical ideas seem unambiguously implicated in his engagement for National Socialism when that movement came to power in 1933.[7] Yet there is more going on with Heidegger than a philosophy of life, death and decision, even if that was what many contemporaries took from his work, and Heidegger's contentions about science grew out of more than sheer romantic resentment, even if his arguments sometimes seem forced.

One task of this essay, at least vis-à-vis the history of science, is to re-situate Heidegger's claims about science within contemporary debate. By this I mean not just the anti-modernist enthusiasms of best-selling authors like Spengler, but academic controversies about discipline and method in the natural sciences.[8] Much of the ground covered by the latter debates is familiar to intellectual historians, and I give only a summary treatment. All the same, this anchoring is critical if we are going to attempt to understand Heidegger's philosophy, rather than just writing it off as dark reactionary mumbling. I also suggest that Heidegger kept his eye on developments in contemporary science, and it was in part by way of these episodes that the particulars of his critique of science were concretely worked out. Starting from problems of temporality, expanding to the *Krisis der Wissenschaft*, and taking up particular issues in contemporary science, physics above all, we can follow his commentary on natural science from its late Wilhelmine origins into the Weimar Republic and through its radicalisation in the Third Reich.

In particular, we can track Heidegger's engagement with 'modern' physics and with the individual he came to take as its iconic stand-in. The Heidegger-Heisenberg exchange, for such it was, gives us insight into the presumptive modernity of modern physics in an era when the modern carried a real charge. At

---

[7] The more sophisticated entries in this debate include Hugo Ott, *Martin Heidegger: Unterwegs zu seiner Biographie* (Frankfurt: Campus, 1988); Michael E. Zimmerman, *Heidegger's Confrontation with Modernity: Technology, Politics, and Art* (Bloomington: Indiana University Press, 1990); Tom Rockmore and Joseph Margolis, eds., *The Heidegger Case: On Philosophy and Politics* (Philadelphia: Temple University Press, 1992); Hans Sluga, *Heidegger's Crisis: Philosophy and Politics in Nazi Germany* (Cambridge, MA: Harvard University Press, 1993); Julian Young, *Heidegger, Philosophy, Nazism* (Cambridge: Cambridge University Press, 1997); Richard Wolin, *Heidegger's Children: Hannah Arendt, Karl Löwith, Hans Jonas, and Herbert Marcuse* (Princeton: Princeton University Press, 2001); Charles Bambach, *Heidegger's Roots: Nietzsche, National Socialism, and the Greeks* (Ithaca, NY: Cornell University Press, 2003); Bernhard Radloff, *Heidegger and the Question of National Socialism: Disclosure and Gestalt* (Toronto. University of Toronto Press, 2007).

[8] I am not trying to draw a boundary around some 'pure' philosophical arena distinct from the culture at large. I am just working from the assumption that historians of science know a lot more about the culture at large than about academic philosophising, and this imbalance shades how they think about Heidegger.

a critical juncture, we see the two men sharing a set of preoccupations: with causality and objectivity, with prediction and control, with the possibility and limits of science. Both marked their distance from nineteenth-century science. Yet they diverged in their reaction: where Heidegger held out against modernity, Heisenberg stayed within its frame. Heisenberg's last-ditch modernism sheds light on what it was that Heidegger believed he was standing against.

Heidegger Read Historically

Heidegger's contemporaries read him in their own ways, for their own reasons. Philosophers today do as well. As a historian, I can only read him by following the conversation of his era, the one in which he was trained and then aimed to amend. There is a massive body of work to draw on. Alongside *Being and Time* and a few famous lectures that popular readers encountered, there is the entire series of course offerings and other materials filling out the projected 102-volume *Gesamtausgabe* edition. As much as anything, reading Heidegger's writings historically gives witness to how he was grounded in something more specific than Weimar's general cultural discussion, and that is the academic philosophy of his day. Heidegger was not just trained on Aristotle, Descartes or Kant as a kind of formal exercise. Rather, he was working out of the middle of traditions of reacting to them, even when non-philosophers rarely caught on.

Only a short-form narrative of Heidegger's philosophical affiliations can be given here, but a short-form version is useful nonetheless. By the mid-1910s, a young Martin Heidegger had emerged from his early studies of Catholic theology and scholastic philosophy. His early writing shows him working his way forward from an original positioning in a three-way force field among contemporary neo-Kantianism, phenomenology and *Lebensphilosophie*. The studies in Freiburg that led to his dissertation *The Doctrine of Judgment in Psychologism: A Critical-Positive Contribution to Logic* and his habilitation thesis on the category problem in Duns Scotus gave him significant exposure to neo-Kantianism. The neo-Kantian problematics of the constitution of the sciences' objects, the question of method in the natural and the historical or cultural sciences, and the relations between philosophy and the special sciences found some place in his early writing and lecturing. As he worked out his own relation to Heinrich Rickert, who directed his habilitation thesis, and Emil Lask, who inspired it in important ways, Heidegger simultaneously came to terms with Edmund Husserl's call to investigate how things appear in consciousness without presuppositions or interpretations. Husserlian phenomenology was a response to primacy claims by empirical psychology and an alternative to neo-Kantian critical idealism. Having been pulled in by Husserl's *Logical Investigations*, published at the turn of the century, Heidegger was now confronted with the *Ideas I*, which, moving in a transcendental direction, appeared as he was starting to define his philosophical stance.[9]

By the end of the war and the dawn of the Weimar period, Heidegger was staking out his own set of concerns. His lecturing in the early Weimar years — teaching first as Husserl's assistant in Freiburg, then moving to Marburg — was marked by a phenomenologist's interest in the structures of experience, albeit without much patience for the transcendental subject to which Husserl had turned. Likewise, it was invested in post-positivist problems of method and the self-positioning of philosophy but sceptical of neo-Kantianism's epistemological starting point and unpersuaded by Rickertian value-philosophy.[10] It was out of this dissatisfaction, running together with his own effort to clarify his personal commitments in philosophising, that we find the phenomenologist taking counsel from *Lebensphilosophie*. The key figure here for Heidegger was Wilhelm Dilthey: Dilthey, who was concerned with the grounding for the human sciences, apart from and co-equal to natural science; Dilthey, who suggested the Cartesian subject was hardly a realistic starting point for the scholarly project of understanding experience; Dilthey, who argued for bringing the temporality and historicity of lived experience back into the picture; Dilthey, whose hermeneutics stood in direct conversation with the neo-Kantians and Husserl.[11] Thus

---

[9] The dissertation was completed in 1913 and published in 1914; the habilitation, in 1915. A rich account of Heidegger's formation is Thomas Sheehan, 'Heidegger's *Lehrjahre*', in *The Collegium Phaenomenologicum: The First Ten Years*, ed. John C. Sallis, Giuseppina Moneta and Jacques Taminiaux (Dordrecht: Kluwer, 1988), 77–137. On the young Heidegger and neo-Kantianism, see Theodore J. Kisiel, 'Why Students of Heidegger Will Have to Read Emil Lask', in *Heidegger's Way of Thought: Critical and Interpretative Signposts* (London: Continuum, 2002), 101–136; and, more generally, Michael Friedman, *A Parting of the Ways: Carnap, Cassirer, and Heidegger* (Chicago: Open Court, 2000). The relation of Heidegger to Husserl is widely discussed; one place to start is Hubert L. Dreyfus, *Being-in-the-World: A Commentary on Heidegger's* Being and Time, *Division I* (Cambridge, MA: MIT Press, 1991). A good overall introduction to Husserl is Barry Smith and David Woodruff Smith, eds., *The Cambridge Companion to Husserl* (New York: Cambridge University Press, 1995).

[10] Many useful materials are collected in Theodore Kisiel and Thomas Sheehan, eds., *Becoming Heidegger: On the Trail of His Early Occasional Writings, 1910–1927* (Evanston, IL: Northwestern University Press, 2007); Theodore Kisiel, *The Genesis of Heidegger's* Being and Time (Berkeley: University of California Press, 1993) works through this period in detail.

[11] On Heidegger and Dilthey, see Thomas R. Wolf, *Hermeneutik und Technik: Martin Heideggers Auslegung des Lebens und der Wissenschaft als Antwort auf die Krise der Moderne* (Würzburg: Königshausen & Neumann, 2005), whose argument is noted in Forman, 'Primacy' (ref. 2), n. 38; István M. Fehér, 'Phenomenology, Hermeneutics, *Lebensphilosophie*: Heidegger's Confrontation with Husserl, Dilthey, and Jaspers', in *Reading Heidegger from the Start: Essays in His Earliest Thought*, ed. Theodore Kisiel and John van Buren (Albany: State University of New York Press, 1994), 73–89; Charles R. Bambach, *Heidegger, Dilthey, and the Crisis of Historicism* (Ithaca, NY: Cornell University Press, 1995); Jeffrey Andrew Barash, *Martin Heidegger and the Problem of Historical Meaning* (Dordrecht: Martinus Nijhoff, 1988). Also useful is Michael Großheim, *Von Georg Simmel zu Martin Heidegger: Philosophie zwischen Leben und Existenz* (Bonn: Bouvier, 1991).

more than Spengler or others in the *lebensphilosophisch* repertoire, it was Dilthey's work that articulated with Heidegger's. And for good reason, since Heidegger was working out of the same problem situation that Dilthey had confronted to start.[12]

In the forms in which Heidegger was drawing on it, *Lebensphilosophie* was not intrinsically distant from the preoccupations of neo-Kantianism and phenomenology. We can say, moreover, that *Lebensphilosophie* helped him bring into focus what he found wanting elsewhere. For Heidegger denied the move to make phenomenology a matter of transcendental consciousness. The later Husserl's recourse to that construct was incompatible with the original impulse of phenomenology, one whose radical reinstatement Heidegger saw himself demanding. This was to get back to the things themselves (*zu den Sachen selbst*) as they in fact showed themselves. The theoretical attitude of disengaged contemplation did not respect the reality of experience: Dilthey had made a similar point in another vein. And the same held of neo-Kantianism. Its epistemological orientation to subjectivity and its objects obscured something that Heidegger came to see as all-important — something that he, teaching on Aristotle no less than on Kant and on phenomenology, was coming to identify as ontology or the question of Being.

Putting his criticisms into concepts, defining his own stance to the problem, Heidegger worked his way in the 1920s from notions such as *Leben* toward *Faktizität* and *Existenz* and on to *Geworfenheit* and *Dasein*. When such terms appeared in *Being and Time* — and with them the horizon of human existence inserted into history, structured by temporality, and circumscribed by the boundaries of factual life — they resonated with Heidegger's situated experience, as well as that of large numbers of readers. At the same time they carried forward a demand to rework existing philosophical conceptions, motivated through recourse to a related intellectual tradition. The lectures and *Being and Time* show a stringency of argument that, even when it strains for solutions, is at odds with the looseness of much popular discussion.

There is *Lebensphilosophie*, and then there is *Lebensphilosophie*. Heidegger's work was received in the space between the two: between rigorous challenges regarding consciousness and method on the one hand, romantic effusions about 'life' on the other. Heidegger was anti-positivist and anti-dualist — for reasons of intellectual consistency, not just anti-scientific resentment. He held out a place for intuition — in the same sense as Husserl, of course. He put the spotlight on life,

---

[12] For Heidegger's reaction to Spengler, see Barash, *Martin Heidegger* (ref. 11), 146–60. Heidegger lectured on Spengler at a popular 'Wissenschaftliche Woche' in 1920, but the manuscript has not been found: Kisiel and Sheehan, *Becoming Heidegger* (ref. 10), lii.

but rejected later existentialist readings of his work as subjectivistic.[13] We would be hard-put to describe him as anti-intellectual, unless the fact that he argued for limits to theoretical thinking is supposed to mark him as such. If he was hard to read, in this period he was not trying to be obscure. His driving persistence, his depth and range, and his ability to push forward in several directions at once: these marked him out for admiration among colleagues, not just the starry-eyed youth.

Science and Crisis

Heidegger's philosophising responded to the moment in other ways, too. He commented in his lectures on the Weimar world around him, tying philosophical abstraction to situated existence.[14] And in another sign of the times, he had his eye on the sciences. This was not as surprising as it may seem. After all, as a student, after he left Catholic theology, he had formally enrolled in Freiburg's scientific-mathematical faculty. He had submitted his philosophy dissertation while holding the status of 'Cand.math'., candidate for a mathematical degree; and while attending Rickert's and others' philosophy courses, he had registered for analytic geometry, differential equations, inorganic chemistry and experimental physics (perhaps even a lab).[15] He left those studies behind, but paid attention to what was coming out of the sciences from a kind of general attentiveness. Besides, he shared the commonplace view that natural science marked the cutting edge of intellectual modernity. It opened up rich opportunities for philosophical reflection — opportunities matched only by the lack of interest among most of its practitioners in carrying out any such thing. What we can look for in his texts is not really a systematic philosophy of science. It is more often occasional comments, typically set out as tangents to some other question and sometimes provoked by contemporary events.[16]

---

[13] Even his thinking of the 1930s was only problematically *lebensphilosophisch* in some well-defined philosophical sense. As Heidegger moved through Ernst Jünger and Nietzsche, his language resonated with popular Nazi-era *Lebensphilosophie*. But he left each of them behind as a nihilistic inversion of modern metaphysics, not a truly new start. For a different emphasis in interpretation, see Wolf, *Hermeneutik und Technik* (ref. 11).

[14] See esp. Hans Ulrich Gumbrecht, *In 1926: Living at the edge of time* (Cambridge, MA: Harvard University Press, 1997).

[15] Sheehan, 'Heidegger's *Lehrjahre*' (ref. 9); Kisiel and Sheehan, *Becoming Heidegger* (ref. 10), xlviii–xlix. His instructor in experimental physics was Franz Himstedt, on whom see Stefan Wolff's paper in this volume. Mathematics was the *Nebenfach* in Heidegger's doctoral examination; the examiner was Lothar Heffter.

[16] A point of entry into the literature is Trish Glazebrook, *Heidegger's Philosophy of Science* (New York: Fordham University Press, 2000).

However, there are common themes to those comments and, more importantly, there is temporal development. Even when we fail, inevitably, to pull out all the references to contemporary science, they come often enough to track against other processes in time.[17]

By the mid-1910s, to start, Heidegger had read up on special relativity. Later he would take on general relativity as well. His habilitation lecture of 1915 was a piece of juvenilia, but it took up questions that would preoccupy him for years, centred on the nature of time. In 'The Concept of Time in the Science of History' ('Der Zeitbegriff in der Geschichtswissenschaft'), physics made its appearance as a foil to something else: in this case, historical studies.[18] The goal of physics since Galileo, so Heidegger suggested, was to develop a generalised dynamics of matter in motion, with time serving as the parameter that allowed that motion to be measured and mathematised. In this respect relativity just continued classical physics.[19] Still, this early move of closure was not Heidegger's last word. Reconsidered as his own thinking unfolded, the theory of relativity came to function as a kind of initial pivot point in his treatment of science.[20]

Heidegger would build a long excursus treating the theory into a famous 1925 lecture series on Dilthey, often seen as a first draft of *Being and Time*. There he argued, in a claim that would ramify widely, that relativity did, after all, revise the time concept. For when the time coordinate became dependent on the state of motion, when time was no longer absolute but local to the place a measurement

---

[17] I take cues from Catherine Chevalley, 'Heidegger and the Physical Sciences', in *Martin Heidegger: Critical Assessments*, ed. Christopher Macann, vol. 4. (New York: Routledge, 1992), 342–64. Since Chevalley's essay was published, however, Heidegger's early development has come into better focus.

[18] This was one of three topics Heidegger proposed for his demonstration lecture, to be chosen by Rickert, as hardly seems a surprise. *Cf.* Heinrich Rickert, *Die Grenzen der naturwissenschaftlichen Begriffsbildung: Eine logische Einleitung in die historischen Wissenschaften* (Tübingen: Mohr, 1902).

[19] In fact, the theory brought out time's relation to measurement even more clearly (particularly in Einstein's operationalist formulations) and its comparability to spatial dimensions in Minkowskian four-dimensional space-time. Martin Heidegger, 'Der Zeitbegriff in der Geschichtswissenschaft', in *Frühe Schriften*, ed. Friedrich-Wilhelm von Herrmann, GA 1 (Frankfurt: Klostermann, 1978), 413–33, on 424; 'The Concept of Time in the Science of History', in Kisiel and Sheehan, *Becoming Heidegger* (ref. 10), 60–72, on 66. See Theodore Kisiel, 'On the Dimensions of a Phenomenology of Science in Husserl and the Young Dr. Heidegger', *Journal of the British Society for Phenomenology* 4 (1973): 217–34.

[20] It is understandable that this point is missed in Klaus Hentschel, *Interpretationen und Fehlinterpretationen der speziellen und der allgemeinen Relativitätstheorie durch Zeitgenossen Albert Einsteins* (Basel: Birkhäuser, 1990), 268–69. The progress of the *Gesamtausgabe* has given us much more evidence to work with.

was made, then the Kantian concept of time was no longer at work.[21] Physics certainly kept its same goal, but the conceptual change within it amounted to a legitimate foundational step — a move of philosophising. Elsewhere, Heidegger would describe that move in terms that made it comparable in consequence to Dilthey's own work:

> [T]he jolt in which a science moves forward always lies in the revision of its fundamental concepts, that is, in the shifting that sets in of its existing propositions and stock of concepts onto new foundations. The upheaval in contemporary physics by Einstein was carried out in this way — not because he started philosophizing about the foundations of physics, but because he went after the fundamental concepts and framing that were involved in particular concrete problems and saw that revision of those concepts was necessary, were the goal of physics to be retained at all.[22]

The passage above, and other comments elsewhere, shows Heidegger understood the construction of Einstein's original paper. He grasped something a 'Cand.math.' could have been expected to appreciate: that the core of the theory was not relativity, but invariance under a transformation group.[23] Staking a claim as an expositor of physics, he cared enough to keep up with at least the semi-popular literature, reading from the series *Fortschritte der mathematischen Wissenschaften in Monographien*, Planck's lectures on theoretical physics and Galileo in Ostwald's *Klassiker*-collection. But it was not simply private interest that underwrote his focus. Since Marburg School Neo-Kantians made mathematical physics central to their epistemology, it was an important field for Heidegger

---

[21] Heidegger offered the following formulation: 'That time is really local time makes sense when one considers that the self is actual time, that it is not something that happens outside of us, a container of being, but we ourselves ... Kant, too, determined time from the apprehension of nature. Now we have to understand time as the reality of ourselves.' Martin Heidegger, 'Wilhelm Diltheys Forschungsarbeit und der gegenwärtige Kampf um eine historische Weltanschauung', *Dilthey-Jahrbuch für Philosophie und Geschichte der Geisteswissenschaften* 8 (1992–93): 143–77, on 172–73; 'Wilhelm Dilthey's Research and the Current Struggle for a Historical Worldview', in Kisiel and Sheehan, *Becoming Heidegger* (ref. 10), 238–74, on 269. What we have of this April 1925 lecture series in Kassel is a *Nachschrift* by Walter Bröcker. All Heidegger translations are my own.

[22] Martin Heidegger, *Logik: Die Frage nach der Wahrheit*, ed. Walter Biemel, GA 21 (Frankfurt: Klostermann, 1976), 16–17. These are lectures from winter semester 1925–26.

[23] He did not draw a clear line, on the other hand, between special and general relativity, and some of his language suggests a one-time student of science putting his vocabulary on display. For further context, including Cassirer, see David Scott, 'The "Concept of Time" and the "Being of the Clock": Bergson, Einstein, Heidegger and the Interrogation of the Temporality of Modernism', *Continental Philosophy Review* 39 (2006): 183–213.

to mark out a position. In any case, he felt comfortable enough concluding early on that theoretical physics was the heart of the discipline, and he stuck with his assessment: 'the foundation of experimental physics is theoretical, i.e., mathematical physics'.[24]

That scarcely meant, on the other hand, that other scientific disciplines were uninteresting. The looming crisis of the sciences tied them all together. This notion of general crisis — not just in philosophy, but throughout the academic disciplines — made its first appearance in Heidegger's work in 1925. Across the board, he suggested in opening his Dilthey lectures, contemporary scholarship stood 'in a great revolution, a productive one, that is opening up new ways of posing questions, new possibilities, new horizons'. This crisis, he continued, 'derives from the period before the war. It has grown out of the continuity of *Wissenschaft* itself, and that is what guarantees the seriousness and sureness of its upheavals'.[25] Alongside physics he had mathematics in view, for instance: the conflict between Hilbertian formalism and Brouwer and Weyl's intuitionism gave insight into just what was at stake. In crisis the disciplines were reflecting on their basic elements and their relation to reality, stripping away presuppositions and established interpretations in order to seek 'an original relationship *zu den Sachen selbst*'. They were carrying out, that is, a phenomenological undertaking. In crisis, the sciences turned philosophical, acknowledging that they could not solve their problems by remaining within their familiar frame.[26]

---

[24] Heidegger, 'Der Zeitbegriff' (ref. 19), 422; 'The Concept of Time' (ref. 19), 65. Commenting on the Marburg neo-Kantians' exclusive attention to mathematical physics, see, e.g., 'Die Idee der Philosophie und das Weltanschauungsproblem', in *Zur Bestimmung der Philosophie*, ed. Bernd Heimbüchel, *GA* 56/57, 1–117 (lecture course for the *Kriegsnotsemester* of 1919), on 80–84; 'The Idea of Philosophy and the Problem of World View', in *Towards the Definition of Philosophy* (London: Athlone, 2000), 1–99, on 67–71.

[25] Heidegger, 'Wilhelm Diltheys Forschungsarbeit' (ref. 21), 144; Wilhelm Dilthey's Research (ref. 21), 242. Kisiel, *Genesis* (ref. 10), 358, points out that this is Heidegger's first invocation of a general crisis of *Wissenschaft*. On Heidegger and crisis, see Bambach, *Heidegger* (ref. 11), which draws on Forman's 'Weimar Culture' on 41, 47. On crisis-talk in Weimar-era science, see Cathryn Carson, 'Method, Moment, and Crisis in Weimar Science', forthcoming in *Weimar Thought: A Critical History*, ed. Peter E. Gordon and John P. McCormack (Princeton: Princeton University Press, 2011).

[26] Martin Heidegger, *Prolegomena zur Geschichte des Zeitbegriffs*, 3rd ed., ed. Petra Jaeger, *GA* 20 (Frankfurt: Klostermann, 1994), 6; *History of the Concept of Time: Prolegomena*, trans. Theodore Kisiel (Bloomington: Indiana University Press, 1985), 4. Heidegger noted the phenomenological inspiration for intuitionism. For the significance of these lectures of summer semester 1925, see Theodore Kisiel, 'On the Way to *Being and Time*; Introduction to the Translation of Heidegger's *Prolegomena zur Geschichte des Zeitbegriffs*', in *Heidegger's Way of Thought: Critical and Interpretative Signposts* (London: Continuum, 2002), 36–63. This material on the crisis of the sciences is carried forward into *Being and Time*, §3.

Could they solve them at all? Heidegger's lecture course 'Fundamental Concepts of Metaphysics: World, Finitude, Solitude' peered at another domain he thought ripe for upheaval. In what are often called his 'biology lectures', in a long excursion on the organism and animality, he worked the ideas of Hans Driesch and Jakob von Uexküll into his reflections on the concepts of world (*Welt*) and life (*Leben*).[27] There was promise, he thought, in the crisis in biology. This was the promise of a discipline that recovered the originality of the organism, a science that acknowledged that the nineteenth-century's physico-chemical reductionism and mechanistically conceived Darwinism took their lead from disciplines dealing with matters other than life. Still, neither Driesch's holistic neovitalism nor von Uexküll's appropriation of the term *Umwelt* was enough. While Heidegger praised the challenge of their views, he did not take them over. Driesch was still trapped in the categories of nineteenth-century mechanism when he tried to create space for a non-mechanical force; 'for biological problems,' Heidegger concluded, 'vitalism is just as dangerous as mechanism'. Von Uexküll's investigations into animal perception did not question thoroughly enough whether animals' existence was more fundamentally divergent from humans', in such a way that notions of world might no longer apply.[28]

Everything depended on individual scientists ready to carry forward such questioning, 'leading researchers ... alongside the indispensable myriad workers and technical experts'.[29] But by the end of the 1920s, Heidegger thought the signs were not good. An intransigently positivist science, facing ill-informed critics appealing to second-hand philosophy and *Weltanschauung*, fell back on the facts and methods of specialised research.[30]

> It certainly seemed for a while that science [*die Wissenschaft*] itself was starting to waver, and out of this arose the slogan of the *Grundlagenkrisis der Wissenschaften*. But the crisis cannot break through in earnest and, above all, it cannot last as needed, for we refuse to let ourselves be shaken up enough to gain

---

[27] Martin Heidegger, *Die Grundbegriffe der Metaphysik: Welt—Endlichkeit—Einsamkeit*, 2nd ed., ed. Friedrich-Wilhelm von Herrmann, GA 29/30 (Frankfurt: Klostermann, 1992); *The Fundamental Concepts of Metaphysics: World, Finitude, Solitude*, trans. William McNeill and Nicholas Walker (Bloomington: Indiana University Press, 1995). For background to the treatment of biology in these lectures of winter semester 1929–30, see Radloff, *Heidegger* (ref. 7), 22–28.

[28] Heidegger, *Grundbegriffe* (ref. 27), 381 (quotation), 379–85; *Fundamental Concepts* (ref. 27), 262 (quotation), 260–64. This disagreement is not picked up in Harrington, *Reenchanted Science* (ref. 6), 53–54.

[29] Heidegger, *Grundbegriffe* (ref. 27), 279; *Fundamental Concepts* (ref. 27), 189.

[30] Martin Heidegger, *Einleitung in die Philosophie*, ed. Otto Saame and Ina Saame-Speidel, GA 27 (Frankfurt: Klostermann, 1996), 26–27, 37. These are lectures from winter semester 1928–29.

the breadth of vision for what is so immense and simultaneously elementary about the new tasks. Science will not go along with it, for it has already been delivered over far too thoroughly to practical-technical subservience.[31]

For philosophy, however, the waning sense of crisis just underlined a fundamental need. That need was for explication of the limits of the sciences and of *Wissenschaft* in general, as Heidegger proceeded to supply.[32]

Quantum Mechanics and the Heidegger-Heisenberg Exchange

Heidegger's emerging critique of *Wissenschaft* made reference to theoretical physics, as commentators have often pointed out. What has not always been appreciated is how implicated in this process was the theoretical physics *of his own day*. Starting in the late 1920s, at a moment when Heidegger's dissatisfaction with the conduct of the sciences was crystallising, we can watch his attention turn from special relativity to quantum mechanics. The shift marks a transition that, in its ramifications in Heidegger's thinking, spills over the boundary closing out the Weimar Republic. It is worth following across that boundary for its relevance to themes that feature prominently in Forman's 'Weimar Culture' — causality and freedom, to start. It points to other issues, moreover — the method and limits of science, objectivity and the subject-object relation — that, while less central in Forman's account of physics, prove to be no less anchored in the intellectual milieu.

With *Being and Time* in print, his triumphant installation accomplished as Husserl's heir-apparent in Freiburg, the no-longer-uncrowned king of contemporary philosophy offered a signal in his introductory course of winter 1928–29. Referring yet again to talk of crisis in physics, Heidegger pinned the crisis this time to something different: the suggestion that 'the concept of causality, of cause and causation, has been made unusable by the newest developments in physics; so, too, the concept of matter'.[33] This meant quantum mechanics (though Heidegger at first did not use the name). Its consequences for causality — or better, physicists' loose philosophising about them — were part of general intellectual discussion; they continued to occupy, and to trouble, Heidegger here and there. His lecture course a year and a half later on human freedom took up Kant's understanding of cause and effect and the apparent opposition between deterministic causation in nature and freedom of the will. To set up Kant's account of causality, preliminary remarks on the natural

---

[31] Heidegger, *Grundbegriffe* (ref. 27), 281; *Fundamental Concepts* (ref. 27), 191.
[32] Heidegger, *Einleitung* (ref. 30), 39.
[33] Ibid., 36.

sciences seemed necessary. (One imagines Heidegger's students perhaps losing the thread, but leave that aside.) Heidegger noted, 'On the one side — that of physics — it is said that we are now finally to the point of understanding that the law of causality is no a priori law of thought, and that only experience and physical thinking can decide about this law'. He illustrated the point with a quotation from a lecture by Pascual Jordan: '"The physicists no longer doubt that the question of whether complete causality obtains can only be decided by experience, that causality is thus not an a priori necessity of thought"'.[34]

However, Heidegger continued with some irritation, 'The last remark naturally refers to the Kantian conception of causality, to which it needs to be said that Kant nowhere and never conceives of or presents the law of causality as an a priori necessity of thought'.[35] When physicists pointed to atomic probabilities to declare the dissolution of causality, breezily proclaiming the ability of empirical science to dictate to philosophy, they misunderstood, first of all, what the philosophers were saying. More importantly, they missed something deeper. What causality was must be defined before its downfall could be entertained. But defined on what basis? Physics might have its own standard practice — for instance, tying it to predicting the future state of a system from complete knowledge of positions and momenta — but the choice of that basis was a question that physics could not answer on its own terms.

Heidegger was evidently following some part of the causality debate in *Die Naturwissenschaften*, possibly in other venues as well. He was less close to the science than he had been with relativity, and to expert readers, if not philosophy students, his lack of familiarity might show.[36] What is crucial is that he kept coming back to the topic: to quantum mechanics, initially meaning statistics and causality, indeterminacy and uncertainty, and increasingly other aspects as well. The theory hardly showed up in his published work, but found its way into occasional lectures and jottings that suggest some ongoing attention. Thus Heidegger read through

---

[34] Martin Heidegger, *Vom Wesen der menschlichen Freiheit: Einleitung in die Philosophie*, ed. Hartmut Tietjen, *GA* 31 (Frankfurt: Klostermann, 1982), 145; *The Essence of Human Freedom: An Introduction to Philosophy*, trans. Ted Sadler (London: Continuum, 2002), 103. These are lectures from summer semester 1930. Heidegger cites P. Jordan, 'Kausalität und Statistik in der modernen Physik', *Die Naturwissenschaften* 15 (1927): 105–110, here 105; as well as M. Born, 'Quantenmechanik und Statistik', *Die Naturwissenschaften* 15 (1927): 238–42.

[35] Heidegger, *Vom Wesen* (ref. 34), 145; *The Essence* (ref. 34), 103.

[36] For instance, in his terminology, or in thinking that the Bohr model of planet like orbits was still used. For the context, see Michael Stöltzner, 'Die Kausalitätsdebatte in den *Naturwissenschaften*: Zu einem Milieuproblem in Formans These', in *Wissensgesellschaft: Transformationen im Verhältnis von Wissenschaft und Alltag*, ed. H. Franz, W. Kogge, T. Möller and T. Wilholt (Bielefeld: Institut für Wissenschafts- und Technikforschung, 2001), 85–128.

Jordan's book *Physik des 20. Jahrhunderts* and made more disparaging comments on it; he attended a lecture by Walther Gerlach and had even sharper things to say. Looking toward lectures, he wrote up notes for himself under such rubrics as 'exact science / measurement / "causality" / Heisenberg's "uncertainty relation"'.[37]

This is the context of Heidegger's dialogue with Heisenberg, an exchange that, while it can only be introduced here, would eventually show up in some of both Heisenberg's and Heidegger's most famous texts.[38] While we want to deal here with the Heidegger-Heisenberg relationship solely through the 1930s, and only in its impacts on Heidegger, we can use it to lay out the latter's incipient critique of natural science. We do not know how the two men's attention was drawn to each other, but we do know that they met for an extended discussion in 1935. Heidegger soon slipped a brief approbation of Heisenberg into his lectures on Kant, whose relation to the science of his day was a central theme of Heidegger's account.

> The greatness and superiority of natural science in the sixteenth and seventeenth century goes back to the fact that those researchers were all philosophers; they understood that there are no plain facts, but that a fact is what it is only in light of the concept that grounds it and according to the reach of that grounding. The hallmark of positivism, by contrast, in which we have stood for decades and still stand more than ever, is believing that we can get by with facts, or other and new facts .... But this attitude rules only where average and follow-on work is done. Where real research is done that opens things up, the situation is no different from three hundred years ago; ... the leading minds of atomic physics today, Niels Bohr and Heisenberg, think philosophically through and through and for that reason alone create new ways of posing questions and above all hold out in the realm of what bears questioning.[39]

---

[37] Martin Heidegger, *Leitgedanken zur Entstehung der Metaphysik, der neuzeitlichen Wissenschaft und der modernen Technik*, ed. Claudius Strube, *GA* 76 (Frankfurt: Klostermann, 2009), 184–85 on Jordan, 188 on Walther Gerlach's 'unexcelled confusion', 173–74 on causality and uncertainty. These appear to be notes connected with Heidegger's lecture 'Die Bedrohung der Wissenschaft', delivered to a small circle of colleagues at Freiburg in 1937.

[38] See Cathryn Carson, 'Science as Instrumental Reason: Heidegger, Habermas, Heisenberg', *Continental Philosophy Review* 42 (2010): 483–509. Other accounts include Hans-Peter Hempel, *Natur und Geschichte: Der Jahrhundertdialog zwischen Heidegger und Heisenberg* (Frankfurt: Anton Hain, 1990); and, more usefully, Otto Pöggeler, 'The Hermeneutics of the Technological World: The Heidegger-Heisenberg Dispute', trans. Michael Kane and Kristin Pfefferkorn-Forbath, *International Journal of Philosophical Studies* 1 (1993): 21–48.

[39] Martin Heidegger, *Die Frage nach dem Ding: Zu Kants Lehre von den transzendentalen Grundsätzen*, ed. Petra Jaeger, *GA* 41 (Frankfurt: Klostermann, 1984), 67; *What is a Thing?*, trans. W.B. Barton, Jr., and Vera Deutsch (Chicago: Regnery, 1967), 67. These lectures of winter semester 1935–36 were given under the title 'Grundfragen der Metaphysik'.

Although Heidegger had written off the sciences in matters of foundational crisis, he made an exception for quantum mechanics. More precisely, he made an exception for leading researchers who did the questioning work of transcendental reflection. Heisenberg seemed a reasonable partner, or so we can speculate, in no small part because he moved beyond simple-minded claims to refute the law of causality. In fact, Heisenberg's philosophising by the early 1930s eschewed grand claims about acausality; the author of the uncertainty principle had little to say about indeterminacy *per se*. Like Bohr, he came to focus on issues that he thought had wider reach: the opening of new possibilities for thought (*Denkmöglichkeiten*) in the transition from classical to quantum physics, the revision of basic physical concepts, the reinsertion of the subject-object relation into the discipline and the feasibility of the scientific enterprise even after a commitment to old-style objectivity was dropped.[40]

Heidegger was making a similar shift — away from causality, with its ties to temporal sequence and freedom, and toward questions of subjects and objects and the security of knowledge. That shift was a critical one in the emergence of new themes in his thinking. He had begun much earlier to think through the meaning of mathematisation, connecting it with Galileo's advance into modernity (*die Neuzeit*) and Descartes' demand for certainty of knowledge. It was in his 1935–36 lectures that Heidegger generalised mathematisation (understood as a particular way of prescribing how nature should make its appearance) into the defining feature of modern science.[41] In the next years he married reflections on Descartes' creation of the modern subject, on the one hand, with Kant's account of the constitution of objects (*Gegenstände*), on the other, to display how subject and object emerged in relation to each other. Mathematisation became modernity's demand that things be represented as objects that were measurable, predictable and predisposed to be brought into line with the subject's will. The world was ordered into an objectified picture, a *Weltbild*, a representation that served the needs of the ordering subject.[42]

---

[40] As argued in Cathryn Carson, *Heisenberg in the Atomic Age: Science and the Public Sphere* (Cambridge: Cambridge University Press), ch. 4.

[41] Heidegger, *Prolegomena* (ref. 26), 245; *History* (ref. 26), 181–82 (1925, on Descartes); Martin Heidegger, *Phänomenologische Interpretation von Kants Kritik der reinen Vernunft*, ed. Ingtraud Görland, *GA* 25 (Frankfurt: Klostermann, 1977), 29–32; *Phenomenological Interpretation of Kant's Critique of Pure Reason*, trans. Parvis Emad and Kenneth Maly (Bloomington: Indiana University Press, 1997), 21–23 (1927–28, on Galileo and Kepler); *Einleitung* (ref. 30), 186–88 (1928–29, on Galileo); *Frage nach dem Ding* (ref. 39), 65–108; *What is a Thing?* (ref. 39), 65–108.

[42] See esp. Martin Heidegger, 'Die Zeit des Weltbildes', in *Holzwege*, ed. Friedrich-Wilhelm von Herrmann, *GA* 5 (Frankfurt: Klostermann, 1977), 75–113; 'The Age of the World Picture', in *The Question Concerning Technology and Other Essays*, trans. William Lovitt (New York: Garland, 1977), 115–54. The reasons for Heidegger's move in this direction have to do with Nietzsche and National Socialism no less than Descartes and Kant, but that exceeds the scope of this essay.

All science was intrinsically oriented to technical application, even and exactly theoretical physics. Theoretical physics, in turn, became the model for *Wissenschaft* at large.

Heidegger's new account of science reinforced his attention to Heisenberg; it may even have drawn from it. For if he had thought in 1935 to praise Heisenberg's openness to transcendental reflection, he decided no later than 1937 that the physicist's reflection did not go far enough. Heidegger wanted to bring into view the process of objectification (*Vergegenständlichung*), understood as natural science's way of taking hold of the world. Heisenberg, in his view, stayed within the subject-object paradigm of modern epistemology — tinkering with it, certainly, when he replaced *Objektivität* by *Objektivierung*, but never escaping it. On point after point Heidegger went after Heisenberg's conceptions. The notion of the measuring instrument influencing the phenomenon under observation was too simple; its interposition in the subject-object relation was ill thought through. The counterposition of subject and object in the interpretation of quantum mechanics was specious, for 'the divide [between them] is never a divide, but precisely a transcendental relation [*Bezug*]'.[43] The very attempt to mark out quantum mechanics as somehow breaking with previous physics was misleading to the core:

> The way in which the doubling of "statistical" and "classical" physics is handled shows exactly that everything remains in the same plane, letting one encroach on the other or coupling them together and recognizing a cut [*Schnitt*] in between. But the question of what kind of objectivity [*Gegenständlichkeit*] corresponds to the statistical way of representation [*Vorstellungsweise*] and how this is to be brought together with the classical one, the question whether a fundamental reflection is needed on the *truth* of the classical projection [*Entwurf*] and its thrownness [*Geworfenheit*], which is slowly becoming manifest, remains unasked.[44]

That is, nothing in quantum mechanics differed essentially from classical physics, itself the fullest expression of the mathematical character of science. If the most reflective of physicists sought to argue the opposite, he remained caught up in the metaphysics of modernity exactly in his inability to see that nothing had changed. And, in fact, Heidegger was confirmed in his conviction that reflection

---

[43] Heidegger, *Leitgedanken* (ref. 37), 180. On Heidegger's appropriation in this period of Kantian transcendental reflection, see Chad Engelland, 'Heidegger on Overcoming Rationalism Through Transcendental Philosophy', *Continental Philosophy Review* 41 (2008): 17–41.

[44] Heidegger, *Leitgedanken* (ref. 37), 179. *Cf.* the remarks on quantum mechanics in Martin Heidegger, *1. Die metaphysischen Grundstellungen des abendländischen Denkens, 2. Einübung in das philosophischen Denken*, ed. Alfred Denker, GA 88 (Frankfurt: Klostermann, 2008), 212–15.

was beyond science's capacity. Its ordering attitude was universal and exceptionless; only philosophy could truly think.

Conclusion

The language, the allusions and the particular foci of attention indicate Heidegger had a specific physicist in mind. Growing from these reflections of the 1930s, the Heidegger-Heisenberg exchange would become a new pivot point in Heidegger's treatment of science. We cannot follow their interaction into the post-World War II era when their paths crossed repeatedly, when Heidegger, in particular, made Heisenberg into a kind of de facto spokesman for science.[45] However, we can use Heidegger's reflections on Heisenberg's physics to highlight some critical features of both men's thinking about science.

Like many others, Heidegger and Heisenberg saw foundational challenges in physics as encapsulating the crisis in science. Heidegger located himself on the very cusp of the crisis, as the self-confident progress of natural science, of *Wissenschaft* and of Western metaphysics was called into question by its own sheer force. Heisenberg, for his part, shared the same narrative of the unfolding of science in modernity and the same inclination to seek out departures from its nineteenth-century form. Moreover, he could serve as the scientific poster child for Weimar's era of crisis. As he would put it after the fact, 'It is not by chance that the development that led to this end [in physics, i.e., quantum mechanics] no longer took place in a time of belief in progress. After the catastrophe of the First World War one understood outside of *Wissenschaft* as well that there were no firm foundations for our existence, secure for all time'.[46]

Yet the driving difference between Heidegger and Heisenberg is where they looked for redemption, and the point where studying their encounter ultimately pays off. Heidegger sought salvation outside of *Wissenschaft* altogether, in a philosophy that thinks in its own way. Heisenberg believed science had resources within itself to face the crisis, even if self-limitation was part of its response. Quantum mechanics might set bounds on objectivity, but — this Heisenberg emphasised — it was still science. Critiquing classical conceptions of science, even replacing them within limits, could open the door to a more differentiated understanding of the scientific project at large.

---

[45] Carson, 'Science' (ref. 38).
[46] Werner Heisenberg, 'Die Beziehungen zwischen Physik und Chemie in den letzten 75 Jahren', *Naturwissenschaftliche Rundschau* 6 (1953): 1–7, on 3, reprinted in *Gesammelte Werke / Collected Works*, ed. Walter Blum, Hans-Peter Dürr and Helmut Rechenberg, vol. C.I (Munich: Piper, 1984), 387–93.

This divergence points us to the different valences of modernity in their thought. Heidegger hardly hesitated to mark his distance from the notion. In his telling, modernity began (proximately) with the Cartesian construction of subject and object and reached a kind of culmination in twentieth-century physics. Heidegger was supremely confident that that tradition had exhausted its internal resources, and so he held out for a kind of ... what shall we call it? Anti-modernity? Non-modernity? The sheer rejection of modernity, after all, 'this negative tendency', as he called it, 'is encumbered with just as great misunderstandings' as the doctrines it reacted against.[47] Rather than just reacting against the modern, and thereby reinforcing it, Heidegger wanted to go back before it and re-think a more original thought. Whatever that meant, it did not involve science. For science was only modern — there was no pre-modern science.

For Heisenberg's part, he, too, claimed the word 'modern' for his science. This was not a trivial move in the 1920s and 1930s, when 'modern' meant so much.[48] The ultimate modernity of modern physics, in his construal, lay in its willingness to expand the possibilities for scientific theory, bringing acausality and limits to objectivity into its frame. He pointed to its openness, that is, to transcendentally reflect on and revise its foundations — though transcendental reflection was just what Heidegger said science could not do. The contrast with Heidegger shows that Heisenberg's was a kind of high modernity that simultaneously built on and cut the ground out from underneath the stylised self-confidence of so-called 'classical' physics, itself modern in a more straightforward sense. Reflexivity meant reconsidering the standards of scientific objectivity, holding open the possibility of epistemic revision, and renewing attentiveness to the conditions of the possibility of knowledge, that foundationally modern step.

Heisenberg's challenges to classical physics, his attempts to push it to the point it broke down, were not intrinsically anti-scientific. Rather, they opened the door (or so he saw it) to new possibilities for thought, new understandings of what counted as science. The exchange between Heisenberg and Heidegger serves as a diagnostic for critiques of scientific forms inherited from the era before Weimar. In this light, the two men's differences show something essential about reactions

---

[47] The context is his lectures on biology: Heidegger, *Grundbegriffe* (ref. 27), 278; *Fundamental Concepts* (ref. 27), 189.

[48] Heidegger's 'modern' science included the physics that Heisenberg called 'classical'. Where Heisenberg saw a break, as we saw above, Heidegger saw none. On modernity in physics, see Richard Staley, 'On the Co-Creation of Classical and Modern Physics', *Isis* 96 (2005): 530–58; Suman Seth, 'Crisis and the Construction of Modern Theoretical Physics', *British Journal for the History of Science* 40 (2007): 25–51; on its meanings in the 1930s, Carson, *Heisenberg* (ref. 40), 49–51.

against scientific modernity: they are intrinsically manifold, defined by what they refuse, not what they are coherently for.

Acknowledgments

The thinking that went into this essay has benefited from many interlocutors. In the context of the Forman Conference I am grateful to Bob Brain and Alan Richardson. I am indebted above all to Paul Forman for his critical engagement and uncompromising posing of questions.